Saponins

T0328847

K. HOSTETTMANN

Institute of Pharmacognosy and Phytochemistry
School of Pharmacy, Lausanne University, Switzerland

and

A. MARSTON

Institute of Pharmacognosy and Phytochemistry
School of Pharmacy, Lausanne University, Switzerland

CAMBRIDGE
UNIVERSITY PRESS

CAMBRIDGE UNIVERSITY PRESS
Cambridge, New York, Melbourne, Madrid, Cape Town, Singapore, São Paulo

Cambridge University Press
The Edinburgh Building, Cambridge CB2 2RU, UK

Published in the United States of America by Cambridge University Press, New York

www.cambridge.org
Information on this title: www.cambridge.org/9780521329705

First published 1995
This digitally printed first paperback version 2005

A catalogue record for this publication is available from the British Library

Library of Congress Cataloguing in Publication data

Hostettmann, K. (Kurt), 1944–
Saponins / K. Hostettmann, A. Marston.
   p.   cm. – (Chemistry and pharmacology of natural products)
Includes bibliographical references and index.
ISBN 0 521 32970 1
1. Saponins.   I. Marston. A. (Andrew), 1953–   . II. Title.
III. Series.
QD325.H73   1995
547.7′83 – dc20   94-688 CIP

ISBN-13 978-0-521-32970-5 hardback
ISBN-10 0-521-32970-1 hardback

ISBN-13 978-0-521-02017-6 paperback
ISBN-10 0-521-02017-4 paperback

Contents

Glossary

Ac	acetyl
ACTH	adrenocorticotrophin, corticotrophin
All	β-D-allose
Api	β-D-apiose
Ara	α-L-arabinose
Bz	benzyl
Dta	6-deoxy-α-L-talopyranose
EBV	Epstein–Barr virus
Fuc	β-D-fucose (6-deoxygalactose)
Gal	β-D-galactose
GalA	β-D-galacturonic acid
GalNAc	N-acetyl-2-amino-2-deoxygalactose
Glc	β-D-glucose
GlcA	β-D-glucuronic acid
GlcNAc	N-acetyl-2-amino-2-deoxyglucose
HI	haemolytic index
HIV	human immunodeficiency virus
5-HT	5-hydroxytryptamine
i.p.	intraperitoneal
i.v.	intravenous
Ma	malonyl
Man	α-L-mannose
Me	methyl
PDE	phosphodiesterase
p.o.	oral
Rha	α-L-rhamnose (6-deoxymannose)
Rib	β-D-ribose

Qui	β-D-quinovose (D-chinovose, 6-deoxyglucose)
s.c.	subcutaneous
Ulo	6-deoxy-β-D-xylo-hexos-4-ulo-pyranose
Xyl	β-D-xylose
f	furanose
p	pyranose

Techniques

APT	attached proton test
CA	collisional activation
CAMELSPIN	cross-relaxation appropriate for minimolecules emulated by locked spins
CD	circular dichroism
CF-FAB	continuous-flow FAB-MS
CI–MS	chemical ionization mass spectrometry
COLOC	^{13}C-^{1}H two-dimensional correlation spectroscopy via long-range couplings
COSY	two-dimensional ^{1}H correlation spectroscopy
CPC	centrifugal partition chromatography
CTLC	centrifugal thin-layer chromatography
DCCC	droplet countercurrent chromatography
D/CI-MS	desorption–chemical ionization mass spectrometry
DEPT	distortionless enhancement by polarization transfer
DQF-COSY (DQ-COSY)	double quantum-filtered, phase-sensitive COSY
EI-MS	electron impact mass spectrometry
ELISA	enzyme-linked immunosorbent assay
FAB-MS	fast atom bombardment mass spectrometry
FD-MS	field desorption mass spectrometry
FLOCK	long-range heteronuclear correlated spectroscopy incorporating bilinear rotation decoupling pulses
FRIT-FAB	frit fast atom bombardment mass spectrometry
GC	gas chromatography
GC–FTIR	gas chromatography–Fourier transform IR
GC–MS	gas chromatography–mass spectrometry

HETCOR	^{13}C-^{1}H heteronuclear correlated spectroscopy
HMBC	^{1}H-detected heteronuclear multiple-bond spectroscopy
HMQC	^{1}H-detected heteronuclear one-bond spectroscopy
HOHAHA	homonuclear Hartmann–Hahn spectroscopy
HPLC	high-performance liquid chromatography
I.D.	internal diameter
INADEQUATE	incredible natural abundance double quantum experiment
INEPT	insensitive nuclei enhanced by polarization transfer
IR	infra-red
LC–MS	liquid chromatography–mass spectrometry
LD	laser desorption
LD/FTMS	laser desorption/Fourier transform mass spectrometry
LPLC	low-pressure liquid chromatography
LSPD	long-range selective proton decoupling
MIKE/CAD	mass-analysed ion kinetic energy/collision-activated dissociation
MPG	microporous glass
MPLC	medium-pressure liquid chromatography
NMR	nuclear magnetic resonance
NOE	nuclear Overhauser enhancement
NOESY	nuclear Overhauser enhancement correlation spectroscopy
ODS	octadecylsilyl
OPLC	overpressure layer chromatography
ORD	optical rotatory dispersion
PD-MS	plasma desorption mass spectrometry
PRFT	partially relaxed Fourier transform spectroscopy
RCT	relayed coherence transfer
RIA	radioimmunoassay
RLCC	rotation locular countercurrent chromatography
ROESY	2-D NOE in a rotating frame
RP	reversed phase

SIMS   secondary ion mass spectrometry
SINEPT   selective INEPT
TLC   thin-layer chromatography
TOCSY   total correlation spectroscopy
TSP   thermospray
UV   ultra-violet
VLC   vacuum liquid chromatography
XHCORR   long-range ^{13}C-^{1}H correlation
                                  spectroscopy

1

Introduction

1.1 General introduction

Saponins are high-molecular-weight *glycosides*, consisting of a sugar moiety linked to a *triterpene* or *steroid* aglycone. The classical definition of saponins is based on their surface activity; many saponins have detergent properties, give stable foams in water, show haemolytic activity, have a bitter taste and are toxic to fish (piscicidal). Such attributes, while not common to all saponins, have frequently been used to characterize this class of natural products. However, because of the numerous exceptions which exist, saponins are now more conveniently defined on the basis of their molecular structure, namely as triterpene or steroid glycosides.

Some saponin-containing plants have been employed for hundreds of years as soaps and this fact is reflected in their common names: soapwort (*Saponaria officinalis*), soaproot (*Chlorogalum pomeridianum*), soapbark (*Quillaja saponaria*), soapberry (*Sapindus saponaria*), soapnut (*Sapindus mukurossi*). Indeed, the name 'saponin' comes from the Latin word *sapo* (soap).

Saponins are constituents of many plant drugs and folk medicines, especially from the Orient. Consequently, great interest has been shown in their characterization and in the investigation of their pharmacological and biological properties.

Since the appearance of Kofler's book *Die Saponine* in 1927, the field has undergone a most remarkable transformation. Kofler described in great detail the properties and pharmacological activities of saponins but at that time not a single saponin had been fully characterized. In contrast, by 1987 the structure of over 360 sapogenins and 750 triterpene glycosides had been elucidated (Bader and Hiller, 1987). This rapid progress is the result of the many dramatic advances in isolation and structure elucidation techniques, a large number of which will be mentioned in this book.

The most common sources of saponins are the higher plants, but increasing numbers are being found in lower marine animals. So far, they have only been found in the marine phylum Echinodermata and particularly in species of the classes Holothuroidea (sea cucumbers) and Asteroidea (starfishes).

1.2   Definitions

The aglycone or non-saccharide portion of the saponin molecule is called the *genin* or *sapogenin*. Depending on the type of genin present, the saponins can be divided into three major classes:

(1) triterpene glycosides
(2) steroid glycosides
(3) steroid alkaloid glycosides.

The genins of these three classes can be depicted as shown in Fig. 1.1.

The aglycones are normally hydroxylated at C-3 and certain methyl groups are frequently oxidized to hydroxymethyl, aldehyde or carboxyl functionalities. When an acid moiety is esterified to the aglycone, the term *ester saponin* is often used for the respective glycosides.

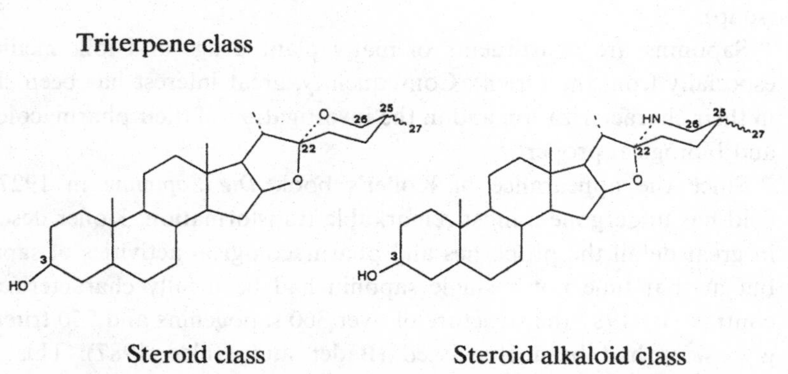

Triterpene class

Steroid class   Steroid alkaloid class

Fig. 1.1. Skeletal types of genin found in the three principal classes of saponin.

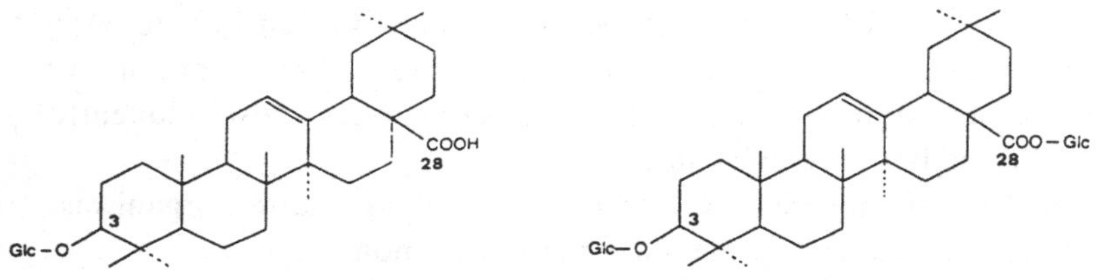

Monodesmosidic   Bidesmosidic

Fig. 1.2. Monodesmosidic and bidesmosidic saponins.

All saponins have in common the attachment of one or more sugar chains to the aglycone. *Monodesmosidic* saponins have a single sugar chain, normally attached at C-3. *Bidesmosidic* saponins have two sugar chains, often with one attached through an ether linkage at C-3 and one attached through an ester linkage (acyl glycoside) at C-28 (triterpene saponins) (Fig. 1.2) or an ether linkage at C-26 (furostanol saponins). *Tridesmosidic* saponins have three sugar chains and are seldom found. Bidesmosidic saponins are easily transformed into monodes-mosidic saponins by, for example, hydrolysis of the esterified sugar at C-28 in triterpene saponins; they lack many of the characteristic properties and activities of monodesmosidic saponins.

The saccharide moiety may be linear or branched, with 11 being the highest number of monosaccharide units yet found in a saponin (Clematoside C from *Clematis manshurica* (Ranunculaceae); Khorlin *et al.* 1965). As a rule, however, most of the saponins so far isolated tend to have relatively short (and often unbranched) sugar chains, containing 2–5 monosaccharide residues. Kochetkov and Khorlin (1966) have introduced the term *oligoside* for those glycosides containing more than 3–4 monosaccharides.

The most common monosaccharide moieties found, and the corre-sponding abbreviations used in this book (according to IUPAC recom-mendations; *Pure Appl. Chem.* (1982) **54**, 1517–1522), are: D-glucose (Glc), D-galactose (Gal), D-glucuronic acid (GlcA), D-galacturonic acid (GalA), L-rhamnose (Rha), L-arabinose (Ara), D-xylose (Xyl) and D-fucose (Fuc). Saponins from marine organisms often contain D-quinovose (Qui) (some-times written as D-chinovose). It is possible that D-fucose is present more often than generally thought, in view of its very similar chromatographic behaviour to L-rhamnose. The glossary contains a full list of the abbrevia-tions used in the book.

Unlike the cardiac glycosides, unusual monosaccharides are seldom found, but it should be noted that uronic acids often occur in triterpene glycosides and that amino sugars may be present. Glucose, arabinose, glucuronic acid and xylose are the monosaccharides most frequently attrached directly to the aglycone.

Acylated sugar moieties are also encountered. In marine organisms, methylated and sulphated sugars are not uncommon.

Configurations of the interglycosidic linkages are given by α and β and the monosaccharides can be in the pyranose (p) or furanose (f) forms.

By virtue of carboxyl groups in the aglycone or sugar parts of a saponin, it can be rendered acidic.

1.3   Biosynthesis

A brief biosynthetic summary of triterpenes and steroids is shown in Fig. 1.3. They are built up of six isoprene units and have a common biosynthetic origin in that they are all derived from *squalene*, presumably via ring opening of squalene-2,3-epoxide (oxidosqualene), followed by a concerted cyclization. It is only recently that the corresponding cyclases have been characterized (Abe *et al.* 1993). While the true triterpenes have 30 carbon atoms, the steroids have only 27 carbons by virtue of the oxidative cleavage of three methyl groups from a C_{30} intermediate. Other details of the biosynthesis of steroids have been reviewed (Heftmann, 1968; Takeda, 1972). Much work remains to be done to clarify the intermediates but at least the distinction between the cholesterol (C_{27}) and the triterpene (C_{30}) pathways is evident.

Figure 1.4 shows the cyclization of the chair–chair–chair–boat conformation of squalene-2,3-epoxide. Following the rearrangement of the tetracyclic carbonium ion, either a pentacyclic (e.g. β-amyrin) or tetracyclic (e.g. dammarenol) triterpene is formed (Luckner, 1990). More details are to be found in a review article by Abe *et al.* (1993). In contrast, cyclization of the chair–boat–chair–boat conformation of squalene-2,3-epoxide, followed by several 1,2-rearrangements, leads to the formation of either cycloartenol or lanosterol (Fig. 1.5).

In the marigold (*Calendula officinalis*, Asteraceae), biosynthesis of oleanolic acid 3-*O*-glycosides occurs in the microsomes. Stepwise oxidation of β-amyrin occurs, giving erythrodiol, oleanolaldehyde and oleanolic acid. Finally the glycosides are formed (Wilkomirski and Kasprzyk, 1979).

Other important classes of secondary metabolites such as phytosterols, cardenolides, cucurbitacins, quassinoids and limonoids are also derived from squalene. There are similarities in the biosynthesis of cardenolides

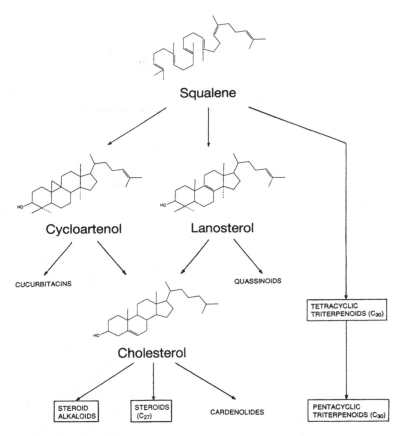

Fig. 1.3. Biosynthesis of triterpenes and steroids.

and steroid saponins. Both can start with cholesterol, both go through a 3-keto intermediate and both processes occur actively in *Digitalis* spp. (Heftmann, 1974). Like the saponins, many cardenolides give foams in aqueous solution but are classified separately because of their different biological activities.

The biosynthetic pathway to cholesterol in algae and green plants is via cycloartenol (Rees *et al.* 1968; Goad and Goodwin, 1972; Goad, 1991), while fungi and non-photosynthetic organisms have lanosterol as intermediate. In keeping with these biosynthetic findings, cycloartane triterpenes are found to be widely distributed throughout the plant kingdom (Boar and Romer, 1975).

At first it was not certain that cholesterol could be a precursor of steroids because its distribution in the plant kingdom was not known. In recent years it has been shown, however, that the occurrence of cholesterol

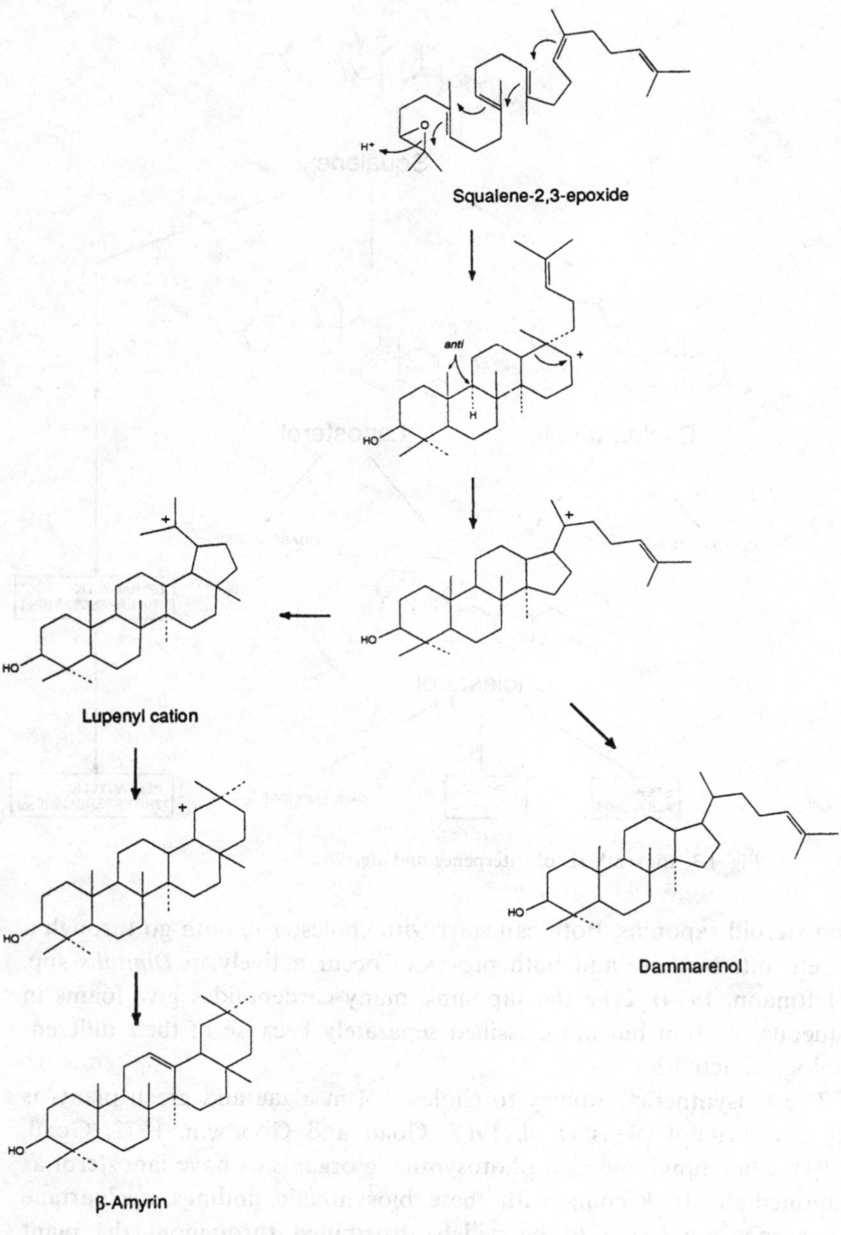
H+
Squalene-2,3-epoxide
anti
H
HO
+
Lupenyl cation
Dammarenol
β-Amyrin

Fig. 1.4. Cyclization of squalene-2,3-epoxide to C_{30} triterpenes.

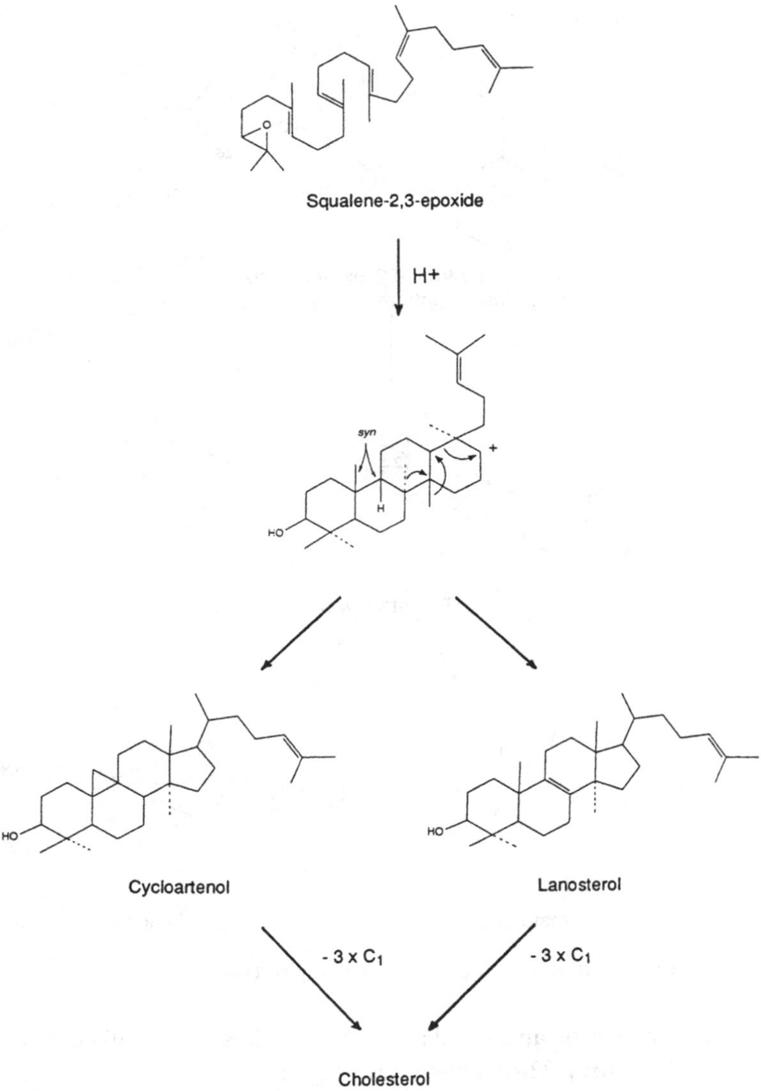

Squalene-2,3-epoxide

H+

Cycloartenol   Lanosterol

- 3 x C₁   - 3 x C₁

Cholesterol

Fig. 1.5. Cyclization of squalene-2,3-epoxide and formation of cholesterol.

is widespread but that it does not occur in large quantities (Heftmann, 1967).

Tschesche and Hulpke (1966) were able to show that cholesterol is a precursor of spirostanols by treating *Digitalis lanata* (Scrophulariaceae) leaves with ^{14}C-labelled cholesterol-β-D-glucoside and showing that radioactive tigogenin and gitogenin were produced. One possibility is that

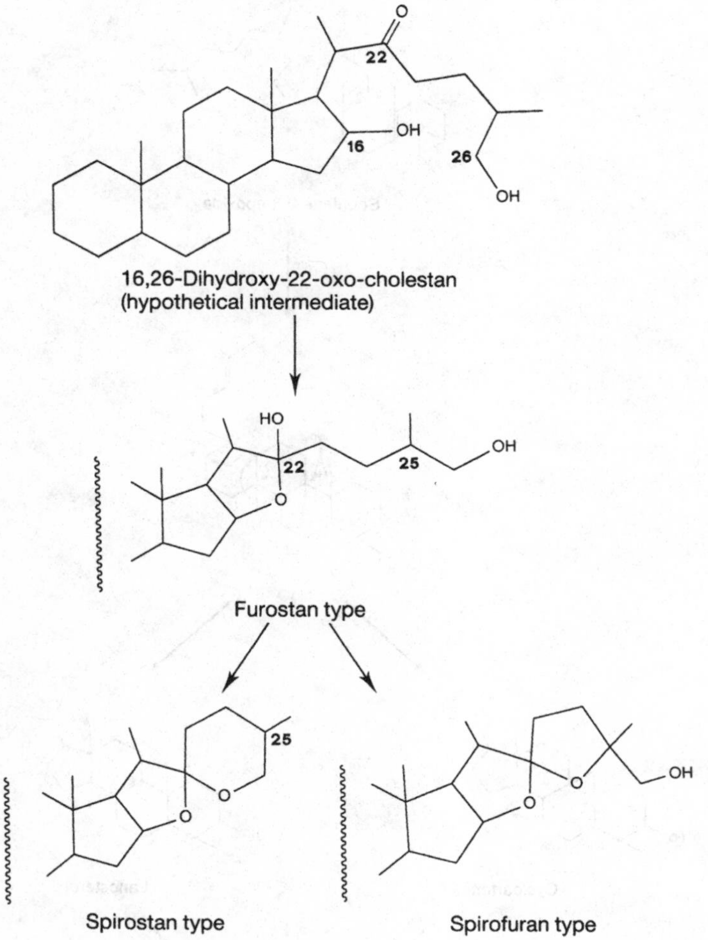

16,26-Dihydroxy-22-oxo-cholestan
(hypothetical intermediate)

Furostan type

Spirostan type

Spirofuran type

Fig. 1.6. Biosynthesis of furostans and spirostans.

furostan and spirostan steroid aglycones are biosynthetically derived from a 16,26-dihydroxy-22-oxo-cholestan (Fig. 1.6).

In tissue cultures from *Dioscorea tokoro* (Dioscoreaceae), diosgenin, yonogenin and tokorogenin can all be derived from cycloartenol, cholesterol, (25R)-cholest-5-en-3β,16β,26-triol and (25R)-cholest-5-en-3β,16β, 22ξ,26-tetraol. The relative efficiencies of their incorporation suggest the following pathway (Tomita and Uomori, 1971):

mevalonic acid → cycloartenol → cholesterol → (25R)-cholest-5-en-3β,16β,26-triol → (25R)-cholest-5-en-3β,16β,22ξ,26-tetraol → diosgenin → yonogenin and tokorogenin.

Certain plants are capable of incorporating a nitrogen atom (from arginine) into the side chain of the steroid moiety, thus giving the steroid alkaloid nucleus (Kaneko *et al.* 1976; Tschesche *et al.* 1976). It is possible that this involves direct replacement of the hydroxyl group in 26-hydroxycholesterol by an amino function but this has not been confirmed (Heftmann, 1967). Spirosolans and solanidans of tuber-bearing *Solanum* species that are genetically closely related to *S. tuberosum* have been postulated to derive from a common intermediate (Fig. 1.7) (Osman, 1984). Without any doubt, however, C_{27} steroid sapogenins and steroid alkaloids, occurring together in plants, are very closely related in terms of biosynthesis and metabolism.

Glycosylation of solanidine has been found to be catalysed by crude enzyme preparations from potato sprouts and tubers, while α-tomatine synthesis and accumulation in the tomato occur mainly in the leaves and roots. The biosynthesis of glycoalkaloids has been reviewed by Vagujfalvi (1965).

Kintia *et al.* (1974) and Wojciechowski (1975) have studied the biosynthesis of the oleanolic acid glycosides from *Calendula officinalis* (Asteraceae) by labelling studies and the use of cell-free enzyme prepar-

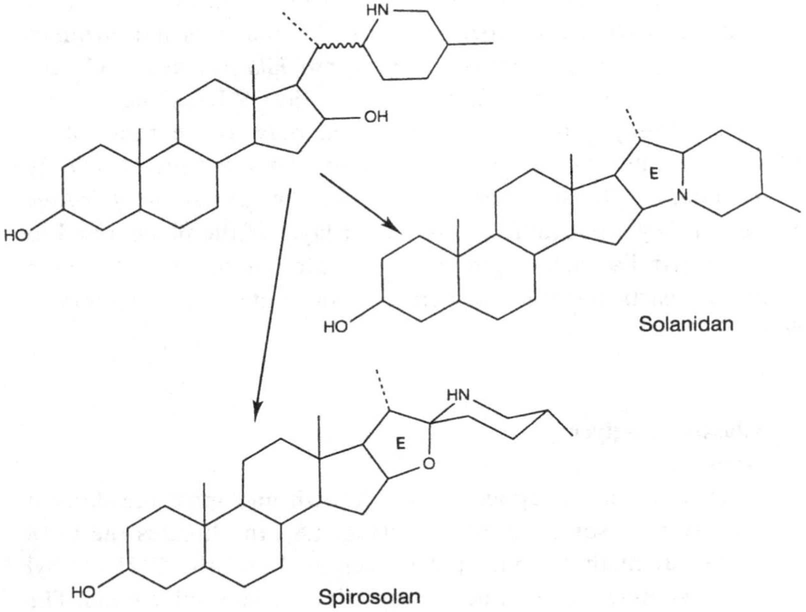

Fig. 1.7. Biosynthetic pathway for the formation of ring E of *Solanum* steroid alkaloids.

ations. Oleanolic acid is first glucuronylated at C-3, to give calendula saponin F. At this stage, the glycoside can either be glucosylated at C-28 (calendula saponin D_2) or galactosylated at the C-3 glucosyl residue (calendula saponin D). Further elongation of the saccharide moiety then occurs, giving calendula saponins B, C and A. The consecutive steps of sugar-chain elongation are catalysed by specific transferases, localized in different cellular compartments; oleanolic acid → calendula saponin F probably occurs in the endoplasmic reticulum; calendula saponin F → D_2 in the cytosol; calendula saponins F → D → B, D_2 → C → A in the Golgi membranes (and/or plasmolemma).

It has been suggested, as a general rule, that introduction of a carbohydrate chain at a hydroxyl group of an aglycone occurs in the stepwise fashion described above, while at the carboxyl group the chain is added in its completed form (Mzhelskaya and Abubakirov, 1968b). Evidence for a stepwise biosynthesis of α-solanine and α-chaconine from solanidine has also been produced (Jadhav *et al.* 1981).

A fair amount of information is available about the location of saponin biosynthesis in plants. While *Calendula officinalis* saponins are bio-synthesized in the green parts of the plant and then transported to the roots (Isaac, 1992), glycyrrhizin is biosynthesized in the roots of *Glycyrrhiza glabra* (Leguminosae) (Fuggersberger-Heinz and Franz, 1984). It is known that glycosylated triterpenes are stored in the roots and rhizomes of *Saponaria officinalis* (Caryophyllaceae), *Gypsophila paniculata* (Caryophyllaceae) and *G. altissima* (Henry *et al.* 1991). Labelling studies have shown that saponins are biosynthesized only in the roots of *G. paniculata* and are not translocated to the shoots. They accumulate mainly in the secondary phloem, while in the roots of *Bupleurum falcatum* (Umbelliferae) they are localized in the outer layer of the phloem and in *Panax ginseng* (Araliaceae) the ginsenosides are found located outside the root cambium, particularly in the periderm and outer cortex (Henry *et al.* 1991).

1.4   Classes of aglycone
1.4.1   *Steroids*
Over 100 steroid sapogenins are known and most are derived from the furostan or spirostan skeleton (Fig. 1.8). In all cases the C-18 and C-19 angular methyl groups are β-orientated and the C-21 methyl group has the α-configuration. There is sometimes a 5,6-double bond. The sapogenins are mostly hydroxylated at C-3.

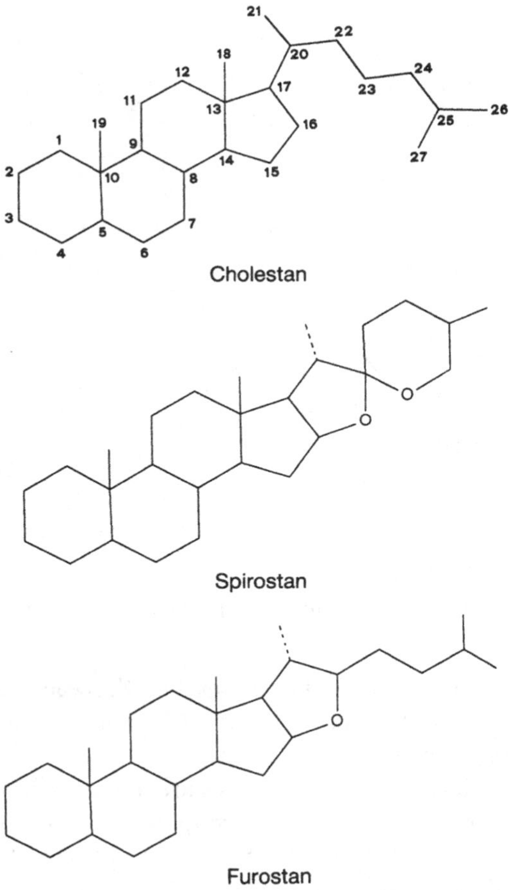

Cholestan

Spirostan

Furostan

Fig. 1.8. The common classes of steroid sapogenin.

1.4.2   Steroid alkaloids

There are two classes of steroid alkaloid sapogenin – the sol-anidans and the spirosolans (Fig. 1.9).

1.4.3   Triterpenes

The pentacyclic triterpenes can be divided into three main classes, depending on whether they have a β-amyrin, α-amyrin or lupeol skeleton (Fig. 1.10). By far the most frequently occurring are saponins belonging to the β-amyrin type (over 50%). Oleanolic acid, one of the β-amyrins, is the most common aglycone of this class (see Chapter 2). The structure elucidation of this triterpene was published in 1949 (Bischof *et al.* 1949).

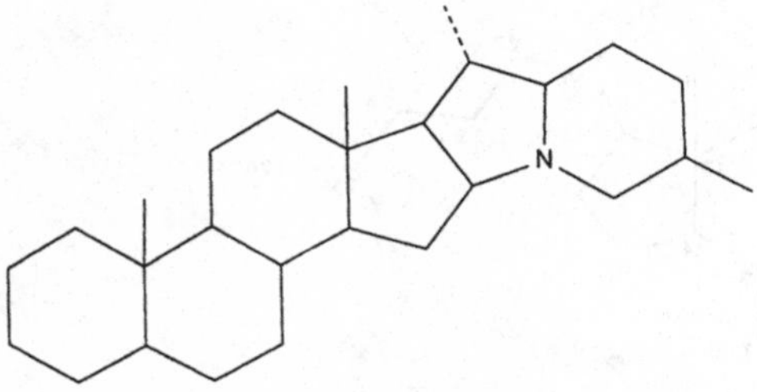

Solanidan

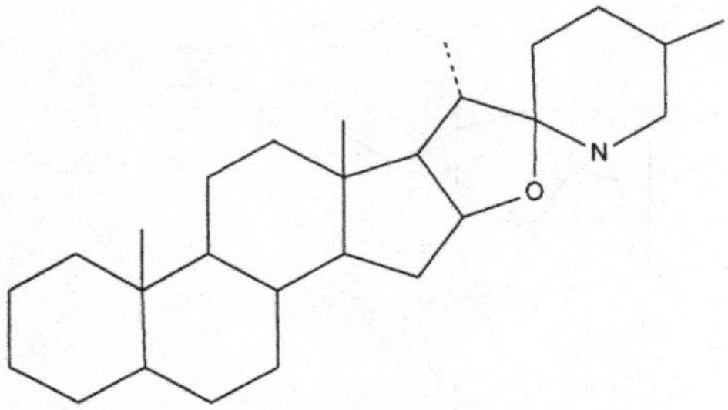

Spirosolan

Fig. 1.9. The two classes of steroid alkaloid sapogenin (conventional represent-ation).

It occurs in the free state in leaves of *Olea europaea* (Oleaceae), leaves of *Viscum album* (Loranthaceae), buds of cloves (*Syzygium aromaticum*, Myrtaceae) and in numerous other plant species.

Important structural elements of these classes are: the unsaturation at C-12(13); the functionalization of the methyl group of C-28, C-23 or C-30; polyhydroxylation at C-2, C-7, C-11, C-15, C-16, C-19, etc. Through etherification or lactonization, formation of an additional ring is possible. Esterification by aliphatic acids sometimes occurs.

Minor modifications of these skeletons are sufficient to obtain the other classes of triterpene (Fig. 1.11).

A certain proportion of the known triterpene saponins belongs to the dammarane class. The aglycones are tetracyclic triterpenes (Fig. 1.12).

Recently, triterpene saponins of the lanostane and holostane type have been isolated from marine organisms (Fig. 1.13). The holostanes are in fact composed of a lanostane skeleton with a distinctive D-ring fused γ-lactone. Lanostanes and cycloartanes (9,19-cyclolanostanes) also occur as aglycones of plant saponins.

1.5   Nomenclature and stereochemistry

The numbering method used for a typical steroid with a hetero-cyclic ring in the side chain is shown in Fig. 1.14: this is a 16,22:22,26-

β-Amyrin type
(oleananes)

α-Amyrin type
(ursanes)

Lupeol type
(lupanes)

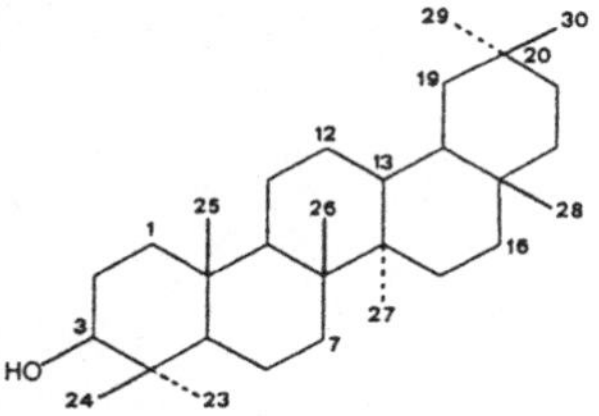

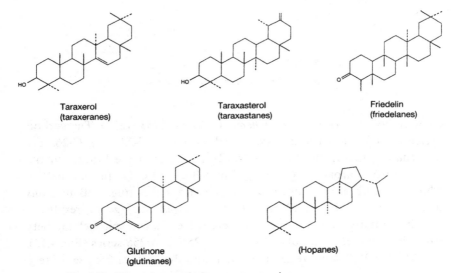

Fig. 1.10. The major classes of triterpene sapogenin.

Taraxerol
(taraxeranes)

Taraxasterol
(taraxastanes)

Friedelin
(friedelanes)

Glutinone
(glutinanes)

(Hopanes)

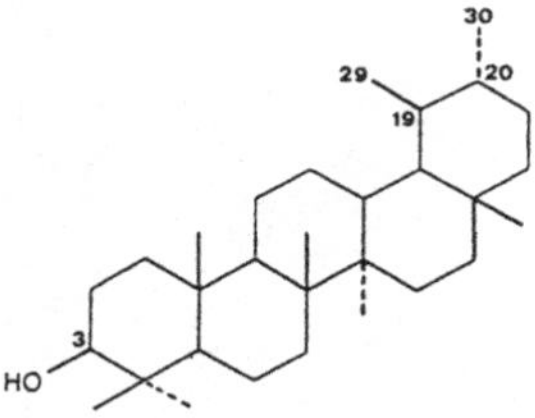

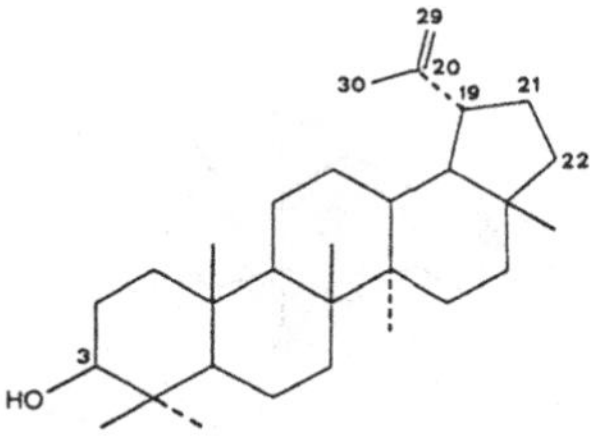

Fig. 1.11. Minor classes of triterpene sapogenin.

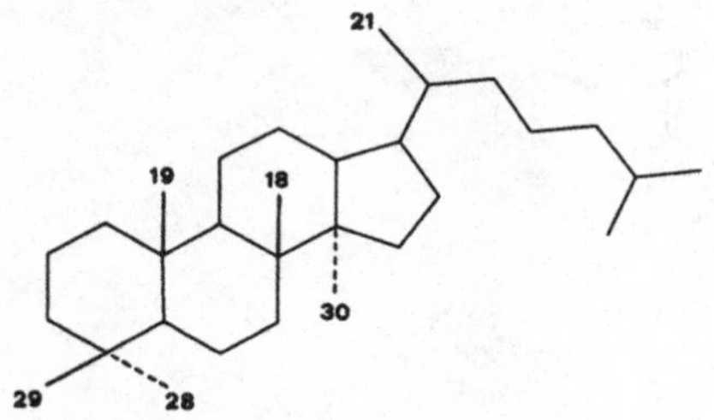

Fig. 1.12. The dammarane skeleton.

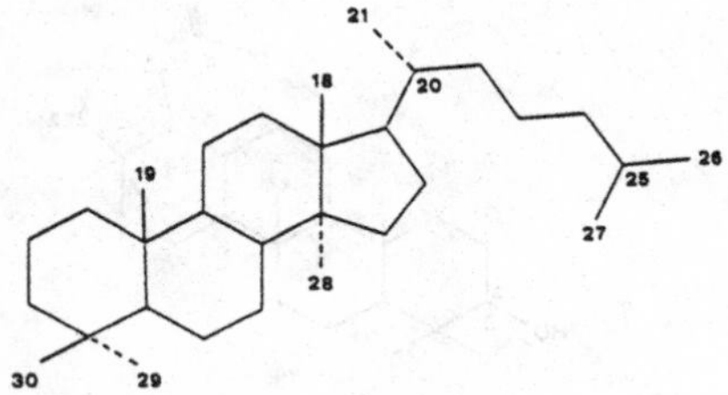

Lanostanes

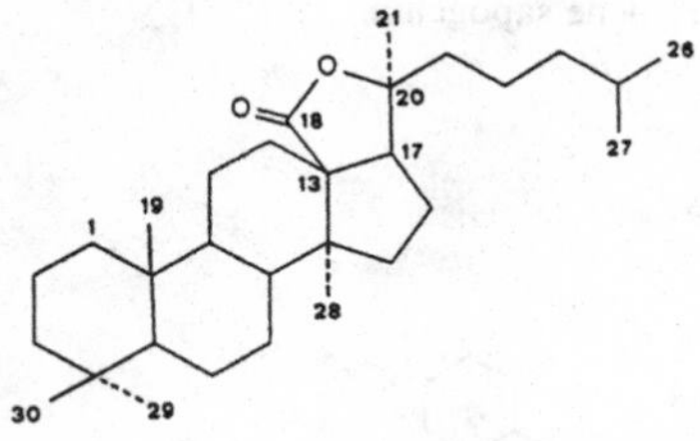

Holostanes

Fig. 1.13. Triterpene sapogenins from marine organisms.

diepoxycholestan (*Pure Appl. Chem.* (1989) **61**, 1783–1822). The carbon skeleton of cholesterol is oxygenated at C-16, C-22 and C-26, the oxygenation taking the form of 16,22- and 22,26-oxide bridges in the characteristic spiroacetal grouping. Rings B/C and C/D are *trans*-linked, while rings D/E are *cis*-linked. Depending on whether rings A/B are *trans* or *cis*, a 5α- (e.g. tigogenin) or a 5β-spirostan (e.g. smilagenin) results.

Spirostans are characterized by the existence of a ketospiroketal moiety (rings E/F) and may be subdivided into a 25R or a 25S series (Fig. 1.15). The 25S (or 25β) series (e.g. yamogenin) and the 25R (or 25α) series (e.g. diosgenin) were formerly referred to as neosapogenins or isosapogenins,

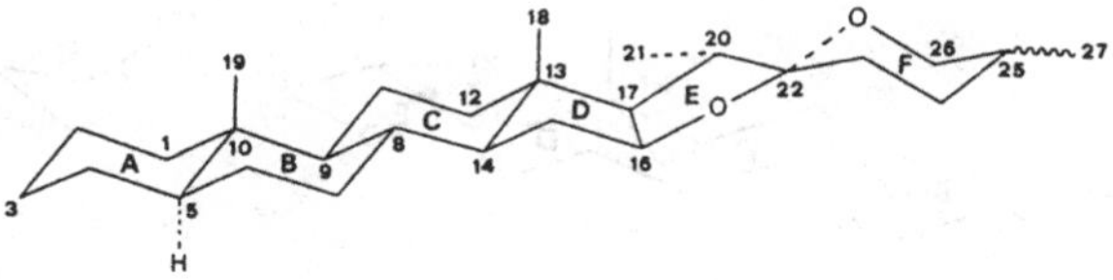

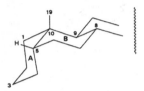

5α-Spirostan

5β-Spirostan

Fig. 1.14. The spirostan aglycone.

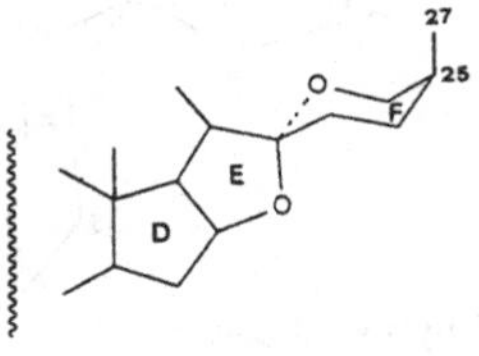

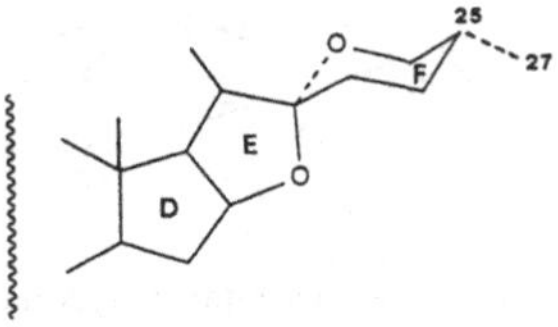

(25S)-   (25R)-

Fig. 1.15. The stereochemistry at C-25 of spirostans.

respectively. The C-25 methyl group is axially orientated in neosapogenins and equatorially orientated in isosapogenins.

Glycosides of the furostan type have ring F open and the sugar moiety attached at C-26.

In the series of steroid alkaloids, the nitrogen atom can be either secondary (*spirosolan* type) (Fig. 1.16) or tertiary (*solanidan* type) (Fig. 1.17). The spirosolans (oxaazaspirodecanes) are the aza-analogues of spirostans and can exist in either the 22R,25R (e.g. solasodine) or the 22S,25S (e.g. tomatidine) configurations (Fig. 1.16). Their numerotation corresponds to that of the spirostans.

In the solanidans, the nitrogen atom belongs to two ring systems simultaneously – hence the chemical name *indolizidines*. The solanidans have a 22R,25S configuration, with the C-25 methyl group in an equatorial position.

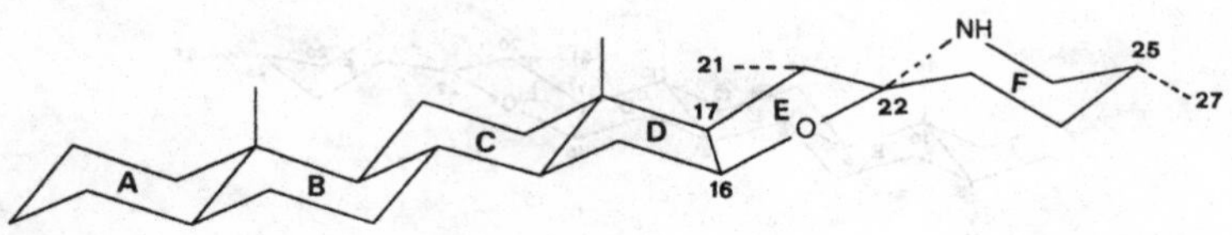

(22*R*,25*R*)-

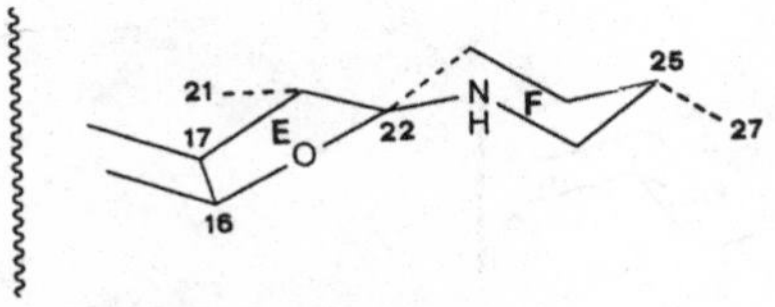

(22*S*,25*S*)-

Fig. 1.16. The spirosolan aglycone.

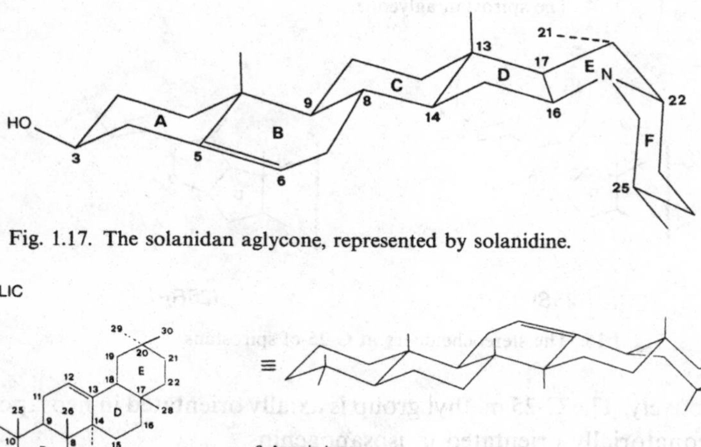

Fig. 1.17. The solanidan aglycone, represented by solanidine.

PENTACYCLIC

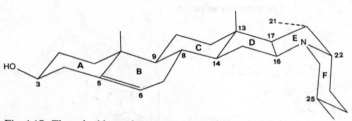

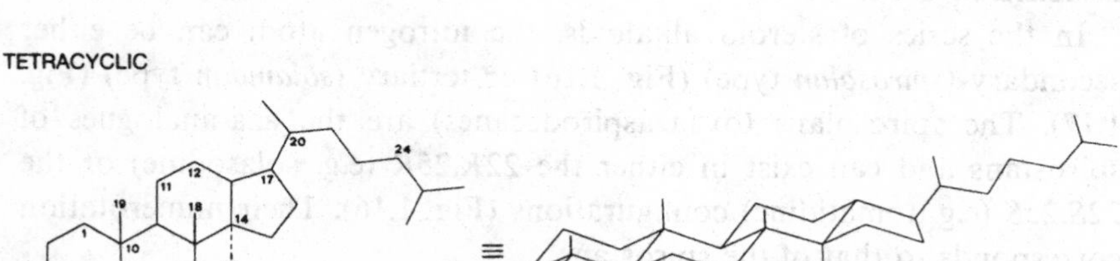

β-Amyrin type
(oleananes)

TETRACYCLIC

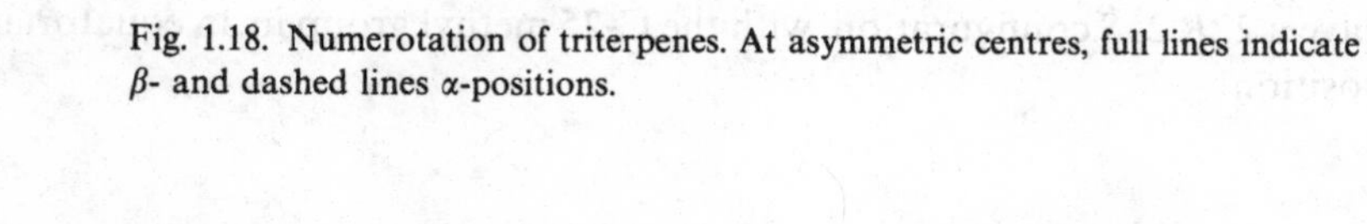

Dammaranes

Fig. 1.18. Numerotation of triterpenes. At asymmetric centres, full lines indicate
β- and dashed lines α-positions.

The situation for the pentacyclic and tetracyclic (dammarane) triterpenes is shown in Fig. 1.18. The pentacyclic triterpene example is from
one of the most frequently encountered classes: the β-amyrin type of
triterpene. Rings A/B, B/C and C/D are generally *trans*-linked, while rings
D/E are *cis*-linked (at asymmetric centres, full lines indicate β- and dashed
lines α-positions). X-ray crystallographic studies indicate a relatively
planar conformation of the triterpene ring system (Roques *et al.* 1978).

2

Occurrence and distribution

Saponins are extremely widely distributed in the plant kingdom. Even by 1927, Kofler had listed 472 saponin-containing plants (Kofler, 1927) and it is now known that over 90 families contain saponins. Gubanov *et al.* (1970) found, in a systematic investigation of 1730 central Asian plant species, that 76% of the families contained saponins.

Saponins occur in plants which are used as human food: soybeans, chick peas, peanuts, mung beans, broad beans, kidney beans, lentils, garden peas, spinach, oats, aubergines, asparagus, fenugreek, garlic, sugar beet, potatoes, green peppers, tomatoes, onions, tea, cassava, yams (Birk and Peri, 1980; Oakenfull, 1981; Price *et al.* 1987) and in leguminous forage species (such as alfalfa). Kitagawa (1984) has described the isolation of soyasaponin I (**454**, Appendix 2) from a number of forage and cover crops, including *Trifolium* spp., *Medicago* spp., *Astragalus* spp. and *Vicia sativa*. Finally, saponins are also present in numerous herbal remedies (see Chapter 7).

Saponin content depends on factors such as the cultivar, the age, the physiological state and the geographical location of the plant. There can be considerable variation in composition and quantity of saponins in vegetable material from different places, as documented for *Lonicera japonica* (Caprifoliaceae) (Kawai *et al.* 1988).

The saponin distribution among the organs of a plant varies considerably. In the garden marigold (*Calendula officinalis*, Asteraceae), for example, saponins with a glucuronic acid moiety at C-3 of oleanolic acid are found in the flowers, while a glucose moiety at the same position is found in the saponins from the roots (Lutomski, 1983; Vidal-Ollivier *et al.* 1989a,b). The flowers contain 3.57% saponins, while the roots have 2.55% of their dry weight in the form of saponins (Isaac, 1992). Ginsenoside levels in *Panax ginseng* (Araliaceae) are lowest in the leafstalks and stem

(0.77%), intermediate in the main root (1.3%) and lateral roots (3.5%) and highest in the leaves (5.2%) and root hairs (6.1%) (Koizumi *et al.* 1982).

In leaf cells of sugar beet and root cells of the pea plant, the triterpene glycosides are present mainly in a bound form, with the highest concentration occurring in the chloroplasts and mitochondria (Anisimov and Chirva, 1980). In young lucerne plants (alfalfa, *Medicago sativa*), they are concentrated in the outer layer of the root bark (Pedersen, 1975). Analysis of soyasaponins in soybeans by TLC, HPLC and GC (see glossary for all abbreviations) has shown that the hypocotyl contains the highest concentrations. The cotyledon also contains soyasaponins but they are absent from the seed coat (Taniyama *et al.* 1988b).

The fact that saponins are localized in organelles which have a high metabolic turnover rate implies that they are not simply ballast material but may be important regulatory substances in the metabolism and development of an organism, i.e. they may be physiologically significant constituents (Anisimov and Chirva, 1980).

2.1   Triterpene saponins

This subdivision forms by far the most extensive collection of saponins, comprising, to date, over 750 triterpene glycosides with 360 sapogenins (Bader and Hiller, 1987). Families and species of plants which have yielded fully and partially characterized triterpene saponins are listed in Appendix 1. Most have either oleanane (β-amyrin) or dammarane skeletons, with ursanes, hopanes, lanostanes or lupanes assuming secondary importance with respect to distribution.

The triterpene saponins can be mono-, bi- or even tridesmosidic. One sugar chain is often attached at C-3 and a second is frequently found to be esterified to the carboxyl group at C-17 of the aglycone (except certain dammarane glycosides and lanostane glycosides which have a second or even a third glycosidically bound sugar chain). The first tridesmosidic triterpene, a 9,19-cyclolanostane (cycloartane) derivative substituted glycosidically at positions C-3, C-6 and C-25, was isolated from the roots of Korean *Astragalus membranaceus* (Leguminosae) (Kitagawa *et al.* 1983e). An example of a tridesmosidic olean-12-en saponin is quinoside A (**354**, Appendix 2), in which sugars are attached at positions C-3, C-23 and C-28 of hederagenin (Meyer *et al.* 1990). A tridesmoside of 16α-hydroxy-medicagenic acid (zahnic acid) has been found in the aerial parts of alfalfa (*Medicago sativa*, Leguminosae) (Oleszek *et al.* 1992).

As monosaccharide units, D-glucose, L-rhamnose, L-arabinose and D-xylose occur widely. Much rarer are D-apiose (saponin **330**),

D-ribose (saponin **206**), D-quinovose (saponin **289**) and D-allose (saponin **311**). (See Appendix 2.)

Among the pteridophytes, there are several species containing triterpene saponins but the gymnosperms are practically free of these compounds. The angiosperms, by comparison, are very rich in saponins. These are found principally in the dicotyledons, in families such as the Leguminosae, Araliaceae, Caryophyllaceae, Asteraceae, Primulaceae and Sapindaceae. The Chenopodiaceae, with 1500 species in 100 genera, are well represented by triterpene saponins, with oleanolic acid as the preponderant aglycone. Dammarane saponins occur especially frequently in the families Araliaceae, Cucurbitaceae and Rhamnaceae.

The main saponin-containing family of the monocotyledons is the Liliaceae, characterized by the predominant occurrence of steroid glycosides in the genera *Allium, Asparagus, Convallaria, Paris, Polygonatum, Ruscus* and *Trillium*. Another excellent source of steroid saponins is the Dioscoreaceae.

Plant material often contains triterpene saponins in considerable amounts. Thus, primula root contains about 5–10% saponin, licorice root between 2% and 12% glycyrrhizin, quillaia bark up to 10% of a saponin mixture and the seeds of the horse chestnut up to 13% aescine. In other words, the concentration of saponins in plants is high when compared with other secondary metabolites.

There are numerous structural variants of the oleanane (or, more precisely, the olean-12-en) skeleton and a number of these are shown in Table 2.1. Other representative triterpene aglycones are listed in Table 2.2. A review article by Tschesche and Wulff (1972) gives references and further examples, together with physical constants, of both oleananes and other triterpenes. Mahato and co-workers (Das and Mahato, 1983; Mahato *et al.* 1992) have published lists of recently isolated triterpenes (not necessarily from saponins) with their physical constants and plant sources. Oleanane triterpenes (and some of their glycosides) have been the subject of an update (Mallavarapu, 1990), covering various aspects of their occurrence and chemistry. Another authoritative source of information on the triterpenes is the book written by Boiteau and colleagues (1964).

The number of known triterpenes is very large (Boiteau *et al.* 1964) and only a small proportion has been characterized in glycosidic form. Triterpenes are frequently isolated only after hydrolysis of plant extracts and it is not always easy to ascertain from published work whether they actually occur in the free or glycosidic forms in the plant itself.

Table 2.1. *Structures of commonly occurring olean-12-en aglycones*

No.	Olean-12-en aglycone	–OH	=O	–COOH	Other
1	β-Amyrin	3β			
2	Oleanolic acid	3β		28	
3	Epikatonic acid	3β		29	
4	α-Boswellic acid	3α		24	
5	Momordic acid	3β	1	28	
6	Glycyrrhetinic acid	3β	11	30	
7	Gypsogenin	3β	23	28	
8	Gypsogenic acid	3β		23,28	
9	Cincholic acid	3β		27,28	
10	Serjanic acid (30-*O*-methyl-spergulagenate)	3β		28	30-COOMe
11	Maniladiol	3β,16β			
12	Sophoradiol	3β,22β			
13	3β,22β-Dihydroxyolean-12-en-29-oic acid	3β,22β		29	
14	2β-Hydroxyoleanolic acid	2β,3β		28	
15	Maslinic acid	2α,3β		28	
16	Echinocystic acid	3β,16α		28	
17	Hederagenin	3β,23		28	
18	Phytolaccagenic acid	3β,23		28	30-COOMe
19	Siaresinolic acid	3β,19α		28	
20	21β-Hydroxyoleanolic acid (machaerinic acid)	3β,21β		28	
21	29-Hydroxyoleanolic acid	3β,29		28	
22	Azukisapogenol	3β,24		29	
23	Soyasapogenol E	3β,24	22		
24	Primulagenin D (28-dehydroprimulagenin)	3β,16α	28		
25	3β,24-Dihydroxyolean-12,15-dien-28-oic acid	3β,24			15-en
26	Soyasapogenol C	3β,24			21-en
27	Glabrinic acid	3β,26	11	30	
28	Quillaic acid	3β,16α	23	28	
29	21β-Hydroxygypsogenin	3β,21	23	28	
30	Barringtogenic acid	2α,3β		23,28	

Table 2.1. (*Cont.*)

No.	Olean-12-en aglycone	–OH	=O	–COOH	Other
31	Medicagenic acid	$2\beta,3\beta$		23,28	
32	Dianic acid	$3\beta,29$		23,28	
33	Soyasapogenol B	$3\beta,22\beta,24$			
34	$3\beta,22\beta,24$-Trihydroxy-olean-12-en-29-oic acid	$3\beta,22\beta,24$		29	
35	Primulagenin A	$3\beta,16\alpha,28$			
36	$2\beta,3\beta,28$-Trihydroxy-olean-12-en	$2\beta,3\beta,28$			
37	Priverogenin A	$3\beta,16\alpha,22\alpha$		28	
38	16α-Hydroxyhederagenin (caulophyllogenin)	$3\beta,16\alpha,23$		28	
39	21β-Hydroxyhederagenin	$3\beta,21\beta,23$		28	
40	$3\beta,21\beta,22\beta$-Trihydroxy-olean-12-en-29-oic acid	$3\beta,21\beta,22\beta$		29	
41	23-Hydroxyimberbic acid	$1\alpha,3\beta,23$		29	
42	Arjunic acid	$2\alpha,3\beta,19\alpha$		28	
43	Arjunolic acid	$2\alpha,3\beta,23$		28	
44	Asterogenic acid	$2\beta,3\beta,16\alpha$		28	
45	Bayogenin	$2\beta,3\beta,23$		28	
46	16-Hydroxy-medicagenic acid	$2\beta,3\beta,16\alpha$		23,28	
47	Presenegenin	$2\beta,3\beta,27$		23,28	
48	Jaligonic acid	$2\beta,3\beta,23$		28,30	
49	Phytolaccagenin	$2\beta,3\beta,23$		28	30-COOMe
50	Belleric acid	$2\alpha,3\beta,23,24$			
51	Barringtogenol A	$2\alpha,3\beta,23,28$			
52	Protobassic acid	$2\beta,3\beta,6\beta,23$		28	
53	Platycogenic acid C	$2\beta,3\beta,16\beta,21\beta$		28	
54	Polygalacic acid	$2\beta,3\beta,16\alpha,23$		28	
55	Tomentosic acid	$2\alpha,3\beta,19\beta,23$		28	
56	Arjungenin	$2\alpha,3\beta,19\alpha,23$		28	
57	Esculentagenic acid	$2\beta,3\beta,23,30$		28	
58	23-Hydroxylongispino-genin	$3\beta,16\beta,23,28$			
59	Cyclamiretin E	$3\beta,16\alpha,28,30$			
60	Soyasapogenol A	$3\beta,21\beta,22\beta,24$			
61	Oxytrogenol	$3\beta,22\beta,24,29$			

(*cont.*)

Table 2.1. (*Cont.*)

No.	Olean-12-en aglycone	–OH	=O	–COOH	Other
62	3α,21β,22α,28-Tetrahy-droxyolean-12-en	3α,21β, 22α,28			
63	3β,23,27,29-Tetrahy-droxyoleanolic acid	3β,23,27, 29		28	
64	Barringtogenol C	3β,16α, 21β,22α, 28			
65	Camelliagenin C	3β,16α, 22α,23, 28			
66	16α-Hydroxyprotobassic acid	2β,3β,6β, 16α,23		28	
67	Platycodigenin	2β,3β,16α, 23,24		28	
68	Protoaescigenin	3β,16α, 21β,22α, 24,28			
69	Theasapogenol A	3β,16α, 21β,22α, 23,28			
70	R$_1$-Barrigenol	3β,15α, 16α,21β, 22α,28			

It is sometimes very difficult to obtain the genuine aglycones from the parent saponins. This problem is especially acute for the triterpenes containing a 13β,28-oxide structure (aglycone skeleton A in Table 2.2). It took a long time, for example, before the aglycone cyclamiretin A (**72**) of cyclamin (from the tubers of *Cyclamen europaeum*, Primulaceae) was completely characterized (Tschesche *et al.* 1966a). As is the case for other 13β,28-oxide aglycones (protoprimulagenin A (**71**), saikogenin F (**76**), etc.), it can be easily ring-opened by acid to the corresponding 12-en-28-alcohol. Primulagenin A (**35**) (Table 2.1) is most probably an artifact produced during hydrolysis of saponins containing protoprimula-genin A (**71**) (Table 2.2) as aglycone (Tschesche *et al.* 1983). It should be noted here, however, that not all 12-en-28-alcohol aglycones are arti-facts. Notable examples are barringtogenol C (**64**) and protoaescigenin (**68**).

Several aglycones exist which deviate from the normal C$_{30}$ skeleton. For example, the nortriterpene pfaffic acid (**125**), parent aglycone of the pfaffosides (Takemoto *et al.* 1983a) has a hexacyclic skeleton.

Table 2.2. *Triterpene aglycones (other than olean-12-en type)*

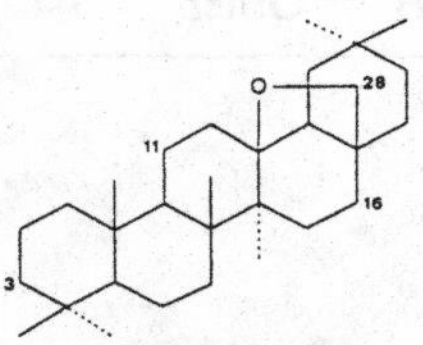

A
(Oleanane)

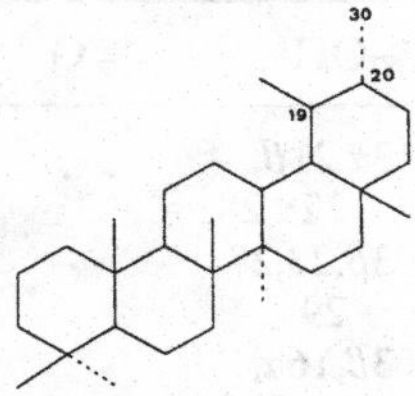

B
(Ursane)

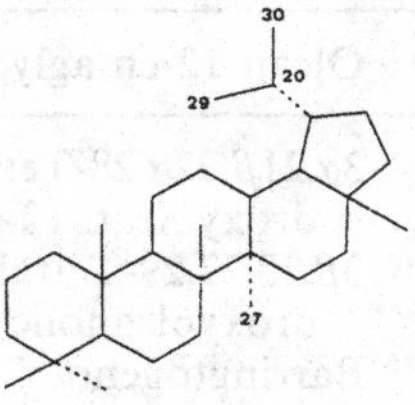

C
(Lupane)

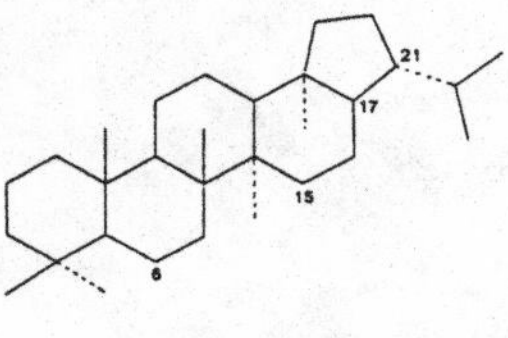

D
(Hopane)

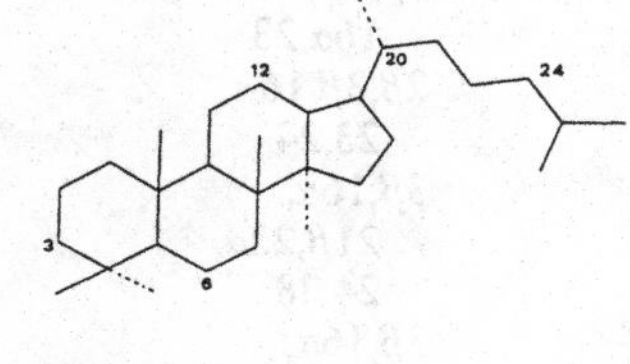

E
(Dammarane)

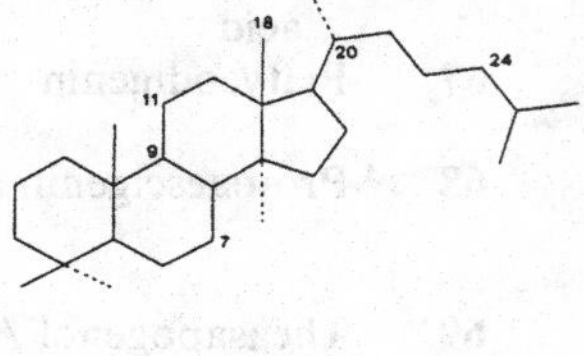

F
(Lanostane)

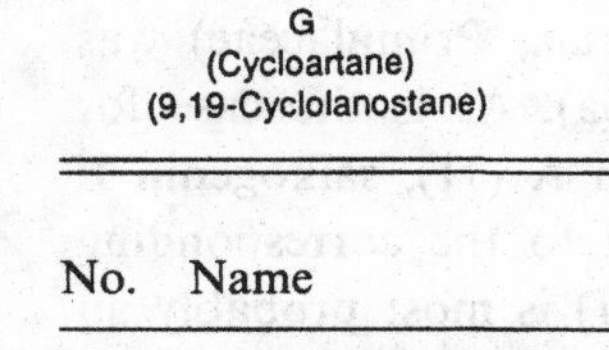

G
(Cycloartane)
(9,19-Cyclolanostane)

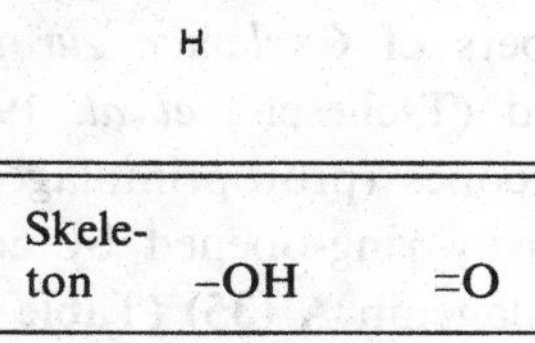

H

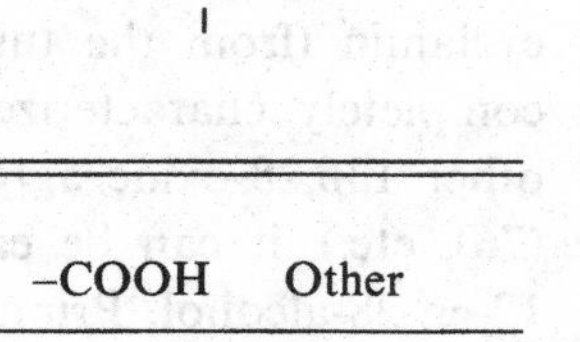

I

No.	Name	Skeleton	–OH	=O	–COOH	Other
71	Protoprimulagenin A	A	$3\beta,16\alpha$			
72	Cyclamiretin A	A	$3\beta,16\alpha$	30		
73	Rotundiogenin A	A	$3\beta,16\alpha$			11-en
74	Saikogenin E	A	$3\beta,16\beta$			11-en
75	Anagalligenone	A	$3\beta,23$	16		
76	Saikogenin F	A	$3\beta,16\beta,23$			
77	Saikogenin G (anagalligenin B)	A	$3\beta,16\alpha,23$			
78	Priverogenin B	A	$3\beta,16\alpha,22\alpha$			

(cont.)

Table 2.2. (*Cont.*)

No.	Name	Skeleton	–OH	=O	–COOH	Other
79	Anagalligenin A	A	3β,16α,22α,28			
80	α-Amyrin	B	3β			12-en
81	Ursolic acid	B	3β		28	12-en
82	Quinovic acid	B	3β		27,28	12-en
83	3β-Hydroxyurs-12,20(30)-dien-27,28-dioic acid	B	3β		27,28	12,20(30)-dien
84	Pomolic acid	B	3β,19α		28	12-en
85	Ilexgenin B	B	3β,19α		28	12-en, (30S)30β
86	Ilexgenin A	B	3β,19α		24,28	12-en
87	21β-Hydroxyursolic acid	B	3β,21β		28	12-en
88	23-Hydroxyursolic acid	B	3β,23		28	12-en
89	3β,23-Dihydroxy-taraxer-20-en-28-oic acid	B	3β,23		28	20-en
90	Rotundic acid	B	3β,19α,23		28	12-en
91	Rotungenic acid	B	3β,19α,24		28	12-en
92	Madasiatic acid	B	2α,3β,6β		28	12-en
93	Asiatic acid	B	2α,3β,23		28	12-en
94	Euscaphic acid	B	2α,3α,19α		28	12-en
95	Tormentic acid	B	2α,3β,19α		28	12-en
96	2α,3β,19α-Trihydroxyurs-12-en-23,28-dioic acid	B	2α,3α,19α		23,28	12-en
97	6β-Hydroxytormentic acid	B	2α,3β,6β,19α		28	12-en
98	7α-Hydroxytormentic acid	B	2α,3β,7α,19α		28	12-en
99	23-Hydroxytormentic acid	B	2α,3β,19α,23		28	12-en
100	24-Hydroxytormentic acid	B	2α,3β,19α,24		28	12-en
101	1α,3β,19α,23-Tetrahydroxyurs-12-en-28-oic acid	B	1α,3β,19α,23		28	12-en
102	Madecassic acid	B	2α,3β,6β,23		28	12-en
103	6β,23-Dihydroxy-tormentic acid	B	2α,3β,6β,19α,23		28	12-en
104	Lupeol	C	3β			20(29)-en
105	Betulin	C	3β,28			20(29)-en
106	Betulinic acid	C	3β		28	20(29)-en

(*cont.*)

Table 2.2. (*Cont.*)

No.	Name	Skele-ton	–OH	=O	–COOH	Other
107	3-*epi*-Betulinic acid	C	3α		28	20(29)-en
108	3β,23-Dihydroxylup-20(29)-en-oic acid	C	3β,23		28	20(29)-en
109	3α-Hydroxylup-20(29)-en-23,28-dioic acid	C	3α		23,28	20(29)-en
110	3α,11α-Dihydroxylup-20(29)-en-23,28-dioic acid	C	3α,11α		23,28	20(29)-en
111	Cylicodiscic acid	C	3β,27α		28	20(29)-en
112	Mollugogenol B	D	3β,6α			15,17(21)-dien
113	(20S)-Protopanaxa-diol	E	3β,12β,20S			24-en
114	(20S)-Protopanaxa-triol	E	3β,6α,12β,20S			24-en
115	Bacogenin A$_1$	E	3β,19,20	16		24-en
116	Seychellogenin	F	3β			7,9(11)-dien 18,20-lactone
117	Mollic acid	G	1α,3β		28	
118	3β,21,26-Trihydroxy-9,19-cyclolanost-24-en	G	3β,21,26			24-en
119	Thalicogenin	G	3β,16β,22,28			24-en
120	3β,16β,24,25-Tetrahy-droxy-9,19-cyclo-lanostane	G	3β,16β,24,25			
121	3β,6α,16β,24,25-Pentahydroxy-9,19-cyclolanostane	G	3β,6α,16β,24,25			
122	Cycloastragenol (astramem-brangenin, cyclo-siversigenin)	H	3β,6α,16β,25			
123	3β-Hydroxy-9,19-cyclolanost-24(28)-en	I	3β			24(28)-en
124	Jessic acid	I	1α,3β	23	29	24(28)-en

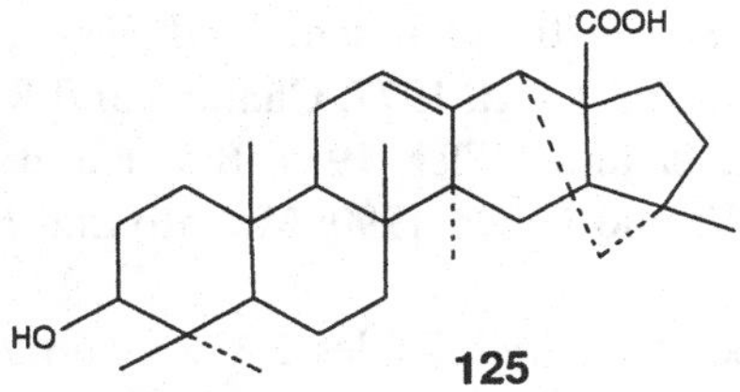

COOH

125

Saponins with nortriterpenoid aglycones are present in *Guaiacum officinale* (Zygophyllaceae) (see Section 2.1.1.) (Ahmad *et al.* 1986a) and certain other plants.

Modified lanostane saponins have been isolated from, for example, *Astragalus* species (Leguminosae) (Kitagawa *et al.* 1983a). Their aglycones contain a cyclopropane moiety and cyclic ethers with 20*S*,24*R* and 20*R*,24*S* conformations, e.g. **122**.

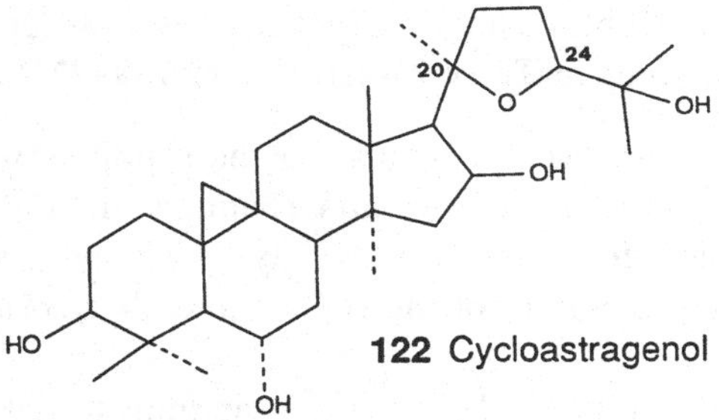

122 Cycloastragenol

Another derivative of a lanostane skeleton occurs in the saponin picfeltarraenin II, which is the 3-*O*-[α-L-rhamnopyranosyl(1→2)]-*β*-D-glucopyranoside of the aglycone picfeltarraegenin II (from *Picria feltarrae*, Scrophulariaceae). This aglycone contains a C-9 methyl substituent and a cyclic ether (**126**) (Jin *et al.* 1987).

A comprehensive listing of the known triterpene saponins is beyond the scope of this book, but for tabulations of over 750 saponins, the reader is referred to a number of excellent review articles in the field: Hiller *et al.*

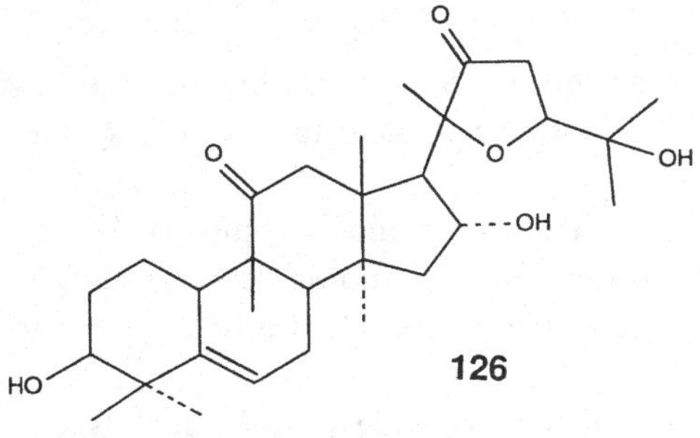

126

1966; Wulff, 1968; Woitke *et al.* 1970; Tschesche and Wulff, 1972; Agarwal and Rastogi, 1974; Hiller and Voigt, 1977; Chandel and Rastogi, 1980; Hiller and Adler, 1982; Adler and Hiller, 1985; Bader and Hiller, 1987; Mahato *et al.* 1988; Schöpke and Hiller, 1990; Mahato and Nandy, 1991.

The selected saponins in Appendix 2 and Tables 2.4, 2.6 and 2.7 (see below) will give an idea of the diversity and distribution of this class of natural product (the condensed system of nomenclature is used, where the common configuration and ring size are implied in the symbol for the monosaccharides; arrows indicating the glycosidic linkages have been replaced by dashes, for simplicity). They are listed according to the structures of the corresponding aglycones in order of increasing sugar-chain length and are divided into three sections:

- triterpene saponins (general) in Appendix 2 (structures **127–644**)
- ester saponins in Tables 2.6 and 2.7 (structures **794–850**)
- dammarane saponins in Table 2.4 (structures **738–771**).

Those documented here correspond mainly to the saponins referred to in other parts of the book. Arabinose and rhamnose in the sugar chains are generally of the L-form and linked α-glycosidically, while the other monosaccharides are generally of the D-form and β-glycosidically linked.

A number of comments can be made about the saponins found in the tables:

- Triterpene glycosides containing oleanolic acid (**2**) and hedera-genin (**17**) as aglycones are found most frequently.
- Clematoside C (**265**), with 11 monosaccharides, has the largest number of sugar residues of any known saponin. A series of saponins with 9 or 10 sugar moieties has been isolated from *Solidago gigantea* (Asteraceae) (Reznicek *et al.* 1989b, 1990a).
- Phaseoloside E (**425**) has the longest single saccharide chain (8 moieties) (Lazurevskii *et al.* 1971). A saponin from *Solidago canadensis* (Asteraceae) has 9 moieties but the interglycosidic linkages have not all been determined yet (Hiller *et al.* 1987b).
- When a sugar chain has an uronic acid moiety, this is almost exclusively glucuronic acid. In saponins containing glucuronic acid, this monosaccharide is generally found adjacent to C-3 of the aglycone.
- There are few examples of monodesmosidic acylglycosides, e.g.

with the sugar chain in position C-28 and no sugar attached at C-3. Asiaticoside (597) is one of the exceptions. It was, in fact, the first such triterpene glycoside to be isolated (Polonsky and Zylber, 1961). These acylglycosides do not show typical monodesmosidic saponin properties and can be classified rather as representatives of bidesmosidic saponins.

- Prosapogenins containing 23-*O*-glycosides of hederagenin and glycosides of 4-*epi*-hederagenin have been described but the true saponins from *Clematis chinensis* (Ranunculaceae) have not yet been isolated (Kizu and Tomimori, 1982).

- An 11α-methoxylated saponin (136) has been isolated from *Acanthopanax hypoleucus* (Araliaceae). This is not an artifact arising from extraction with methanol, as might be supposed, but a true natural product since it is obtained after ethanol extraction of the plant material (Kohda *et al.* 1990).

Bidesmosides form the largest single group of known triterpene glycosides. This is probably because of their abundance in individual plants and their widespread distribution. One of the most frequently observed of these saponins is gypsoside A (292, Appendix 2), first isolated from *Gypsophila pacifica* (Caryophyllaceae). It has since been found in 44 out of 62 species of the Caryophyllaceae and, while present in most of the genera *Gypsophila* (24 species) and *Acanthophyllum* (eight species), it is completely lacking in *Dianthus* species (Yukhananov *et al.* 1971). These bidesmosidic glycosides present a formidable challenge for their isolation and structure elucidation because of very complex (and often highly branched) sugar chains.

A number of novel cyclic bidesmosidic saponins have recently been isolated from the tubers of *Bulbostemma paniculatum* (Cucurbitaceae). The tubers, 'Tu Bei Mu', are used in Chinese folk medicine for their anti-inflammatory and verrucidal properties. Tubeimoside I (645) has at its core a bayogenin moiety. Two of the monosaccharides (arabinose and rhamnose) are bridged by 3-hydroxy-3-methyl glutaric acid (Kasai *et al.* 1986a; Kong *et al.* 1986). Tubeimosides II and III are glycosides of polygalacic acid. In tubeimoside III, the arabinosyl moiety of tubeimosides I and II is replaced by a glucosyl moiety (Kasai *et al.* 1988a).

A dimeric saponin (648) has been isolated from the leaves of the Korean medicinal plant *Rubus coreanus* (Rosaceae). The aglycone of this saponin was named coreanogenoic acid and consists of two units of oxygenated 19α-hydroxyursolic acid (Ohtani *et al.* 1990).

An *N*-acetylated glycoside of oleanolic acid, aridanin (649), has been

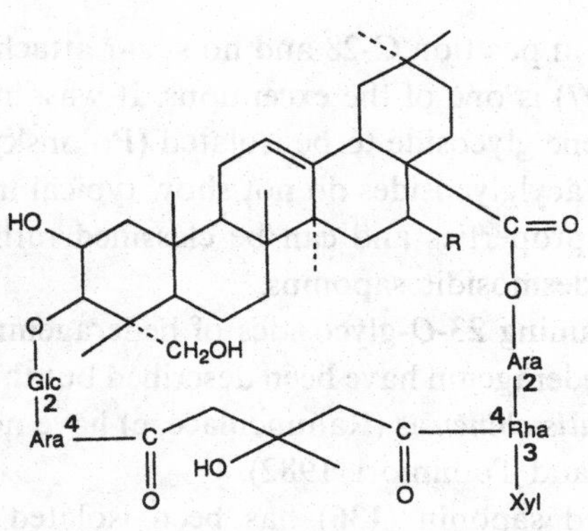

645 Tubeimoside I   R = -H

646 Tubeimoside II   R = -OH

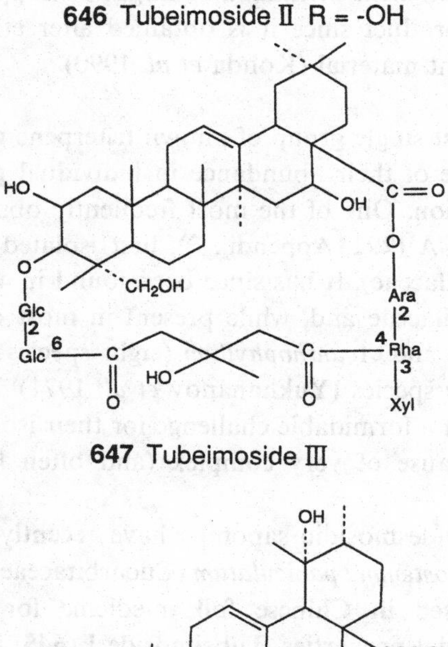

647 Tubeimoside III

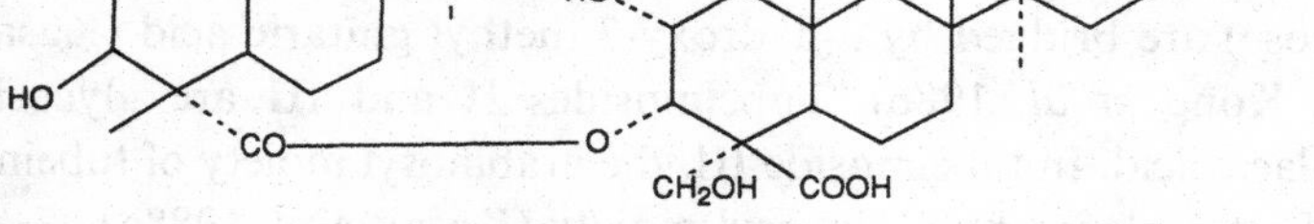

648

isolated from the leaves of *Pithecellobium cubense* and *P. arboreum* (Leguminosae) (Ripperger *et al.* 1981) and from the fruits of *Tetrapleura tetraptera* (Leguminosae) (Adesina and Reisch, 1985). Further phytochemical work on the fruits of *T. tetraptera* has shown the presence of two further molluscicidal *N*-acetylated glycosides of oleanolic acid (**650, 651**) and one of echinocystic acid (**652**) (Maillard *et al.* 1989).

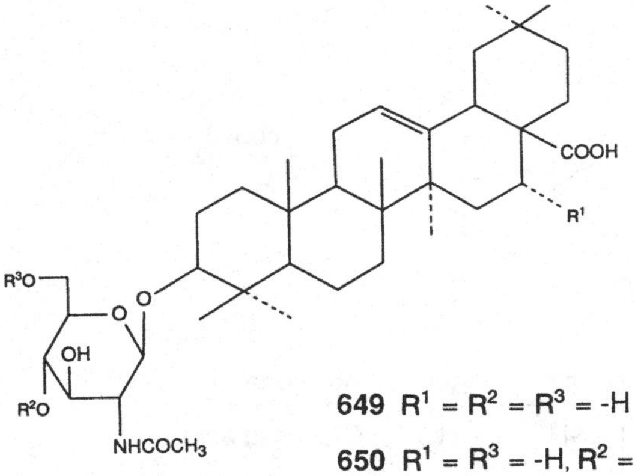

649 $R^1 = R^2 = R^3 = $ -H

650 $R^1 = R^3 = $ -H, $R^2 = $ -Gal

651 $R^1 = R^2 = $ -H, $R^3 = $ -Glc

652 $R^1 = $ -OH, $R^2 = R^3 = $ -H

Similarly, 2-acetamido-2-deoxy-β-D-glucopyranosyl (*N*-acetyl-D-glucosamine) derivatives of echinocystic acid have been obtained from the root bark of *Albizzia anthelmintica* (Leguminosae) (**653–655**) (Carpani *et al.* 1989) and from the seeds of *A. lucida* (**656**) (Orsini *et al.* 1991).

Three further saponins containing *N*-acetyl-D-glucosamine and apiose moieties, from the bark of *Entada phaseoloides* (Leguminosae), are shown

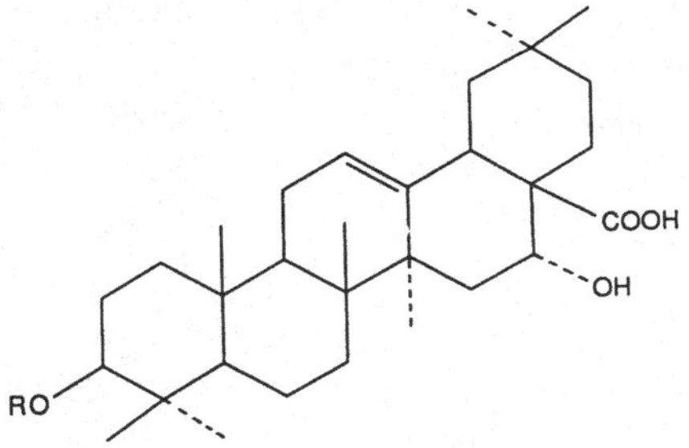

653 R = -Glc^6-Ara^2-Ara^3-Glc
　　　　　|^2
　　　NHCOCH$_3$

655 R = -Glc^6-Ara
　　　　　|^2
　　　NHCOCH$_3$

654 R = -Glc^6-Ara^2-Ara
　　　　　|^2
　　　NHCOCH$_3$

656 R = -Glc^6-Fuc^2-Xyl
　　　　　|^2
　　　NHCOCH$_3$

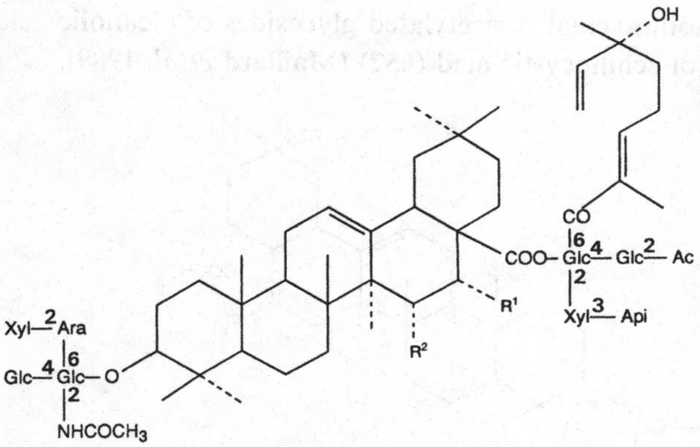

	R^1	R^2	Aglycone
657	-H	-H	Oleanolic acid
658	-OH	-H	Echinocystic acid
659	-OH	-OH	Entagenic acid

Fig. 2.1. Saponins of *Entada phaseoloides* (Leguminosae).

in Fig. 2.1. They are all octasaccharides and esterified with the monoterpene 2,6-dimethyl-6-hydroxy-2-*trans*-2,7-octadienoic acid and acetic acid (Okada *et al.* 1987, 1988a,b).

Unsaturated sugars occur very rarely; one example of a saponin with an unsaturated sugar moiety is kotschyioside (**660**), or 3β-*O*-(4-deoxy-α-hex-4-enopyranosyluronic   acid)-(3→1)-α-L-rhamnopyranosyl   28-*O*-β-D-glucopyranosyl oleanolic acid. This was isolated from the leaves of *Aspilia kotschyi* (Asteraceae) (Kapundu *et al.* 1988).

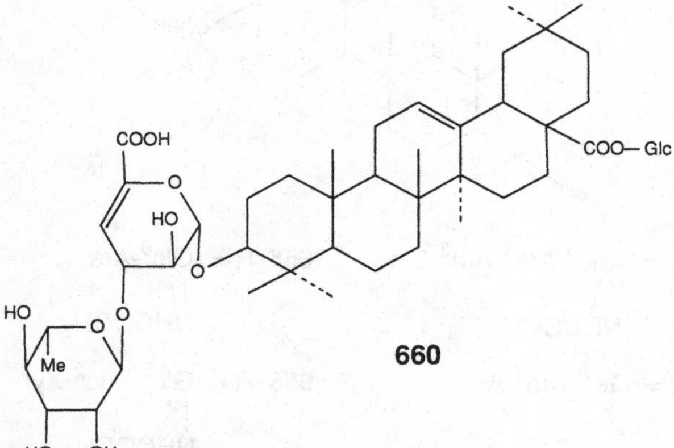

660

Very recently, Nishioka and co-workers (Ageta *et al.* 1988) have isolated a new class of compound, consisting of ellagitannins with a triterpene glycoside core, from *Castanopsis cuspidata* (Fagaceae) leaves. They consist of a mixture of 16 oleanane-type and ursane-type triterpenoids, with galloyl moieties attached at different hydroxyl substituents of the triterpene aglycone, e.g. **661–664**.

661 R^1 = H, R^2 = -CH$_3$

662 R^1 = -CH$_3$, R^2 = -H

663 R^1 = H, R^2 = -CH$_3$

664 R^1 = -CH$_3$, R^2 = -H

Desfontainoside (**665**) is a triterpenic glycoside with an ester-linked seco-iridoid congener found in the stems of *Desfontainia spinosa* (Logania-ceae). Secoxylogenin is conjugated with the C-24 hydroxyl group of 24-hydroxytormentic acid, which is in turn glucosylated at C-28 (Houghton and Lian, 1986b).

The first triterpene saponins sulphated at C-3 were isolated from *Bupleurum rotundifolium* (Umbelliferae) leaves. Rotundiosides B (**667**, Fig. 2.2) and C (**668**) are tetrasaccharides of oleanolic acid, while rotundioside A

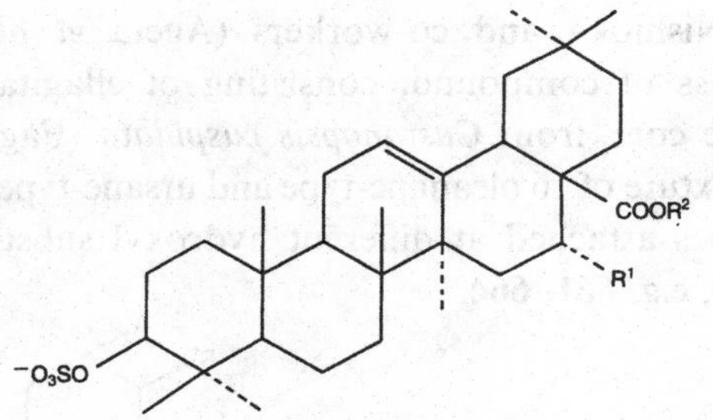

	R¹	R²
666 Rotundioside A	-OH	-Glc⁶-Glc²-Glc⁶-Glc
667 Rotundioside B	-H	-Glc⁶-Glc²-Glc⁶-Glc
668 Rotundioside C	-H	-Glc²-Glc²-Glc⁶-Glc

Fig. 2.2. Sulphated saponins from *Bupleurum rotundifolium* (Umbelliferae).

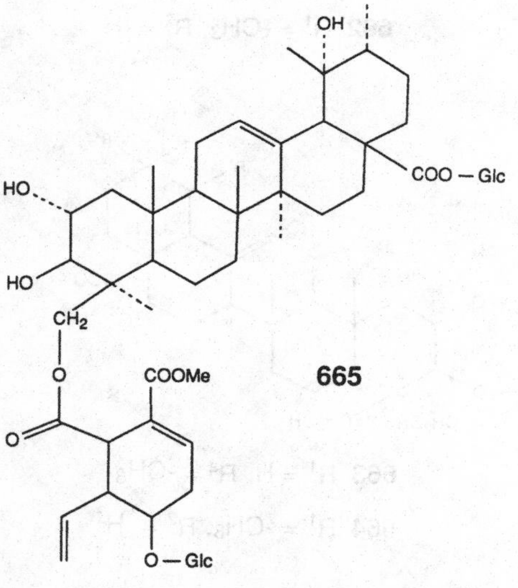

(**666**) is a tetrasaccharide of echinocystic acid. These saponins are monodesmosides acylated at C-28 (Akai *et al.* 1985a). Another C-3 sulphated saponin occurs in the leaves of *Hedera helix* (Araliaceae) (Elias *et al.* 1991) and a sulphate of 3-*epi*-betulinic acid (**669**) was obtained from the leaves of *Schefflera octophylla* (Araliaceae) (Sung and Adam, 1991). Saponins sulphated at position C-23 have been isolated from *Patrinia scabiosaefolia* seeds (Valerianaceae). As in the case of the sulphated glycosides from *B. rotundifolium*, they are monodesmosidic, ester-linked

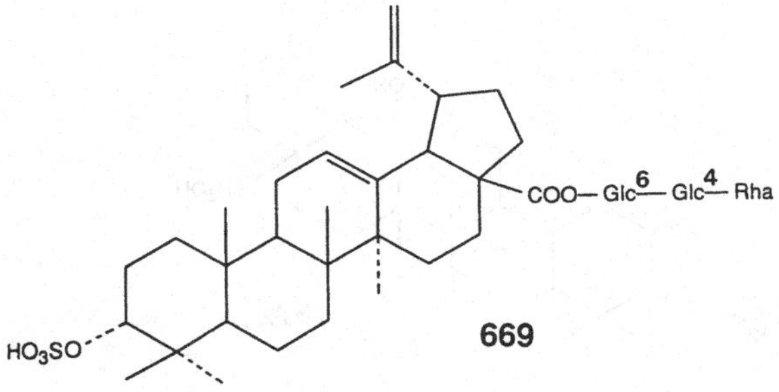

669

saponins. One (**670**) is a gentiobioside of 23-hydroxyursolic acid and one (**671**) a gentiobioside of hederagenin (Inada *et al.* 1988).

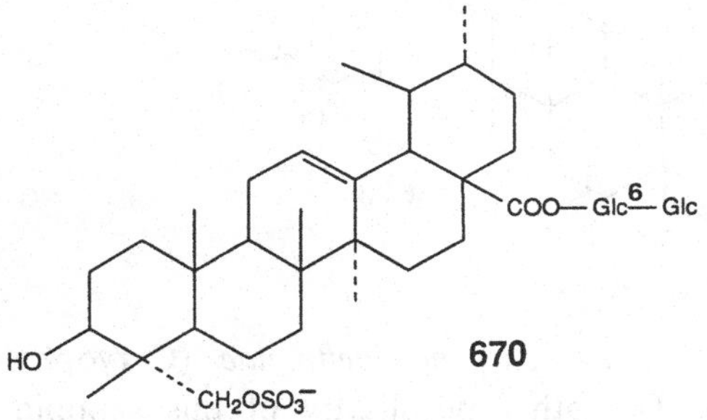

670

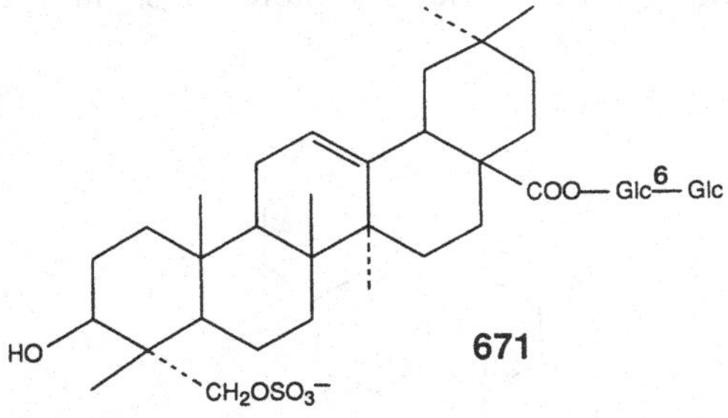

671

2.1.1   *Triterpene saponins with modified aglycone skeletons*

Some examples exist of saponins which possess *modified* oleanolic acid skeletons. One of these is the olean-11-en-3,13,23,28-tetraol verbasco-genin, found in verbascosaponin (**672**) from the flowers of *Verbascum phlomoides* (Scrophulariaceae) (Tschesche *et al.* 1980).

Saikosaponin BK1 (**673**) from *Bupleurum kunmingense* (Umbelliferae) has as aglycone 16-*epi*-saikogenin C, in which the 13,28-oxy bridge of saikogenins is replaced by a 13(18)-en structure (Luo *et al.* 1987).

Saikogenin D, the parent aglycone of polycarponoside A (**674**), has a similar olean-11,13(18)-diene skeleton. Polycarponoside A was isolated

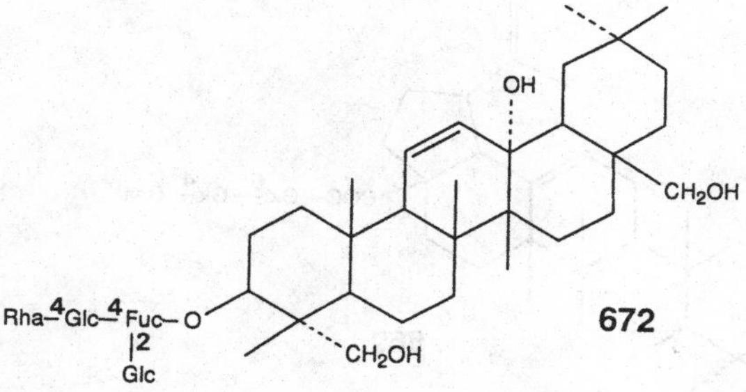
OH
CH2OH
Rha-4-Glc-4-Fuc-O
2
Glc
CH2OH

672

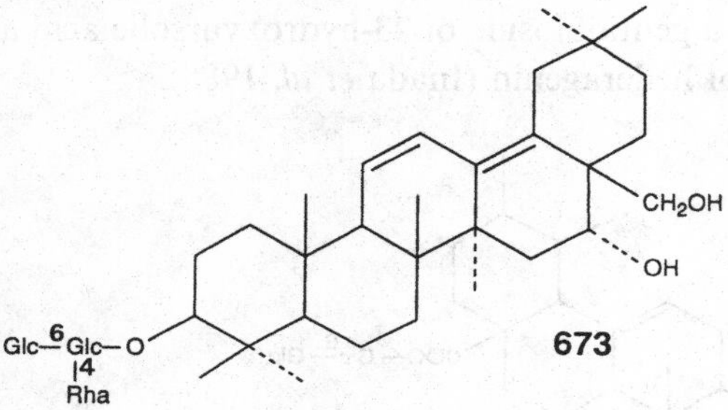
CH2OH
OH
Glc-6-Glc-O
4
Rha

673

from the aerial parts of *Polycarpone loeflingiae* (Caryophyllaceae) (Bhandari *et al.* 1990). The other peculiarity of this saponin is the presence of both arabinopyranose and arabinofuranose in the same molecule.

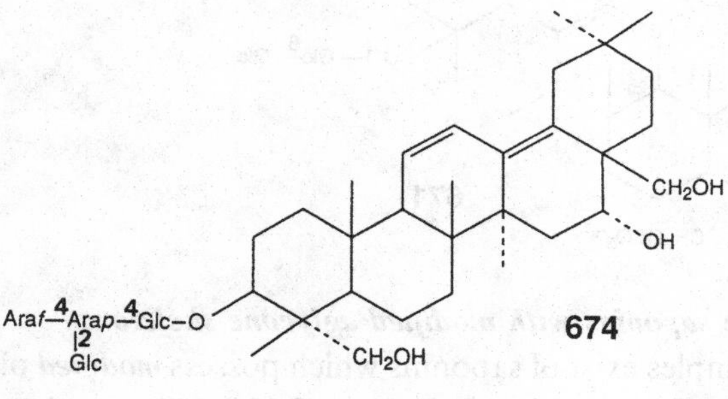
CH2OH
OH
Araf-4-Arap-4-Glc-O
2
Glc
CH2OH

674

The glycoside acaciaside (**675**) from the seeds of *Acacia auriculiformis* (Leguminosae) is a trisaccharide of acacic acid lactone and has a terminal arabinosyl moiety (Mahato *et al.* 1989).

A series of sweet-tasting glycosides has been found in the roots of *Periandra dulcis* (Leguminosae) from Brazil. One of these, periandrin I (**676**), is a 25-al-olean-18(19)-en-30-oic acid glucuronide (Hashimoto *et al.* 1983).

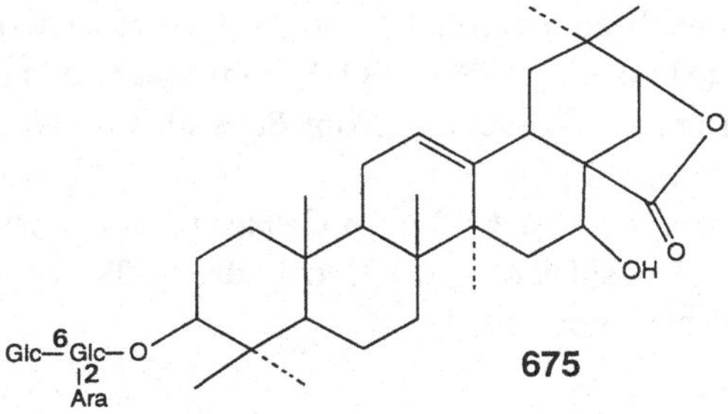

675

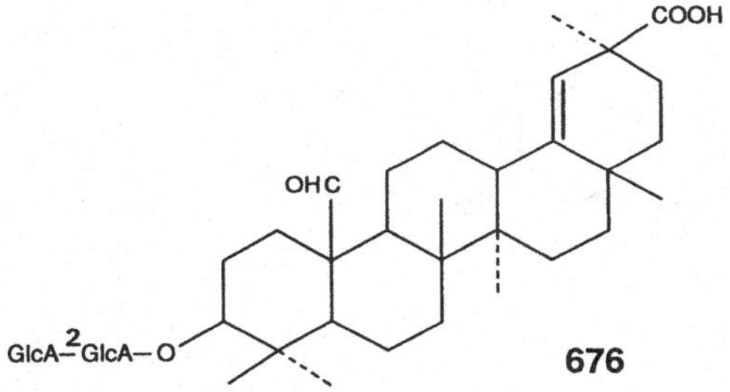

Glc—^{6}Glc—O
|2
Ara

676

GlcA—^{2}GlcA—O

The stem, bark and leaves of *Guaiacum officinale* (Zygophyllaceae) contain nortriterpenoid saponins (**677–685**) (Ahmad *et al.* 1986a,b, 1988a, 1989a–d). The aglycone is 30-norolean-12,20(29)-dien-28-oic acid, a C_{29} triterpene.

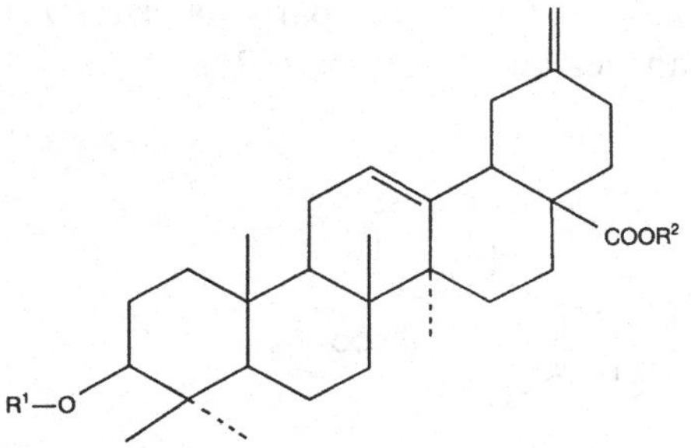

R^1—O

677 R^1 = -Ara, R^2 = -H

678 R^1 = -Ara^3-Glc, R^2 = -Glc

679 R^1 = -Ara^2-Rha, R^2 = -H

680 R^1 = -Ara^3-Rha, R^2 = -H

681 R^1 = -Ara^3-Glc^3-Rha, R^2 = -H

682 R^1 = -Ara^3-Glc^3-Rha, R^2 = -Glc

683 R^1 = -Ara^3-Glc, R^2 = -H
|2
Rha^3-Rha

684 R^1 = -Ara^3-Glc, R^2 = -Glc
|2
Rha^3-Rha

685 R^1 = -Ara^3-Glc, R^2 = -Glc^6-Glc
|2
Rha^3-Rha

Related glycosides have been isolated from the leaves of *Acanthopanax senticosus* (Araliaceae) (Shao *et al.* 1988, 1989a), from *Stauntonia chinensis* (Lardizabalaceae) (Wang *et al.* 1989c) and from *Boussingaultia baselloides* (Basellaceae) (Espada *et al.* 1990, 1991).

Two nortriterpene saponins (**686, 687**) from *Celmisia petriei* (Asteraceae) have been identified as glycosides of $2\beta,17,23$-trihydroxy-28-norolean-12-en-16-one (Rowan and Newman, 1984).

686 R = -Arap^2-Glc^6-Ara$p$

687 R = -Arap^2-Glc^6-Ara$p$
$$\begin{array}{c} | 2 \\ Ac \end{array}$$

Pyrocincholic acid is a nortriterpene with a 13,14 double bond. A 3β-*O*-β-6-deoxy-D-glucopyranosyl-28-*O*-β-D-glucopyranoside (**688**) of this sapogenin has been found in the aerial parts of *Isertia haenkeana* (Rubiaceae) (Rumbero-Sanchez and Vazquez, 1991).

688

A glycoside containing as aglycone the 3,4-secoderivative of gypsogenic acid (**689**) occurs in *Dianthus superbus* (Caryophyllaceae). This co-exists with glycosides of gypsogenic acid in the plant (Oshima *et al.* 1984c).

A new 3,4-secolupane triterpene divaroside (**690**) has been isolated from

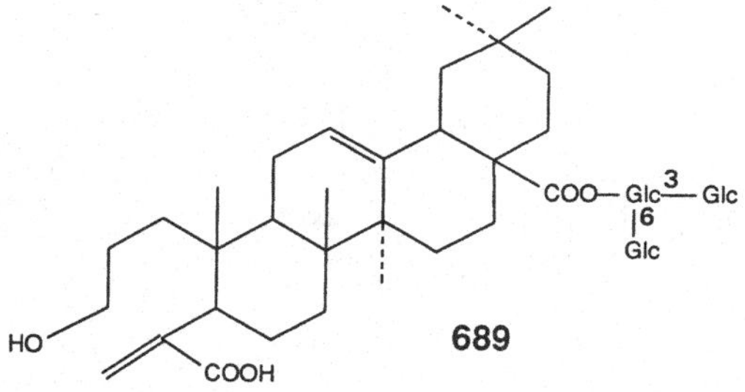

689

the leaves of *Acanthopanax divaricatus* (Araliaceae). The structure of the new compound was established as the β-gentiobiosyl ester of chiisanogenin (Matsumoto *et al.* 1987).

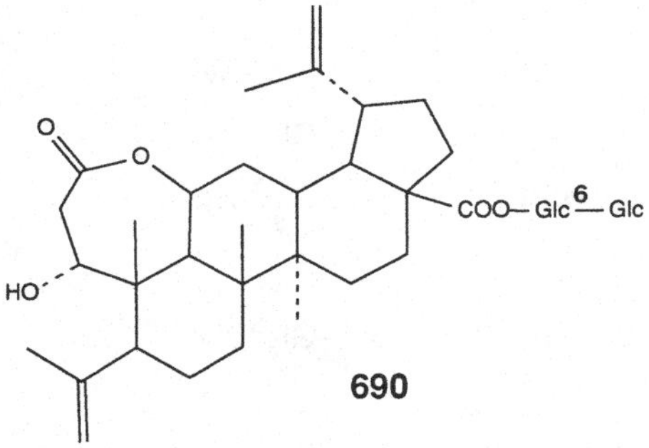

690

Aerial parts of *Sesamum alatum* (Pedaliaceae) contain alatosides A–C (**691–693**), glycosides of a novel 18,19-seco-urs-12-en aglycone called alatogenin. Alatosides B and C are bidesmosides with two osidic linkages at C-3 and C-19 (Potterat *et al.* 1992).

The 18,19-secoursane glycosides ilexosides A–F (Fig. 2.3, **694–699**)

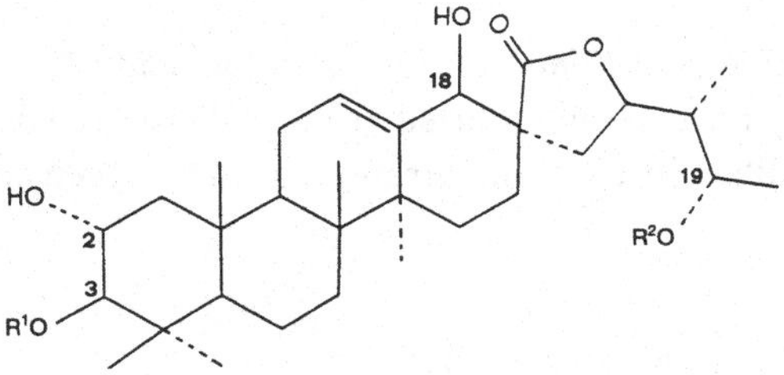

691 R^1 = -Xyl^2-Rha, R^2 = -H

692 R^1 = -Xyl^2-Rha, R^2 = -Glc

693 R^1 = -Xyl^2-Glc, R^2 = -Glc

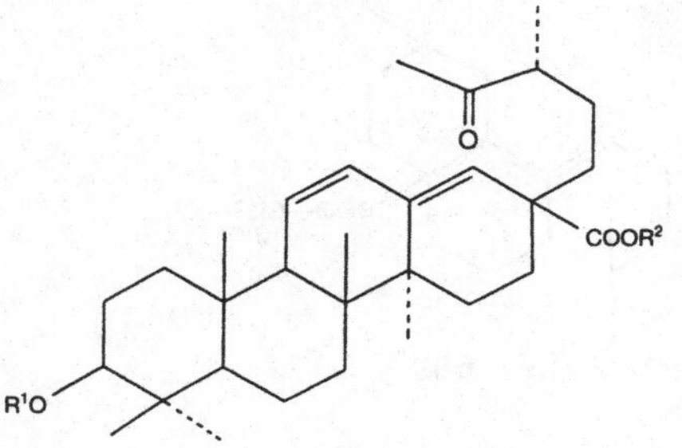

		R^1	R^2
694	Ilexoside A	-Ara	-H
695	Ilexoside B	-Ara^3-Glc	-H
696	Ilexoside C	-H	-Glc^6-Xyl
697	Ilexoside D	-Ara	-Glc^6-Xyl
698	Ilexoside E	-Ara	-Glc^6-Xyl^2 Rha
699	Ilexoside F	-Ara^3-Glc	-Glc^6-Xyl^2 Rha

Fig. 2.3. 18,19-Seco-ursane glycosides of *Ilex crenata*.

are present in the fruit of *Ilex crenata* (Aquifoliaceae). The aglycone is α-ilexanolic acid (3β-hydroxy-19-oxo-18,19-seco-11,13(18)-ursadien-28-oic acid). Because of the conjugated diene system, these glycosides exhibit strong UV absorption in the 240–250 nm region (Kakuno *et al.* 1991a,b).

Glucosides of ceanothic acid (**700**) and isoceanothic acid (**701**), modified lupane triterpenes, have been reported from *Paliurus ramosissimus* (Rhamnaceae). These differ in the configuration of the oxygen function at C-3 (Lee *et al.* 1991).

2.1.2   *Licorice saponins* (see Chapter 7)

Glycyrrhizin (**269**, Appendix 2) from *Glycyrrhiza glabra* (Leguminosae) is a rather special saponin, not only from the point of view of the unusual aglycone skeleton. The glycoside contains two glucuronic acid units, is highly polar and can be relatively easily purified

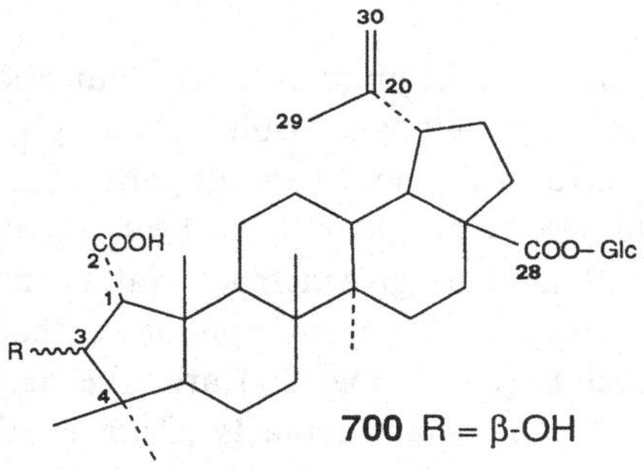

700 R = β-OH

701 R = α-OH

and crystallized. However, it lacks most of the characteristic properties of saponins. In addition to glycyrrhizin, a very large number of other glycosides exist in the very complex saponin mixture from the roots of the plant (Russo, 1967). Recently, a series of glucuronyl saponins have been characterized from Chinese *Glycyrrhizae radix* (*Glycyrrhiza uralensis* and *G. inflata*) (Kitagawa *et al.* 1988c, 1989b, 1991). These include the bidesmosidic licorice-saponin A3 (**270**, Appendix 2), the ester saponin **702** (licorice-saponin D3) and the olean-11,13-dien glycoside **703** (licorice-saponin K2).

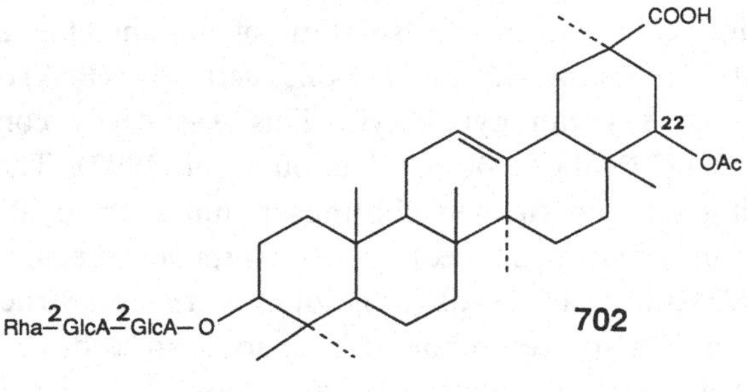

702

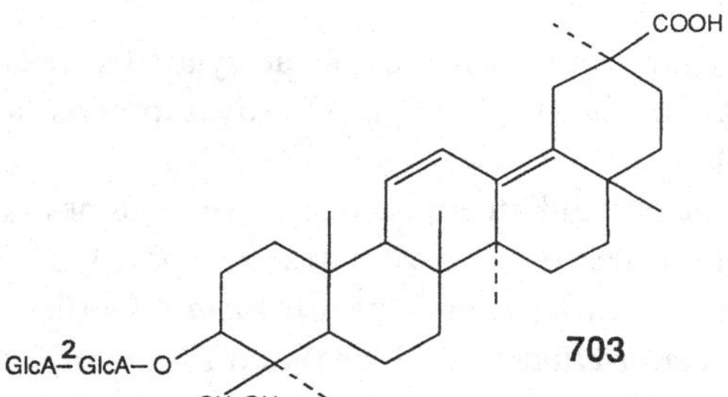

703

2.1.3   Soyasaponins

Soybean seeds (*Glycine max*, Leguminosae) contain about 2%
of glycosides, in the form of saponins and isoflavonoid glycosides.
The saponins of soybeans have been under investigation since the
beginning of the century. Numerous problems have been encountered
because their hydrolysis has tended to give artifacts rather than the
true aglycones (Price *et al.* 1987). The current opinion is that soyasapogenols A (**60**), B (**33**) and E (**23**) (Table 2.1) are true aglycones,
while soyasapogenols C (**26**), D and F are artifacts (Price *et al.* 1987,
Kitagawa *et al.* 1988a). The corresponding glycosides are the so-called
'group A saponins', 'group B saponins' and 'group E saponins', respectively. Group B saponins, soyasaponins I–V, are shown in Fig. 2.4
and Appendix 2, while group A saponins, soyasaponins A_1 (**518**), A_2
(**517**) and A_3 (**519**) are shown in Appendix 2. Group B saponins
have a widespread occurrence in edible beans (Price *et al.* 1986), while
group A saponins are present only in the seed hypocotyl of soybeans (Shiraiwa *et al.* 1991). Very recently, a series of acetylated
saponins, acetyl-soyasaponins A_1–A_6 and saponins Ac and Ad, have been
isolated from soybeans (Fig. 2.4). In these glycosides of soyasapogenol
A, the terminal monosaccharide moiety of the C-22 sugar chain is
acetylated (Kitagawa *et al.* 1988b; Taniyama *et al.* 1988a, 1988b; Shiraiwa
et al. 1991). It is very likely, therefore, that soyasaponins A_1, A_2 and A_3
are not actually present in soybean seeds (Shiraiwa *et al.* 1991). Support
for this supposition comes from the isolation of an unstable 22-(2,3-
dihydro-2,5-dihydroxy-6-methyl-4*H*-pyran-4-one) derivative (BeA) of soyasaponin I (Bb) from soybean hypocotyls. This was easily converted
into soyasaponin I (**454**) above 40 °C (Kudou *et al.* 1992). The same
γ-pyrone-containing saponin (named chromosaponin I or soyasaponin
VI) was also found in etiolated pea seedlings (*Pisum sativum*, Leguminosae)
(Tsurumi *et al.* 1992) and alfalfa seeds (Massiot *et al.* 1992b). Structurally
similar conjugates have also been reported from soybean seeds and other
legumes such as mung bean (*Vigna mungo*) and cowpea (*Vigna sinensis*)
(Kudou *et al.* 1993).

An interesting observation is that whereas the acetylated soyasaponins
are bitter and astringent, the non-acetylated soyasaponins are not
(Kitagawa *et al.* 1988b).

Soyasaponin I has been found in numerous other legumes, such as
Phaseolus vulgaris (kidney bean), *P. aureus* (runner bean), *P. lunatus*
(butter bean), *P. coccineus* (scarlet runner bean), *Vicia faba* (field bean),
Lens culinaris (lentil), *Cicer arietinum* (chick pea) and *Pisum sativum* (pea)
(Price *et al.* 1987).

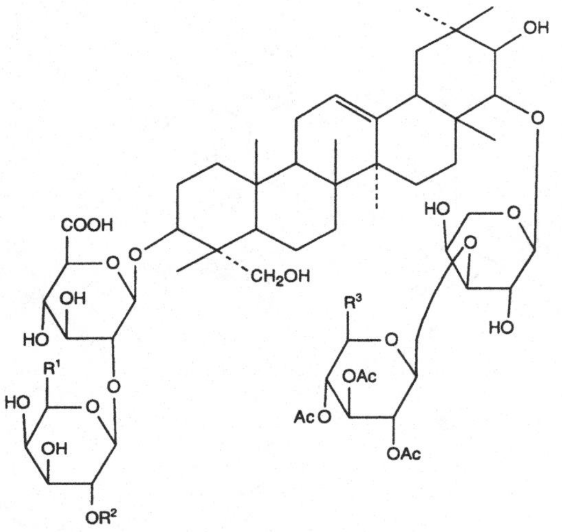

Group A

		R^1	R^2	R^3
704	Acetyl-soyasaponin A_1	$-CH_2OH$	-Glc	$-CH_2OAc$
705	Acetyl-soyasaponin A_2	$-CH_2OH$	-H	$-CH_2OAc$
706	Acetyl-soyasaponin A_3	-H	-H	$-CH_2OAc$
707	Acetyl-soyasaponin A_4	$-CH_2OH$	-Glc	-H
708	Acetyl-soyasaponin A_5	$-CH_2OH$	-H	-H
709	Acetyl-soyasaponin A_6	-H	-H	-H
710	Saponin A_c	$-CH_2OH$	-Rha	$-CH_2OAc$
711	Saponin A_d	-H	-Glc	$-CH_2OAc$

Fig. 2.4. Structures of saponins from soybean seed.

2.1.4   Cycloartane glycosides

A number of cycloartane (9,19-cyclolanostane) glycosides have been isolated (notably from *Astragalus* and *Thalictrum* species) by Japanese and Russian groups (see, for example, **633**, Appendix 2). Many of the glycosides (e.g. **633–642** in Appendix 2) from the genus *Astragalus* have cycloastragenol (**122**, Table 2.2) as aglycone.

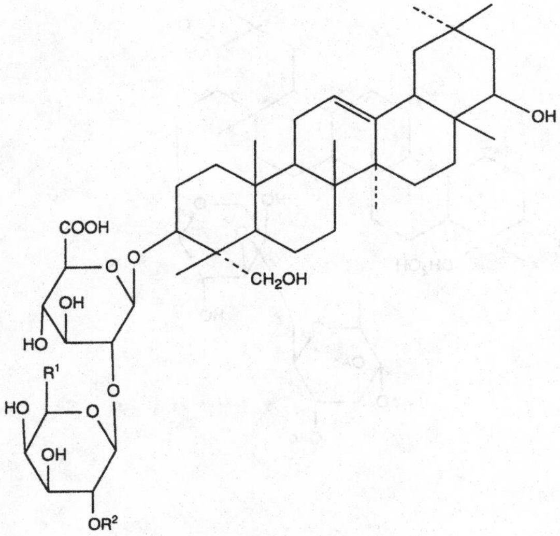

Group B

		R^1	R^2
454	Soyasaponin I	-CH$_2$OH	-Rha
451	Soyasaponin II	-H	-Rha
448	Soyasaponin III	-CH$_2$OH	-H
447	Soyasaponin IV	-H	-H
453	Soyasaponin V	-CH$_2$OH	-Glc

Fig. 2.4. (*Cont.*)

Different sapogenins, on the contrary, are found in *Thalictrum* species (Ranunculaceae). The corresponding saponins include cyclofoetoside A (**631**, Appendix 2) and cyclofoetoside B (**712**) from *T. foetidum* (Ganenko *et al.* 1986b). *Thalictrum squarrosum* has yielded squarroside A1 (**713**) and B1 (**714**). These have a five-membered ring in the side chain (Khamidullina *et al.* 1989). Very similar glycosides, thalictosides I (**715**) and II (**716**), have been isolated from the aerial parts of an unidentified *Thalictrum* species growing in Japan. It has been stated that these acetals could possibly be formed from the corresponding hemiacetals during the extraction procedure (Yoshimitsu *et al.* 1993).

712 Cyclofoetoside B

713 Squarroside A1 R = -Glc
714 Squarroside B1 R = -Man^6-Glc

715 Thalictoside I R^1 = -OCH$_3$, R^2 = -H
716 Thalictoside II R^1 = -H, R^2 = -OCH$_3$

In addition to cycloastragenol derivatives, glycosides (**717, 718**) of a remarkable hexacyclic heterocycle have been reported from *Astragalus orbiculatus* (Leguminosae) (Isaev *et al.* 1986; Agzamova *et al.* 1987a,b). Another *Astragalus* species, *A. trigonus*, contains a diglucoside (**719**) of 6-oxo-cycloartenol (El-Sebakhy and Waterman, 1985).

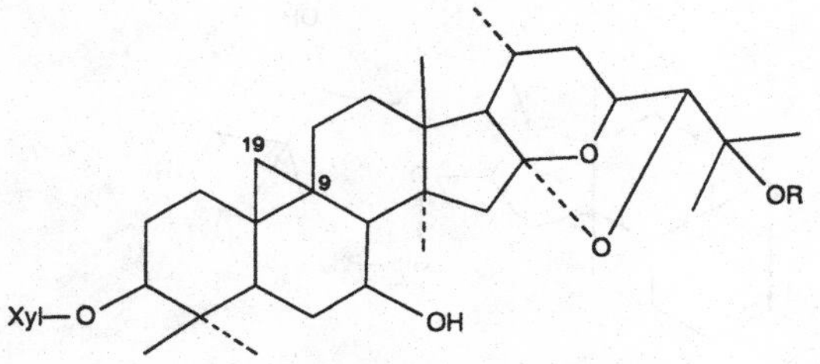

717 Cycloorbicoside A R = -H
718 Cycloorbicoside G R = -Glc

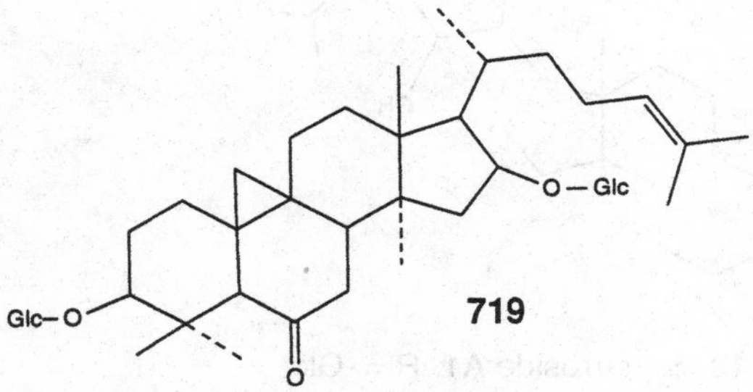

719

Novel cycloartenol saponins containing a peptide bond have been found in the plant *Mussaenda pubescens* (Rubiaceae). One of these (**720**) is a 3-*O*-glucoside (Xu *et al.* 1991).

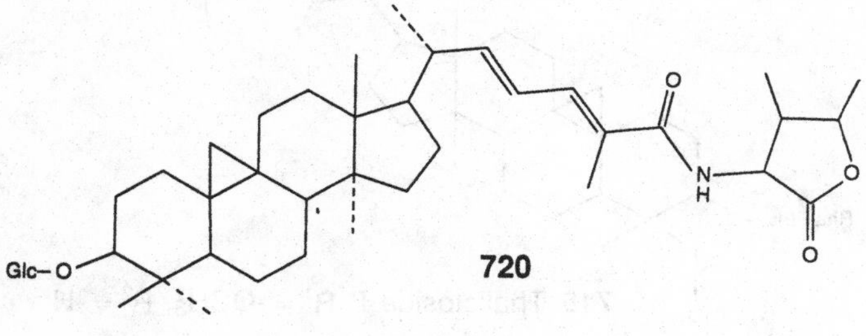

720

A member of the Passifloraceae, *Passiflora quadrangularis*, contains two cycloartane glycosides (**721**) and (**722**) in the leaves, in addition to quadranguloside (**629**, Appendix 2) (Orsini *et al.* 1987).

Passiflorine (**723**) from the leaves of *Passiflora edulis* (Passifloraceae) is the *β*-glucosyl ester of (22*R*,24*S*)-22,31-epoxy-24-methyl-1*α*,3*β*,24,31-tetrahydroxy-9,19-*cyclo*-9*β*-lanostan-29-oic acid (Bombardelli *et al.* 1975).

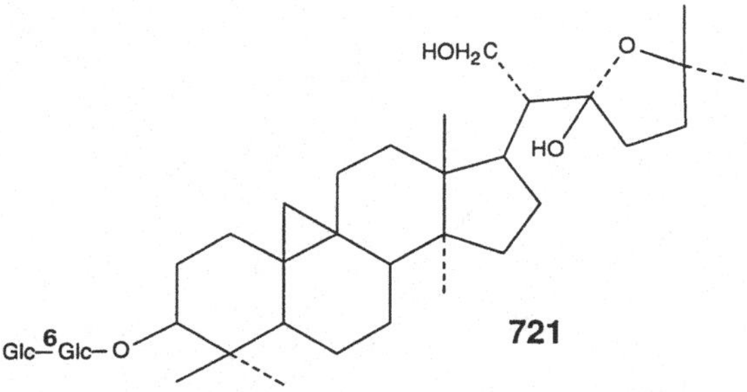

721

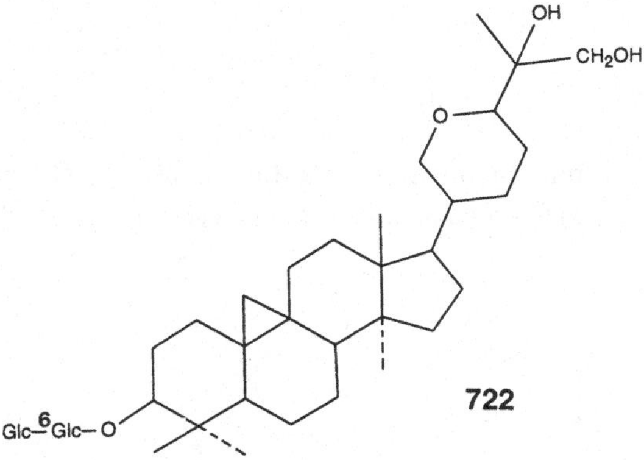

722

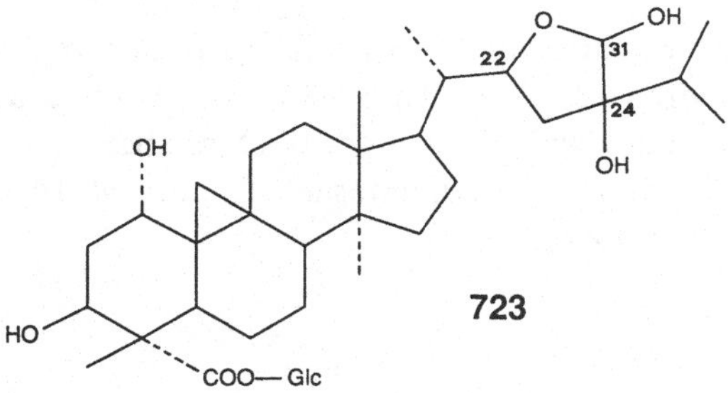

723

Other cycloartane glycosides have been found in the genera *Cimicifuga*, *Actaea*, *Beesia* and *Souliea*, all members of the Ranunculaceae. For example, the xylosides beesiosides I–IV were discovered in the rhizomes of *Beesia calthaefolia* (Inoue *et al.* 1985; Sakurai *et al.* 1986). Beesiosides III and IV were also found in the rhizomes of *Souliea vaginata* (Inoue *et al.* 1985). Beesioside II (**724**) has the structure 3-*O*-β-D-xylopyranosyl-(20*S*,24*R*)-16β,18-diacetoxy-20,24-epoxy-3β,15α,25-trihydroxy-9,19-cyclo-lanostane (Sakurai *et al.* 1986).

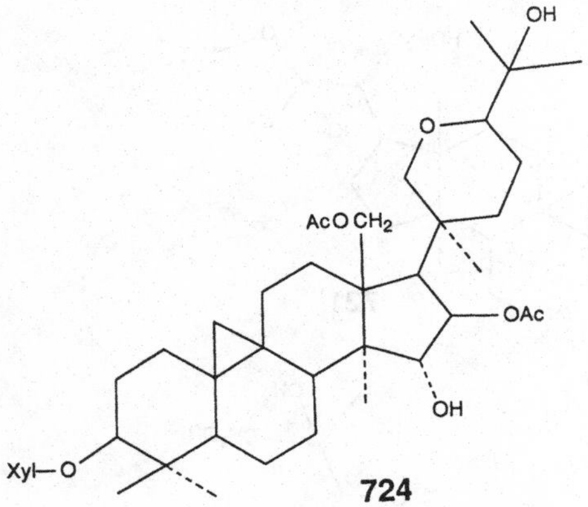

724

The xylosides are often chemically unstable. Such is the case for shengmanol xyloside (**725**) from *Cimicifuga japonica* (Kimura *et al.* 1982).

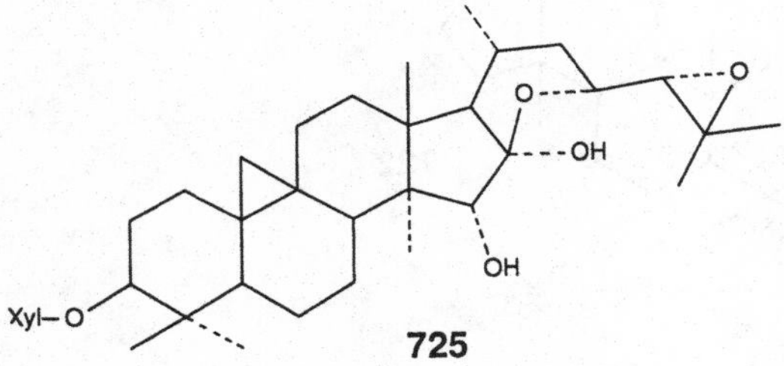

725

A sweet-tasting glycoside, abrusoside A (**726**), the 3β-D-glucopyranoside of (20*S*,22*S*)-3β,22-dihydroxy-9,19-cyclolanost-24-en-26,29-dioic acid δ-lactone (abrusogenin), together with abrusosides B–D, has been discovered in the leaves of *Abrus precatorius* (Leguminosae) (Choi *et al.* 1989) and *A. fruticulosus* (Fullas *et al.* 1990).

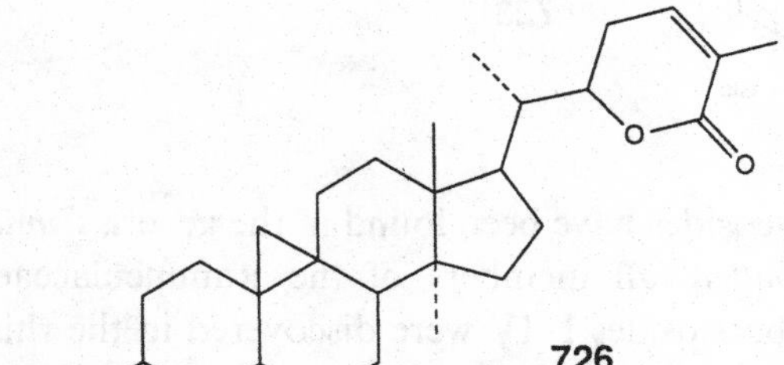

726

The saponin 3-*O*-[β-D-glucopyranosyl-(1→2)-β-D-glucopyranosyl]-25-*O*-α-L-rhamnopyranosyl-(20*S*,24*S*)-3β,16β,20,24,25-pentahydroxy-9,19-

cyclolanostane (**727**) from *Oxytropis bicolor* (Leguminosae) is a bidesmoside with three sugar moieties (Sun *et al.* 1991).

727

2.1.5   *Dammarane saponins*

Dammarane saponins are derived from lanostane-type saponins
and form a large proportion of known triterpene saponins. They occur
especially frequently in plants of the families Araliaceae, Cucurbitaceae
and Rhamnaceae.

Besides various *Panax* (Araliaceae) species, dammarane saponins have
also been isolated, for example, from *Gynostemma pentaphyllum* (Cucurbitaceae) (Takemoto *et al.* 1983b; Yoshikawa *et al.* 1987a). This plant has
also furnished saponins with different dammarane aglycones. One example
is gypenoside LX (**728**), which is 2α,3β,12β,(20S),25-pentahydroxydammar-
23-ene 20-*O*-β-primeveroside (Takemoto *et al.* 1986). Gypenoside LXX
(**729**) has a 3β,12β,(20S),26-tetrahydroxydammar-24-ene aglycone and
gypenoside LXXI (**730**) has a 3β,12β,(20S),(24S)-tetrahydroxydammar-
25-ene aglycone (Yoshikawa *et al.* 1987a). Other dammarane glycosides
from *G. pentaphyllum* include aglycones hydroxylated at C-19: gypenoside
LXV   (3β,12β,19,(20S)-tetrahydroxydammar-24-ene 20-*O*-β-primevero-
side) (**731**) is one of these (Yoshikawa *et al.* 1987b).

Glycosides of alnustic acid, a C_{31}-secodammarane-type triterpenoid,
have been found in female flowers of *Alnus serrulatoides* (Betulaceae). The
structures of the five saponins **732–734**, **736**, **737** are shown (Aoki *et al.*
1982, 1990). An additional glycoside, **735**, is present in the male flowers
of *A. serrulatoides* and in the flowers of *A. pendula* (Aoki *et al.* 1990).
Three of the saponins (**733**, **735**, **737**) are acylated at C-2 of the
monosaccharide moiety.

Ginseng saponins

The triterpene glycosides from ginseng ('ginsenosides') are named
according to their mobility on TLC plates: their polarity decreases from

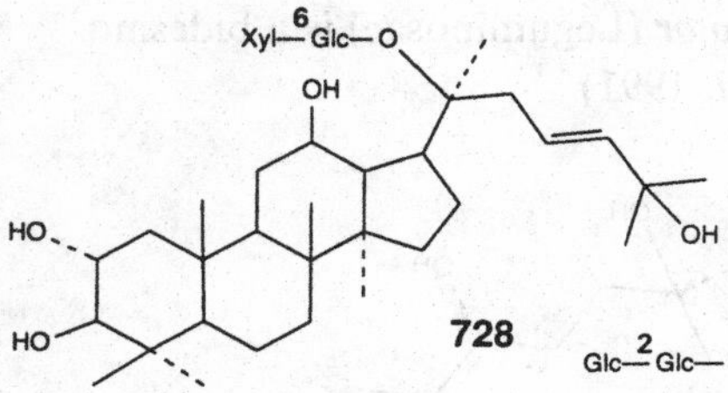

Xyl—⁶Glc—O

728

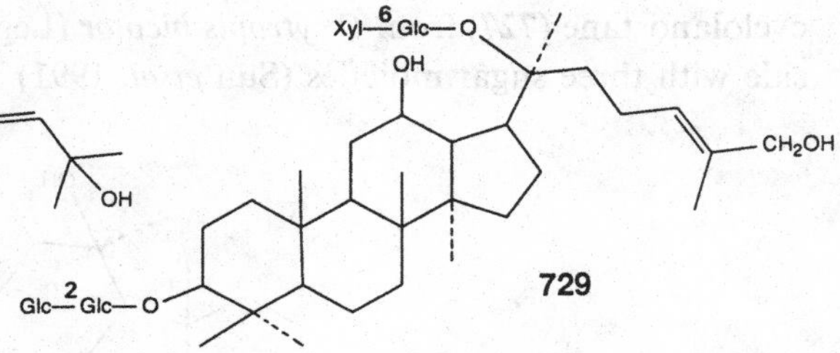

Xyl—⁶Glc—O

729

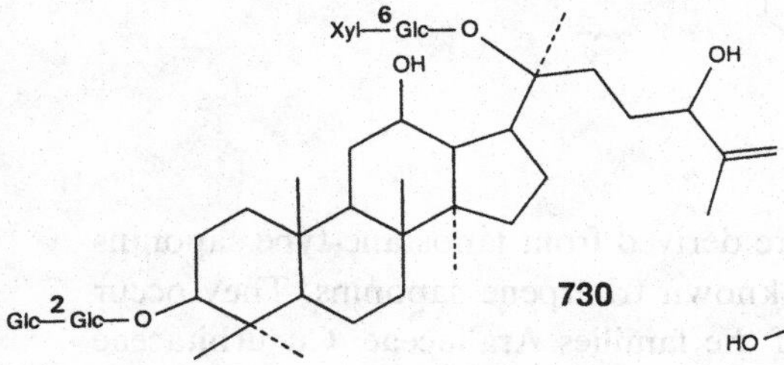

Xyl—⁶Glc—O

730

Xyl—⁶Glc—O

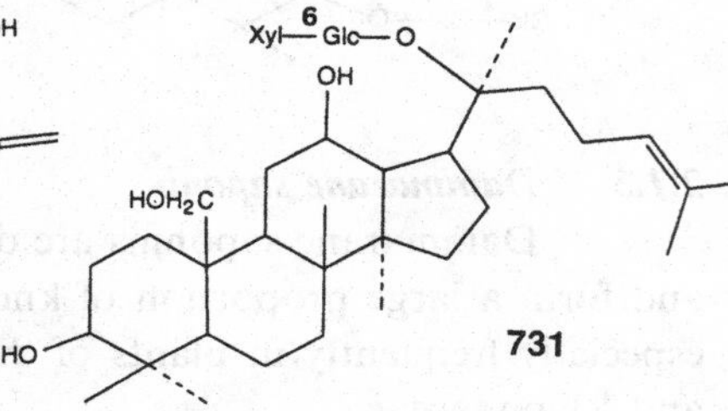

731

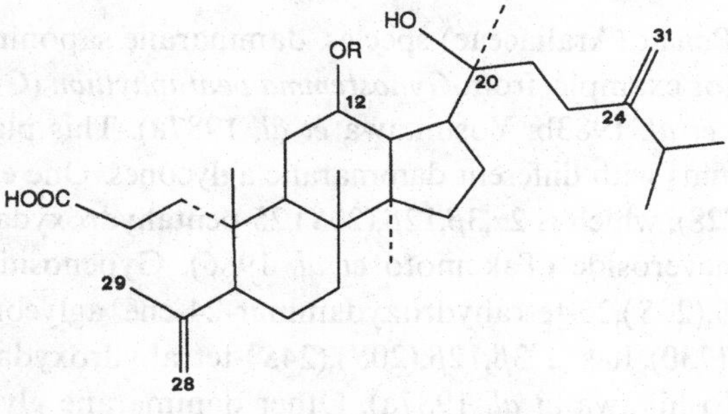

732 R = -Ara   **735** R = -Xyl²-Ac

733 R = -Ara²-Ac **736** R = -Glc

734 R = -Xyl   **737** R = -Glc²-Ac

index 'a' to index 'h'. This property is, of course, a function of the number of monosaccharide residues in the sugar chain (Table 2.3).

Many of the ginsenosides are neutral bidesmosidic saponins (Rb$_1$, Rb$_2$, Rc, Rd, Re, Rg$_1$) and some are monodesmosidic (Rf, Rg$_2$) (Table 2.4).

Up to 1984, 18 saponins had been isolated from white ginseng: ginsenosides Ro, Ra$_1$, Ra$_2$, Ra$_3$, Rb$_1$, Rb$_2$, Rb$_3$, Rc, Rd, Re, Rf, Rg$_1$, Rg$_2$, Rg$_3$, Rh$_1$, 20-glucoginsenoside Rf, notoginsenoside R$_1$, quinquenoside R$_1$ (Tanaka and Kasai, 1984) and by 1986 the total was already 25 (Shibata, 1986).

Table 2.3. *The naming of ginsenosides*

Ginsenoside	No. of sugar units
Ra	5
Rb, Rc	4
Rd, Re	3
Rf, Rg	2
Rh	1

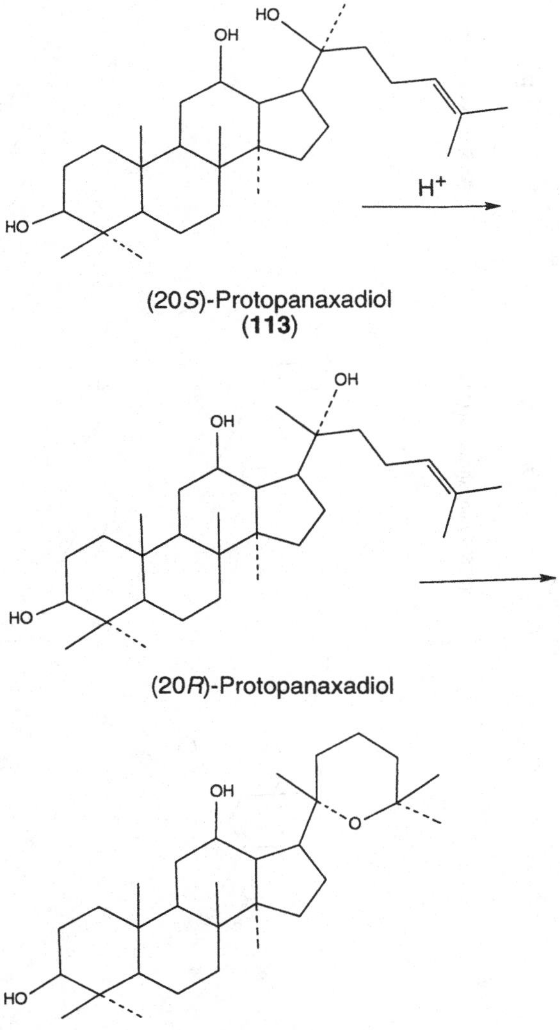

(20*S*)-Protopanaxadiol
(113)

(20*R*)-Protopanaxadiol

(20*R*)-Panaxadiol

Fig. 2.5. Transformation of protopanaxadiol in acidic media.

Table 2.4. *Structures of dammarane saponins*

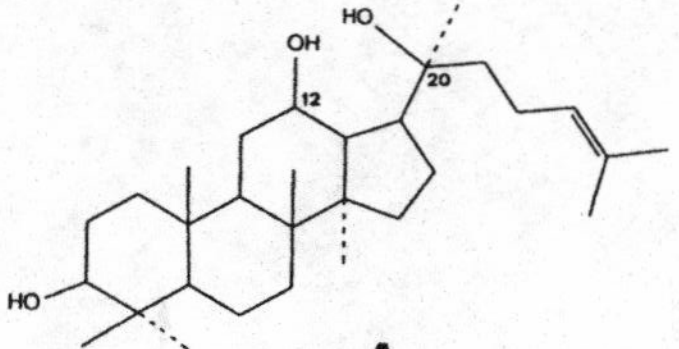

A

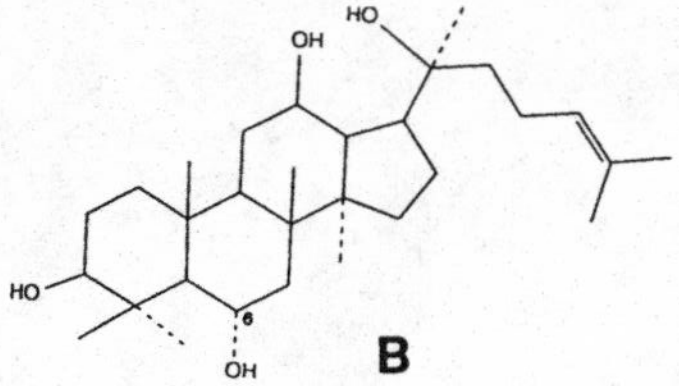

(20S)-Protopanaxadiol (113)

B

C

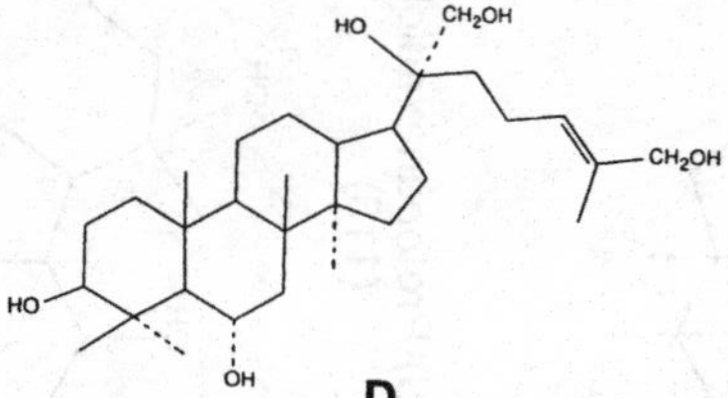

D

(20S)-Protopanaxatriol (114)

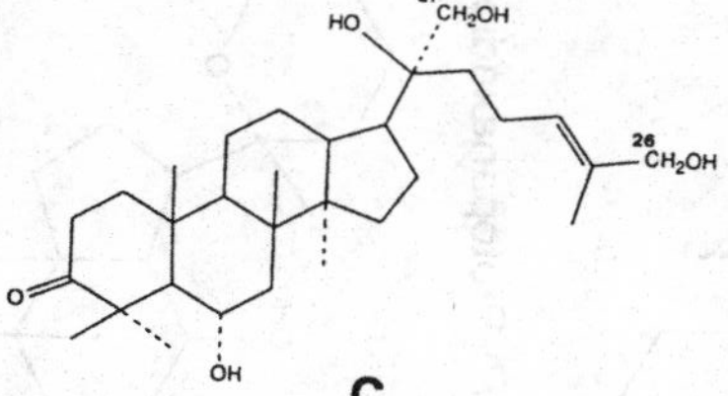

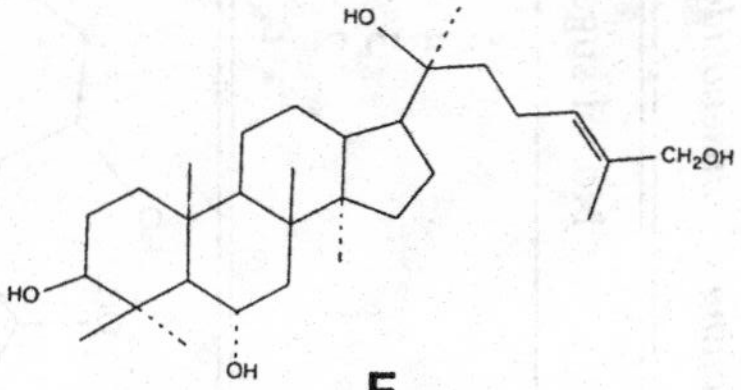

E

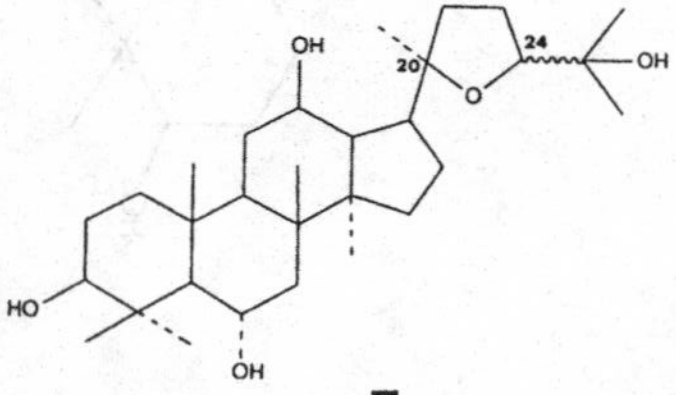

F

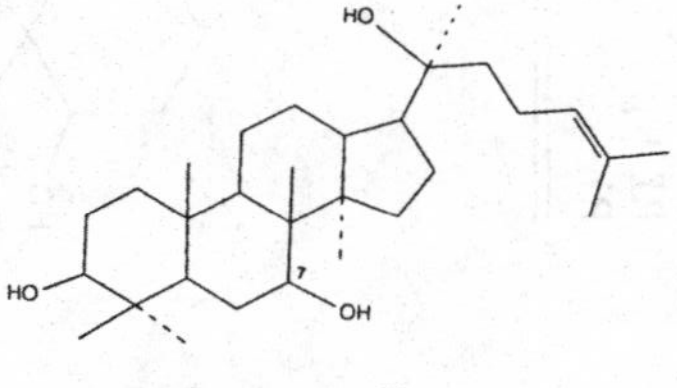

G

No.	Name	Aglycone	Structure	Plant	Reference
738	Ginsenoside Rh$_2$	A	3-Glc	*Panax ginseng* (Araliaceae)	Chen *et al.*, 1986
739	Notoginsenoside Fe	A	3-Glc 20-Glc^6-Ara*f*	*P. trifolius* (Araliaceae)	Lee and der Marderosian, 1988
740	Gypenoside XVII	A	3-Glc 20-Glc^6-Glc	*Gynostemma pentaphyllum* (Cucurbitaceae)	Takemoto, 1983
				Panax notoginseng (Araliaceae)	Tanaka and Kasai, 1984
				P. quinquefolius	Tanaka and Kasai, 1984
741	Gypenoside IX	A	3-Glc 20-Glc^6-Xyl	*Gynostemma pentaphyllum* (Cucurbitaceae)	Takemoto *et al.*, 1983b
				Panax notoginseng (Araliaceae)	Tanaka and Kasai, 1984
742	Ginsenoside Rg$_3$	A	3-Glc^2-Glc	*P. ginseng* (Araliaceae)	Tanaka and Kasai, 1984
				P. quinquefolius	Xu *et al.*, 1987
743	Ginsenoside Rd	A	3-Glc^2-Glc 20-Glc	*P. ginseng* (Araliaceae)	Tanaka and Kasai, 1984
				P. notoginseng	Tanaka and Kasai, 1984
				P. quinquefolius	Tanaka and Kasai, 1984
				P. pseudoginseng	Tanaka and Kasai, 1984
				P. japonicus	Tanaka and Kasai, 1984
				P. trifolius	Lee and der Marderosian, 1988
744	Malonyl-ginsenoside Rd	A	3-Glc^2-Glc^6-Ma 20-Glc	*P. ginseng* (Araliaceae)	Kitagawa *et al.*, 1983b
				Gynostemma pentaphyllum (Cucurbitaceae)	Kuwahara *et al.*, 1989
745	Ginsenoside Rb$_2$	A	3-Glc^2-Glc 20-Glc^6-Arap	*Panax ginseng* (Araliaceae)	Tanaka and Kasai, 1984
				P. notoginseng	Tanaka and Kasai, 1984
				P. quinquefolius	Tanaka and Kasai, 1984

(cont.)

Table 2.4 (*Cont.*)

No.	Name	Aglycone	Structure	Plant	Reference
746	Malonyl-ginsenoside Rb$_2$	A	3-Glc^2-Glc^6-Ma 20-Glc^6-Ara$p$	*P. ginseng* (Araliaceae)	Kitagawa *et al.*, 1983b
747	Ginsenoside Rc	A	3-Glc^2-Glc 20-Glc^6-Ara$f$	*P. ginseng* (Araliaceae) *P. notoginseng* *P. japonicus* *P. quinquefolius* *P. trifolius*	Tanaka and Kasai, 1984 Tanaka and Kasai, 1984 Tanaka and Kasai, 1984 Tanaka and Kasai, 1984 Lee and der Marderosian, 1988
748	Malonyl-ginsenoside Rc	A	3-Glc^2-Glc^6-Ma 20-Glc^6-Ara$f$	*P. ginseng* (Araliaceae)	Kitagawa *et al.*, 1983b
749	Ginsenoside Rb$_1$	A	3-Glc^2-Glc 20-Glc^6-Glc	*P. ginseng* (Araliaceae) *P. notoginseng* *P. quinquefolius* *P. pseudoginseng* *P. japonicus*	Tanaka and Kasai, 1984 Tanaka and Kasai, 1984 Tanaka and Kasai, 1984 Tanaka and Kasai, 1984 Tanaka and Kasai, 1984
750	Malonyl-ginsenoside Rb$_1$	A	3-Glc^2-Glc^6-Ma 20-Glc^6-Glc	*P. ginseng* (Araliaceae) *Gynostemma pentaphyllum* (Cucurbitaceae)	Kitagawa *et al.*, 1989 Kuwahara *et al.*, 1989
751	Ginsenoside Rb$_3$	A	3-Glc^2-Glc 20-Glc^6-Xyl	*Panax ginseng* (Araliaceae) *P. notoginseng* *P. quinquefolius* *P. pseudoginseng* *P. japonicus* *P. trifolius*	Tanaka and Kasai, 1984 Tanaka and Kasai, 1984 Tanaka and Kasai, 1984 Tanaka and Kasai, 1984 Tanaka and Kasai, 1984 Lee and der Marderosian, 1988
752	Ginsenoside Ra$_1$	A	3-Glc^2-Glc 20-Glc^6-Arap^4-Xyl	*P. ginseng* (Araliaceae)	Tanaka and Kasai, 1984

No.	Name	Type	Structure	Source	Reference
753	Ginsenoside Ra$_2$	A	3-Glc^2-Glc 20-Glc^6-Araf^2-Xyl	_P. ginseng_ (Araliaceae)	Tanaka and Kasai, 1984
754	Ginsenoside Ra$_3$	A	3-Glc^2-Glc 20-Glc^6-Glc^3-Xyl	_P. ginseng_ (Araliaceae)	Tanaka and Kasai, 1984
755	Ginsenoside RA$_0$	A	3-Glc^2-Glc^2-Glc 20-Glc^6-Glc	_P. quinquefolius_ (Araliaceae)	Xu _et al._, 1987
756	Chikusetsusaponin III	A	3-Glc^6-Xyl $\vert^2$ Glc	_P. japonicus_ (Araliaceae)	Tanaka and Kasai, 1984
757	Ginsenoside Rh$_1$	B	6-Glc	_P. ginseng_ (Araliaceae) _P. notoginseng_	Tanaka and Kasai, 1984 Tanaka and Kasai, 1984
758	Ginsenoside Rg$_1$	B	6-Glc 20-Glc	_P. ginseng_ (Araliaceae) _P. notoginseng_ _P. quinquefolius_ _P. japonicus_ _P. zingiberensis_ _Luffa cylindrica_ (Cucurbitaceae)	Tanaka and Kasai, 1984 Tanaka and Kasai, 1984 Tanaka and Kasai, 1984 Tanaka and Kasai, 1984 Tanaka and Kasai, 1984 Takemoto _et al_, 1984
759	Ginsenoside Rf	B	6-Glc^2-Glc	_Panax ginseng_ (Araliaceae) _P. trifolius_	Tanaka and Kasai, 1984
760	20-Glucoginsenoside Rf	B	6-Glc^2-Glc 20-Glc	_P. ginseng_ (Araliaceae) _P. notoginseng_ _P. japonicus_	Tanaka and Kasai, 1984 Tanaka and Kasai, 1984 Tanaka and Kasai, 1984 Tanaka and Kasai, 1984
761	Pseudoginsenoside RC$_1$	B	3-Glc^2-Glc^6-Ac 20-Glc	_P. pseudoginseng_ (Araliaceae)	Namba _et al._, 1986
762	Ginsenoside Rg$_2$	B	6-Glc^2-Rha	_P. ginseng_ (Araliaceae) _P. notoginseng_ _P. quinquefolius_	Tanaka and Kasai, 1984 Tanaka and Kasai, 1984 Tanaka and Kasai, 1984

(cont.)

Table 2.4 (Cont.)

No.	Name	Aglycone	Structure	Plant	Reference
763	Ginsenoside Re	B	6-Glc^2-Rha 20-Glc	*P. japonicus* *P. trifolius* *P. ginseng* (Araliaceae) *P. notoginseng* *P. quinquefolius* *P. pseudoginseng* *P. japonicus* *P. trifolius* *Luffa cylindrica* (Cucurbitaceae)	Tanaka and Kasai, 1984 Tanaka and Kasai, 1984 Tanaka and Kasai, 1984 Tanaka and Kasai, 1984 Tanaka and Kasai, 1984 Tanaka and Kasai, 1984 Tanaka and Kasai, 1984 Tanaka and Kasai, 1984 Takemoto et al., 1984
764	Kizuta saponin K$_{7A}$	C	26-Glc	*Hedera rhombea* (Araliaceae)	Kizu et al., 1985a
765	Kizuta saponin K$_{7B}$	D	26-Glc	*H. rhombea* (Araliaceae)	Kizu et al., 1985a
766	Kizuta saponin K$_9$	E	3-Glc 26-Glc	*H. rhombea* (Araliaceae)	Kizu et al., 1985a
767	Kizuta saponin K$_{13}$	E	3-Glc^2-Glc 26-Glc	*H. rhombea* (Araliaceae)	Kizu et al., 1985a
768	24S-Pseudoginsenoside F$_{11}$	F	6-Glc^2-Rha	*Panax pseudoginseng* (Araliaceae)	Namba et al., 1986
769	24S-Majonoside-R2	F	6-Glc^2-Xyl	*P. japonicus* (Araliaceae) *P. pseudoginseng*	Morita et al., 1982 Namba et al., 1986
770	Luperoside A	G	3-Glc 20-Glc	*Luffa operculata* (Cucurbitaceae)	Kusumoto et al., 1989
771	Luperoside F	G	3-Glc^2-Rha 20-Glc^6-Glc	*L. operculata* (Cucurbitaceae)	Kusumoto et al., 1989

Structure elucidation of saponins from ginseng roots (*Panax ginseng* and *P. japonica*) has proved very awkward as a result of the extreme sensitivity of the aglycones protopanaxadiol and protopanaxatriol to acid hydrolysis (Fig. 2.5) and the consequent formation of artifacts. On removal of the sugars from ginsenosides, the corresponding aglyca cyclize very readily, via addition of the 20-OH to the 23,24-double bond, giving pyran derivatives (Tanaka and Kasai, 1984).

The vine *Thladiantha grosvenorii* (syn. *Momordica grosvenorii*; Cucurbitaceae) contains an intensely sweet glycoside, mogroside V (**772**). The aglycone mogrol is a tetracyclic triterpene hydroxylated at C-11 (Takemoto *et al.* 1983c).

Jujubogenin saponins

Jujubogenin (**773**) is a dammarane-type triterpene whose structure has been confirmed by X-ray crystallography (Kawai *et al.* 1974). It is the parent aglycone of a number of saponins obtained from members of the Rhamnaceae (Table 2.5). Two of the jujubogenin saponins isolated (**782**, **784**) contain the rare monosaccharide 6-deoxytalose (Okamura *et al.* 1981).

Ziziphin (**788**), the structure of which has recently been revised (Yoshikawa *et al.* 1991b), from *Ziziphus jujuba*, is a sweetness-inhibiting substance (Kurihara *et al.* 1988) containing an acetylated sugar residue.

Sulphated jujubogenin saponins (**792**, **793**) have been isolated from the bark of *Ziziphus joazeiro* (Higuchi *et al.* 1984).

Table 2.5. *Glycosides of jujubogenin*

No.	Name	Structure	Plant	Reference
774	Hovenoside D	$3\text{-Ara}^3\text{-Glc}^6\text{-Glc}$ $\quad\ \vert^2\qquad\ \vert^2$ $\quad \text{Xyl}\quad\ \text{Xyl}$	*Hovenia dulcis* (Rhamnaceae)	Inoue et al., 1978
775	Hovenoside G	$3\text{-Ara}^3\text{-Glc}$ $\quad\ \vert^2\qquad\ \vert^2$ $\quad \text{Xyl}\quad\ \text{Xyl}$	*H. dulcis*	Inoue et al., 1978
776	Hovenoside I	$3\text{-Ara}^3\text{-Glc}$ $\qquad\ \vert^2$ $\qquad \text{Xyl}$	*H. dulcis*	Inoue et al., 1978
777	Saponin C2	$3\text{-Ara}^3\text{-Glc}$ $\qquad\ \vert^2$ $\qquad \text{Rha}$	*H. dulcis*	Kimura et al., 1981
778	Saponin D	$3\text{-Glc}^2\text{-Rha}$ 20-Rha	*H. dulcis*	Kimura et al., 1981
779	Saponin G	3-Glc 20-Rha	*H. dulcis*	Kimura et al., 1981
780	Jujuboside A	$3\text{-Ara}^3\text{-Glc}^6\text{-Glc}$ $\quad\ \vert^2\qquad\ \vert^2$ $\quad \text{Rha}\quad\ \text{Xyl}$	*Ziziphus jujuba* (Rhamnaceae)	Otsuka et al., 1978a
781	Jujuboside B	$3\text{-Ara}^3\text{-Glc}$ $\quad\ \vert^2\qquad\ \vert^2$ $\quad \text{Rha}\quad\ \text{Xyl}$	*Z. jujuba*	Otsuka et al., 1978a
782	Ziziphus saponin I	$3\text{-Ara}^3\text{-Glc}$ $\qquad\ \vert^2$ $\qquad \text{Dta}$	*Z. jujuba*	Okamura et al., 1981

No.	Name	Structure	Species	Reference
783	Ziziphus saponin II	3-Ara^3-Glc $\mid^2$ Rha	*Z. jujuba*	Okamura *et al*, 1981
784	Ziziphus saponin III	3-Ara^3-Glc $\mid^2$ Dta Xyl	*Z. jujuba*	Okamura *et al*, 1981
785	Jujuba saponin I	3-Ara^2-Rha 20-Rha	*Z. jujuba*	Yoshikawa *et al*, 1991b
786	Jujuba saponin II	3-Ara^2-Rha 20-Rha^2-Ac	*Z. jujuba*	Yoshikawa *et al*, 1991b
787	Jujuba saponin III	3-Ara^2-Rha 20-Rha^3-Ac	*Z. jujuba*	Yoshikawa *et al*, 1991b
788	Ziziphin	3-Ara^2-Rha 20-Rha^3-Ac $\mid^2$ Ac	*Z. jujuba*	Kurihara *et al*, 1988; Yoshikawa *et al*, 1991b
789	Colubrin	3-Ara^3-Glc^2-Xyl	*Colubrina asiatica* (Rhamnaceae)	Wagner *et al*, 1983
790	Colubrinoside	3-Ara^3-Glc^3-Xyl $\mid^2$ $\mid^2$ Ac Ac	*C. asiatica*	Wagner *et al*, 1983
791		3-Arap^3-Glc $\mid^2$ Ara$f$	*Ziziphus joazeiro* (Rhamnaceae)	Higuchi *et al*, 1983
792		3-Arap^3-Glc^4-SO$_3^-$ $\mid^2$ Ara$f$	*Z. joazeiro*	Higuchi *et al*, 1983
793		3-Arap^3-Glc^4-SO$_3^-$ $\mid^2$ Araf^3-SO$_3^-$	*Z. joazeiro*	Higuchi *et al*, 1983

2.1.6   *Ester saponins*

The term 'ester saponin' is used here to describe those glycosides which are acylated on the aglycone or on the sugar chain with an acid moiety. Representatives of these saponins are very important in phytotherapeutical preparations and are found in aescine, *Senegae radix* and the like.

The ester saponins belong to the group of the most complicated saponins, the isolation of which is very difficult as a result of easy acyl migration or deacylation (Nakayama *et al.* 1986a; Jimenez *et al.* 1989). The formation of artifacts during hydrolysis reactions is also common. In fact, numerous ester saponins have only been incompletely characterized and the locations of their acyl groups still remain to be determined (see, for example, the saponins from *Sanicula europaea* (Umbelliferae) (Kühner *et al.* 1985) and from *Bellis perennis* (Asteraceae) (Hiller *et al.* 1988)). Higuchi *et al.* (1987) were only able to separate saponins from the bark of *Quillaja saponaria* (Rosaceae) after deacylation with 6% sodium bicarbonate solution. Similarly, desacylmasonosides, polygalacic acid glycosides from the corms of *Crocosmia masoniorum* (Iridaceae), were isolated after treatment of the crude saponin mixture with 1% sodium bicarbonate–ethanol (1 : 1) (Asada *et al.* 1989b). It is also certain that a number of characterized saponins are actually ester saponins and that during the extraction/isolation process the ester linkage is cleaved.

Structure elucidation is often problematic and substitution positions of acyl groups are difficult to establish with certainty, as is the case with the acetylated saponins from *Sapindus rarak* (Sapindaceae) (Hamburger *et al.* 1992). Extensive use of 1-D and 2-D NMR techniques (Chapter 4) may be necessary for unambiguous structure determination (Massiot *et al.* 1992a).

Figure 2.6 shows some of the organic acids found in combination with triterpene saponins. A selection of triterpene ester saponins is presented in Tables 2.6 and 2.7. Since few acylated steroid saponins have as yet been discovered, these are included in Section 2.2. The known acylated dammarane saponins are listed in Table 2.4.

A series of antifungal compounds, avenacins A-1 (**851**), A-2 (**852**), B-1 (**853**) and B-2 (**854**), has been found in oat roots (*Avena sativa*, Gramineae). It was many years before their structures were definitively assigned (Begley *et al.* 1986; Crombie *et al.* 1987), but it is now known that they are composed of complex aglycones containing either a benzoate or an anthranilate ester linked to a trisaccharide moiety.

Gymnemic acids I, II, III and IV (Fig. 2.7) are antisweet (sweetness-

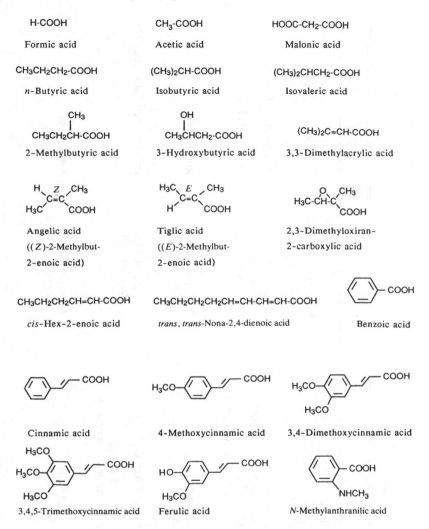

Fig. 2.6. Acid moieties of ester saponins.

inhibiting) substances from the leaves of *Gymnema sylvestre* (Asclepiadaceae) (Yoshikawa *et al.* 1989a). They all contain a glucuronic acid
moiety, and the gymnemagenin aglycone (Tsuda *et al.* 1989) is esterified
at positions C-21 and C-28. The substitution pattern was established by
long-range selective proton decoupling (LSPD). It should be noted that
alternative structures for gymnemic acids I and II have been presented,
without the acetyl group at position C-28 (Maeda *et al.* 1989).

A second series of gymnemagenin glycosides has been isolated from
Gymnema sylvestre: gymnemic acids V–VII. Gymnemic acid VII is the

Table 2.6. *Ester saponins with acylated aglycones*

No.	Name	Structure	Plant	Reference
21β-Hydroxyoleanolic acid aglycone (20)				
794		3-Glc	*Enterolobium*	Delgado *et al.*, 1986
		21-Cinnamate	*contorstisiliquum*	
			(Leguminosae)	
795		3-Glc^2-GlcNAc	*Mussatia* sp.	Jimenez *et al.*, 1989
		21-(4-Methoxycinnamate)	(Bignoniaceae)	
796		3-Glc^2-Glc^3-Xyl	*Mussatia* sp.	Jimenez *et al.*, 1989
		21-(3,4-Dimethoxycinnamate)		
23-Hydroxyimberbic acid aglycone (41)				
797		1-Acetate	*Combretum indicum*	Rogers, 1988
		23-Rha	(Combretaceae)	
Presenegenin aglycone (47)				
798		3-Rha^4-Rha^4-Xyl^4-Glc	*Polygala chinensis*	Brieskorn and Kilbinger, 1975
		$\mid^2$	(Polygalaceae)	
		Rha^3-Glc		
		27-(4-Methoxycinnamate)		
Barringtogenol C aglycone (64)				
799		3-GlcA^2-Glc	*Aesculus indica*	Sati and Rana, 1987
		21-Angelate	(Hippocastanaceae)	
		22-Angelate		

Protoaescigenin aglycone (68)

800	Aescine	3-GlcA^4-Glc $\mid^2$ Glc 21-Angelate (tiglate) 22-Acetate	*A. hippocastanum* (Hippocastanaceae)	Wulff and Tschesche, 1969
801	Aesculuside-A	3-GlcA^4-Glc $\mid^2$ Glc 21-Angelate	*A. indica* (Hippocastanaceae)	Singh *et al.*, 1986
802	Bunkankasaponin A	3-GlcA^2-Glc 21-Fuc^4-Ac $\mid^3$ Angelate 22-Acetate	*Xanthoceras sorbifolia* (Sapindaceae)	Chen *et al.*, 1985)
803	Bunkankasaponin C	3-GlcA^2-Glc 21-Fuc^4-Ac $\mid^3$ Angelate 28-Acetate	*X. sorbifolia* (Sapindaceae)	Chen *et al.*, 1985

Theasapogenol A aglycone (69)

804	Theasaponin	3-GlcA^3-Gal $\mid^2$ Ara^2-Xyl 21-Angelate (tiglate) 22-Acetate	*Thea sinensis* (Theaceae)	Tschesche *et al.*, 1969a

(*cont.*)

Table 2.6 (*Cont.*)

No.	Name	Structure	Plant	Reference
R$_1$-Barrigenol aglycone (70)				
805	Dodonoside A	3-GlcA-Gal | Ara*f* 21-2-Methylbutyrate 22-2,3-Dimethyloxiran-2-carboxylate or 21-2,3-Dimethyloxiran-2-carboxylate 22-2-Methylbutyrate	*Dodonaea viscosa* (Sapindaceae)	Wagner *et al.*, 1987
806	Dodonoside B	3-GlcA-Gal | Ara*f* 21-Angelate 22-2,3-Dimethyloxiran-2-carboxylate or 21-2,3-Dimethyloxiran-2-carboxylate 22-Angelate	*D. viscosa* (Sapindaceae)	Wagner *et al.*, 1987
807		3-GlcA^2-Glc + C$_7$ monoester	*Eryngium planum* (Umbelliferae)	Voigt *et al.*, 1985
Cyclamiretin A aglycone (72)				
808	Saxifragifolin C	3-Ara^4-Glc^2-Xyl |^2 Glc^4-Glc 22-Acetate	*Androsace saxifragifolia* (Primulaceae)	Pal and Mahato, 1987

809 Saxifragifolin A

$3\text{-Ara}^4\text{-Glc}^2\text{-Xyl}$
$|^2$
Glc
22-Acetate

A. saxifragifolia
(Primulaceae)

Waltho *et al.*, 1986

Priverogenin B aglycone (78)

810 Anagalloside C

$3\text{-Ara}^4\text{-Glc}^4\text{-Glc}$
$|^2 \quad |^2$
Glc   Xyl
22-Acetate

Anagallis arvensis
(Primulaceae)

Glombitza and Kurth, 1987a,b

Anagalligenin A aglycone (79)

811 Anagalloside A

$3\text{-Ara}^4\text{-Glc}^4\text{-Glc}$
$|^2 \quad |^2$
Glc   Xyl
22-Acetate

A. arvensis
(Primulaceae)

Glombitza and Kurth, 1987a,b

812 Desglucoanagalloside A

$3\text{-Ara}^4\text{-Glc}$
$|^2 \quad |^2$
Glc   Xyl
22-Acetate

A. arvensis
(Primulaceae)

Glombitza and Kurth, 1987a,b

Table 2.7. *Ester saponins with acylated saccharides*

No.	Name	Structure	Plant	Reference
Oleanolic acid aglycone (2)				
813		3-GlcA²-Xyl⁴-Ac	*Panax pseudoginseng* ssp. *himalaicus* var. *angustifolius* (Araliaceae)	Shukla and Thakur, 1988
814	Silphioside A	3-Glc⁶-Ac 28-Glc	*Silphium perfoliatum* (Asteraceae)	Davidyants *et al.*, 1986
815	Silphioside C	3-Glc⁶-Ac |² Glc	*S. perfoliatum* (Asteraceae)	Davidyants *et al.*, 1985
Gypsogenin aglycone (7)				
816	Lucyoside M	3-Glc⁶-Ac 28-Glc	*Luffa cylindrica* (Cucurbitaceae)	Takemoto *et al.*, 1985
Echinocystic acid aglycone (16)		3-GlcA		
817	Tragopogonsaponin D	28-Xyl²—CO—CH=CH	*Tragopogon porrifolius* (Asteraceae)	Warashina *et al.*, 1991
Hederagenin aglycone (17)				
818	Kizuta saponin K₈	3-Ara 28-Glc⁶-Glc⁴-Rha |⁶ Ac	*Hedera rhombea* (Araliaceae)	Kizu *et al.*, 1985b

For compound 817, the 28-Xyl²—CO—CH=CH group connects to:

No.	Name / Structure	Source (Family)	Reference
819	3-Ara^2-Glc 28-Glc^6-Glc	*Patrinia scabiosaefolia* (Valerianaceae)	Choi and Woo, 1987
820	3-Glc^6-Ac	*Schefflera impressa* (Araliaceae)	Srivastava, 1989
821	3-Ara^2-Rha 28-Glc^6-Glc^6-Ac	*Lonicera japonica* (Caprifoliaceae)	Kawai et al., 1988

822 Kizuta saponin K$_{11}$	3-Ara^2-Rha
	28-Glc^6-Glc^4-Rha
	    |^6
	    Ac
		Hedera rhombea
		(Araliaceae)
		H. nepalensis
			Kizu et al., 1985b
			
			Kizu et al., 1985c

823	3-Ara^2-Rha 28-Glc^6-Glc^4-Rha^2-Ac	*Astrantia major* (Umbelliferae)	Hiller et al., 1990b
824	3-Ara^2-Rha 28-Glc^6-Glc^4-Rha^3-Ac	*A. major*	Hiller et al., 1990b
825 Mukurozi saponin E$_1$	3-Ara^2-Rha^3-Xyl^4-Ac	*Sapindus mukurossi* (Sapindaceae)	Kimata et al., 1983

826	3-Ara^2-Rha^3-Xyl^4-Ac
	    |^3
	    Ac
		S. delavayi
		S. delavayi
		S. rarak
			Nakayama et al., 1986a
			Nakayama et al., 1986a
			Hamburger et al., 1992

827 Mukurozi saponin G	3-Ara^2-Rha^3-Ara^4-Ac
	    |^3
	    Ac
		S. mukurossi
		S. delavayi
			Kimata et al., 1983
			Nakayama et al., 1986a

29-Hydroxyoleanolic acid aglycone (21)

No.	Name / Structure	Source (Family)	Reference

828 Ciwujianoside D$_3$	3-Ara
	28-Glc^6-Glc^4-Rha
	    |^6
	    Ac
		Acanthopanax senticosus
		(Araliaceae)
			Shao et al., 1989a

(cont.)

Table 2.7. (*Cont.*)

No.	Name	Structure	Plant	Reference
829	Ciwujianoside A₄	3-Ara²-Glc 28-Glc⁶-Glc⁴-Rha ‚⁶ Ac	*A. senticosus* (Araliaceae)	Shao *et al.*, 1989a
	Presenegenin aglycone (47)			
830	Senegin II	3-Glc 28-Fuc²-Rha⁴-Xyl⁴-Gal ‚⁴ 3,4-dimethoxycinnamate	*Polygala senega* (Polygalaceae)	Shoji *et al.*, 1971; Tsukitani *et al.*, 1973

831	Senegin III
	(onjisaponin B)
		3-Glc
		  Rha
		  ‚³
		28-Fuc²-Rha⁴-Xyl⁴-Gal
		‚⁴
		4-methoxycinnamate
			P. senega
			(Polygalaceae)
			P. tenuifolia
				Shoji *et al.*, 1971; Tsukitani
				et al., 1973
				Sakuma and Shoji, 1982

832	Onjisaponin A	3-Glc
		  Rha Api
		  ‚³ ‚³
		28-Fuc²-Rha⁴-Xyl⁴-Gal
		‚⁴
		4-methoxycinnamate
			P. tenuifolia
			(Polygalaceae)
				Sakuma and Shoji, 1982

833	Onjisaponin C	3-Glc 28-Fuc²-Rha⁴-Xyl⁴-Gal ‚⁴ 3,4,5-trimethoxycinnamate	*P. tenuifolia* (Polygalaceae)	Sakuma and Shoji, 1982

834	Senegin IV	3-Glc Rha Rha |3 |3 28-Fuc²-Rha⁴-Xyl⁴-Gal |⁴ 4-methoxycinnamate	*P. senega* (Polygalaceae)	Shoji *et al.*, 1971; Tsukitani and Shoji, 1973

Polygalacic acid aglycone (54)

835	Bellissaponin BA₁	3-Rha 28-Fuc²-Rha⁴-Xyl³-Rha |⁴ *E*-buta-2-enoate	*Bellis perennis* (Asteraceae)	Schöpke *et al.*, 1991
836	Bellissaponin BA₂	3-Glc 28-Fuc²-Rha⁴-Xyl³-Rha |⁴ *E*-buta-2-enoate	*B. perennis* (Asteraceae)	Schöpke *et al.*, 1991
837	Solidagosaponin II	16-Glc²-Ara⁴-(3-hydroxybutyrate)	*Solidago virgaurea* (Asteraceae)	Inose *et al.*, 1991
838	Solidagosaponin IV	16-Glc²-Ara⁴-(3-hydroxybutyrate) 28-Rha	*S. virgaurea* (Asteraceae)	Inose *et al.*, 1991
839	Solidagosaponin VI	16-Glc²-Ara⁴-(3-hydroxybutyrate) 28-Xyl	*S. virgaurea* (Asteraceae)	Inose *et al.*, 1991
840	Solidagosaponin VIII	16-Glc²-Ara⁴-(3-hydroxybutyrate) 28-Ara	*S. virgaurea* (Asteraceae)	Inose *et al.*, 1991

Soyasapogenol A aglycone (60)

704	Acetyl-soyasaponin A₁	3-GlcA²-Gal²-Glc 22-Ara³-Glc(Ac)₄	*Glycine max* (Leguminosae)	Kitagawa *et al.*, 1988b
707	Acetyl-soyasaponin A₄	3-GlcA²-Gal²-Glc 22-Ara³-Xyl(Ac)₃	*G. max* (Leguminosae)	Taniyama *et al.*, 1988a

(cont.)

Table 2.7. (Cont.)

No.	Name	Structure	Plant	Reference
Ursolic acid aglycone (81)				
841	Sanguisorba saponin	3-Ara 28-Glc³-Ac	*Sanguisorba officinalis* (Rosaceae)	Bukharov and Karneeva, 1970a
Pomolic acid aglycone (84)				
842		3-Ara²-Ac	*Ilex cornuta* (Aquifoliaceae)	Qin et al, 1986
3β,6α,16β,24,25-Pentahydroxy-9,19-cyclolanostane aglycone (121)				
843	Askendoside A	3-Xyl²-Ara $\mid$³ Ac	*Astragalus taschkendicus* (Leguminosae)	Isaev et al., 1983d
Cycloastragenol aglycone (122)				
844	Astragaloside II	3-Xyl²-Ac	*A. membranaceus* (Leguminosae)	Kitagawa et al., 1983a
845	Cyclosieversioside A	3-Xyl³-Ac $\mid$² Ac	*A. galegiformis* *A. sieversianus* (Leguminosae)	Alaniya et al., 1984 Svechnikova et al., 1982
846	Askendoside B	3-Xyl²-Ara $\mid$³ Ac 6-Xyl	*A. taschkendicus* (Leguminosae)	Isaev et al., 1983c

847	Astrasieversianin IX	$3\text{-Xyl}^2\text{-Rha}$ $\overset{	^3}{\text{Ac}}$ 6-Xyl	*A. sieversianus* (Leguminosae)	Gan *et al.*, 1986a
848	Astrasieversianin XI	$3\text{-Xyl}^2\text{-Rha}$ $\overset{	^4}{\text{Ac}}$ 6-Xyl	*A. sieversianus* (Leguminosae)	Gan *et al.*, 1986a
849	Asernestioside B	$3\text{-Xyl}^2\text{-Rha}$ $\overset{	^3}{\text{Ac}}$ 25-Glc	*A. ernestii* (Leguminosae)	Wang *et al.*, 1989a
850	Asernestioside C	$3\text{-Xyl}^2\text{-Rha}$ $\overset{	^4}{\text{Ac}}$ 25-Glc	*A. ernestii* (Leguminosae)	Wang *et al.*, 1989a

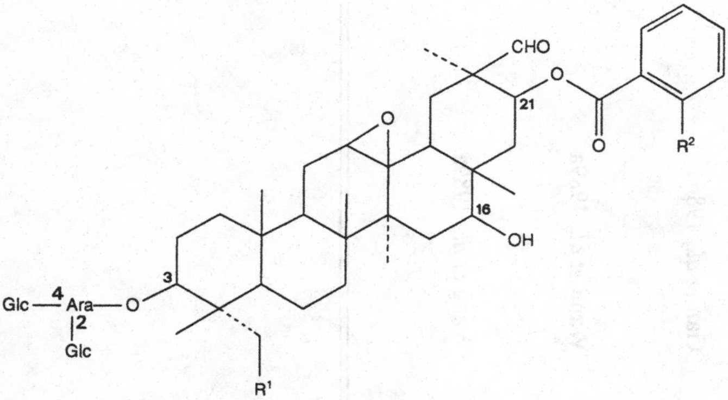

851 Avenacin A-1 R^1 = -OH, R^2 = -NHCH$_3$

852 Avenacin A-2 R^1 = -OH, R^2 = -H

853 Avenacin B-1 R^1 = -H, R^2 = -NHCH$_3$

854 Avenacin B-2 R^1 = -H, R^2 = -H

3-*O*-glucuronide of gymnemagenin, and gymnemic acid V has a 3-*O*-glucuronyl-21,22-bis-*O*-tigloyl substitution pattern (Yoshikawa *et al.* 1989b). Gymnemic acids VIII and IX, also ester saponins, have an oxoglycoside (hexulopyranoside) moiety attached to the glucuronic acid residue (Kiuchi *et al.* 1990).

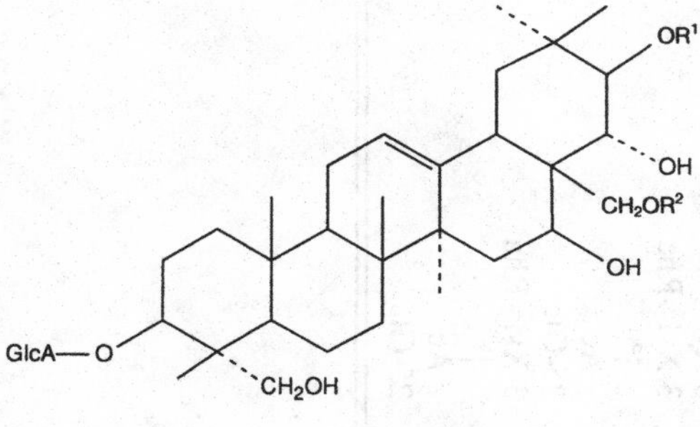

	R^1	R^2
855 Gymnemic acid I	Tigloyl	Acetyl
856 Gymnemic acid II	2-Methylbutyroyl	Acetyl
857 Gymnemic acid III	2-Methylbutyroyl	Acetyl
858 Gymnemic acid IV	Tigloyl	Acetyl

Fig. 2.7. Gymnemic acids from *Gymnema sylvestre*.

The saponin mixtures aescine (from *Aesculus hippocastanum*, Hippocastanaceae) and theasaponin (from tea seeds, *Thea sinensis*, Theaceae) have the same or similar aglycones to the *Gymnema* saponins. Aescine is now well characterized but the components of theasaponin have still not been completely analysed (Chapter 7).

Eighteen acylated glycosides of echinocystic acid, tragopogonsaponins A–R, have been isolated from the roots of *Tragopogon porrifolius* (Asteraceae). They contain *p*-coumaric acid, ferulic acid, 4-hydroxyphenyl propionic acid or 4-hydroxy-3-methoxyphenyl propionic acid as the ester moiety. Certain pairs of these saponins, e.g. **859** and **860**, could not be separated, even by semi-preparative HPLC. The structures of these mixtures were determined by NMR spectroscopy, FAB-MS and comparison with related glycosides (Warashina *et al.* 1991).

859 Tragopogonsaponin I R = -CO-CH$_2$-CH$_2$-⟨benzene⟩-OH

860 Tragopogonsaponin J R = -CO-CH$_2$-CH$_2$-⟨benzene⟩-OH, OCH$_3$

Two of the rare examples of triterpene saponins from the monocotyledons are the acylated saponins crocosmiosides A (**861**) and B (**862**) from the corms of montbretia (*Crocosmia crocosmiiflora*, Iridaceae). Structure elucidation of these hydroxylated palmitic acid esters required extensive use of FAB-MS, ^{13}C-NMR spectroscopy and hydrolysis experiments (Furuya *et al.* 1988; Asada *et al.* 1989a). The aglycone is polygalacic acid.

An olean-18(19)-en triterpene glycoside from *Periandra dulcis* (Leguminosae) contains a 3,5-dimethoxy-4-hydroxybenzoyl (syringyl) moiety esterified at C-22 (**863**) (Ikeda *et al.* 1990, 1991).

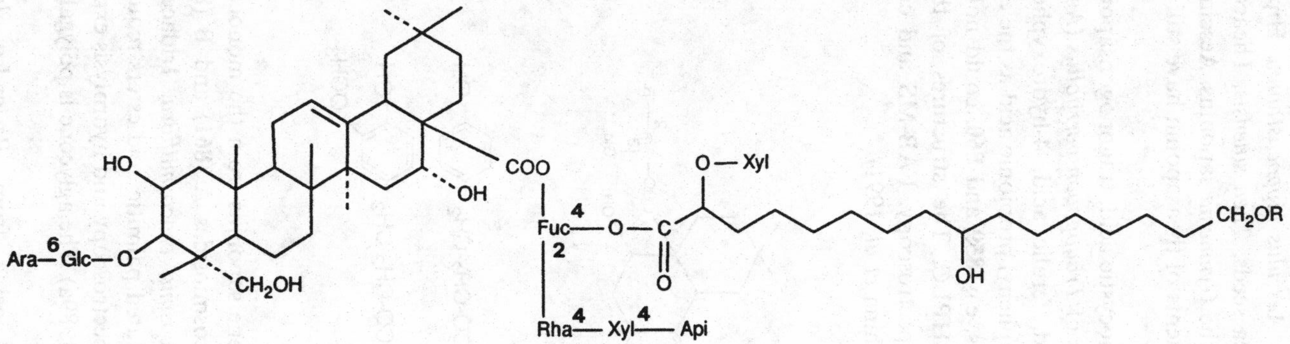

CH_2OR

OH

O—Xyl

C

O

Fuc

Rha—Xyl—Api

Rha

COO

OH

CH_2OH

HO

Ara—Glc

861 Crocosmioside A R = -Rha

862 Crocosmioside B R = -H

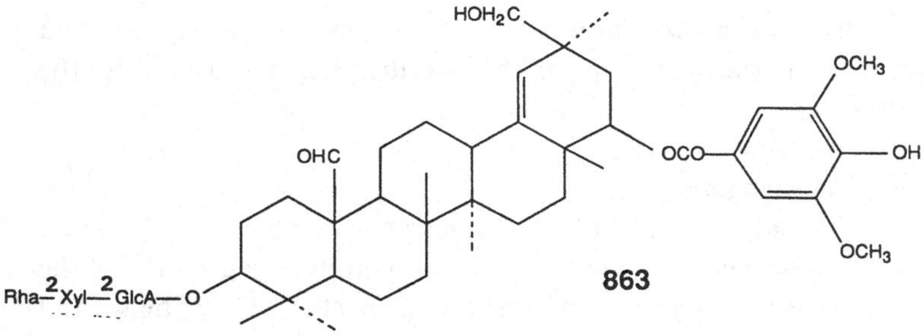

863

The aglycone of glochidioside R (**866**) and glochidiosides N (**864**) and Q (**865**) is 16-*O*-benzoyl-gymnestrogenin. The glycosides were isolated from *Glochidion heyneanum* (Euphorbiaceae) (Srivastava and Kulshreshtha, 1986, 1988).

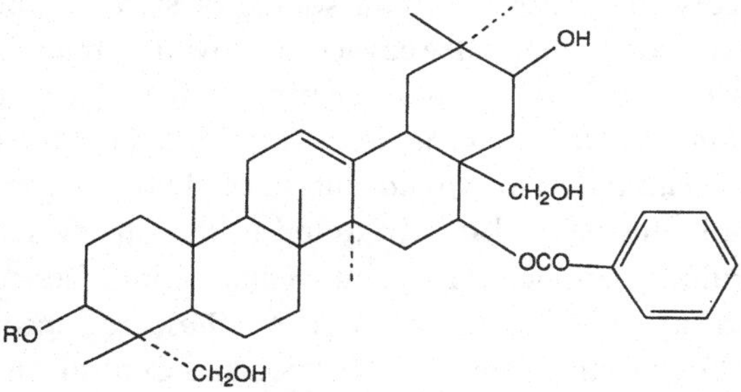

864 Glochidioside N R = -Glc

865 Glochidioside Q R = -Glc^2-Glc

866 Glochidioside R R = -Ara^3-Glc

There are one or two examples of saponins which have a methylated glucuronic acid moiety in the sugar chain. Such is the case for momordin Ia (**867**) from the roots of the Chinese medicinal plant *Momordica cochinchinensis* (Cucurbitaceae) (Kawamura *et al.* 1988) and for saponin

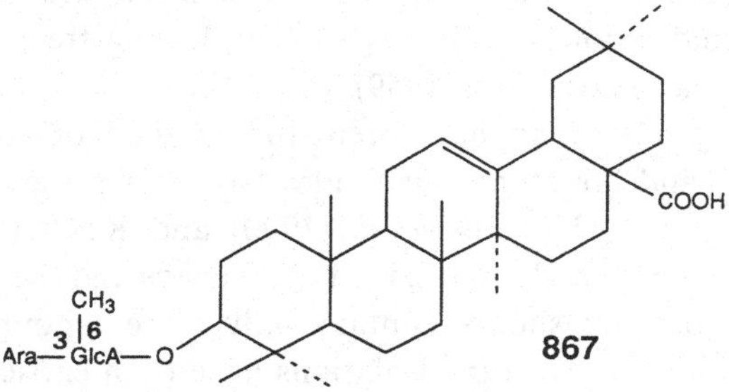

867

358 (Appendix 2) from *Schefflera impressa* (Araliaceae) (Srivastava and Jain, 1989). The carboxyl group for the esterification is provided by the saponin itself.

2.2   Steroid saponins

The steroid saponins are divided into two main groups. The first (and largest) group, the *spirostanol* glycosides, comprises aglycones of the spirostan type, with a sugar chain generally at position C-3. The second group, the *furostanol* glycosides, has a second sugar moiety at position C-26 and can therefore be classified as bidesmosidic. They consequently lack most of the characteristic saponin properties. Spirostanol glycosides are found mainly in the seeds, roots or bulbs of plants, whereas furostanol glycosides are in the assimilatory parts.

Steroid saponins are less widely distributed in nature than the triterpenoid type. Monocotyledons are the main source of steroid saponins, with the Liliaceae, Dioscoreaceae and Agavaceae providing many representatives. The most important saponin-bearing genera are: in the Liliaceae family *Trillium, Chlorogalum, Smilax, Nolina, Agapanthus*; in the Agavaceae the *Agave, Yucca* and *Manfreda*; and in the Dioscoreaceae the *Dioscorea*. Steroid saponins have also been found in the Amaryllidaceae, Bromeliaceae and Palmae families. The genus *Smilax* comprises mainly climbing plants, with saponins in the woody roots. *Dioscorea* species, of which about 250 are known, grow mainly in Mexico and Central America. Their bulbous rhizomes are excellent sources of steroid saponins and may weight up to 50 kg. Those which are used as foodstuffs (yams) include *D. alata, D. bulbifera, D. cayennensis, D. esculenta, D. opposita* (syn. *D. batatas*). The dried underground parts of *D. villosa* (from North America) have been used in folk medicine for the treatment of rheumatoid arthritis, intestinal cholic, cholecystitis and dysmenorrhoea (*British Herbal Pharmacopoeia*, 1983). Agaves are tropical and subtropical plants, whose fleshy leaves are used both for obtaining sisal fibre and saponins. Species of *Yucca* are characterized by long stems and a palm-like crown of leaves. They are found in tropical America and the USA; the leaves are sources both of fibres and steroid saponins. The seeds of the Joshua tree, *Yucca brevifolia*, contain 18% saponins (Wall, 1959).

The saponins of the genus *Allium* have been studied: the work on the constituent spirostanol and furostanol glycosides has been reviewed by Fenwick and Hanley (1985), Kravets *et al.* (1990) and Koch (1993). Glycosides are found in the epigeal organs, the flower heads and the seeds; the leaves and flower-bearing shoots contain only trace amounts of glycosides and sapogenins. Many of the isolations have been carried out by the group of Abubakirov (Kravets *et al.* 1986a,b and references cited therein) and the species investigated include the garden onion, *A. cepa* (Appendix 3), *A. turcomanicum, A. karataviense, A. erubescens, A. giganteum* and *A. waldstenii.* The most widespread genin is diosgenin, found in 18 *Allium* species (Kravets *et al.* 1990). The fruit and seeds of onion contain glycosides but these are absent in the bulbs, leaves and flower-bearing shoots (Kravets *et al.* 1990). The tigogenin constituents of garlic (*A. sativum*) have been characterized by Matsuura *et al.* 1989a (Appendix 3).

Of the dicotyledons, the genera *Digitalis* (Scrophulariaceae) and *Trigonella* (Leguminosae) have been found to contain steroid saponins. Further examples have been found in other species of the Scrophulariaceae and Leguminosae, together with the families Solanaceae, Simaroubaceae, Cornaceae, Zygophyllaceae, Ranunculaceae, Asteraceae and Cruciferae. Some species of *Strophanthus* (Apocynaceae) and *Digitalis* contain both steroid saponins and cardenolides.

It should also be noted that simple steroid glycosides (sitosterol-3-*O*-β-D-glucopyranoside, for example) are widespread in the plant kingdom. Certain of these, such as tenuifolioside A (**868**) from *Lilium tenuifolium* (Liliaceae) (Mimaki *et al.* 1989), bear a close resemblance to starfish saponins (see Section 2.4).

868

Very bitter steroid glycosides, for example vernonioside A_1 (**869**), have been isolated from the leaves of *Vernonia amygdalina* (Asteraceae) (Ohigashi *et al.* 1991; Jisaka *et al.* 1992).

2.2.1   Spirostanol glycosides

The correct structure of the aglycone part of these glycosides remained elusive for many years (Table 2.8). It was not until 1939 that Marker proposed the novel spiroketal arrangement, labile under certain acidic conditions, for sarsasapogenin (Marker and Rohrmann, 1939).

Table 2.8. *Structures of selected spirostanol aglycones*

No.	Name	Skeleton	C-25	OH
870	Sarsasapogenin	A	S	3β
871	Smilagenin	A	R	3β
872	12β-Hydroxysmilagenin	A	R	3β,12β
873	Rhodeasapogenin	A	S	1β,3β
874	Isorhodeasapogenin	A	R	1β,3β
875	Samogenin	A	R	2α,3β
876	12β-Hydroxysamogenin	A	R	3β,12β
877	Markogenin	A	S	2β,3β
878	Yonogenin	A	R	2β,3α
879	Convallagenin A	A	S	1β,3β,5β
880	Convallagenin B	A	S	1β,3β,4β,5β
881	Tokorogenin	A	R	1β,2β,3α
882	Tigogenin	B	R	3β
883	Neotigogenin	B	S	3β
884	Gitogenin	B	R	2α,3β
885	Agigenin	B	R	2α,3β,6β
886	Digitogenin	B	R	2α,3β,15β
887	Chlorogenin	B	R	3β,6α
888	Paniculogenin	B	S	3β,6α,23β
889	(25R)-Spirostan-3β,17α,21-triol	B	R	3β,17α,21
890	Alliogenin	B	R	2α,3β,5α,6β
891	(25R)-5α-Spirostan-2α,3β,5α,6α-tetraol	B	R	2α,3β,5α,6α
892	(24S,25R)-5α-Spirostan-2α,3β,5α,6β,24-pentaol	B	R	2α,3β,5α,6β,24
893	Yamogenin (neodiosgenin)	C	S	3β
894	Diosgenin	C	R	3β
895	Yuccagenin	C	R	2α,3β
896	Lilagenin	C	S	2α,3β
897	Ruscogenin	C	R	1β,3β
898	(25S)-Ruscogenin	C	S	1β,3β
899	Neoprazerigenin	C	S	3β,14α
900	Pennogenin	C	R	3β,17α
901	Isonuatigenin	C	S	3β,25S
902	Cepagenin	C	R	1β,3β,24
903	24α-Hydroxypennogenin	C	R	3β,17α,24α
904	Ophiogenin	C	R	3β,14α,17α
905	Sibiricogenin	C	R	3β,14α,23
906	Convallamarogenin	D		1β,3β
907	Neoruscogenin	E		1β,3β
908	Hecogenin	F	R	3β
909	Neohecogenin	F	S	3β
910	Manogenin	F	R	2α,3β
911	Sisalagenin	F	S	3β
912	Nuatigenin	G	S	3β
913	Hispigenin	H		3β,6β
914	Solagenin	I	S	3β,6α

Structures **915–1058** are listed in Appendix 3, p. 422.

869

Although over 100 different spirostanol aglycones are known (see Table 2.8 and Patel *et al.* 1987), the number of chemically defined glycosides is relatively small for two reasons: (i) purification problems are commonplace and (ii) many of the isolated compounds are quite possibly artifacts arising from the extraction process. One source of artifacts is the ready conversion of furostanol glycosides into spirostanols in the presence of plant enzymes. The best source of genuine spirostanol saponins is the seeds, where they are thought to have some inhibitory function on the growth of fungi (Tschesche and Wulff, 1972).

Several combinations of isomers are possible for the aglycones, depending on the chiral centres C-20, C-22 and C-25, but most native sapogenins appear to have the 20*S*,20*R*-configuration, as illustrated by digitogenin (**886**). By comparison, both of the C-25 epimers are possible in the plant and mixtures of these, e.g. diosgenin ((25*R*)-spirost-5-en-3β-ol, **894**) and yamogenin ((25*S*)-spirost-5-en-3β-ol, **893**) do occur; their ratio is dependent upon factors such as the morphological part and the stage of development of the plant. The equatorial (25*R*) conformation is more stable than the axial (25*S*) conformation and the axial isomer is converted into the equatorial isomer by the action of hydrochloric acid in alcohol to give an equilibrium.

886 Digitogenin: (25*R*)-5α-spirostan-2α,3β,15β-triol

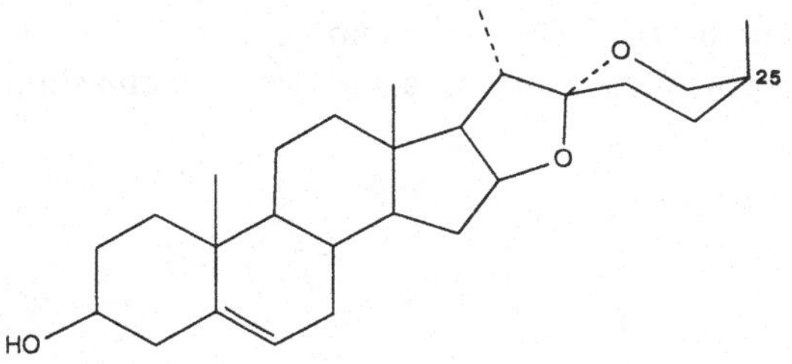

893 Yamogenin (25*S*)

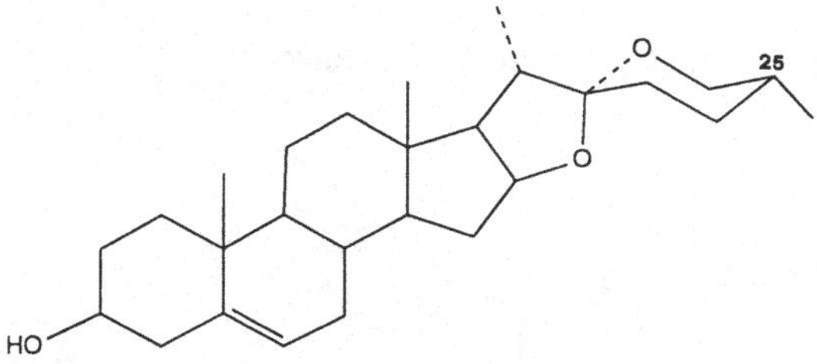

894 Diosgenin (25*R*)

The chromatographic separation of glycosides with the same sugar moiety and aglycones differing only in the structure at C-25 can be extremely difficult (Miyahara *et al.* 1983). It cannot, therefore, be excluded that some steroid saponins which are claimed to be mono-substances are, in fact, mixtures of 25*R* and 25*S* epimers and even the 25(27)-enes.

In most cases, the sugar chain is affixed to position C-3 of the aglycone but the hydroxyl groups at C-1, C-2, C-5, C-6 and C-11 can also be involved. Substitution in the F ring is also known. The sapogenins of the Liliaceae are often oxidized at C-1, C-2, C-5 or C-6, those of the Amaryllidaceae at C-12 and those of the Scrophulariaceae at C-2 and/or C-15.

The structures of a number of spirostanol aglycones are shown in Table 2.8. Certain glycosides are listed in Appendix 3 (structures **915–1058**). More comprehensive compilations are to be found in reviews by Takeda (1972), Mahato *et al.* (1982), and Voigt and Hiller (1987). An extensive list of aglycones and glycosides is provided in dictionary form (Hill *et al.* 1991).

Spirostanol glycosides with a lactone ring in the sapogenol have been found in the aerial parts of *Solanum dulcamara* (Solanaceae). One of these, soladulcoside B (**1060**), has a 3-*O*-β-chacotrioside moiety attached to a tigogenin-like aglycone, while soladulcoside A (22*R*,25*R*)-3β,15α,23αtrihydroxy-5α-spirostan-26-one   3-*O*-α-L-rhamnopyranosyl-(1→2)-β-D-glu-copyranoside) (**1059**) possesses a disaccharide chain (Yamashita *et al.* 1991).

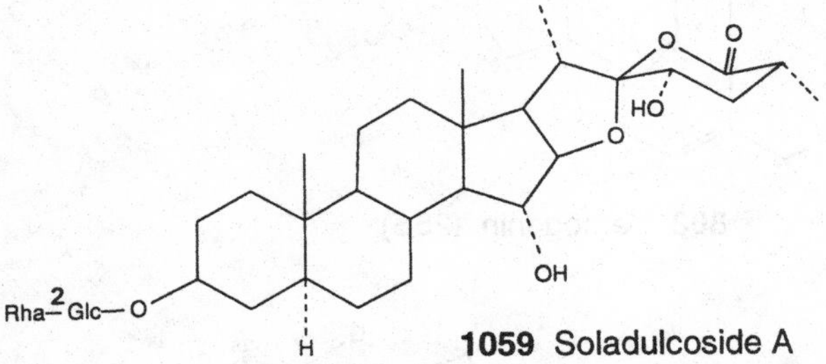

1059 Soladulcoside A

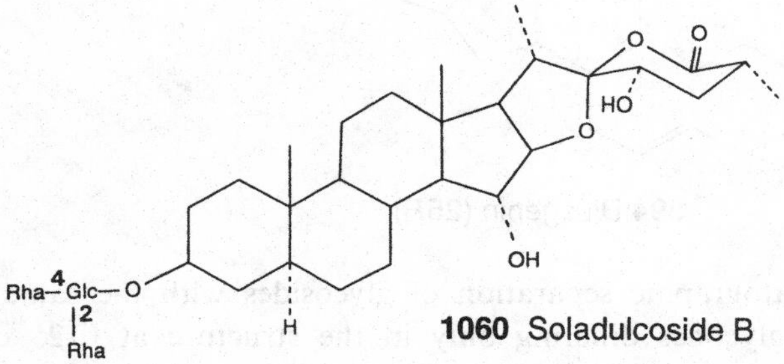

1060 Soladulcoside B

Nuatigenin (**912**, Table 2.8) has a furospirostan structure, in which ring F becomes a five-membered furan ring instead of a six-membered pyran ring as in spirostans.

A number of the glycosides have been isolated from *Digitalis*, in addition to the numerous cardiac glycosides from this genus. Similar spirostanol saponins are found both in *D. purpurea* and *D. lanata* (Appendix 3). Historically, *Digitalis purpurea* was the first source of steroid saponins. Schmiedeberg (1875) isolated a material from commercial digitalis extracts that he named digitonin; this was subsequently shown to contain 10–20% of a second glycoside, named gitonin (Windaus and Schneckenburger, 1913).

Of great commercial interest are the diosgenin glycosides from *Dioscorea* species. Diosgenin (**894**) was first isolated from Japanese *Dioscorea tokoro* in 1936 (Tsukamoto and Ueno, 1936; Fujii and Matsukawa, 1936) and was named after the source plant. It is common to all members of the Dioscoreaceae which contain steroid sapogenins and, in fact, consultation of Appendix 3 will show that glycosides of diosgenin and its C-25 isomer yamogenin are among the most frequently documented spirostanol saponins. Gracillin (**995**) and dioscin (**1002**) differ from each other only in the nature of the third sugar in the saccharide moiety and are important as starting materials in the synthesis of steroid hormones (see Chapter 7).

A characteristic feature of the Dioscoreaceae is the existence of sapogenins in which the hydroxyl group at the C-3 position has an α-configuration, e.g. yonogenin (**878**) and tokorogenin (**881**, Table 2.8).

Parillin (**923**, Appendix 3) has been isolated from the roots and rhizomes of *Smilax* species, including *Sarsaparillae radix* (*Smilax aristolochiaefolia*, Liliaceae), but is most probably an artifact resulting from the furostanol bidesmoside sarsaparilloside.

Spirostanol saponins from *Convallaria* (Liliaceae) possess sugar chains attached not only at position C-3 but also at C-1 and C-5 (Appendix 3). Similarly, ruscin (**1040**) and tokoronin (**941**) have the saccharide moiety at C-1. The bulbs of *Allium giganteum* (Liliaceae) contain alliogenin (**890**, Table 2.8) with a monosaccharide attached at position C-2 (Sashida *et al.* 1991a). Saponins containing an aglycone (paniculogenin, **888**, Table 2.8) hydroxylated at C-3, C-6 and C-33 have been isolated from *Solanum paniculatum* (Solanaceae). Paniculonin A (**977**) and paniculonin B (**978**) both have an α-oriented disaccharide moiety at C-6 with a D-quinovose (6-deoxyglucose) residue (Ripperger and Schreiber, 1968).

Saponins of chlorogenin (**887**, Table 2.8) glycosylated only at C-6 are present in the bitter-tasting bulbs of *Camassia cusickii* (Liliaceae). They also contain glycosides of (25*R*)-6α-hydroxy-5α-spirostan-3-one (**1061**) and (25*R*)-3,3-dimethoxy-5α-spirostan-6α-ol (**1062**). This latter is a ketal form of the carbonyl compound **1061** and is claimed not to be an artifact of the isolation process (Mimaki *et al.* 1991).

An *allomethylose* moiety is attached to apiofuranose in a heloniogenin glycoside (**1063**) from *Heloniopsis orientalis* (Liliaceae). This unusual sugar moiety has previously only been found in cardiac glycosides (Nakano *et al.* 1989a).

The fresh bulbs of *Lilium brownii* (Liliaceae) contain two novel steroid saponins. One of these, the 27-acyloxyspirostanol saponin brownioside (**1064**), contains a 3-hydroxy-3-methylglutaroyl moiety at position C-27. The sapogenin corresponds to isonarthogenin (27-hydroxydiosgenin). Glycoside **1054** (see Appendix 3) is a furospirostanol saponin (Mimaki and Sashida, 1990a). Acid hydrolysis of this nuatigenin glycoside gave an isomeric artifact isonuatigenin (see also Tschesche and Richert, 1964; Kesselmeier and Budzikiewicz, 1979).

Glucoconvallasaponin B (**940**) and convallamarogenin triglycoside (**1039**) from *Convallaria majalis* both possess two sugar moieties attached to their aglycones, at C-3, C-5 and C-1, C-3, respectively (Appendix 3).

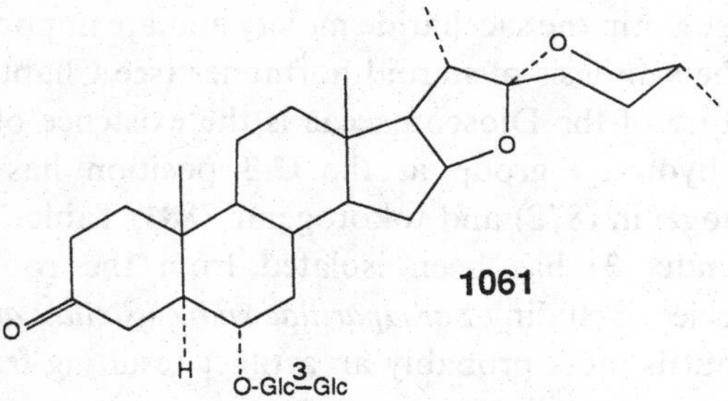

1061

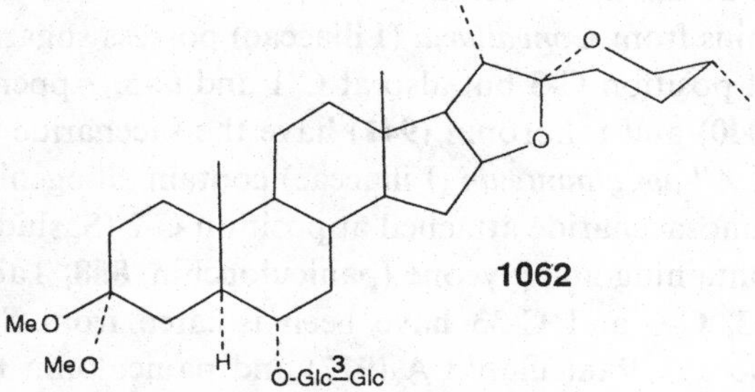

1062

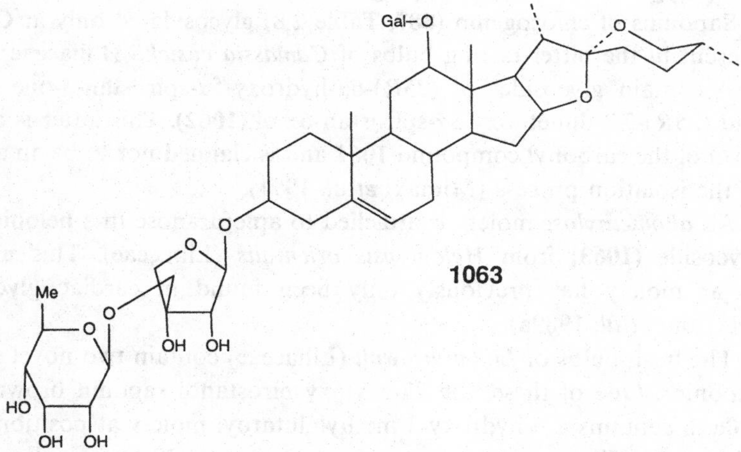

1063

The occurrence of sulphated steroid glycosides is very rare in the plant kingdom and among the few examples known is a ruscogenin glycoside (**1016**) (Appendix 3) from the tubers of *Ophiopogon planiscapus* (Liliaceae) (Watanabe *et al.* 1983).

A group of 18-norspirostanol glycosides has been found in *Trillium kamtschaticum, T. tschonoskii* and *Paris quadrifolia* (all Liliaceae). One of these is a penta-*O*-acetyl derivative epitrillenoside CA (**1065**) from *T. tschonoskii* (Ono *et al.* 1986), in which the saccharide moiety is attached

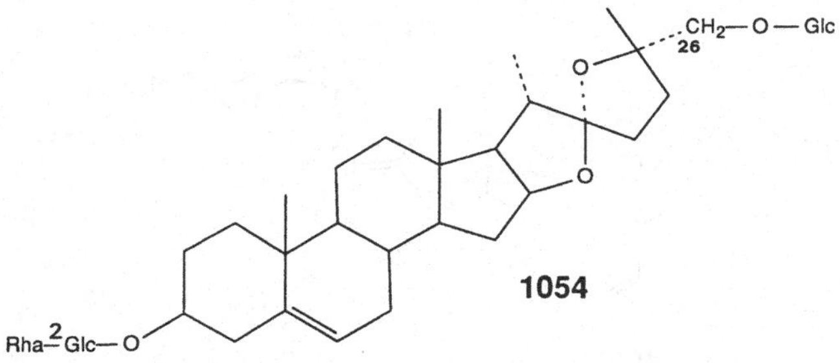

1054

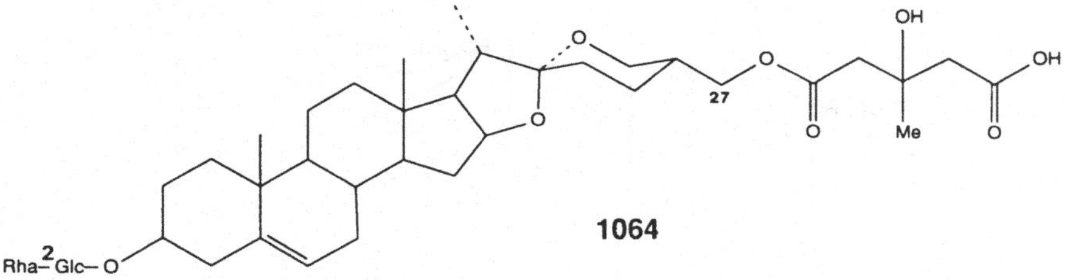

1064

at C-1 of epitrillenogenin. Another is trillenoside A (**1066**), a glycoside of trillenogenin with an apioside moiety in the sugar chain (Nohara *et al.* 1980; Fukuda *et al.* 1981). Deoxytrillenoside A, from the underground parts of *T. kamtschaticum*, is the 21-deoxy compound of **1066**, the aglycone being 21-deoxytrillogenin (Fukuda *et al.* 1981).

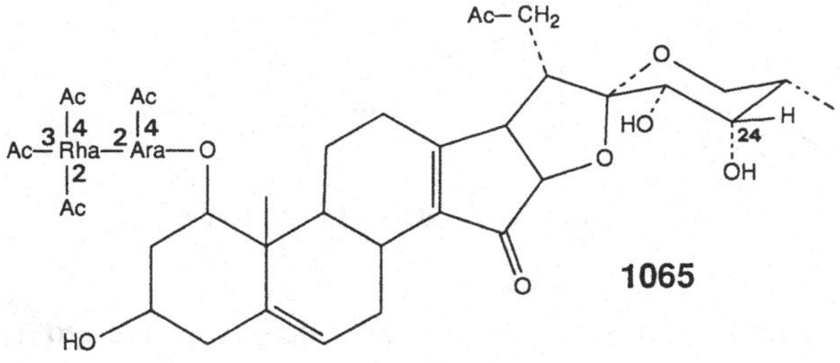

1065

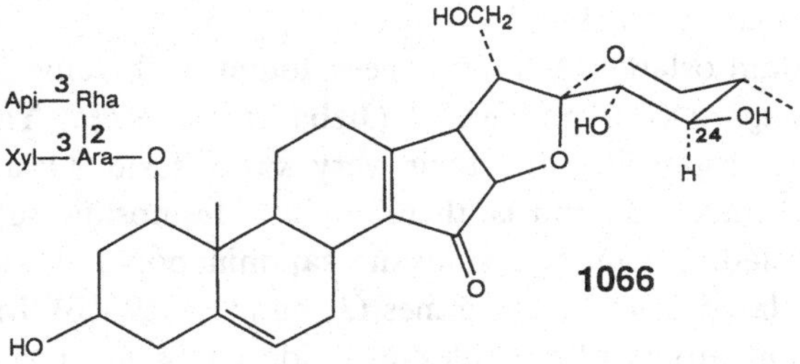

1066

Kryptogenin glycosides, in which both rings E and F of the spirostan skeleton are open, have similarly been isolated from *T. kamtschaticum*. Kryptogenin 3-*O*-β-D-glucopyranoside (**1067**) when first reduced with

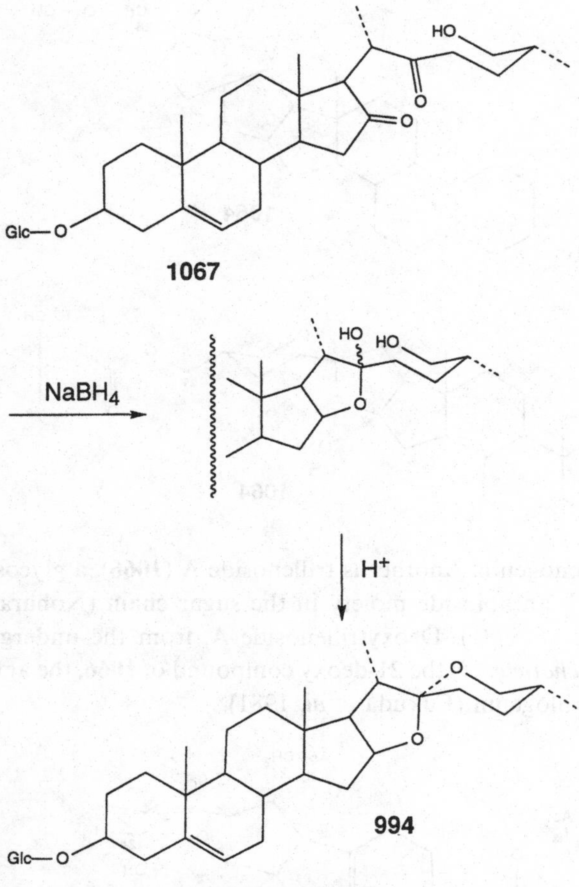

Fig. 2.8. Conversion of kryptogenin glucoside into trillin.

sodium borohydride and subsequently acidified is converted to trillin (**994**) (Fig. 2.8) (Nohara *et al.* 1975).

An unusual saponin osladin (**1068**) has been found in rhizomes of the fern *Polypodium vulgare* (Polypodiaceae) (Jizba *et al.* 1971a; Yamada *et al.* 1992) and is responsible for their very *sweet* taste. Osladin is glycosylated at C-3 and C-26 and is, therefore, a bidesmosidic saponin.

Together with osladin, a monodesmosidic saponin polypodosaponin (**1069**) has been isolated from the rhizomes (Jizba *et al.* 1971b). In both saponins five carbon atoms of a cholesterol side chain form a cyclo-hemiacetal. This cyclic ether formation bears some resemblance to the aglycones of the astragalosides. Other closely related compounds (**1070–1072**) have been isolated from the rhizomes of *Polypodium glycyrrhiza* (licorice fern) (Kim *et al.* 1988; Kim and Kinghorn, 1989). One of these, polypodoside A (**1070**), is 600 times sweeter than a 6% w/v sucrose solution (Kim *et al.* 1988). Polypodoside C (**1072**) contains the unusual monosaccharide *acofriose* at the C-26 position.

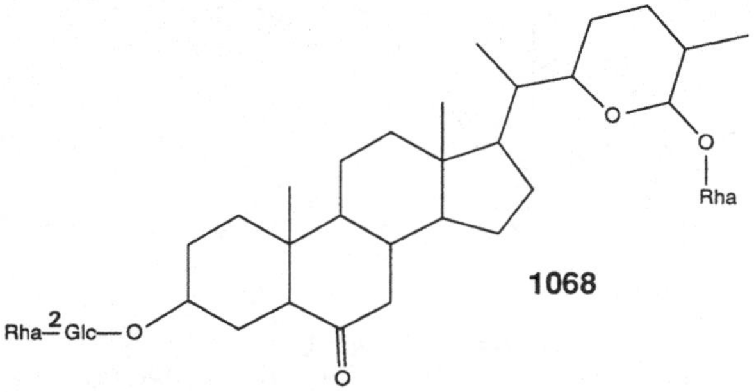

1068

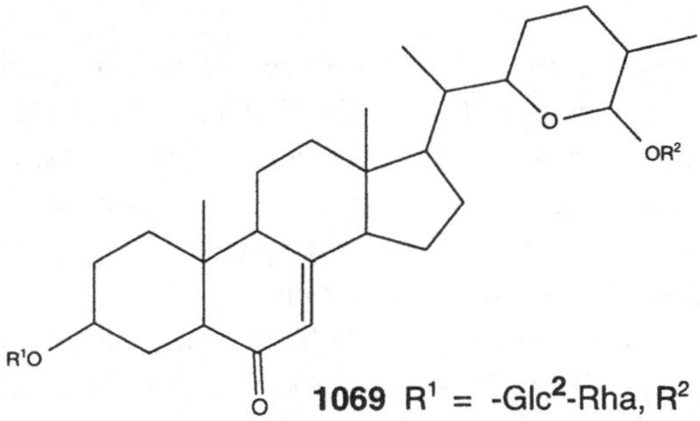

Rha$\underline{^2}$Glc—O

R'O

1069 R^1 = -Glc^2-Rha, R^2 = -H

1070 R^1 = -Glc^2-Rha, R^2 = -Rha

1071 R^1 = -Glc, R^2 = -Rha

1072 R^1 = -Glc, R^2 = -Rha^3-CH$_3$

2.2.2 *Furostanol glycosides*

The existence of open-chain glycosides of this type was first postulated by Marker and Lopez in 1947 and the first example, jurubine (**1073**), was isolated (from *Solanum torvum*) by Schreiber and Ripperger

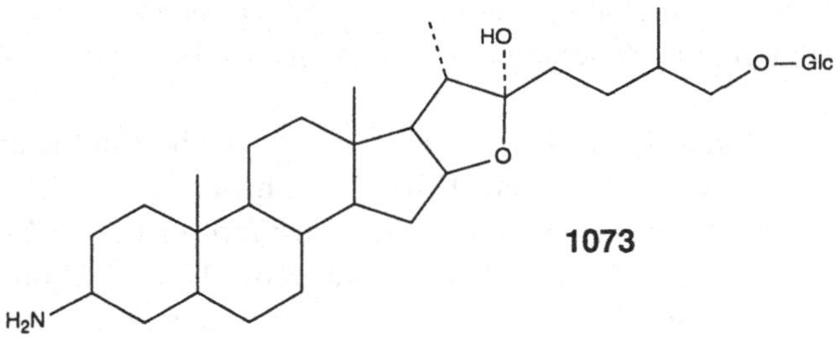

1073

(1966). The simplest member of the series is asparagoside B (**1074**), the 26-*O*-glucoside of (25*S*)-5β-furostan-3β,22α,26-triol (Goryanu *et al.* 1976; Sharma *et al.* 1982).

1074 Asparagoside B
((25*S*)-5β-furostan-3β,22α,26-triol)

The classical representative of furostanol glycosides is *sarsaparilloside* (**1075**) from the roots of *Sarsaparillae radix* (*Smilax aristolochiaefolia*, Liliaceae) and, as is the case for other members of this class, acid or enzymatic hydrolysis of the sugar in position 26 leads to spontaneous ring closure, producing the spirostanol derivative parillin, a glycoside of sarsasapogenin (Fig. 2.9) (Tschesche *et al.* 1969b).

The difficulty of obtaining sarsaparilloside from plant extracts can be explained by the fact that the C-26 sugar moiety is readily cleaved by enzymes present in the plant itself and enzyme-inhibitory conditions are necessary. This holds true for other furostanol glycosides and explains why in, for example, *Digitalis* and *Dioscorea* species earlier work only revealed the spirostanol glycosides which are partially or wholly derived from furostanol glycosides in the plants. The co-occurrence of spirostanol saponins with the corresponding furostanol saponins in numerous plants is the logical consequence of this ready conversion.

While action of glycosidases on furostanol saponins usually produces spirostanol glycosides after cleavage of the C-26 sugar moiety (Tschesche and Gutwinski, 1975; Yahara *et al.* 1985), it has been claimed that hydrolysis with the β-glucosidase enzyme avenacosidase results in the formation of an intact 26-desglucofurostanol glycoside (Grünweller *et al.* 1990).

Another problem is the ready formation of methyl ethers in the presence of methanol as extraction solvent. Etherification occurs at C-22 and in certain cases (e.g. Gupta *et al.* 1985) the furostanol glycoside has been characterized as its 22-*O*-methyl derivative. Konishi and Shoji (1979) reported the formation of 22-methoxyfurostanols from a methanol extract

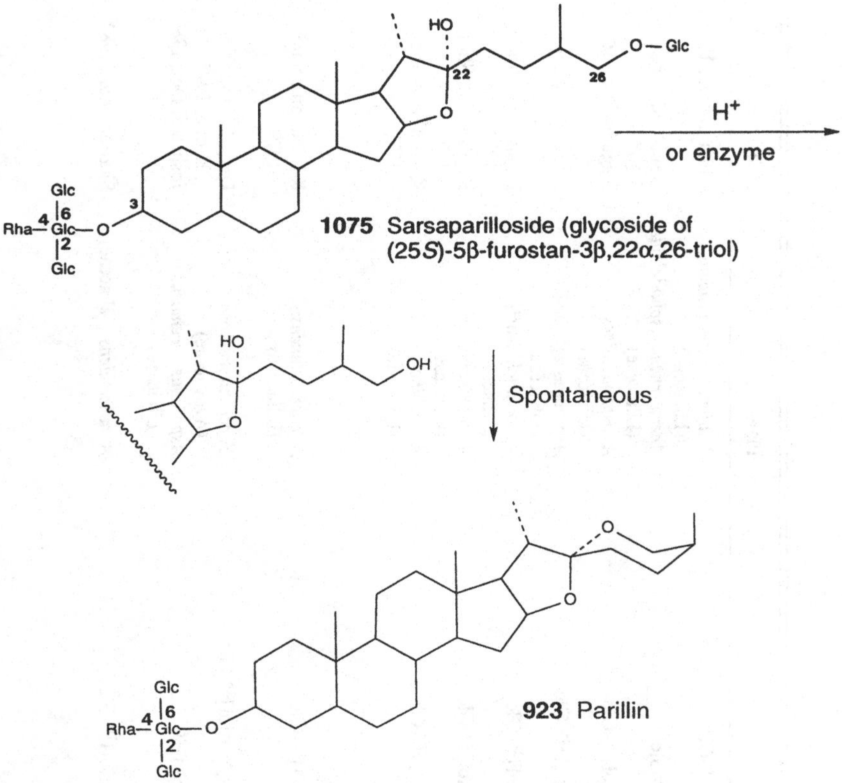

1075 Sarsaparilloside (glycoside of (25*S*)-5β-furostan-3β,22α,26-triol)

923 Parillin

Fig. 2.9. Conversion of sarsaparilloside into parillin.

of *Asparagus cochinchinensis* (Liliaceae) roots. The genuine 22-hydroxyfurostanols were obtained by changing the extraction solvent to pyridine. Boiling the 22-methoxy derivatives with aqueous acetone also regenerates the free hydroxyl group (Hirschmann and Hirschmann, 1958; Kamel *et al.* 1991).

A list of furostanol glycosides is shown in Table 2.9. This table is arranged in the order of increasing number of hydroxyl substituents. Further examples are to be found in a review article by Voigt and Hiller (1987). Although numerous plants contain bidesmosidic (and tridesmosidic) furostanol saponins, as shown by colour reactions, the list of known saponins is small because of the problems of isolation. The known glycosides of the furostan series all have a D-glucose residue at C-26, except for afromontoside (**1099**) from the flowers of *Dracaena afromontana* (Agavaceae) (Reddy *et al.* 1984), which is a furostanol 26-*O*-α-L-rhamnopyranoside.

Table 2.9. *Structures of furostanol saponins*

No.	Name	Structure	Plant	Reference
1076		Rha-^{2}Glc-^{3}[(25R)-furost-5-en-3β,26-diol]^{26}-Glc	*Trillium kamtschaticum* (Liliaceae)	Nohara et al., 1975
1077	Anemarsaponin B	Glc-^{2}Gal-^{3}[furost-20(22)-en-3β,26-diol]^{26}-Glc	*Anemarrhena asphodeloides* (Liliaceae)	Dong and Han, 1991
1078		Glc-^{2}Gal-^{3}[(25S)-5β-furostan-3β,22,26-triol]^{26}-Glc	*A. asphodeloides* (Liliaceae)	Nagumo et al., 1991
1079	Officinalisnin-I	Glc-^{2}Glc-^{3}[(25S)-5β-furostan-3β,22α,26-triol]^{26}-Glc	*Asparagus officinalis* (Liliaceae)	Kawano et al., 1975
1080	Asp-IV'	Xyl-^{4}Glc-^{3}[(25S)-5β-furostan-3β,22,26-triol]^{26}-Glc	*A. cochinchinensis* (Liliaceae)	Konishi and Shoji, 1979
1081	Asp-V'	Rha-^{6}Glc-^{3}[(25S)-5β-furostan-3β,22,26-triol]^{26}-Glc	*A. cochinchinensis* (Liliaceae)	Konishi and Shoji, 1979
1082	Asp-VI'	Rha-^{6}Glc-^{3}[(25S)-5β-furostan-3β,22,26-triol]^{26}-Glc $\vert^4$ Xyl Rha $\vert^6$	*A. cochinchinensis* (Liliaceae)	Konishi and Shoji, 1979
1083	Asp-VII'	Xyl-^{4}Glc-^{3}[(25S)-5β-furostan-3β,22,26-triol]^{26}-Glc $\vert^2$ Glc	*A. cochinchinensis* (Liliaceae)	Konishi and Shoji, 1979
1084	Protoyuccoside C	Gal-^{2}Glc-^{4}Glc-^{3}[(25S)-5β-furostan-3β,22α,26-triol]^{26}-Glc	*Yucca filamentosa* (Agavaceae)	Dragalin and Kintia, 1975
1085	Shatavarin-I	Glc-^{4}Glc-^{3}[(25S)-5β-furostan-3β,22α,26-triol]^{26}-Glc $\vert^2$ Rha	*Asparagus racemosus* (Liliaceae)	Joshi and Dev, 1988
1086	Asparanin B	Rha-^{4}Glc-^{3}[(25S)-5β-furostan-3β,22α,26-triol]^{26}-Glc $\vert^2$ Glc	*A. adscendens* (Liliaceae)	Sharma et al., 1982

No.	Saponin	Structure	Source	Reference
1087		$\begin{array}{l}\quad\ \ \text{Ara}\\\quad\ \ |^6\\ \text{Rha-}^4\text{Glc-}^3[(25S)\text{-}5\beta\text{-furostan-}3\beta,22\alpha,26\text{-triol}]^{26}\text{-Glc}\\ \quad\ \ |^2\\ \quad\ \ \text{Glc}\\ \quad\ \ |^6\\ \quad\ \ \text{Glc}\end{array}$	*A. curillus* (Liliaceae)	Sati and Sharma, 1985
1075	Sarsaparilloside	$\begin{array}{l}\text{Rha-}^4\text{Glc-}^3[(25S)\text{-}5\beta\text{-furostan-}3\beta,22\alpha,26\text{-triol}]^{26}\text{-Glc}\\ \quad\ \ |^2\\ \quad\ \ \text{Glc}\end{array}$	*Smilax aristolochiaefolia* (Liliaceae)	Tschesche et al., 1969b
1088	Melongoside N	$\text{Glc-}^2\text{Glc-}^3[(25R)\text{-}5\alpha\text{-furostan-}3\beta,22\alpha,26\text{-triol}]^{26}\text{-Glc}$	*Solanum melongena* (Solanaceae) / *Lycopersicon lycopersicum* (Solanaceae)	Kintia and Shvets, 1985b / Sato and Sakamura, 1973
1089		$\text{Glc-}^2\text{Glc-}^4\text{Gal-}^3[(25R)\text{-}5\alpha\text{-furostan-}3\beta,22\alpha,26\text{-triol}]^{26}\text{-Glc}$		
1090	Melongoside P	$\begin{array}{l}\text{Rha-}^3\text{Glc-}^3[(25R)\text{-}5\alpha\text{-furostan-}3\beta,22\alpha,26\text{-triol}]^{26}\text{-Glc}\\ \quad\ \ |^2\\ \quad\ \ \text{Glc}\end{array}$	*Solanum melongena* (Solanaceae)	Kintia and Shvets, 1985a
1091	Lanatigoside	$\begin{array}{l}\text{Xyl-}^3\text{Glc-}^4\text{Gal-}^3[(25R)\text{-}5\alpha\text{-furostan-}3\beta,22\alpha,26\text{-triol}]^{26}\text{-Glc}\\ \quad\ \ |^2\\ \quad\ \ \text{Glc-}^3\text{Gal}\end{array}$	*Digitalis lanata* (Scrophulariaceae)	Tschesche et al., 1972
1092	Sativoside-B1	$\begin{array}{l}\text{Glc-}^3\text{Glc-}^2\text{Glc-}^4\text{Gal-}^3[(25R)\text{-}5\alpha\text{-furostan-}3\beta,6\beta,26\text{-triol}]^{26}\text{-Glc}\\ \quad\ \ |^3\\ \quad\ \ \text{Glc}\end{array}$	*Allium sativum* (Liliaceae)	Matsuura et al., 1989a
1093	Sativoside-R1	$\begin{array}{l}\text{Glc-}^3\text{Glc-}^2\text{Glc-}^4\text{Gal-}^3[(25R)\text{-}5\alpha\text{-furostan-}3\beta,6\beta,26\text{-triol}]^{26}\text{-Glc}\\ \quad\ \ |^3\\ \quad\ \ \text{Xyl}\end{array}$	*A. sativum* (Liliaceae)	Matsuura et al., 1989a
1094	Chinenoside I	$\begin{array}{l}\text{Ara-}^6\text{Glc-}^3[(25R)\text{-}5\alpha\text{-furostan-}6\text{-one-}3\beta,22,26\text{-triol}]^{26}\text{-Glc}\\ \quad\ \ |^4\\ \quad\ \ \text{Xyl}\end{array}$	*A. chinense* (Liliaceae)	Matsuura et al., 1989b
1095	Chloromaloside B	$\begin{array}{l}\qquad\qquad\qquad\qquad\qquad\qquad{}^{26}\text{Glc}\\ \qquad\qquad\qquad\qquad\qquad\quad\ \ |\ \ {}^{22}\text{OCH}_3\\ \text{Glc-}^2\text{Glc-}^4\text{Gal}\!-\!\!-\!^3[(25S)\text{-}5\text{-furostan-}12\text{-one-}3\beta,22,26\text{-triol}]\\ \quad\ \ |^3\\ \quad\ \ \text{Xyl}\end{array}$	*Chlorophytum malayense* (Liliaceae)	Li et al., 1990a

(cont.)

Table 2.9. (Cont.)

No.	Name	Structure	Plant	Reference
1096	Melongoside O	Glc-^{2}Glc-^{3}[(25R)-furost-5-en-3β,22α,26-triol]^{26}-Glc	*Solanum melogena* (Solanaceae)	Kintia and Shvets, 1985b
1097		Rha-^{2}Glc-^{3}[(25R)-furost-5-en-3β,22,26-triol]^{26}-Glc	*Ophiopogon planiscapus* (Liliaceae)	Watanabe et al., 1983
1098	Trigofoenoside A	Rha-^{2}Glc-^{3}[(25S)-furost-5-en-3β,22,26-triol]^{26}-Glc	*Trigonella foenum-graecum* (Leguminosae)	Hardman et al., 1980
1099	Afromontoside	Rha-^{4}Glc-^{3}[(25R)-furost-5-en-3β,22α,26-triol]^{26}-Rha	*Dracaena afromontana* (Agavaceae)	Gupta et al., 1985; Reddy et al., 1984

1100	Polyphyllin H	Ara-^{4}Glc-^{3}[(25R)-furost-5-en-3β,22α,26-triol]^{26}-Glc
		      $\vert^3$
		      Rha
			Paris polyphylla (Liliaceae)	Singh et al, 1982a

1101	Solayamocinoside C	Gal-^{2}Glc-^{6}Glc-^{3}[(25S)-furost-5-en-3β,22α,26-triol]^{26}-Glc	*Solanum dulcamara* (Solanaceae)	Willuhn and Köthe, 1983
1102	Solayamocinoside B	Rha-^{2}Glc-^{6}Glc-^{3}[(25S)-furost-5-ene-3β,22α,26-triol]^{26}-Glc	*S. dulcamara* (Solanaceae)	Steinegger and Hänsel, 1988

1103	Methyl protogracillin	Glc—^{3}Glc—^{3}[furost-5-en-3β, 22α, 26-triol]^{26}-Glc ; ^{22}-OCH$_3$
		   $\vert^2$
		   Rha
			Paris vietnamensis (Liliaceae)	Namba et al, 1989

1104	Protogracillin	Glc-^{3}[(25S)-furost-5-en-3β,22α,26-triol]^{26}-Glc $\vert^2$ Rha	*Dioscorea gracillima* (Dioscoreaceae)	Kawasaki and Yamauchi, 1962

1105	Deltoside	Glc-^{4}Glc-^{3}[(25R)-furost-5-en-3β,22,26-triol]^{26}-Glc
		    $\vert^2$
		    Rha
			Ophiopogon planiscapus (Liliaceae)	Watanabe et al., 1983

1106		Glc-^{3}GlcA-^{3}[(25R)-furost-5-en-3β,22α,26-triol]^{26}-Glc
		     $\vert^2$
		     Rha
			Solanum lyratum (Solanaceae)	Yahara et al, 1985

1107	Protodioscin	Rha-⁴Glc-³[(25R)-furost-5-en-3β,22α,26-triol]²⁶-Glc
		        |²
		       Rha
			Dioscorea septemloba
			(Dioscoreaceae)
			Trigonella foenum-graecum
			(Leguminosae)
			Balanites aegyptiaca
			(Balanitaceae)
				Kiyosawa et al., 1968
				Bogacheva et al., 1976
				Kamel et al., 1991

1108	Asparasaponin I	Rha-⁴Glc-³[(25S)-furost-5-en-3β,22α,26-triol]²⁶-Glc
		        |²
		       Rha
			Asparagus officinalis
			(Liliaceae)
				Kawano et al., 1977

1109	Methylproto-
	dioscin
		Glc
		|
		Rha—⁴Glc—³[(25R)-furost-5-en-3β, 22α, 26-triol]——OCH₃
		       |²
		      Rha
			Trachycarpus fortunei
			(Palmae)
				Hirai et al., 1984

1110	Sibiricoside A	²⁶Glc
		Xyl-³Glc-⁴Gal—³[(25S)-furost-5-en-3β, 22, 26-triol]——²²OCH₃
		      |²
		     Glc
		     Rha
		     |⁴
			Polygonatum sibiricum
			(Liliaceae)
				Son et al., 1990

1111	Polyfuroside	Xyl-³Glc-³[(25R)-furost-5-en-3β,22,26-triol]²⁶-Glc
		       |²
		      Rha
			Balanites aegyptiaca
			(Balanitaceae)
				Kamel et al., 1991

1112	Polyfuroside	Glc-³Glc-⁴Gal-³[(25R)-furost-5-en-3β,22α,26-triol]²⁶-Glc
		       |²
		      Glc
		      Gal
		      |⁴
			Polygonatum officinale
			(Liliaceae)
				Janeczko et al., 1987

1113	Solayamocinoside
	A
		Gal-³Rha-³[(25S)-furost-5-en-3β,22α,26-triol]²⁶-Glc
		       |²
		      Ara
			Solanum dulcamara
			(Solanaceae)
				Willuhn and Köthe, 1983
				Steinegger and Hänsel, 1988

1114	Asperoside	Rha-⁴Rha-⁴Glc-³[(25R)-furost-5-en-3β,22α,26-triol]²⁶-Glc
		       |²
		      Glc
			Smilax aspera (Liliaceae)	Petricic and Radosevic, 1969

(cont.)

Table 2.9. (Cont.)

No.	Name	Structure	Plant	Reference
1115		$Rha\text{-}^4Rha\text{-}^4Glc\text{-}^3[(25R)\text{-furost-5-en-}3\beta,22\alpha,26\text{-triol}]^{26}\text{-}Glc$	*Paris quadrifolia* (Liliaceae)	Nohara *et al.*, 1982
1116	Trigofoenoside G	$Xyl\text{-}^4Glc\text{-}^6Glc\text{-}^3[(25R)\text{-furost-5-en-}3\beta,22,26\text{-triol}]^{26}\text{-}Glc$ $\qquad\quad \|^2$ $\qquad\quad Rha$	*Trigonella foenum-graecum* (Leguminosae)	Gupta *et al.*, 1984
1117	Trigofoenoside B	$Rha\text{-}^4Glc\text{-}^3[(25S)\text{-}5\alpha\text{-furostan-}2\alpha,3\beta,22,26\text{-tetraol}]^{26}\text{-}Glc$	*T. foenum-graecum* (Leguminosae)	Gupta *et al.*, 1986
1118	Lanagitoside	$Xyl\text{-}^3Glc\text{-}^4Gal\text{-}^3[(25R)\text{-}5\alpha\text{-furostan-}2\alpha,3\beta,22\alpha,26\text{-tetraol}]^{26}\text{-}Glc$ $\quad \|^2$ $Glc\text{-}^3Gal$	*Digitalis lanata* (Scrophulariaceae)	Tschesche *et al.*, 1972
1119	YG-4	$Xyl\text{-}^4Xyl\text{-}^3Glc\text{-}^4Gal\text{-}^3[(25R)\text{-}5\alpha\text{-furostan-}2\alpha,3\beta,22,26\text{-tetraol}]^{26}\text{-}Glc$ $\qquad\qquad\quad \|^2$ $\qquad\qquad\quad Glc$	*Yucca gloriosa* (Agavaceae)	Nakano *et al.*, 1988
1120	Capsicoside A	$Glc\text{-}^3Glc\text{-}^4Gal\text{-}^3Glc\text{-}^3[(25S)\text{-}5\alpha\text{-furostan-}2\alpha,3\beta,22\alpha,26\text{-tetraol}]^{26}\text{-}Glc$ $\qquad\qquad\quad \|^2$ $\qquad\qquad\quad Glc$	*Capsicum annum* (Solanaceae)	Tschesche and Gutwinski, 1975
1121	Protoeruboside-B	$Glc\text{-}^3Glc\text{-}^4Gal\text{-}^3Glc\text{-}^3[(25R)\text{-}5\alpha\text{-furostan-}3\alpha,6\beta,22\alpha,26\text{-tetraol}]^{26}\text{-}Glc$ $\qquad\qquad\quad \|^2$ $\qquad\qquad\quad Glc$	*Allium sativum* (Liliaceae)	Matsuura *et al.*, 1989a
1122	Pardarinoside B	$\qquad\qquad\qquad\qquad Ac \qquad\qquad OCH_3$ $\qquad\qquad\qquad\qquad \|^{26}\qquad\qquad \|^{22}$ $Rha—^2Glc—^3[(25R)\text{-}5\alpha\text{-furostan-}3\beta,17\alpha,26\text{-triol}]$	*Lilium pardarinum* (Liliaceae)	Shimomura *et al.*, 1989
1123		$Rha\text{-}^2Glc\text{-}^3[(25R)\text{-furost-5-en-}3\beta,17\alpha,22,26\text{-tetraol}]^{26}\text{-}Glc$	*Trillium kamtschaticum* (Liliaceae)	Nohara *et al.*, 1975
1124		$Rha\text{-}^4Glc\text{-}^3[(25R)\text{-furost-5-en-}3\beta,17\alpha,22,26\text{-tetraol}]^{26}\text{-}Glc$ $\quad \|^2$ $\quad Rha$	*T. kamtschaticum* (Liliaceae)	Fukuda *et al.*, 1981

No.	Name	Structure	Source	Reference
1125		Rha-^{2}Ara-^{1}[(25R)-furost-5-en-1β,3β,22,26-tetraol]^{26}-Glc $\quad\quad$ |^4 $\quad\quad$ SO$_3$	*Ophiopogon planiscapus* (Liliaceae)	Watanabe *et al.*, 1983
1126 1127	Alliofuroside A Spicatoside B	Rha-^{2}Gal-^{1}[furost-5-en-1β,3β,22,26-tetraol]^{26}-Glc	*Allium cepa* (Liliaceae) *Liriope spicata* (Liliaceae)	Kravets *et al.*, 1986a Lee *et al.*, 1989
1128		Xyl-^{3}Fuc-^{1}[(25S)-furost-5-en-1β,3β,22,26-tetraol]$\begin{smallmatrix}26\\22\end{smallmatrix}$ $\quad$ |^2 $\quad\quad$ OCH$_3$ $\quad$ Glc $\quad$ Glc	*Polygonatum sibiricum* (Liliaceae)	Son *et al.*, 1990
1129	Convallamaroside	Rha-^{2}Qui$\begin{smallmatrix}1\\3\end{smallmatrix}$[5$\beta$-furost-25-en-1$\beta$,3$\beta$,22$\alpha$,27-tetraol]^{27}-Glc	*Convallaria majalis* (Liliaceae)	Tschesche *et al.*, 1973
1130	Deglucoruscoside	Glc-^{4}Rha Rha-^{2}Ara-^{1}[furost-5,25(27)-dien-1β,3β,22α,26-tetraol]^{26}-Glc	*Ruscus ponticus* (Liliaceae) *R. aculeatus*	Korkashvili *et al.*, 1985 Bombardelli *et al.*, 1975
1131	Ruscoside	Glc-^{3}Rha-^{2}Ara-^{1}[furost-5,25(27)-dien-1β,3β,22α,26-tetraol]^{26}-Glc	*R. aculeatus* (Liliaceae)	Bombardelli *et al.*, 1972
1132	Ampeloside Bf$_1$	Glc-^{3}Glc-^{4}Gal-^{3}[(25R)-5α-furostan-2α,3β,6β,22,26-pentaol]^{26}-Glc	*Allium ampeloprasum* (Liliaceae)	Morita *et al.*, 1988
1133	Ampeloside Bf$_2$	Glc-^{4}Gal-^{3}[(25R)-5α-furostan-2α,3β,6β,22,26-pentaol]^{26}-Glc	*A. ampeloprasum* (Liliaceae)	Morita *et al.*, 1988
1134	Pardarinoside A	Rha-^{2}Glc-^{3}[(25R)-5α-furostan-3β,14α,17α,22α,26-pentaol]$\begin{smallmatrix}26\\22\end{smallmatrix}$ $\quad\quad$ Ac $\quad$ OCH$_3$	*Lilium pardarinum* (Liliaceae)	Shimomura *et al.*, 1988
1135	Pardarinoside C	Glc-^{4}Glc-^{3}[(25R)-5α-furostan-3β,14α,17α,22α,26-pentaol]$\begin{smallmatrix}26\\22\end{smallmatrix}$ $\quad$ |^2 $\quad\quad$ Ac $\quad$ OCH$_3$ $\quad$ Rha	*L. pardarinum* (Liliaceae)	Shimomura *et al.*, 1988

Furostanol saponins have been found in the leaves of *Digitalis lanata* (see Table 2.9), whereas the seeds provide a reservoir of spirostanol saponins. The furostanol glycosides sarsaparilloside, protogracillin and protodioscin have all been obtained from the roots of the plants concerned, while the corresponding spirostanol saponins gracillin and dioscin have been isolated from the leaves.

A sulphated pentahydroxyfurostanol glycoside has been isolated from *Aspidistra elatior* (Liliaceae) with the sulphate group on the aglycone (**1136**) (Konishi *et al.* 1984).

1136

A protoruscogenin-type furostanol glycoside containing a sulphate group conjugated with the hydroxyl group of a sugar moiety (**1125**) (Table 2.9) has been identified in the tubers of *Ophiopogon planiscapus* (Liliaceae) (Watanabe *et al.* 1983).

The underground parts of *Trachycarpus wagnerianus* (Palmae) furnished, in addition to the diosgenin spirostanol tetraglycoside Pb (**1007**) (Appendix 3), the furostanol 22-*O*-methyl derivative methylproto-Pb (**1137**) and pseudoproto-Pb (**1138**), which contains a 20(22)-en moiety (obtained also by refluxing methylproto-Pb in acetic acid) (Hirai *et al.* 1986).

The aerial parts of *Solanum lyratum* (Solanaceae) contain not only furostanol and spirostanol glucuronides but also a steroid glucuronide, (22*R*)-cholest-5-en-3β,16β,22,26-tetraol rhamnosyl-glucuronopyranoside (**1139**). It is therefore quite possible that this compound is a biogenetic precursor of the spirostanol and furostanol glycosides (Yahara *et al.* 1989b).

2.3   Steroid alkaloid glycosides (glycoalkaloids, azasteroids)

The number and variety of naturally occurring steroid alkaloid glycosides can in no way be compared with the steroid glycosides. Even their distribution is limited, the Solanaceae family (which includes many

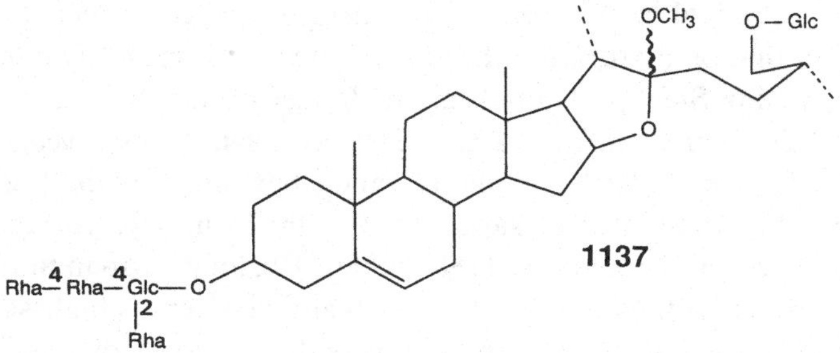

OCH₃ shown as OCH_3, O—Glc

1137

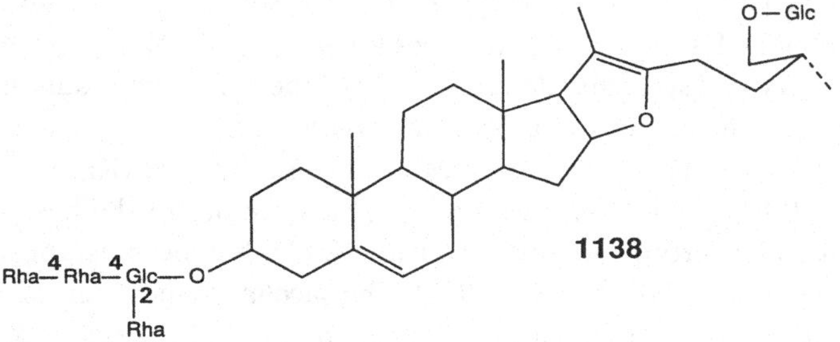

Rha—⁴Rha—⁴Glc—O
        |2
       Rha

1138

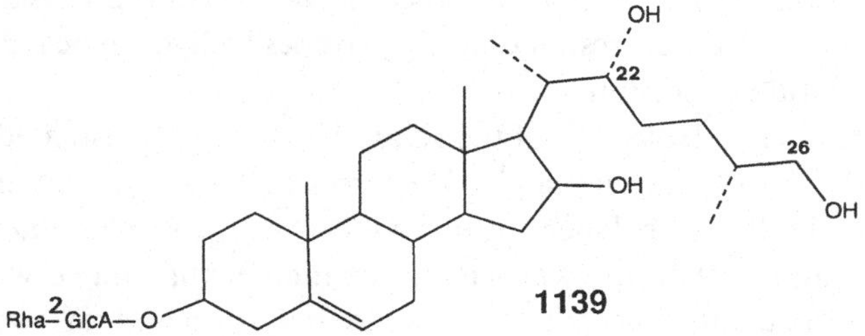

O—Glc

Rha—⁴Rha—⁴Glc—O
        |2
       Rha

1139

OH   22

OH   26

OH

Rha—²GlcA—O

agricultural crop plants important to humans, such as potato, tomato, eggplant, capsicum and naranjilla) constituting the largest source. According to a compilation (Schreiber, 1968), solasonine (**1147**, Table 2.10) and related glycosides are known from approximately 200 *Solanum* species.

Glycoalkaloids are usually found in all organs of the plant, with the highest concentrations in regions of high metabolic activity, such as flowers, unripe berries, young leaves and sprouts.

In the Solanaceae family, although steroid alkaloids have been isolated from all organs, the complex glycoalkaloid mixtures are generally found in larger amounts in the roots than in the aerial parts. In the fruits, there is usually a gradual metabolization to nitrogen-free constituents during ripening, so that on maturation they contain almost exclusively non-toxic neutral saponins. Steroid alkaloids are no longer present in ripe tomatoes (*Lycopersicon lycopersicum* (L.) KARST EX FARW.; synonyms: *Lycopersicon esculentum*, *Solanum lycopersicum* L.) and they are metabolized into products other than neutral saponins. On the contrary, young green tomatoes are rich in α-tomatine (**1158**, Table 2.10), found also in the leaves, stems and roots. In the potato, the glycoalkaloid content is highest in the flowers and sprouts (0.2–0.4%); in the tubers the glycoalkaloids are found mainly in the peel and around the eyes (Jadhav *et al.* 1981).

There are two basic skeletal types: the *spirosolans* and the *solanidans* (see Chapter 1). Examples of both types are given in Tables 2.10 and 2.11. The chemistry and structure determination of the aglycones is summarized in a comprehensive review (Schreiber, 1968), while the occurrence of steroid alkaloids and their glycosides has been tabulated (Ripperger and Schreiber, 1981 and references therein). The structure of the best-known of the solanidan group, solanidine (Table 2.11), has been established as solanid-5-en-3β-ol (Jadhav *et al.* 1981). The saccharide portions of certain glycoalkaloids have been assigned trivial names. Thus, the trioside in α-solanine is *β-solatriose* and that in α-chaconine is *β-chacotriose*. The tetrasaccharide moiety of tomatine and demissine is called *β-lycotetraose*. It is possible in certain cases that the so-called *β-*, *γ*- and *δ*-glycosides are artifacts arising from the corresponding α-glycosides by hydrolytic reactions (enzymes, isolation procedure, etc.).

The first glycoside ('solanée') of the solanidan group was isolated from black nightshade *Solanum nigrum* by Desfosses in 1821. The presence of a compound similar to 'solanée' in the potato plant (*S. tuberosum*) was reported by Baup in 1826 and was named solanine. Kuhn and co-workers (1955a,b) subsequently showed solanine to be composed of two glycosides of solanidine, α-chaconine (**1177**) and α-solanine (**1174**). These two alkaloids represent up to 95% of the total alkaloids in leaves of *S. tuberosum* and *S. chacoense*. In addition, *β*- and *γ*-forms of solanine and chaconine possessing a shortened chain are present, together with α-solamarine, *β*-solamarine and demissidine. More than 95% of the total glycoalkaloid content of commercially available potato tubers consists of α-solanine and α-chaconine (Morgan *et al.* 1985).

The stems of bittersweet, or *Solanum dulcamara* (Solanaceae), are used in Russia for their antiallergic, antiphlogistic, analgesic, cardiotonic and anti-shock activities. They contain glycoalkaloids of solasodine, soladulcidine and 5,6-dehydrotomatidine but also spirostanol glycosides and furostanol glycosides of proto-yamogenin (see Table 2.9). The latter

Table 2.10. *Structures of spirosolan glycosides*

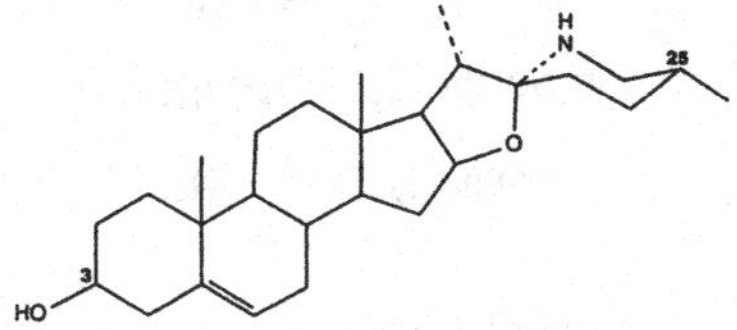

1141 Solasodine

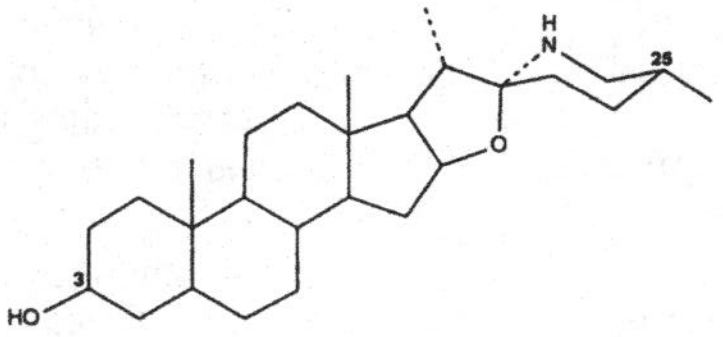

1142 Soladulcidine

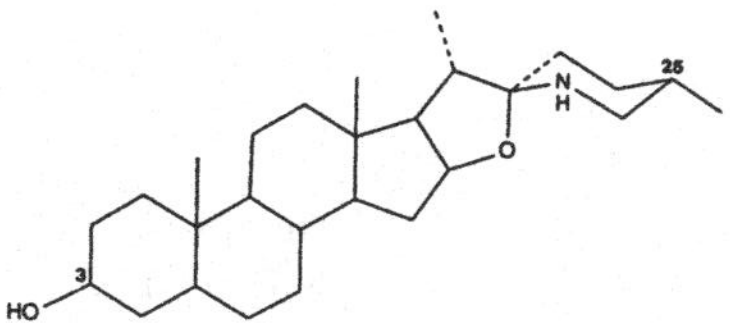

1143 Tomatidine

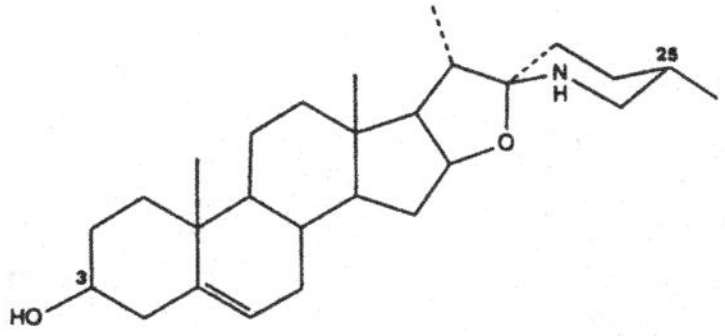

1144 5-Dehydrotomatidine (tomatidenol)

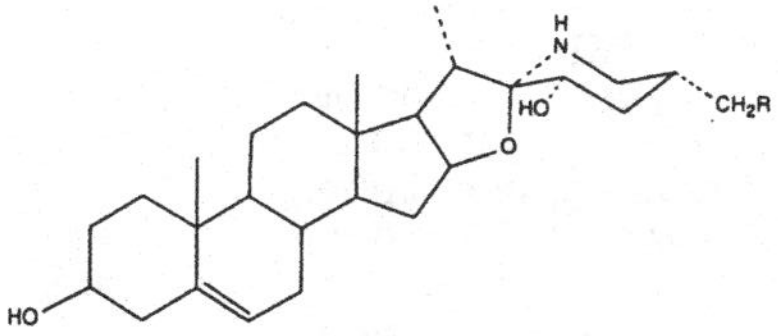

1145 Solaverol A R = H

1146 Solaverol B R = OH

No.	Name	Structure	Plant	Reference
1147	Solasonine (α-solasonine)	Glc-³Gal-³[solasodine] |² Rha	*Solanum tuberosum* (Solanaceae) *S. erianthum* *S. ptycanthum*	Schreiber, 1968 Moreira *et al.*, 1981 Eldridge and Hockridge, 1983

Table 2.10. (*Cont.*)

No.	Name	Structure	Plant	Reference
			S. dasyphyllum	Adesina, 1985
			S. khasianum	Mahato et al., 1980a
			S. sanctae-catharinae	Leonart et al., 1982
			S. xanthocarpum	Siddiqui et al., 1983
			S. incanum	Lin et al., 1986
			S. melongena	El-Khrisy et al., 1986
			S. scabrum	Adesina and Gbile, 1984
			S. nigrum	Ridout et al., 1989
1148	Solamargine (α-solamargine)	Rha-^{4}Glc-^{3}[solasodine] $\mid^2$ Rha	*S. erianthum* (Solanaceae)	Moreira et al., 1981
			S. japonense	Murakami et al., 1984
			S. ptycanthum	Eldridge and Hockridge, 1983
			S. dasyphyllum	Adesina, 1985
			S. xanthocarpum	Siddiqui et al., 1983
			S. khasianum	Mahato et al., 1980a
			S. incanum	Lin et al., 1986
			S. melongena	El-Khrisy et al., 1986
			S. scabrum	Adesina and Gbile, 1984
			S. nigrum	Ridout et al., 1989
1149	β_1-Solamargine	Rha-^{2}Glc-^{3}[solasodine]	*S. ptycanthum* (Solanaceae)	Eldridge and Hockridge, 1983
1150		Xyl-^{3}Glc-^{4}Gal-^{3}[solasodine] $\mid^2$ Glc (Rha, Glc)-^{3}[solasodine]	*Lilium brownii* (Liliaceae) *Solanum japonense* (Solanaceae)	Mimaki and Sashida, 1990c Murakami et al., 1984
1151	Solasurine	(Rha, Glc)-^{3}[solasodine]	*S. xanthocarpum* (Solanaceae)	Siddiqui et al., 1983
1152	Khasianine (β_2-solamargine)	Rha-^{4}Glc-^{3}[solasodine]	*S. khasianum* (Solanaceae)	Mahato et al., 1980a
			S. incanum	Lin et al., 1990
1153	Solaplumbine	Glc-^{4}Rha-^{3}[solasodine]	*Nicotiana plumbaginifolia* (Solanaceae)	Singh et al, 1974

No.	Name	Structure	Source	Reference
1154	Solaplumbinin	Rha-³[solasodine]	*N. plumbaginifolia* (Solanaceae)	Singh *et al.*, 1974

1155		Glc-⁴Glc-³[solasodine]
		  |²
		  Rha
			Lilium brownii (Liliaceae)	Mimaki and Sashida, 1990c

1156	Incanumine	Xyl-⁴Rha-⁴Glc-³[solasodine]
		       |³
		       Xyl
			Solanum incanum (Solanaceae)	Lin *et al.*, 1990

1157		Xyl-³Glc-⁴Gal-³[soladulcidine]
		     |²
		     Glc
			S. japonense (Solanaceae)	Murakami *et al.*, 1984

1158	Tomatine (α-tomatine)	Xyl-³Glc-⁴Gal-³[tomatidine]
		     |²
		     Glc
			Lycopersicon lycopersicum (Solanaceae)
			Solanum pennelli
			S. demissum
			S. lycopersicoides
			S. demissum (Solanaceae)
				Fontaine *et al.*, 1948
				Juvik *et al.*, 1982
				Osman and Sinden, 1982
				Oleszek *et al.*, 1986
				Osman and Sinden, 1982

1159	Neotomatine	Glc-³Glc-⁴Gal-³[tomatidine]
		     |²
		     Glc
			S. tuberosum (Solanaceae)	Maga, 1980

1160	α-Solamarine	Glc-³Gal-[5-dehydrotomatidine]
		     |²
		     Rha
			S. tuberosum (Solanaceae)	Maga, 1980

1161	β-Solamarine	Rha-⁴Glc-[5-dehydrotomatidine]
		     |²
		     Rha		

1162	Solaverine I	Rha-⁴Glc-³[solaverol A]
		     |²
		     Rha
			S. toxicarum (Solanaceae)
			S. verbascifolium
				Yamashita *et al.*, 1990
				Yamashita *et al.*, 1990

1163	Solaverine II	Glc-³Gal-³[solaverol A]
		     |²
		     Rha
			S. toxicarum (Solanaceae)
			S. verbascifolium
				Yamashita *et al.*, 1990
				Yamashita *et al.*, 1990

1164	Solaverine III	Rha-⁴Glc-³[solaverol B]
		     |²
		     Rha
			S. toxicarum (Solanaceae)
			S. verbascifolium
				Yamashita *et al.*, 1990
				Yamashita *et al.*, 1990

Table 2.11. *Structures of solanidan glycosides*

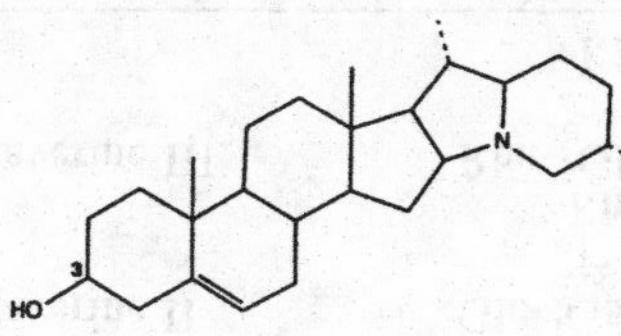

1165 Solanidine

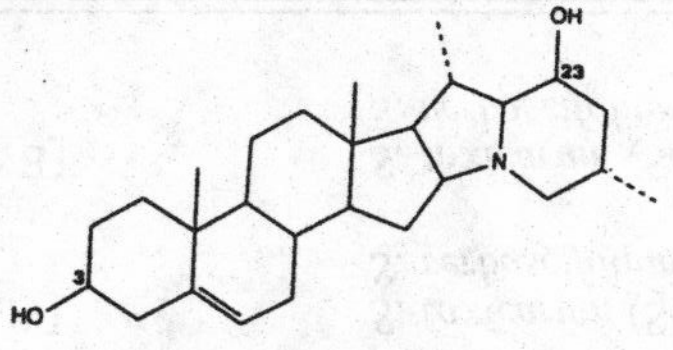

1166 Leptinidine (23β-hydroxysolanidine)

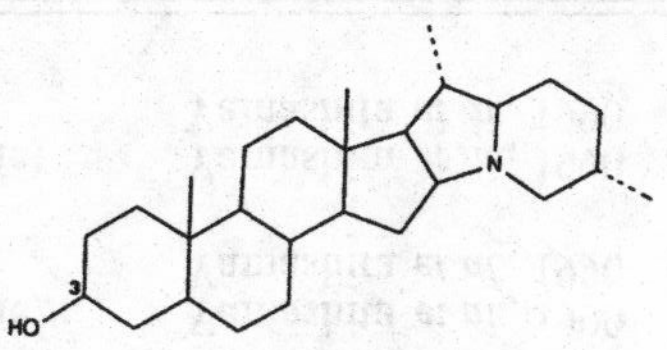

1167 Demissidine

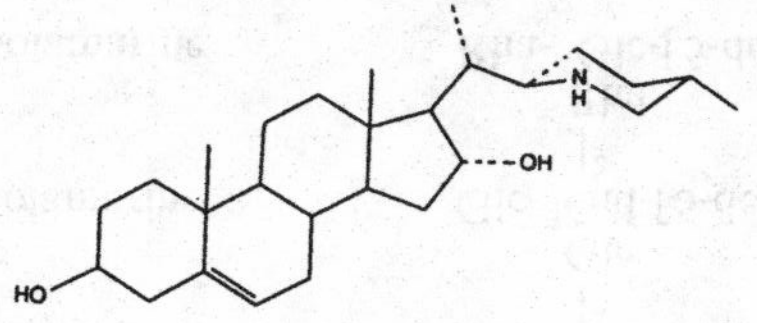

1168 Isoteinemine

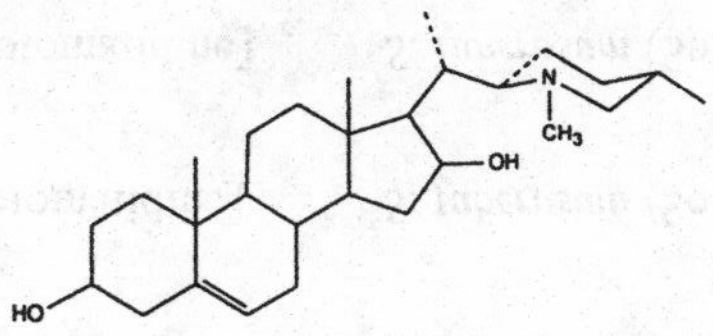

1169 Hapepunine

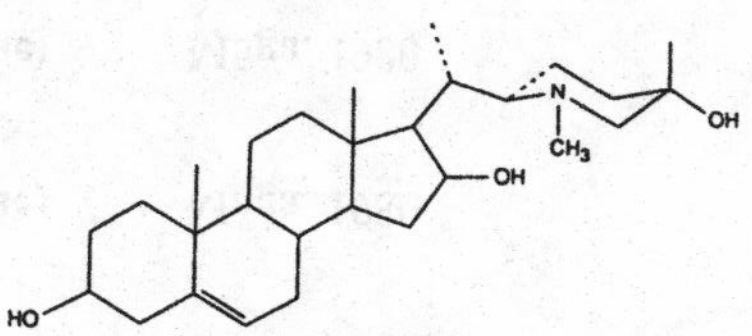

1170 Pingbeinine

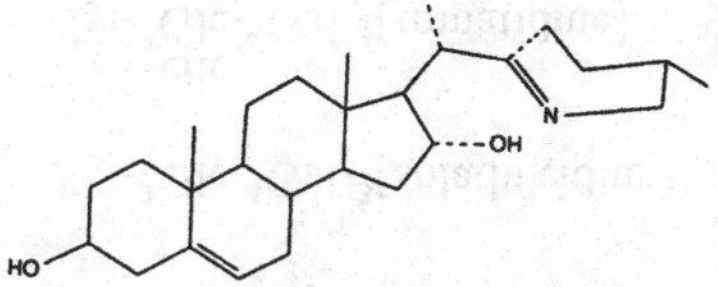

1171 Etioline

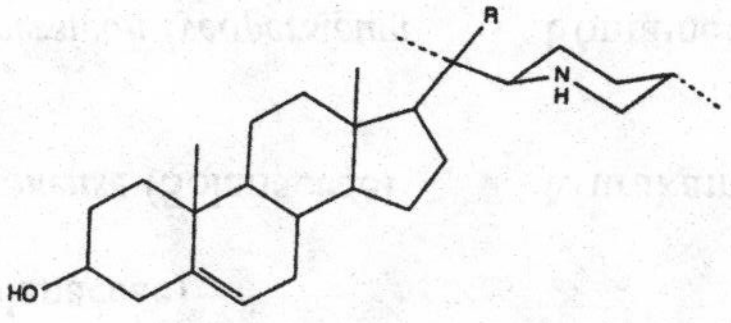

1172 Veramiline R = H

1173 Stenophylline B R = OH

No.	Name	Structure	Plant	Reference
1174	Solanine (α-solanine)	Glc-³Gal-³[solanidine] |² Rha	*Solanum erianthum* (Solanaceae) *S. tuberosum* *S. ptycanthum* *S. dasyphyllum* *Capsicum annuum* (Solanaceae)	Moreira *et al.*, 1981 Hwang and Lee, 1983 Eldridge and Hockridge, 1983 Adesina, 1985 Gutsu *et al.*, 1984
1175	β-Solanine	Glc-³Gal-³[solanidine]	*Solanum tuberosum* (Solanaceae)	Jadhav *et al.*, 1981
1176	γ-Solanine	Gal-³[solanidine]	*S. tuberosum*	Jadhav *et al.*, 1981
1177	Chaconine (α-chaconine)	Rha-⁴Glc-³[solanidine] |² Rha	*S. tuberosum* *S. chacoense* *S. ptycanthum*	Hwang and Lee, 1983 Kuhn *et al.*, 1955b Eldridge and Hockridge, 1983
1178	β₁-Chaconine	Rha-²Glc-³[solanidine]	*S. chacoense* *S. tuberosum* *Fritillaria thunbergii* (Liliaceae) *Notholirion hyacinthinum* (Liliaceae)	Kuhn *et al.*, 1955b Jadhav *et al.*, 1981 Kitajima *et al.*, 1982 Xu and Xue, 1986
1179	β₂-Chaconine	Rha-²Glc-³[solanidine]	*Solanum tuberosum*	Jadhav *et al.*, 1981
1180	γ-Chaconine	Glc-³[solanidine]	*S. tuberosum*	Jadhav *et al.*, 1981
1181		Glc-⁴Glc-³[solanidine] |² Rha	*Fritillaria thunbergii* (Liliaceae) *Notholirion bulbuliferum* (Liliaceae) *N. hyacinthinum* *Fritillaria camtschatcensis* (Liliaceae)	Kitajima *et al.*, 1982 Qiu *et al.*, 1982 Xu and Xue, 1986 Mimaki and Sashida, 1990b

(cont.)

Table 2.11. (*Cont.*)

No.	Name	Structure	Plant	Reference
1182	Dehydrocommersonine	Glc-³Glc-⁴Gal-³[solanidine] |² Glc	*Solanum chacoense* (Solanaceae)	Zacharius and Osman, 1977
1183	Hyacinthoside	Glc-⁴Glc-⁴Glc-³[solanidine] |² Rha	*Notholirion hyacinthinum* (Liliaceae)	Xu and Xue, 1986
1184	Neohyacinthoside	Glc-⁶Glc-³Glc-⁴Glc-³[solanidine] |² Rha	*N. hyacinthinum* (Liliaceae)	Xu and Xue, 1988
1185	Leptine I	Rha-Glc-³[23-acetylleptinidine] ⌐ Rha	*Solanum chacoense* (Solanaceae)	Kuhn and Löw, 1961
1186	Leptine II	Rha-Gal-³[23-acetylleptinidine] ⌐ Glc	*S. chacoense*	Kuhn and Löw, 1961
1187		Glc-²Glc-³Glc-³[(23S,25ε)- solanidine-3β,23-diol]	*S. lyratum*	Murakami *et al.*, 1985
1188	Demissine	Xyl-³Glc-⁴Gal-³[demissidine] |² Glc	*S. demissum* *S. chacoense*	Osman and Sinden, 1982
1189	Commersonine	Glc-³Glc-⁴Gal-³[demissidine] |² Glc	*S. demissum* *S. commersonii*	Osman and Sinden, 1982 Osman *et al.*, 1976
1190 1191	Capsicastrine Isocapsicastrine	Gal-³[isoteinemine] Glc-³[isoteinemine]	*S. capsicastrum* *S. capsicastrum*	Lin *et al.*, 1987 Lin and Gan, 1989
1192		Rha-²Glc-³[hapepunine]	*Fritillaria thunbergii* (Liliaceae)	Kitajima *et al.*, 1982
1193 1194	Pingbeininosine Havanine	Glc-³[pingbeinine] Glc-³[etioline]¹⁶-Ac	*F. ussuriensis* *Solanum havanense* (Solanaceae)	Xu *et al.*, 1990 Basterrechea *et al.*, 1984
1195 1196	Etiolinine	Glc-⁴Glc-³[etioline] Glc-³[veramiline]	*S. havanense* *Veratrum taliense* (Liliaceae)	Basterrechea *et al.*, 1986 Mizuno *et al.*, 1990
1197		Glc-³[stenophylline B]	*V. taliense*	Mizuno *et al.*, 1990

are responsible for the bitter taste of the drug (Steinegger and Hänsel, 1988).

Variations on the two skeletal types (spirosolans and solanidans) and a number of unusual aglycones have been found in members of the Liliaceae, such as the genera *Veratrum* and *Fritillaria*. The 3-*O*-glycoside of imperialine (**1140**) from *Fritillaria ussuriensis* (Xu *et al.* 1983) and *Petilium eduardi* (Shakirov *et al.* 1965) is a representative of the fritillaria glycoalkaloids. The genus *Veratrum*, on the other hand, is a source of the ceveratum class of glycoalkaloids.

1140

Another liliaceous plant, *Lilium brownii*, contains the steroid alkaloid β_1-solamargine and several steroid saponins in the bulbs (Mimaki and Sashida, 1990c).

2.4   Saponins from marine organisms

Since the early 1970s, a new field of saponin research has emerged – namely the investigation of saponins from marine organisms. Although they often possess typical saponin characteristics, their aglycones are difficult to classify with those found in the plant kingdom and hence a special section is necessary for this category.

In 1960, Hashimoto and Yasumoto noticed, after an unsuccessful attempt to keep starfishes in an aquarium, that the water containing the dead starfishes foamed readily. Further investigation led to the isolation of a saponin mixture which was both haemolytic and toxic. Since this discovery, a large number of saponins from marine organisms have been characterized, although there are only half a dozen groups specializing in this isolation work.

At first it was not certain whether these glycosides were assimilated from food sources or whether the organism was capable itself of their biosynthesis. However, it is now known that marine organisms can biosynthesize at least a number of the saponins. The monosaccharides are most probably taken up with the diet.

Starfish belong to the echinoderms. The phylum Echinodermata is divided into two subphyla one of which, the Eleutherozoa, is itself subdivided into four classes:

- Holothuroidea (sea cucumbers)
- Echinoidea (sea urchins)
- Asteroidea (starfishes)
- Ophiuroidea (brittle stars).

Saponins have been isolated from two classes of Echinodermata: the Holothuroidea and the Asteroidea (Yasumoto *et al.* 1966). Those from sea cucumbers (holothurins) are triterpene glycosides, based on the holostane γ-lactone skeleton (derived from lanostane) (see Tables 2.12 and 2.13, structures **1212–1242**), whereas those from starfishes (asterosaponins) are steroid glycosides (Minale *et al.* 1982) (see Table 2.15, structures **1266–1299**).

More recently, sulphated polyhydroxylated steroids have been isolated from Halichondriidae sponges (Stonik, 1986).

A recent review of toxins from echinoderms contains details of the many saponins isolated from these organisms (Habermehl and Krebs, 1990), while the numerous structures of steroid glycosides from the Asteroidea and the Ophiuroidea have been listed by D'Auria *et al.* (1993). The oligoglycosides and polyhydroxysteroids from echinoderms have been reviewed by Minale *et al.* (1993).

One of the most notable features of many of the glycosides from marine organisms is the sulphation of the aglycone or sugar moiety. Furthermore, a 6-deoxyhexose monosaccharide is frequently present. Whereas, in plants, rhamnose (6-deoxymannose) is common, saponins from echinoderms have fucose (6-deoxygalactose) and quinovose (6-deoxyglucose) units; 3-*O*-methylglucose and 3-*O*-methylxylose moieties are also found.

Extracts and pure saponins from the Echinodermata have a broad spectrum of biological activities (see Chapters 5 and 6). Starfishes produce a slimy secretion from their skin glands with which they hunt and paralyse their prey. Contact dermatitis can result from touching certain asteroids. At least one of the functions of saponins in the tube feet and epidermis of starfish such as *Asterias rubens* and *Marthasterias glacialis* is as a defence against predators, parasites and microorganisms

(Mackie *et al.* 1977). They are lytic to mammalian and fish erythrocytes and toxic to fish.

The toxins of sea cucumbers are also used in defence. They are produced in the skin, for example, and when the animal is attacked they are ejected from the body cavity through the anus. Even in a dilution of 1 : 100 000 the crude saponin kills fish within a few minutes.

The first triterpene glycosides from sea cucumbers to be fully characterized were the holotoxins A (**1227**) and B (**1228**) (Kitagawa *et al.* 1976a, 1978a), although Yamanouchi (1943) had obtained a crystalline surface-active and haemolytic toxin from the same species, *Holothuria leucospilota*, in 1942. Early work was complicated by the fact that hydrolysis gave artifactual aglycones and this explains why the true structures were not elucidated until recently.

1227 Holotoxin A R = -Xyl^4-Glc^3-Glc^3-OCH$_3$
$$\quad\quad\quad\quad\quad\quad\quad\quad\quad\quad\quad |2$$
Qui^4-Glc^3-Glc^3-OCH$_3$

1228 Holotoxin B R = -Xyl^4-Glc^3-Glc
$$\quad\quad\quad\quad\quad\quad\quad\quad\quad\quad\quad |2$$
Qui^4-Glc^3-Glc^3-OCH$_3$

In the last few years nearly 50 saponins, with sugar chains composed of up to six monosaccharide units, have been isolated, all derivatives of holostanol with different side chains and the sugar moiety attached at C-3. Elyakov and co-workers found that the triterpene glycosides holothurins A (**1241**) and B (**1240**) (and related holothurins) were the major saponins in all 25 species of the genera *Holothuria* and *Actinopyga* that they examined (Burnell and ApSimon, 1983). *Curcumaria japonica* (order: Dendrochirota) is one of the most economically important holothurians and is used as food in Japan and other countries. The glycoside cucumarioside A$_2$-2 (**1232**) isolated from this sea cucumber has a labile 7(8)-ene aglycone and a pentasaccharide sugar chain (Avilov *et al.* 1984).

Table 2.12. *Aglycones of sea cucumber saponins*

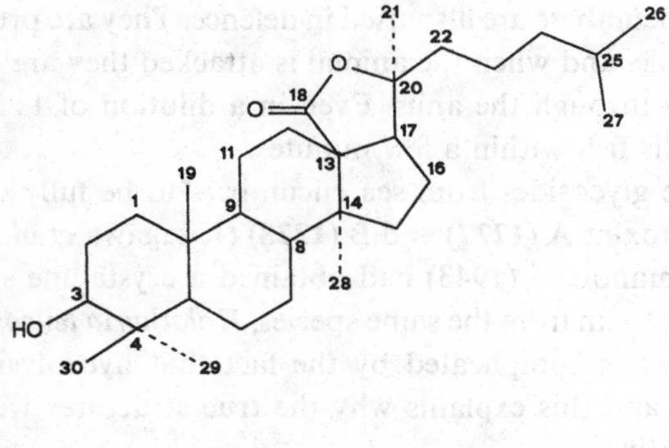

A

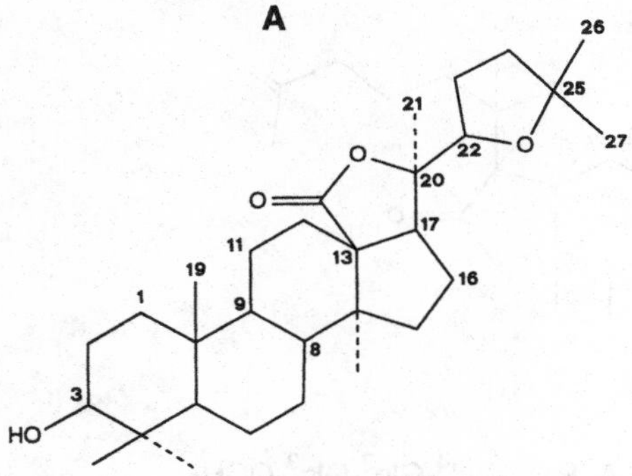

B

No.	Skeleton	Substituents
1198	A	12α-H, 17α-H, 9(11)-en
1199	A	12α-OH, 17α-H, 9(11)-en
1200	A	12α-OH, 17α-H, 9(11)-en, 25-OAc
1201	A	12α-OH, 17α-OH, 9(11)-en
1202	A	12α-OH, 17α-OH, 9(11),22(23)-dien, 25-OAc
1203	A	12α-OH, 17α-OH, 22-OH, 9(11)-en
1204	A	12α-OH, 17α-OH, 9(11), 24(25)-dien
1205	A	12α-H, 17α-H, 9(11),25(26)-dien, 16-one
1206	A	12α-H, 17α-H, 7(8),25(26)-dien, 16-one
1207	A	12α-H, 17α-H, 7(8),25(26)-dien, 23-OAc
1208	A	12α-H, 17α-H, 7(8),24(25)-dien, 16-OAc
1209	A	12α-H, 17α-H, 7(8)-en, 23-OAc
1210	B	12α-OH, 17α-H, 9(11)-en
1211	B	12α-OH, 9(11)-en

Table 2.13. *Triterpene glycosides from marine organisms*

No.	Name	Aglycone	Carbohydrate moiety	Organism	Reference
1212	Bivittoside D (bohadschioside A$_1$)	**1198**	$3\text{-Xyl}^4\text{-Glc}^3\text{-Glc}^3\text{-OMe}$ $\quad\vert^2$ $\text{Qui}^4\text{-Glc}^3\text{-Glc}^3\text{-OMe}$	*Bohadschia bivittata* *B. argus* *B. marmorata* *B. vitiensis*	Kitagawa *et al.*, 1989a Stonik, 1986 Stonik, 1986 Stonik, 1986
1213	Bivittoside A	**1199**	$3\text{-Xyl}^2\text{-Qui}$	*B. bivittata*	Kitagawa *et al.*, 1989a
1214	Bivittoside B	**1199**	$3\text{-Xyl}^4\text{-Glc}^3\text{-Glc}^3\text{-OMe}$ $\quad\vert^2$ Qui	*B. bivittata*	Kitagawa *et al.*, 1989a
1215	Bivittoside C (bohadschioside A)	**1199**	$3\text{-Xyl}^4\text{-Glc}^3\text{-Glc}^3\text{-OMe}$ $\quad\vert^2$ $\text{Qui}^4\text{-Glc}^3\text{-Glc}^3\text{-OMe}$	*B. bivittata* *B. argus* *B. marmorata* *B. vitiensis*	Kitagawa *et al.*, 1989a Stonik, 1986 Stonik, 1986 Stonik, 1986
1216	Pervicoside C	**1199**	$3\text{-Xyl}^2\text{-Qui}^4\text{-Glc}^3\text{-Glc}^3\text{-OMe}$ $\quad\vert^4$ $\text{OSO}_3^-\text{Na}^+$	*Holothuria pervicax*	Kitagawa *et al.*, 1989c
1217	Neothyoside A (pervicoside A)	**1200**	$3\text{-Xyl}^2\text{-Qui}^4\text{-Glc}^3\text{-Glc}^3\text{-OMe}$ $\quad\vert^4$ $\text{OSO}_3^-\text{Na}^+$	*Neothyone gibbosa* *Holothuria pervicax*	Encarnacion *et al.*, 1989 Kitagawa *et al.*, 1989c
1218	Echinoside B (holothurin B$_1$)	**1201**	$3\text{-Xyl}^2\text{-Qui}$ $\quad\vert^4$ $\text{OSO}_3^-\text{Na}^+$	*Actinopyga echinites* *Holothuria floridana*	Kitagawa *et al.*, 1985b Stonik, 1986
1219	Echinoside A (holothurin A$_2$)	**1201**	$3\text{-Xyl}^2\text{-Qui}^4\text{-Glc}^3\text{-Glc}^3\text{-OMe}$ $\quad\vert^4$ $\text{OSO}_3^-\text{Na}^+$	*H. edulis* *H. floridana* *Bohadschia graeffei* *Actinopyga echinites*	Stonik, 1986 Stonik, 1986 Stonik, 1986 Kitagawa *et al.*, 1985b

Table 2.13. (*Cont.*)

No.	Name	Aglycone	Carbohydrate moiety	Organism	Reference
1220	Holothurinoside B	1202	3-Xyl^2-Qui^4-Glc^3-Glc^3-OMe $\underset{\mid^2}{}$ Glc	*Holothuria forskalii*	Rodriguez et al., 1991
1221	Holothurin A$_1$	1203	3-Xyl^2-Qui^4-Glc^3-Glc^3-OMe $\underset{\mid^4}{}$ OSO$_3^-$X$^+$	*H. floridana* *H. grisea*	Stonik, 1986 Stonik, 1986
1222	Pervicoside B	1204	3-Xyl^2-Qui^4-Glc^3-Glc^3-OMe $\underset{\mid^4}{}$ OSO$_3^-$Na$^+$	*H. pervicax*	Kitagawa et al., 1989c
1223	Caudinoside A	1205	3-Xyl^2-Qui^4-Glc^3-Glc^3-OMe	*Paracaudina ransonetti*	Kalinin et al., 1986
1224	Cladoloside A	1205	3-Xyl^2-Qui^4-Xyl^3-Glc^3-OMe	*Cladolabes* sp.	Avilov and Stonik, 1988
1225	Cladoloside B	1205	3-Xyl^2-Qui^4-Xyl^3-Glc^3-OMe	*Cladolabes* sp.	Avilov and Stonik, 1988
1226	Neothyonidioside	1205	3-Xyl^2-Qui^4-Xyl^3-Glc^3-OMe $\underset{\mid^4}{}$ OSO$_3^-$Na$^+$	*Neothyonidium magnum*	Bedoya Zurita et al., 1986
1227	Holotoxin A	1205	3-Xyl^4-Glc^3-Glc^3-OMe $\underset{\mid^2}{}$ Qui^4-Glc^3-Glc^3-OMe	*Stichopus japonicus*	Stonik, 1986 Kitagawa, 1988
1228	Holotoxin B	1205	3-Xyl^4-Glc^3-Glc $\underset{\mid^2}{}$ Qui^4-Glc^3-Glc^3-OMe	*S. japonicus*	Stonik, 1986 Kitagawa, 1988
1229	Holotoxin A$_1$	1205	3-Xyl^4-Glc^3-Glc^3-OMe $\underset{\mid^2}{}$ Qui^4-Xyl^3-Glc^3-OMe	*S. japonicus*	Stonik, 1986
1230	Holotoxin B$_1$	1205	3-Xyl^4-Glc^3-Glc $\underset{\mid^2}{}$ Qui^4-Xyl^3-Glc^3-OMe	*S. japonicus*	Stonik, 1986

1231	Psolothurin A	1205	$3\text{-}Xyl^2\text{-}Glc^4\text{-}Qui^4\text{-}Glc^3\text{-}OMe$ $\quad\mid^6\quad\quad\mid^6$ $OSO_3^-X^+\ OSO_3^-Na^+$ Xyl $\mid^2$	*Psolus fabricii*	ApSimon *et al.*, 1984
1232	Cucumarioside A_2-2	1206	$3\text{-}Xyl^2\text{-}Qui^4\text{-}Glc^3\text{-}OMe$ $\quad\quad\mid^4$ $OSO_3^-X^+$	*Cucumaria japonica*	Stonik, 1986
1233	Stichoposide A	1207	$3\text{-}Xyl^2\text{-}Qui$	*Stichopus chloronotus*	Stonik, 1986
1234	Stichoposide B	1207	$3\text{-}Xyl^2\text{-}Glc$	*S. chloronotus*	Stonik, 1986
1235	Stichoposide C (stichloroside C_1)	1207	$3\text{-}Xyl^2\text{-}Qui^4\text{-}Xyl^3\text{-}Glc^3\text{-}OMe$ $\quad\quad\mid^4$ $Glc^3\text{-}Glc^3\text{-}OMe$	*Cucumaria fraudatrix*	Stonik, 1986
1236	Cucumarioside C_1	1208	$3\text{-}Xyl^2\text{-}Qui^4\text{-}Glc^3\text{-}Glc^3\text{-}OMe$ $\quad\mid^6$ $OSO_3^-X^+$	*Cucumaria fraudatrix*	Stonik, 1986
1237	Thelenotoside A	1209	$3\text{-}Xyl^2\text{-}Qui^4\text{-}Xyl^3\text{-}Glc^3\text{-}OMe$	*Thelenota ananas*	Stonik *et al.*, 1982
1238	Holothurinoside D	1210	$3\text{-}Xyl^2\text{-}Qui$	*Holothuria forskalii*	Rodriguez *et al.*, 1991
1239	Holothurinoside C	1210	$3\text{-}Xyl^2\text{-}Qui^4\text{-}Glc^3\text{-}Glc^3\text{-}OMe$	*H. forskalii*	Rodriguez *et al.*, 1991
1240	Holothurin B	1211	$3\text{-}Xyl^2\text{-}Qui$ $\quad\mid^4$ $OSO_3^-X^+$	*H. leucospilota* *H. edalis* *H. atra* *Actinopyga flammea* and others	Stonik, 1986 Kitagawa, 1988 Kitagawa, 1988 Kitagawa, 1988
1241	Holothurin A	1211	$3\text{-}Xyl^2\text{-}Qui^4\text{-}Glc^3\text{-}Glc^3\text{-}OMe$ $\quad\quad\mid^4$ $OSO_3^-X^+$	*Holothuria leucospilota* *H. squamifera* *Bohadschia graeffi* *Actinopyga agassizi* *A. flammea* and others	Stonik, 1986 Kitagawa, 1988 Ivanova and Kuznetsova, 1985 Kitagawa, 1988 Kitagawa, 1988
1242	Holothurinoside A	1211	$3\text{-}Xyl^2\text{-}Qui^4\text{-}Glc^3\text{-}Glc^3\text{-}OMe$ $\quad\quad\mid^4$ Glc	*Holothuria forskalii*	Rodriguez *et al.*, 1991

The first pure starfish saponin was not isolated until 1960, from *Asterina pectinifera* (Hashimoto and Yasumoto, 1960), and it was not until the mid-1970s that the chemistry of the marine saponins became properly known.

The sulphated steroid glycosides (asterosaponins) from starfishes (see Table 2.15, **1266–1299**) are characterized by aglycones possessing a 3β,6α-diol pattern, a 9,11-double bond and often a 23-oxo function. The sugar moiety is frequently attached at C-6 and the sulphate at C-3. The oligosaccharide portion includes fucose (6-deoxygalactose), quinovose (6-deoxyglucose) and often xylose, galactose or glucose. The saponins are fragile molecules and they often give artifacts on acid hydrolysis instead of the true aglycones. It is known, for example, that 3β,6α-dihydroxy-5α-pregn-9(11)-en-20-one (asterone) is frequently obtained after acid hydrolysis of asterosaponins (Habermehl and Krebs, 1990). This may mean that some of the steroid aglycones so far characterized are actually artifacts.

Saponins with thornasterol A (**1243**) as aglycone are widely distributed among the starfishes (Table 2.14). This steroid probably has a 20S-configuration (α) for the alcohol group at C-20 (Riccio *et al.* 1985a).

1243

Ovarian asterosaponin-1 (OA-1, **1300**) was isolated from the ovary of *Asterias amurensis* and was shown to contain a 6-deoxy-β-D-xylo-hexos-4-ulo-pyranose (Ulo) residue which was readily hydrated in media containing water (Okano *et al.* 1985). This glycoside has also been found in *A. forbesi* (forbeside C; Findlay *et al.* 1987). A second *A. amurensis* saponin (Co-ARIS II, **1301**), this time from the egg jelly, contains the same carbohydrate moiety but has a 17(20),24-dien-23-one side chain (Fujimoto *et al.* 1987).

Steroid cyclic glycosides have been isolated from starfish of the genus *Echinaster*. Sepositoside A (**1302**) from *Echinaster sepositus* has a cyclic trisaccharide moiety which includes a glucuronic acid residue (de Simone

Table 2.14. *Aglycones of starfish saponins*

No.	R	Other substituents
1243 (Thornasterol A)	A	6α-OH, 9(11)-en
1244	A	6α-OH, 12α-OH, 9(11)-en
1245	B	6α-OH, 9(11)-en
1246 (Dihydromartasterone)	C	6α-OH, 9(11)-en
1247	D	6α-OH, 9(11)-en
1248	E	6α-OH, 9(11)-en
1249	F	6α-OH, 9(11)-en
1250	G	5α-OH, 6β-OH, 8β-OH, 15α-OH

(Cont.)

Table 2.14. (*Cont.*)

No.	R	Other substituents
1251	G	6α-OH, 8β-OH, 15β-OH
1252	G	4β-OH, 6α-OH, 8β-OH, 15β-OH
1253	H	6α-OH, 15α-OH
1254	H	6β-OH, 8β-OH, 15α-OH
1255	H	4β-OH, 6β-OH, 8β-OH, 15α-OH
1256	I	6β-OH, 8β-OH, 15α-OH, 16β-OH
1257	J	4β-OH, 6α-OH, 8β-OH, 15β-OH, 16β-OH
1258	K	6α-OH, 7α-OH, 8β-OH, 15β-OH, 16β-OH
1259	K	6β-OH, 8β-OH, 15α-OH, 16β-OH, 4(5)-en
1260	L	6α-OH, 7α-OH, 8β-OH, 15β-OH
1261	M	6α-OH, 8β-OH, 15β-OH, 16β-OH
1262	N	6α-OH, 8β-OH, 15β-OH, 16β-OH
1263	O	6α-OH, 8(9)-en
1264	P	4β-OH, 6β-OH, 8β-OH, 15α-OH
1265	Q	6β-OH, 8β-OH, 15α-OH, 16β-OH, 4(5)-en

1300 OA-1 (Co-ARIS I) R = -Ulo^3-Qui^4-Fuc^2-Fuc
$|^2$
Qui

1301 Co-ARIS II R = -Ulo^3-Qui^4-Fuc^2-Fuc
$|^2$
Qui

et al. 1981). Very mild acid hydrolysis with 1 M HCl at 50 °C for approximately 1 min gives ring opening.

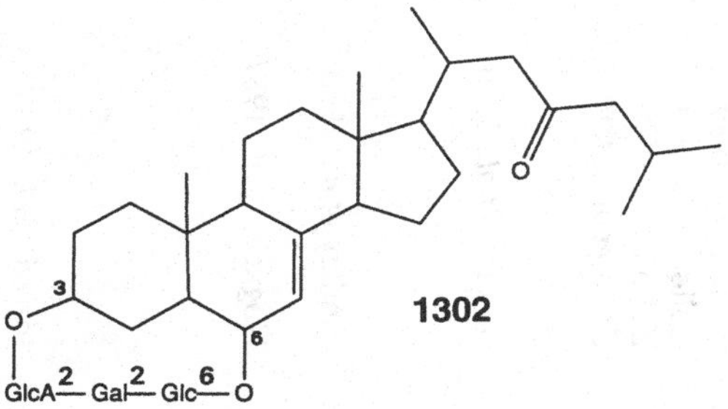

1302

GlcA$\frac{2}{}$Gal$\frac{2}{}$Glc$\frac{6}{}$O

Polyhydroxysteroids glycosylated in the side chains (e.g. amurensoside A, **1286**, Table 2.15) occur only in small amounts and are present in both sulphated and non-sulphated forms. The aglycone can be hydroxylated at C-3, C-6, C-8, C-15, C-24 and sometimes at C-4, C-5, C-7 and C-16. Steroids glycosylated at C-24 exist in the starfish genera *Protoreaster*, *Hacelia*, *Patiria* and *Culcita* (see Table 2.15). Cholestane (e.g. halityloside E, **1285**) and stigmastane (e.g. halityloside A, **1291**) skeletons are both found as aglycones of starfish saponins.

The absolute configuration at C-24 of asterosaponin P_1 (**1283**) from the starfish *Asterina pectinifera* has been confirmed by X-ray analysis of the desulphated derivative as (24S) (Cho *et al.* 1992).

More recently, a weakly cytotoxic and piscicidal saponin with amino sugars has been isolated from a sponge (*Asteropus sarasinosum*). This saponin, sarasinoside A_1 (**1303**), has 3β-hydroxy-4,4-dimethylcholesta-8,24-dien-23-one (a norlanostane) as aglycone and the monosaccharides xylose, glucose, *N*-acetyl-2-amino-2-deoxygalactose (GalNAc) and *N*-acetyl-2-amino-2-deoxyglucose (GlcNAc) (Kitagawa *et al.* 1987a; Schmitz *et al.* 1988). The carbon skeleton of the norlanostane aglycone is considered to be an intermediate in the biogenetic pathway from lanosterol to cholesterol (Kobayashi *et al.* 1991).

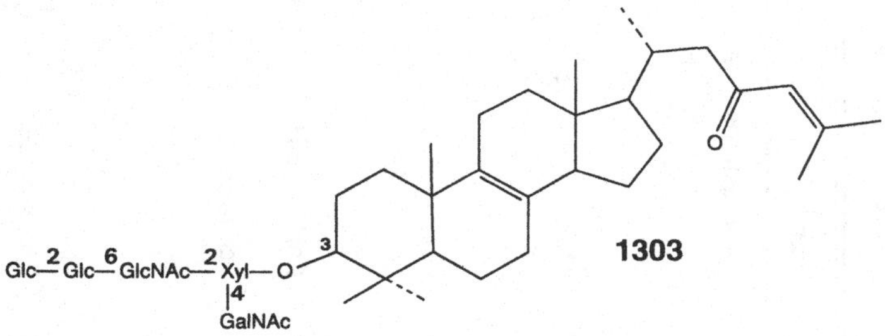

Glc$\frac{2}{}$Glc$\frac{6}{}$GlcNAc$\frac{2}{}$Xyl—O
|4
GalNAc

1303

Table 2.15. *Steroid glycosides from starfish*

No.	Name	Aglycone	Structure	Organism	Reference
1266	Thornasteroside A	1243	3-OSO$_3^-$X$^+$ 6-Qui^3-Xyl^4-Gal^2-Fuc $\mid^2$ Qui	*Acanthaster planci* *Halityle regularis* *Nardoa gomophia* *Pisaster ochracens* *P. brevispinus* *P. ochraceus* *P. brevispinus*	Kitagawa and Kobayashi, 1978 Riccio et al., 1985a Riccio et al., 1986a Zollo et al., 1989 Zollo et al., 1989 Zollo et al., 1989 Zollo et al., 1989
1267	Marthasteroside A$_1$	1243	3-OSO$_3^-$X$^+$ 6-Qui^3-Xyl^4-Gal^2-Fuc^3-Fuc $\mid^2$ Qui		
1268	Glycoside B$_2$	1243	3-OSO$_3^-$X$^+$ 6-Qui^3-Xyl^4-Gal^2-Fuc^3-Qui $\mid^2$ Qui	*Asterias amurensis*	Ikegami et al., 1979
1269	Forbeside H	1243	3-OSO$_3^-$Na$^+$ 6-Qui^3-Xyl^{4}-^{2}-Qui $\mid^2$ Qui	*A. forbesi*	Findlay et al., 1990
1270	Pectinioside A	1243	3-OSO$_3^-$Na$^+$ 6-Qui^3-Qui^4-Glc^2-Fuc $\mid^2$ Qui	*Asterina pectinifera*	Noguchi et al., 1987
1271	Pectinioside G	1243	3-OSO$_3^-$Na$^+$ 6-Qui^3-Qui^4-Glc^4-Ara $\mid^2 \quad \mid^2$ Qui $\quad$ Fuc	*A. pectinifera*	Iorizzi et al., 1990

1272	Tenuispinoside C	**1244**	$3\text{-OSO}_3^-\text{Na}^+$ $6\text{-Qui}^3\text{-Xyl}^4\text{-Fuc}^2\text{-Fuc}$ $\mid^2$ Qui	*Coscinasterias tenuispina*	Riccio *et al.*, 1986c
1273	Asterosaponin A	**1245**	$3\text{-OSO}_3^-\text{X}^+$ $6\text{-Qui}^4\text{-Qui}^4\text{-Fuc}^4\text{-Fuc}$	*Asterias amurensis*	Ikegami *et al.*, 1972
1274	Saponin C	**1246**	$3\text{-OSO}_3^-\text{X}^+$ 6-Glc-Qui-Fuc-Qui $\mid^2$ Fuc	*Marthasterias glacialis*	Minale *et al.*, 1982
1275	Acanthaglycoside A	**1247**	$3\text{-OSO}_3^-\text{Na}^+$ $6\text{-Qui}^3\text{-Xyl}^4\text{-Qui}^2\text{-Fuc}$ $\mid^2$ Qui	*Acanthaster planci*	Itakura and Komori, 1986b
1276	Versicoside B	**1248**	$3\text{-OSO}_3^-\text{Na}^+$ $6\text{-Qui}^3\text{-Xyl}^4\text{-Gal}^2\text{-Fuc}^3\text{-Gal}$ $\mid^2$ Qui	*Asterias amurensis*	Itakura and Komori, 1986a
1277	Versicoside C	**1248**	$3\text{-OSO}_3^-\text{Na}^+$ $6\text{-Qui}^3\text{-Xyl}^4\text{-Gal}^2\text{-Fuc}$ $\mid^2$ Qui	*A. amurensis*	Itakura and Komori, 1986a
1278	Regularoside A	**1249**	$3\text{-OSO}_3^-\text{Na}^+$ $6\text{-Glc}^3\text{-Qui}^4\text{-Qui}^2\text{-Fuc}$ $\mid^2$ Qui	*Halityle regularis*	Riccio *et al.*, 1985a
1279	Henricoside A	**1249**	$3\text{-OSO}_3^-\text{Na}^+$ $6\text{-Qui}^3\text{-Xyl}^4\text{-Gal}^2\text{-Ara}$ $\mid^2$ Qui	*Henricia laeviuscola*	D'Auria *et al.*, 1990

(*cont.*)

Table 2.15. (Cont.)

No.	Name	Aglycone	Structure	Organism	Reference
1280	Nodososide	1250	24-Ara^2-Xyl^2-OMe	Protoreaster nodosus Poraster superbus Pentaceraster alveolatus	Minale et al., 1982 Riccio et al., 1985b Zollo et al., 1986
1281		1251	24-Ara^2-Xyl^2-OMe	Hacelia attenuata	Minale et al., 1982
1282		1251	24-Ara^3-OSO$_3^-$Na$^+$	Oreaster reticulatus	Segura de Correa et al., 1985
1283	Asterosaponin P$_1$	1252	24-Ara^5-OSO$_3^-$Na$^+$ $\mid^3$ OMe	Patiria pectinifera Oreaster reticulatus	Stonik, 1986 Segura de Correa et al., 1985
1284	Culcitoside C	1252	24-Ara^2-Xyl^4-OMe $\mid^2$ OMe	Asterina pectinifera Culcita novaguinea	Iorizzi et al., 1990 Stonik, 1986
1285	Halityloside E	1252	24-Ara^2-Xyl^2-OMe	Sphaerodiscus placenta Halityle regularis Nardoa gomophia Porania pulvillus	Zollo et al., 1987 Iorizzi et al., 1986 Riccio et al., 1986a Andersson et al., 1987
1286	Amurensoside A	1253	15-OSO$_3^-$Na$^+$ 24-Xyl	Asterias amurensis	Riccio et al., 1988
1287	Pisasteroside B	1254	3-Xyl	Pisaster ochracens	Zollo et al., 1989
1288	Forbeside I	1255	24-OSO$_3^-$Na$^+$ 3-Xyl^4-OMe $\mid^2$ OMe	Asterias forbesi	Findlay and He, 1991
1289	Forbeside J	1255	3-Xyl^4-OMe $\mid^2$ OMe 24-Ara	A. forbesi	Findlay and He, 1991

1290		3-Xyl^2-OMe	*Poraster superbus*	Riccio et al., 1985b
1291	Halityloside A	24-Xyl^2-Xyl^2-OMe	*Sphaerodiscus placenta*	Zollo et al., 1987
			Halityle regularis	Iorizzi et al., 1986
			Nardoa novacaledonia	Riccio et al., 1986a
			N. gomophia	Riccio et al., 1986a
1292	Coscinasteroside A	3-Xyl 26-OSO$_3^-$Na$^+$	*Coscinasterias tenuispina*	Riccio et al., 1987a
1293	Echinasteroside A	3-Xyl^2-OMe	*Henricia laeviuscola*	D'Auria et al., 1990
1294	Indicoside A	28-Galf^5-OMe	*Astropecten indicus*	Riccio et al., 1987b
1295	Pisasteroside A	28-Glc^6-OSO$_3^-$Na$^+$	*Pisaster ochraceus*	Zollo et al., 1989
1296	Pisasteroside C	29-Xyl^4-OSO$_3^-$Na$^+$	*P. brevispinus*	Zollo et al., 1989
1297	Laeviuscoloside A	3-Glc^2-Gal	*P. brevispinus*	Zollo et al., 1989
1298	Forbeside K	3-Xyl^2-OMe 27-Ara	*Henricia laeviuscola*	D'Auria et al., 1990
			Asterias forbesi	Findlay and He, 1991
1299	Forbeside L	3-Xyl^2-OMe	*A. forbesi*	Findlay and He, 1991

The ophiuroid *Ophioderma longicaudum* has yielded two steroid glyco-side sulphates, longicaudoside-A **(1304)** and longicaudoside-B **(1305)**. These differ from other echinoderm saponins in that the monosaccharide is attached at position C-12 (Riccio *et al.* 1986b).

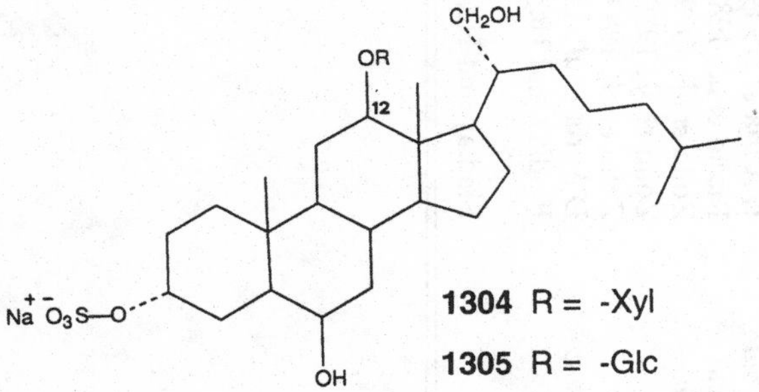

1304 R = -Xyl
1305 R = -Glc

Eryloside A **(1306)** is an exception. This antifungal and cytotoxic steroid glycoside has not been isolated from a starfish but from the Red Sea sponge Erylus lendenfeldii (Carmely *et al.* 1989). It has a rare 8,14-diene structure and a 4α-methyl substituent. Erylosides C **(1307)** and D

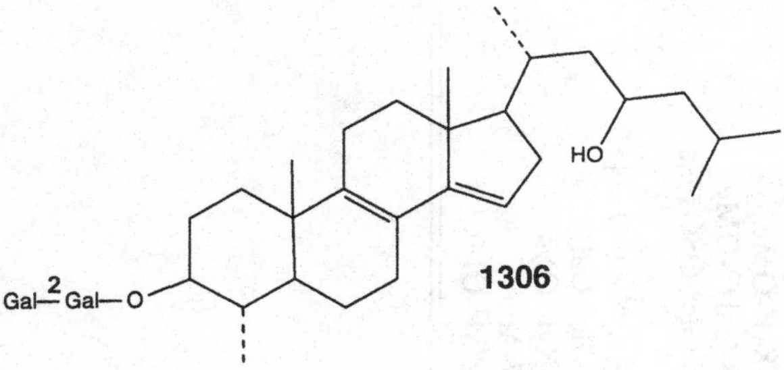

1306

are 14-carboxy-24,25-dimethyl-lanosta-8(9),24(31)-dienes from a Pacific *Erylus* sponge. Their structures were elucidated by a range of 2D-NMR techniques, including HOHAHA, HMBC and HMQC (D'Auria *et al.* 1992).

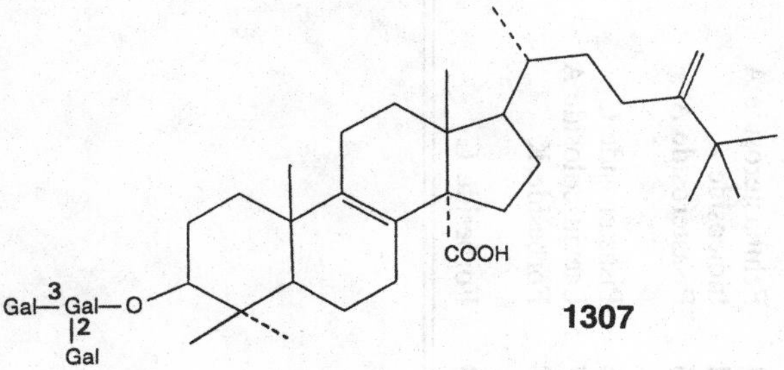

1307

A steroid glycoside (**1308**) bearing a strong resemblance to starfish saponins has also been isolated from the Sri Lankan soft coral *Sinularia crispa* (Alcyonaceae). This glycoside contains an acetylated α-fucose moiety and is spermatostatic at 0.5 mg/ml on rat cauda epididymal spermatozoa (Tillekeratne *et al.* 1989).

1308

3

Analysis and isolation

An important aspect of the modern use of plant extracts as pharmaceutical preparations is the characterization and determination of the individual active constituents. Such is also the case for saponin preparations, which often require sophisticated techniques for the isolation, structure elucidation and analysis of their component triterpene and steroid glycosides. When biological testing of the pure compounds is to be performed, it is necessary to isolate them in sufficient quantity and purity. As many foodstuffs contain saponins, their isolation and characterization is vital in order to investigate their biological activities and possible toxic effects.

A large body of work on the isolation of saponins comes from Japan, where there is great interest in the constituents and pharmacological properties of Oriental drugs, many of which contain either triterpene glycosides or steroid glycosides.

Techniques of isolation (to be found in Section 3.2, together with extraction methods) and structure elucidation described in this book put an emphasis on the triterpene glycosides. The other classes of saponins (steroid glycosides, steroid alkaloid glycosides) rely on the same or similar methods of chromatography and spectroscopic/chemical analysis (e.g. MS, ^{13}C-NMR, acid hydrolysis, enzymatic hydrolysis, alditol acetate formation, etc.). As far as steroid alkaloid glycosides are concerned, alternative isolation methods relying on the basic nature of the nitrogen atom can be employed (see below). Exceptions and special cases are mentioned in the relevant sections.

3.1   Analysis and quantitative determination

Different methods have been employed for the qualitative and quantitative determination of saponins: haemolysis, piscicidal activity, gravimetry, spectrophotometry, TLC, GC, HPLC, etc. (Hiller *et al.* 1966; Hiller and Voigt, 1977; Price *et al.* 1987). Determinations based on classical properties of saponins (haemolysis, surface activity, fish toxicity) have largely been replaced by photometric methods such as densitometry, colorimetry of derivatives and, more recently, by GC and HPLC. The quantitative analysis of *Ginseng radix* in the *Pharmacopoea Helvetica VII*, for example, relies on reaction with glacial acetic acid/sulphuric acid and spectrophotometry at 520 nm of the red colour formed. The β-aescine component of horse chestnut (*Aesculus hippocastanum*, Hippocastan-aceae) saponin can be spectrophotometrically determined after treatment with a mixture of iron(III) chloride, acetic acid and sulphuric acid (Schlemmer and Bosse, 1963). Spectrophotometric methods are very sensitive (Steinegger and Marty, 1976) but not suitable for estimating saponins in crude plant extracts since the reactions are not specific and coloured products may form with compounds which accompany the saponins, such as phytosterols and flavonoids. Another problem, common to much of the analytical work on saponins, is their incomplete extraction from the vegetable material.

A very simple but inadequate test is based, of course, on the foam-forming properties of saponins (Birk, 1969).

Analysis of extracts and quantitative determination of ginsenosides in ginseng roots has even been successfully performed by droplet countercurrent chromatography (DCCC) with chloroform–methanol–*n*-propanol–water (45:60:6:40) as solvent system and the upper layer as the mobile phase. The separated ginsenosides were treated with phenol-sulphuric acid reagent and determined colorimetrically (Otsuka *et al.* 1977).

Because of the commercial and nutritive aspects of soya and its products, a large proportion of the quantitative studies on saponins have been conducted with this legume (Price *et al.* 1987).

3.1.1   *Colour reactions*

(*1*) *Aromatic aldehydes*. Anisaldehyde, vanillin and other aro-matic aldehydes in strong mineral acid (for example, sulphuric, phos-phoric, perchloric acids) give coloured products with aglycones. The absorption maxima of these entities lie between 510 and 620 nm. A dehydration reaction probably occurs, forming unsaturated methylene groups which give coloured condensation products with the aldehydes. These reactions are, however, not very specific and a number of other classes of substance can react. With vanillin–sulphuric acid, spirostan saponins give two visible absorptions, one of them located around the

455 to 460 nm region; triterpene saponins with a C-23 hydroxyl group have a peak located between 460 and 485 nm (Hiai *et al.* 1976).

(*2*) *Liebermann–Burchard test*. Unsaturated and hydroxylated triterpenes and steroids give a red, blue or green coloration with acetic anhydride and sulphuric acid (Abisch and Reichstein, 1960). Since terpenoid saponins tend to produce a pink or purple shade and steroid saponins a blue-green coloration, differentiation of the two classes is possible.

(*3*) *Cerium(IV) sulphate or iron(III) salts and inorganic acids, such as sulphuric acid*. This gives a violet-red coloration of the solution.

(*4*) *A 30% solution of antimony(III) chloride in acetic anhydride–acetic acid*. This reagent gives colour reactions with hydroxytriterpenes and hydroxysteroids.

(*5*) *Antimony(III) chloride in nitrobenzene–methanol*. The differentiation of 5,6-dehydro-derivatives of steroid glycosides (diosgenin and solasodine glycosides) and 5α- or 5β-H-derivatives (e.g. tomatine) is accomplished with antimony(III) chloride in nitrobenzene–methanol: the 5,6-dehydro-derivatives give a red colour at room temperature.

(*6*) *Ehrlich reagent*. Furostanol derivatives give a red coloration with Ehrlich reagent (1 g *p*-dimethylaminobenzaldehyde + 50 ml 36% HCl + 50 ml ethanol; spray and heat) and a yellow colour with anisaldehyde (Kawasaki, 1981). Spirostanol derivatives do not react (Kiyosawa *et al.* 1968) and pennogenin glycosides are also negative. Another reagent specific for furostanol derivatives has been described by Sannié *et al.* (1951).

(*7*) *Carbazole*. The presence of uronic acids can be established by reaction with carbazole, in the presence of borate and concentrated sulphuric acid (Bitter and Muir, 1962).

Several of the reagents used for colour reactions are similarly employed in TLC detection and for the spectrophotometric and colorimetric determination of saponins (Hiai *et al.* 1976).

3.1.2   *Haemolysis*

Although the more important techniques for the qualitative and quantitative analysis of saponins are TLC and HPLC, for historical reasons a short treatment of haemolytic methods is necessary.

The ability of saponins to rupture erythrocytes has been used for decades as a detection and quantification method. Various quantitative methods using haemolysis have been reviewed by Birk (1969) and the older methods were described in detail by Kofler (1927) and Boiteau *et al.* (1964). The procedure of Büchi and Dolder (1950) has been improved and modified several times (Wasicky and Wasicky, 1961; van Kampen and Zijlstra, 1961), but the most frequently measured parameter is the change in absorbance of the supernatant of an erythrocyte suspension after haemolysis by a saponin or by a saponin-containing mixture. Various amounts of the saponin-containing material are mixed with a suspension of washed erythrocytes in isotonic buffer at pH 7.4. After 24 h, the mixture is centrifuged and haemolysis is indicated by the presence of haemoglobin (red) in the supernatant.

The *European Pharmacopoeia* uses as a unit the quantity in ml of ox blood (diluted 1 : 50) which is totally haemolysed by 1 g of test substance. As a standard, the saponin mixture from the roots of *Gypsophila paniculata* L. (Caryophyllaceae) has by definition an activity of 30 000. The haemolytic index (HI) is calculated as follows:

$$HI \text{ (haemolytic index)} = 30\,000 \times a/b$$

where a is the quantity of standard saponin (g) required for complete hydrolysis and b is the quantity of substance (g) required for complete hydrolysis.

Other methods include that of Mackie *et al.* (1977) which measures the absorbance at 545 nm of the supernatant after haemolysis and defines one unit of activity as the quantity of haemolytic material that causes 50% haemolysis. Reichert *et al.* (1986) have described a visual method which does not require spectrophotometry for the determination of saponins in quinoa grain. The assay was standardized against a commercial saponin preparation. A more extensive discussion of haemolytic indices and structure–activity considerations is given in Chapter 5.

Haemolytic methods have the disadvantage that they rely on the complete absence of other surface-active compounds which may also be haemolytic. Furthermore, there is often no correlation between the haemolytic activity and the pharmacological activity of the plant drug (Wagner and Reger, 1986b). Numerous 'typical' saponins, such as glycyrrhizin or sarsaparilloside, show very weak or no haemolytic activity. This is because haemolytic activity depends greatly on the arrangement of the sugar moieties and also on the presence of polar substituents (carboxyl or hydroxyl groups) on the aglycone, especially rings D and E.

Table 3.1. *Visualization reagents for the TLC of triterpene saponins*

Reagent	Reference
Vanillin–sulphuric acid	Godin, 1954
Vanillin–phosphoric acid	Oakenfull, 1981
Liebermann–Burchard (acetic anhydride–	Abisch and Reichstein, 1960
sulphuric acid)	Wagner *et al.* 1984a
1% Cerium sulphate in 10% sulphuric acid	Kitagawa *et al.* 1984b
10% Sulphuric acid in ethanol	Price *et al.* 1987
50% Sulphuric acid	Price *et al.* 1987
p-Anisaldehyde–sulphuric acid	Wagner *et al.* 1984a
Komarowsky (*p*-hydroxybenzaldehyde–	Wagner *et al.* 1985
sulphuric acid)	
Antimony(III) chloride	Wagner *et al.* 1984a
Blood	Wagner *et al.* 1984a
Water	

3.1.3   Thin-layer chromatography (TLC)

The qualitative analysis of saponins by TLC is of great import-ance for all aspects of saponin investigations. TLC plates (usually silica gel) can handle both pure saponins and crude extracts, are inexpensive, rapid to use and require no specialized equipment. A number of visual-ization reagents are available for spraying onto the plates (Table 3.1). Methods of preparation of the most common reagents are as follows:

- Vanillin–sulphuric acid (Godin reagent). A 1% solution of vanillin in ethanol is mixed in a 1:1 ratio with a 3% solution of perchloric acid in water and sprayed onto the TLC plate. This is followed by a 10% solution of sulphuric acid in ethanol and heating at 110 °C.
- Liebermann–Burchard reagent. Concentrated sulphuric acid (1 ml) is mixed with acetic anhydride (20 ml) and chloroform (50 ml). Heating at 85–90 °C gives the required coloration on the TLC plate.
- Antimony(III) chloride. The TLC plate is sprayed with a 10% solution of antimony chloride in chloroform and heated to 100 °C.
- Anisaldehyde–sulphuric acid. Anisaldehyde (0.5 ml) is mixed with glacial acetic acid (10 ml), methanol (85 ml) and con-centrated sulphuric acid (5 ml). This solution is sprayed onto the TLC plate, which is then heated at 100 °C.

Details of the preparation of other reagents are given in the literature references cited. Spraying with vanillin–sulphuric acid in the presence of ethanol and perchloric acid (Godin, 1954), for example, gives a blue or violet coloration with triterpene saponins. With anisaldehyde–sulphuric acid, a blue or violet–blue coloration is produced on heating the TLC plate. A systematic study of aldehyde–sulphuric acid spray reagents has been carried out and some useful conclusions on their compositions have been drawn (Stahl and Glatz, 1982). Spraying TLC plates with a solution of cerium sulphate in sulphuric acid gives violet–red, blue or green fluorescent zones under 365 nm UV light (Kitagawa *et al.* 1984b). Steroid saponins can be identified by their characteristic colour reaction with antimony (III) trichloride (Takeda *et al.* 1963). Occasionally, simply spraying the plates with water is sufficient to reveal the saponins present. Further precisions and additional spray reagents are to be found in a very comprehensive book on TLC (Stahl, 1969). The TLC analysis of aglycones formed after hydrolysis of saponins is described in Chapter 4.

The most frequently used solvent for TLC is chloroform–methanol–water (65 : 35 : 10) (Kawasaki and Miyahara, 1963), but other solvents can be found in a review by Hiller *et al.* (1966). The system *n*-butanol–ethanol–ammonia (7 : 2 : 5) is also popular, especially for glycosides containing uronic acid residues; for very polar mixtures *n*-butanol–acetic acid–water (4 : 1 : 5; upper layer) or chloroform–methanol–acetic acid–water (60 : 32 : 12 : 8) are used.

Systems employed for the TLC of glycoalkaloids include ethyl acetate–pyridine–water (30 : 10 : 30; upper phase). Visualization is with steroid reagents (anisaldehyde–sulphuric acid) or with alkaloid reagents (Dragendorff reagent, cerium(IV) sulphate). Other TLC solvents and visualization reagents are given by Jadhav *et al.* (1981) and Baerheim Svendsen and Verpoorte (1983).

TLC analysis of ginseng saponins is conveniently performed with the upper phase of the solvent system *n*-butanol–ethyl acetate–water (100 : 25 : 50) (spray reagent: anisaldehyde–sulphuric acid).

Numerous quantitative determinations have also been possible by TLC (Price *et al.* 1987). Here the density of spots obtained with a suitable spray reagent can be measured directly using a densitometer. For example, ginsenoside determination in *Panax ginseng* spray-dried extracts was performed at 530 nm, after spraying the TLC plates with vanillin–sulphuric acid (Vanhaelen and Vanhaelen-Fastré, 1984). Alternatively, quantitative determinations are possible by carrying out TLC separations, scraping the relevant band off the plates (located, for example, with iodine vapour), eluting the saponin and measuring the UV absorbance after addition of a suitable reagent (e.g. concentrated sulphuric acid) (Kartnig *et al.* 1972).

Reversed-phase TLC plates have now been commercially introduced and these provide an excellent analytical method for saponins, complementary to TLC on silica gel plates. Almost exclusive use of methanol–water and acetonitrile–water mixtures is made for developing reversed-phase (RP-8 or RP-18) plates. For hederagenin 3-*O*-glucoside (**350**, Appendix 2), a saponin with only one sugar moiety, TLC on Merck RP-8 HPTLC plates with methanol–water 7:3 gives R_f 0.18. The bidesmosidic hederagenin saponin **399** (Appendix 2), which has a total of four glucose units, gives R_f 0.54 under the same TLC conditions. Another possibility is to use DIOL HPTLC glass-backed plates. These can be used with normal silica gel TLC-type solvents or with methanol–water and acetonitrile–water solvents, as for RP-TLC.

3.1.4   High-performance liquid chromatography (HPLC)

HPLC, because of its speed, sensitivity and adaptability to non-volatile, polar compounds, is ideal for the analysis of saponins and sapogenins. The single largest difficulty is the lack of a suitable chromophore for UV detection in most saponins but there are ways of overcoming this problem by employing:

- refractive index detection
- mass detection
- derivatization.

However, assuming gradient changes are small, UV detection at around 203–210 nm with suitably pure solvents is perfectly feasible. A number of successful separations have been carried out with acetonitrile–water gradients, using UV detection (Table 3.2). Acetonitrile is preferred to methanol at low wavelengths because of its smaller UV absorption. Most of the work has been performed on reversed-phase C-8 and C-18 columns, although a DIOL support gives very satisfactory results with acetonitrile–water as eluent (Domon *et al.* 1984). If the polarity difference is not too great within a series of saponins under test (only small changes in the sugar chain, for example), isocratic elution is possible.

A mixture of eight olean-12-en saponins has been separated on an octyl-bonded column using gradient elution with aqueous acetonitrile (methanol could only be used under isocratic conditions). The quantity of acetonitrile was increased from 30% to 40% over 20 min and relatively little baseline drift was observed at 206 nm. More polar bidesmosidic saponins eluted much quicker than monodesmosidic saponins and

Table 3.2. *HPLC analyses of saponins*

Sample	Column	Solvent	Detection	Reference
Oleanane saponins	C-8	CH_3CN-H_2O	206 nm	Domon et al., 1984
	C-18	CH_3CN-H_2O	206 nm	Slacanin et al., 1988
	DIOL	CH_3CN-H_2O (85:15)	206 nm	Domon et al., 1984
Saponins	C-18	$CH_3CN-H_2O-TFA^a$	210 nm	Burnouf-Radosevich and Delfel, 1986
Triterpene saponins	Hydroxyapatite	CH_3CN-H_2O (87:13) (80:20)	210 nm	Kasai et al., 1987b
Soyasaponins	Anion exchange	$CH_3CN-0.4$ M H_3BO_3 (20:80)	UV	Yamaguchi et al., 1986
	Silica gel	$CHCl_3-MeOH-H_2O-HOAc$	Mass detection	Ireland and Dziedzic, 1986b
	C-18	CH_3CN-H_2O	Fluorescence	Kitagawa et al., 1985a,c
				Tani et al., 1985
	C-18	CH_3CN-n-PrOH$-H_2O-HOAc$ (32.3:4.2:63.4:0.1)	205 nm	Shiraiwa et al., 1991
Saikosaponins	Silica gel	$CHCl_3-MeOH-H_2O$ (30:10:1)	Refractive index	Kaizuka and Takahashi, 1983
	C-18	CH_3CN-H_2O (47:53)	210 nm	Shimizu et al., 1983
	C-18	$MeOH-H_2O-HOAc-Et_3N$	254 nm	Kimata et al., 1979
	Hydroxyapatite	CH_3CN-H_2O	205 nm	Okuyama et al., 1989
Glycyrrhizin	C-18	$MeOH-H_2O-TBA^b-H_3PO_4$	254 nm	Sagara et al., 1985
	C-18	$CH_3CN-H_2O-HOAc$	UV	Hiraga et al., 1984
	C-18	$MeOH-H_2O-HClO_4$	254 nm	Tsai and Chen, 1991
Gypsogenin glucuronide	C-18	$MeOH-H_2O-TBA^b-H_3PO_4$	206 nm	Henry et al., 1989
Ginsenosides	Silica gel	n-Heptane-n-BuOH$-CH_3CN-H_2O$	207 nm	Sticher and Soldati, 1979
	Silica gel	$CHCl_3-MeOH-H_2O$ (30:17:2)	Refractive index	Kaizuka and Takahashi, 1983
	C-18	$MeOH-H_2O$	203 nm	Sticher and Soldati, 1979
	C-18	CH_3CN-H_2O	203 nm	Soldati and Sticher, 1980
	C-18	CH_3CN-H_2O	203 nm	Sollorz, 1985
	C-18	CH_3CN-H_2O	203 nm	Pietta et al., 1986

(cont.)

Table 3.2. (Cont.)

Sample	Column	Solvent	Detection	Reference
	C-18	CH_3CN–50 mM KH_2PO_4	202 nm	Yamaguchi et al., 1988a,b
	C-18	CH_3CN–H_2O–H_3PO_4	202 nm	Yamaguchi et al., 1988a
	C-18	CH_3CN–H_2O–H_3PO_4	205 nm	Guédon et al., 1989
	C-18	CH_3CN–H_2O	203 nm	Petersen and Palmqvist, 1990
	NH_2	CH_3CN–H_2O–H_3PO_4	202 nm	Yamaguchi et al., 1988a,b
	Anion exchange	CH_3CN–0.25 M H_3BO_3 (12.5:87.5)	UV	Yamaguchi et al., 1986
				Yamaguchi et al., 1988b
	Hydroxyapatite	CH_3CN–H_2O (80:20)	210 nm	Kasai et al., 1987b
				Yamaguchi et al., 1988b
Hedera helix saponins	C-18	CH_3CN–H_2O	205 nm	Wagner and Reger, 1986a
Aesculus saponins	C-18	CH_3CN–H_2O–H_3PO_4	210 nm	Wagner et al., 1985
Aesculus saponins	C-18	CH_3CN–H_2O–H_3PO_4	205 nm	Pietta et al., 1989
Primula saponins	C-18	CH_3CN–H_2O–H_3PO_4	195 nm	Wagner and Reger, 1986b
Calendula saponins	C-18	$MeOH$–H_2O–H_3PO_4	210 nm	Vidal-Ollivier et al., 1989a
Steroid saponins	Silica gel	Hexane–EtOH–H_2O	208 nm	Xu and Lin, 1985
	NH_2	CH_3CN–H_2O	208 nm	Xu and Lin, 1985
	C-18	CH_3CN–H_2O	202 nm	Rauwald and Janssen, 1988
Glycoalkaloids	C-18	CH_3CN–H_2O–ethanolamine (45:55:0.1)	210 nm	Sinden et al., 1986
	C-18	CH_3CN–H_2O (1:1) (+5 mM sodium lauryl sulphate +5 mM sodium sulphate decahydrate)	200 nm	Friedman and Dao, 1992
Solasodine glycosides	C-18	MeOH–0.001 M Tris buffer	205 nm	Crabbe and Fryer, 1980
	C-18	CH_3CN–0.01 M Tris buffer	205 nm	Crabbe and Fryer, 1980
Potato glycoalkaloids	C-18	MeOH–H_2O–H_3PO_4	206 nm	Kajiwara et al., 1984
Potato glycoalkaloids	Carbohydrate	CH_3CN–THF–H_2O	215 nm	Bushway and Storch, 1982
Potato glycoalkaloids	NH_2	CH_3CN–KH_2PO_4–H_2O	208 nm	Saito et al., 1990a

[a] Trifluoroacetic acid.
[b] Tetra-n-butylammonium hydroxide.

glucuronides were less retained than the other glycosides. The apolar octylsilyl support was selective for the lipophilic part of the saponins; glycosides of hederagenin were eluted before the same glycosides of the less polar oleanolic acid (Domon *et al.* 1984). A crude methanol extract of *Phytolacca dodecandra* (Phytolaccaceae) berries contained a complex mixture of bidesmosidic triterpene glycosides which could be separated under similar conditions on a C-8 column with a gradient of 28% to 38% acetonitrile over 20 min. When using a DIOL column with acetonitrile–water (85:15) isocratically, there was an efficient separation of the *P. dodecandra* bidesmosidic saponins but the order of elution changed (Domon *et al.* 1984).

Detection at low wavelengths, which leads to problems of unstable baselines, interference by traces of highly UV-active material, etc., can conveniently be avoided by carrying out HPLC analyses with *derivatized* saponins. One possibility is to functionalize free carboxyl groups found in the saponin, as has been reported for the quantitative determination of monodesmosidic saponins from *P. dodecandra*. Treatment of oleanolic acid glycosides with 4-bromophenacyl bromide in the presence of potassium bicarbonate and a crown ether results in the formation of bromophenacyl derivatives (Fig. 3.1). The 4-bromophenacyl derivatives strongly absorb at 254 nm and detection can be performed at this wavelength without interference from solvent (Slacanin *et al.* 1988).

Using a solvent gradient, monodesmosidic saponins in an aqueous extract of *P. dodecandra* berries are rather difficult to quantify by HPLC (Fig. 3.2). Derivatization of the extract with 4-bromophenacyl bromide, however, allows straightforward analysis by HPLC (Fig. 3.3). The monodesmosidic saponins are responsible for the molluscicidal activity of the water extract, so their quantitative determination is most important when screening different batches of the plant (Slacanin *et al.* 1988).

Alfalfa root saponins have also been determined by this technique (Oleszek *et al.* 1990), as has a zanhic acid tridesmoside from the aerial parts of alfalfa (Nowacka and Oleszek, 1992).

An alternative determination method is to prepare fluorescent coumarin derivatives by esterification of the carboxylic acid moiety. By this means, soyasaponins were analysed and determined quantitatively in different

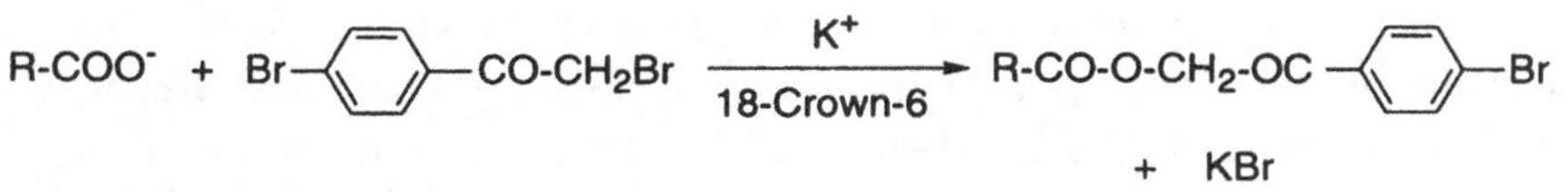

$$R\text{-}COO^- + Br\text{-}\langle\bigcirc\rangle\text{-}CO\text{-}CH_2Br \xrightarrow[\text{18-Crown-6}]{K^+} R\text{-}CO\text{-}O\text{-}CH_2\text{-}OC\text{-}\langle\bigcirc\rangle\text{-}Br$$

$$+ \quad KBr$$

Fig. 3.1 Derivatization of oleanolic acid glycosides.

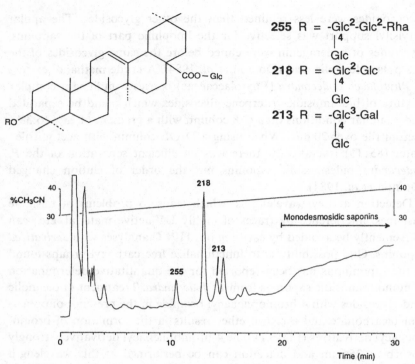

Fig. 3.2. HPLC analysis of an aqueous extract of *Phytolacca dodecandra* berries. Column: NovaPak C-18, 4 µm (150 × 3.9 mm); eluent: 30 → 40% acetonitrile over 30 min; flow rate: 1.5 ml/min; detection: 206 nm.

varieties and different organs of soybeans, with anthracene as internal standard (Kitagawa *et al.* 1984a; Tani *et al.* 1985).

One of the earliest derivatization methods for the HPLC of saponins involved reacting benzoyl chloride with ginsenosides in pyridine. This led to benzoylation of hydroxyl groups and was more likely to afford a single reaction product than 4-nitrobenzoylation, 3,5-dinitrobenzoylation or β-naphthoylation. By this means, ginsenosides Rb_1, Rb_2, Rc, Rd, Re and Rg_1 were distinguished in red ginseng and white ginseng (Besso *et al.* 1979). However, ginsenosides contain 9-15 hydroxyl groups and are difficult to quantitatively benzoylate, especially if steric problems occur. For this reason, the method has not been exploited any further.

In order to remove interfering (often highly UV-absorbing) material, a pre-purification step may be necessary. This can take the form of a clean-up on Sep-Pak[R] C_{18} (Wagner *et al.* 1985; Wagner and Reger, 1986a,b; Pietta *et al.* 1986; Ireland and Dziedzic, 1986b; Bushway *et al.* 1986; Guédon *et al.* 1989) or Extrelut[R] (Soldati and Sticher, 1980; Sollorz,

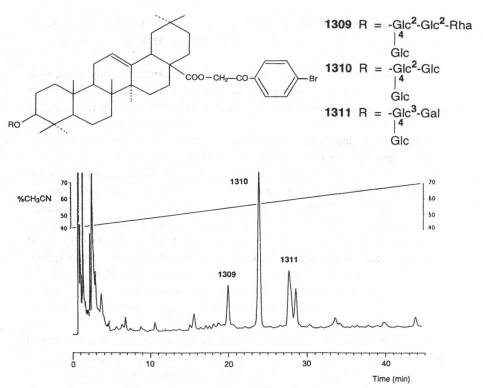

Fig. 3.3. HPLC analysis of a derivatized aqueous extract from *Phytolacca dodecandra* berries. Column: NovaPak C-18, 4 µm (150 × 3.9 mm), eluent: 40 → 70% acetonitrile over 45 min; flow rate: 1 ml/min; detection: 254 nm.

1985) cartridges. For glycoalkaloids, sample preparation with Sep-Pak[R] C_{18} and Sep-Pak[R] NH_2 cartridges has been reported (Saito *et al.* 1990a).

In the case of ionic compounds (those containing a free carboxyl group on the aglycone or glucuronic acid moieties, for example), some method of suppressing ion formation is required if peak broadening is to be avoided (Henry *et al.* 1989). This can be achieved by addition of a low UV-absorbing acid to the eluent, such as phosphoric acid (Wagner *et al.* 1985; Wagner and Reger, 1986b) or trifluoroacetic acid (Burnouf-Radosevich and Delfel, 1986). Another possibility is to use ion-pair HPLC, with a counter-ion added to the mobile phase. The capacity factor of the ionic compounds is increased by forming ion complexes with the pairing reagent. Thus, a quantitative method for the determination of glycyrrhizin in *Glycyrrhiza radix* (*Liquiritiae radix*) has been described, in which the mobile phase consisted of aqueous methanol with tetra-*n*-butyl-ammonium hydroxide and phosphoric acid added to pH 6 (Sagara *et al.*

1985). Derivatization of carboxyl groups (as mentioned above) is an alternative to additives in the mobile phase; peak resolution is considerably enhanced.

Many of the analyses listed in Table 3.2 are quantitative methods, with UV detection at low wavelengths (e.g. Soldati and Sticher, 1980; Wagner and Reger, 1986a). The advantage of quantitative HPLC over photometric methods is that the amounts of the individual saponins in a mixture or extract can be determined. Adulterations are easier to discern. Quantitative determination of ginsenosides by HPLC, for example, is now the favoured method and gives better results than those obtained by colorimetric, gas chromatographic and TLC–fluorimetric techniques (Soldati and Sticher, 1980). Both isocratic (Soldati and Sticher, 1980) and gradient elution (Sollorz, 1985) have been employed. Most of the gradient HPLC methods for the determination of ginsenosides fail to separate Rg_1 and Re. However, it is claimed that the most important ginsenosides, including the last two, can be separated on an octadecylsilyl column with an acetonitrile–water gradient (Petersen and Palmqvist, 1990). In this particular study, several commercially available reversed-phase C-18 columns were tested for their selectivity.

Shiraiwa *et al.* (1991) have analysed the 'group A' saponin content of soybean (*Glycine max*, Leguminosae) seed hypocotyls from 457 different varieties. Six closely related acetyl-soyasaponins were separated by HPLC on a YMC-PACK R-ODS-5 column (5 μm, 250 × 4.6 mm) with acetonitrile–*n*-propanol–water–acetic acid (32.3:4.2:63.4:0.1), at a flow-rate of 0.5 ml/min.

In the HPLC of *Sapindus rarak* (Sapindaceae) fruits, the acetylated hederagenin saponins were only poorly separated from other saponins and quantitative analysis was achieved by hydrolysis and measurement of the hederagenin (the sole aglycone) formed (Hamburger *et al.* 1992).

The glycyrrhizin contents of 26 types of licorice confectionery and health products (including cough mixtures and herbal teas), six batches of licorice root and 10 samples of commercial raw material have been determined by the method of Vora (1982), using a Spherisorb C-18 (5 μm) column and acetonitrile–water–acetic acid gradients (Spinks and Fenwick, 1990).

Several approaches have been used for the analysis of glycoalkaloids. Bushway *et al.* (1979) tested C-18, NH_2 and carbohydrate columns for the separation of α-chaconine, β-chaconine and α-solanine. More recently, the addition of ethanolamine to the solvent system has been reported, using a C-18 column (Sinden *et al.* 1986), while Saito *et al.* (1990a) performed separations of potato glycoalkaloids on an NH_2

column with the solvent combination acetonitrile–water–potassium dihydrogen phosphate.

One of the problems lies in the residual silanol sites in C-8 or C-18 packing materials. With basic compounds like glycoalkaloids, tailing and excessive elution times are a consequence. The separation of mixtures of α-chaconine, α-solanine, β-chaconines, β-solanines, γ-chaconine, δ-solanine and aglycones has therefore been attempted on different commercially available C-18 columns (Friedman and Levin, 1992). Development of a single assay for both glycoalkaloids and aglycones proved unrealistic. Aglycones required a non-acidic column, while glycoalkaloids were best separated on a very acidic column with a solvent such as 35% acetonitrile and 100 mM ammonium phosphate (monobasic) adjusted to pH 3.5 with phosphoric acid (Friedman and Levin, 1992).

Analytical HPLC is sometimes capable of resolving mixtures of very closely related saponins. For example, mixtures of steroid saponins containing $25R,25S$ and $25(27)$-en isomers can be detected by ^{13}C-NMR (Miyahara *et al.* 1983) or by HPLC methods. Miyahara *et al.* (1983) succeeded in separating glycosides of rhodeasapogenin, its $25R$-epimer (isorhodeasapogenin) and the $25(27)$-dehydro-derivative (convallamarogenin) by HPLC, using a TSK-GEL LS-410 C-18 column at low temperature (0 °C), with 90% methanol at 0.5 ml/min, and a Radial Pak C-18 column with 80% methanol at 1 ml/min.

However, the peak resolution of saponin mixtures on reversed-phase HPLC columns is sometimes insufficient and other methods have of necessity been employed:

- hydroxyapatite columns
- HPLC of borate complexes
- chemically modified porous glass columns
- silica gel columns.

Hydroxyapatite $Ca_{10}(PO_4)_6(OH)_2$. Hydroxyapatite is more hydrophilic than silica gel and can be used with simple binary aqueous solvent systems, thus facilitating detection by UV. It is stable in neutral and alkaline media. Recently, hard spherical particles of hydroxyapatite which are resistant to high pressure (up to 150 kg/cm²) have been prepared, opening the way to HPLC applications. Saponins of *Sapindus mukurossi* (Sapindaceae) (Fig. 3.4) and *Anemone rivularis* (Ranunculaceae) have been separated on hydroxyapatite columns (PENTAX PEC 101; 100 × 7.5 mm; Asahi Optical Co., Tokyo) with acetonitrile–water (87:13) as mobile phase for monodesmosidic saponins and acetonitrile–

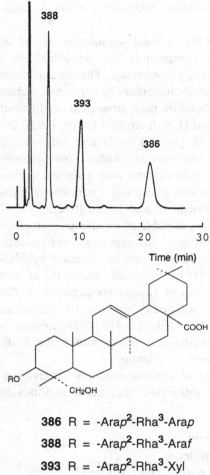

386 R = -Arap^2-Rha^3-Arap
388 R = -Arap^2-Rha^3-Araf
393 R = -Arap^2-Rha^3-Xyl

Fig. 3.4. HPLC of saponins of *Sapindus mukurossi* on a hydroxyapatite column. Eluent: acetonitrile–water (87:13); flow rate: 2 ml/min; detection: UV 202 nm.

water (80:20) for bidesmosidic saponins (flow rate 2 ml/min). Saponins differing only in the terminal pentose unit and which could not be separated by RP-HPLC were resolved by this method (Kasai *et al.* 1987b). The separation of ginsenosides from *Panax ginseng* (Araliaceae) was achieved in the isocratic mode (acetonitrite–water, 80:20) or, better, with a linear gradient (acetonitrile–water 90:10 → 70:30) (Kasai *et al.* 1987b). As is observed for silica gel, the glycosides are eluted in order of increasing polarity, i.e. the opposite of RP-HPLC.

Borate ion-exchange HPLC. This method has found application in the analysis of mono- and oligosaccharides. Tanaka and collaborators (Yamaguchi *et al.* 1986) have extended the technique to the separation of isomeric glycosides from *Sapindus mukurossi*. The best results were obtained with an anion exchange column (Asahipak ES-502N; Asahi Kasei Kogyo Co.; 100 × 7.6 mm) and 0.4 M H_3BO_3 in 20% (v/v) acetonitrile (pH 8) at 75 °C. Similar conditions were employed for separations of ginseng and *Anemone rivularis* saponins. Chromatographic characteristics depend on the formation of borate complexes with *cis*-diols in the saccharide moiety. After separations, borate can be removed as volatile methyl borate by repeated co-distillation of the eluate with methanol.

Chemically modified porous glass. Microporous glass (MPG) has a high chemical resistance and is stable between pH 2 and 12. Octadecyl porous glass (MPG-ODS) has been prepared as a packing for reversed-phase HPLC and used for the rapid and efficient separation of ginsenosides (Kanazawa *et al.* 1987). Furthermore, it is possible to separate both ginsenosides and saikosaponins simultaneously from extracts of combination drugs ('Shosaiko-to') containing ginseng and bupleurum root. A simple acetonitrile–water (25.5 : 74.5) mixture is sufficient for the separation (Kanazawa *et al.* 1990a). Comparison of MPG-ODS and silica-ODS columns for the HPLC of ginseng extract and for mixtures of ginsenosides has shown that the retention behaviour was similar but that capacity factors were smaller on an MPG-ODS column. The resolution of certain pairs of ginsenosides was better on MPG-ODS columns (Kanazawa *et al.* 1993).

Silica gel. The use of water-containing mobile phases is often unavoidable for the separation of saponins, and silica gel HPLC does not normally lend itself to such eluents. However, a modification of the column packing has made possible the separation of water-soluble glycosides without column deterioration. The procedure involves first washing the column with methanol, then with the mixture chloroform–methanol–ethanol–water (62 : 16 : 16 : 6) and finally the solvent system to be used for the separation (Kaizuka and Takahashi, 1983). By this means, efficient analyses of ginseng saponins and saikosaponins from *Bupleurum falcatum* could be achieved. Neutral *Dioscorea* steroid saponins have also been separated on 5 μm silica gel columns with a water-containing eluent: hexane–ethanol–water (8 : 2 : 0.5). This system permitted the separation of the C-25 epimers collettiside IV and gracillin but not the epimers collettiside III and dioscin. Better results were not achieved with a C-18 column (methanol–water 78:22) or an NH_2 column (acetonitrile–water 85:15) (Xu and Lin, 1985).

3.1.5 *Liquid chromatography–mass spectrometry (LC–MS)*

The combination of HPLC and mass spectrometry (MS) has only recently been successfully applied to the analysis of non-volatile, thermally labile molecules such as saponins because of the difficulties involved with the interfacing of the systems. However, several types of efficient interface for direct and indirect introduction of HPLC column effluent have now been developed (Arpino, 1990). Satisfactory results have been obtained for the qualitative analysis of crude saponin fractions of *Panax ginseng*, *P. japonicus* and *Bupleurum falcatum* by combining semi-micro HPLC with a frit-fast atom bombardment (FRIT-FAB) interface (Hattori *et al.* 1988). For this purpose, an NH_2 column (μS-Finepak SIL NH_2, Jasco; 25 cm × 1.5 mm internal diameter (I.D.)) was used, rather than an octadecyl silica column, with a 1:20 split ratio of effluent (100 μl/min → 5 μl/min). Elution with a linear gradient of acetonitrile and water, containing 1% glycerol allowed a better peak sharpness than that obtained by isocratic elution. Negative FAB mass spectra were recorded for saponins with a molecular weight of up to 1235 (the bidesmoside cauloside F, **363**, Appendix 2). Pseudomolecular $[M − 1]^-$ ions as well as fragment ions ascribed to the cleavage of sugar moieties were observed (Hattori *et al.* 1988).

A FRIT-FAB LC–MS system was also used to separate a mixture of the isomeric saponins rosamultin and arjunetin (both molecular weight 650) from *Rosa rugosa* (Rosaceae). Rosamultin (an ursane glycoside) and arjunetin (an oleanane glycoside) both have a single glucose residue at C-28 and were analysed in both the negative and positive FAB modes with xenon as neutral gas. HPLC was performed on an octadecylsilica column (250 × 1.5 mm) with acetonitrile–water (7:3, containing 0.5% glycerol) as solvent at a flow rate of 1 ml/min. Pseudomolecular $[M − 1]^-$ and $[M + 1]^+$ ions were observed, together with strong peaks caused by the parent aglycones in the negative FAB mass spectra (Young *et al.* 1988).

It is also possible to detect saponins by dynamic secondary ion mass spectroscopy (SIMS) (similar to dynamic FAB interfacing, in which eluent is passed directly into the source). Thus, HPLC combined with UV (206 nm) and SIMS detection has been employed to analyse a mixture of one mono- and two bidesmosidic triterpene glycosides (Fig. 3.5) (Marston *et al.* 1991).

A disadvantage with interfaces of the FRIT-FAB and CF-FAB type is the low flow rates required (around 5 µl/min). After HPLC separation, effluent splitting is necessary. The thermospray (TSP) interface (Blackley and Vestal, 1983), however, is characterized by its simplicity and its ability to handle flow rates of 1–2 ml/min. This makes the technique more attractive for problems involving the analysis of plant constituents. At the heart of the TSP technique is a soft ionization of molecules, similar to chemical ionization MS. This allows analysis of non-volatile and thermally labile mono-, di- and even triglycosides. Information is provided about the molecular weight of the saponin and the nature and sequence of the sugar chains. TSP LC–MS has been used for the analysis of molluscicidal saponins in a methanol extract of *Tetrapleura tetraptera* (Leguminosae) fruits (Maillard and Hostettmann, 1993). With post-column addition of 0.5 M ammonium acetate (0.2 ml/min) to provide the volatile buffer for ion evaporation ionization, the TSP LC–MS total ion current (mass range 450 to 1000 a.m.u) corresponded well with HPLC–UV analysis at 206 nm (Fig. 3.6a). Ion traces at m/z 660, 676, 880 and 822 gave signals representing the pseudomolecular $[M + H]^+$ ions of the major saponins **649–652**. The TSP mass spectrum acquired for each saponin in the extract displayed a major peak for the pseudomolecular $[M + H]^+$ ion. Fragmentations of the sugar moieties were observed, as shown in Fig. 3.6b for the principal molluscicidal saponin aridanin (**649**), where loss of a *N*-acetylglucosyl moiety gave rise to an $[A + H]^+$ peak for the aglycone (Maillard and Hostettmann, 1993).

The technique of LC–MS, as applied to the investigation of saponins, is still in its infancy, but the potential of the method is enormous, especially as GC–MS is of little practical use and in HPLC alone the identities of peaks can only be confirmed by their retention times. Not only is LC–MS amenable to the analysis of these glycosides in plants, foodstuffs and medicines but it will also be of value, via MS–MS, to the structure determination of individual saponins in a mixture.

3.1.6   Radioimmunoassay (RIA)

A radioimmunoassay has been developed for the determination of picogram amounts of aescine. This allows quantification of aescine in extracts of *Aesculus hippocastanum* (Hippocastanaceae) and in non-purified biological samples. As virtually no interference is observed for various potentially cross-reacting compounds, RIA provides a highly sensitive and selective method for the analysis of aescine (Lehtola and Huhtikangas, 1990).

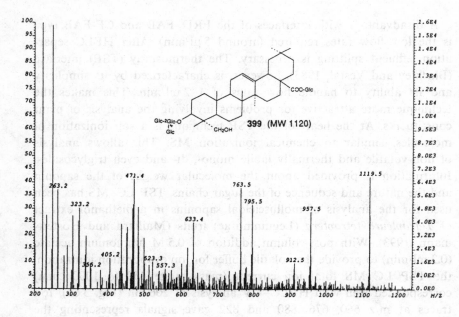

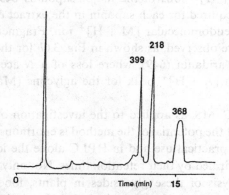

Fig. 3.5. HPLC combined with UV (206 nm) and SIMS detection of a saponin mixture. HPLC conditions: column: RP-18, 50 × 1 mm, eluent: water + 1% glycerol (solvent A), acetonitrile + 9% methanol + 1% glycerol (solvent B), 0 → 100% B in 20 min, flow rate: 1 ml/min. MS conditions: VG AutoSpec Q, dynamic SIMS negative ion mode, 1% glycerol matrix. (*cont.*)

3.1.7   *Enzyme-linked immunosorbent assay (ELISA)*

Glycoalkaloids do not have a suitable UV chromophore for HPLC analysis and therefore absorbance is measured at around 210 nm. Relatively large sample sizes and extensive sample clean-up are needed to overcome background noise. In addition, UV absorbance will not

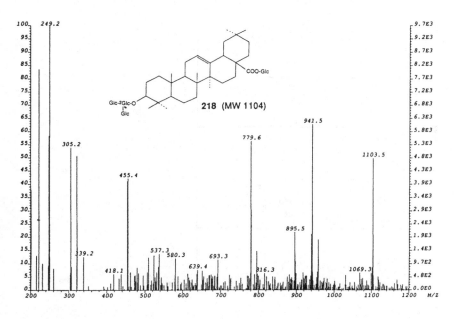

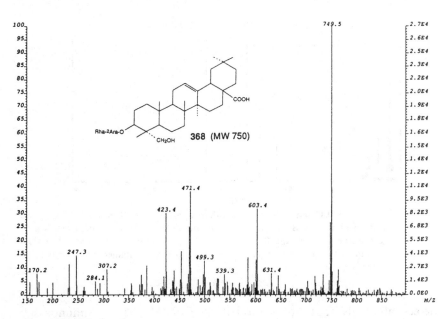

Fig. 3.5. (*cont.*)

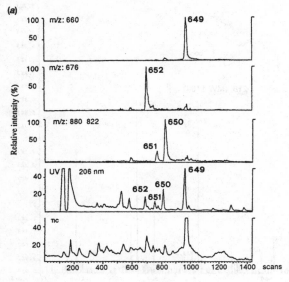

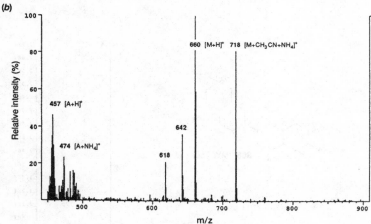

Fig. 3.6. HPLC analysis of a methanol extract of *Tetrapleura tetraptera* fruits. Column: μ Bondapak C-18, 10 μm (300 × 3.9 mm); eluent: 30 → 80% acetonitrile over 30 min; flow-rate: 1 ml/min. (*a*) TSP LC-MS and HPLC-UV 206 nm; (*b*) TSP mass spectrum of aridanin (**649**) obtained after on-column analysis. (*cont.*)

detect alkaloids that are saturated at C-5. However, immunoassay methods are specific, sensitive and useful for rapid analysis of large numbers of samples. An ELISA for the total glycoalkaloid content of commercially available potato tubers was both technically simple and cheap, requiring minimal sample preparation. The antisera allowed

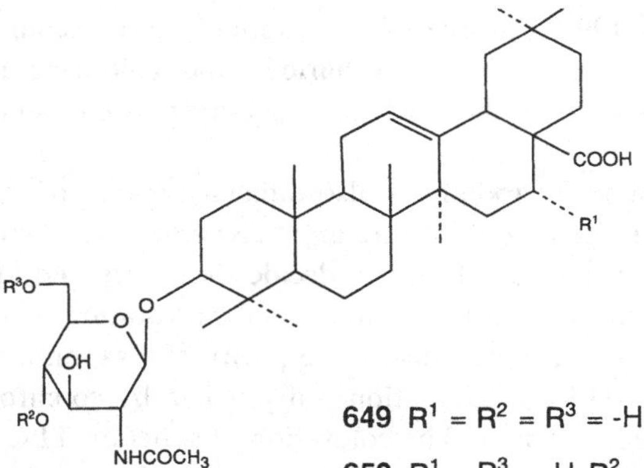

649 $R^1 = R^2 = R^3 = -H$

650 $R^1 = R^3 = -H, R^2 = -Gal$

651 $R^1 = R^2 = -H, R^3 = -Glc$

652 $R^1 = -OH, R^2 = R^3 = -H$

Fig. 3.6. (*cont.*)

quantification of solanidan-based glycosides and the sensitivity achieved was 2 µg/100 g potato tissue (Morgan *et al.* 1985).

An improved enzyme immunoassay with bovine serum albumin coating conjugates has been used to quantify the major solanidine glycoalkaloids (α-solanine and α-chaconine) in *Solanum tuberosum* cultivars. Demissidine glycoalkaloids saturated at C-5 can also be quantified by this procedure (Plhak and Sporns, 1992).

An enzyme immunoassay for ginsenoside Rb_1 in *Panax ginseng* (Araliaceae) has been introduced. This is capable of detecting 0.04 ng of ginsenoside Rb_1, but the anti-ginsenoside Rb_1 antiserum cross-reacts with ginsenosides Rb_2 (21.8%) and Rc (10.6%) (Kanaoka *et al.* 1992).

3.2   Isolation

Since saponins have relatively large molecular weights and are of high polarity, their isolation poses quite a challenge. Another of the major problems involved in the isolation of pure saponins is the presence of complex mixtures of closely related compounds, differing subtly either in the nature of the aglycone or the sugar part (nature, number and positions of attachment of the monosaccharides). Difficulties are also encountered with labile substituents such as esters. The major genuine soybean saponin, a γ-pyrone derivative (BeA), is only extracted by aqueous ethanol at room temperature. Extraction with heating (80 °C) leads to fission of the ester moiety and formation of soyasaponin I

(Bb) (Kudou *et al.* 1992). In the plant, saponins are accompanied by very polar substances such as saccharides and colouring matter (phenolic compounds and the like), are not easily crystallized and can be hygroscopic.

Characterization of pure saponins is also difficult because of the lack of crystalline material. Melting points are imprecise and often occur with decomposition. It is, therefore, difficult to decide the purity and identity of samples simply from the melting point, the optical rotation value and other physical constants. A better idea of the purity of a saponin can be obtained by TLC or HPLC examination – if possible by co-chromatography with an authentic sample. The coloration of spots on TLC plates after spraying with suitable reagents is an additional indicator of identity (see Table 3.1).

3.2.1 *Extraction and preliminary purification*

Extraction procedures have to be as mild as possible because certain saponins can undergo the following transformations:

- enzymatic hydrolysis during water extraction (Wulff, 1968; Domon and Hostettmann, 1984; Kawamura *et al.* 1988 – see below)
- esterification of acidic saponins during alcohol treatment
- hydrolysis of labile ester groups
- transacylation.

Therefore, care has to be taken to follow the individual steps by TLC (Section 3.1).

Although numerous variations are possible, current general procedures for obtaining crude saponin mixtures are as follows:

- extraction with methanol, ethanol, water or aqueous alcohol
- a defatting step (generally with petroleum ether); this can either be performed before the extraction step or on the extract itself
- extractives are dissolved or suspended in water and shaken with *n*-butanol saturated with water (Rothman *et al.* 1952a,b).
- precipitation (optional) of saponins with diethyl ether or acetone.

A dialysis step can be included (see, for example, Zhou *et al.* 1981; Massiot *et al.* 1988a) in order to remove small water-soluble molecules such as sugars.

The most efficient extraction of dry plant material is achieved with methanol or aqueous methanol. Methanol is also used for fresh plant material but an alternative is to extract frozen plant parts with propanol–water in an Ultra-Turrax mixer (Hiller and Voigt, 1977). Water is a less efficient extraction solvent for saponins (unless specifically water-soluble glycosides are desired) but has the advantage of being easily lyophilized and gives a cleaner extract. Depending on the proportion of water used for extraction, either monodesmosidic or bidesmosidic saponins may be obtained (Domon and Hostettmann, 1984; Kawamura *et al.* 1988). Fresh vegetable material contains active enzymes (esterases) which, when homogenized with a solvent, are able to convert bidesmosides into monodesmosides. Even dry material may contain esterases which are activated in the presence of water. Kawamura *et al.* (1988) studied the formation of momordin I (a monodesmosidic oleanolic acid saponin) from momordin II (the corresponding bidesmoside) in dried roots of *Momordica cochinchinensis* (Cucurbitaceae), after homogenization in water or aqueous methanol. The conversion takes place in water and in 30% and 60% methanol solutions but not in 80% and 100% methanol solutions. On the contrary, homogenates of the fresh roots in methanol retained enzyme activity. However, the enzymes could be inactivated by first soaking the fresh roots in 4% hydrochloric acid and the bidesmoside was then shown to be the major component. It is, therefore, obvious that the correct choice of extraction procedure is an extrememely important first step.

In the past, precipitation with lead acetate, barium hydroxide, tannic acid and magnesium oxide, and salting-out with ammonium sulphate or magnesium sulphate were practised (Kofler, 1927); precipitation with cholesterol is still reported from time to time (Tschesche and Ziegler, 1964). Other, more classical, extraction details have been described by Boiteau *et al.* (1964).

Methods typically used to purify proteins, such as dialysis, ion-exchange chromatography and size-exclusion chromatography, are useful in partially separating saponins in aqueous solution from non-saponin components but are ineffective in separating individual saponins because of the tendency of saponins to form mixed micelles. Hence, effective separation requires the use of organic solvents or solvent/water systems that solubilize the amphiphilic saponins as monomers so that the formation of mixed micelles does not interfere with separation.

Furostanol saponins. A common phenomenon observed for furostanol saponins is the formation of 22-OCH$_3$ derivatives during extraction with methanol. The genuine 22-hydroxyfurostanols can either be obtained by extraction with another solvent (e.g. pyridine) or by treatment of the methoxylated artifacts with boiling aqueous acetone (Konishi and Shoji, 1979).

Steroid alkaloids. For the isolation of steroid alkaloids, advantage can be taken of the fact that they form water-soluble salts. Crude glycoalkaloid preparations can therefore be obtained by precipitation from weakly acidic plant extracts with ammonia (Jadhav *et al.* 1981). Otherwise, purification techniques are much the same as for the other classes of saponin.

3.2.2   Chromatography

The isolation of pure saponins requires one or (as is almost always the case) more chromatographic separation steps in order to remove other polar constituents of alcoholic or aqueous plant extracts.

A variety of modern separation techniques (Hostettmann *et al.* 1986, 1991; Marston and Hostettmann, 1991b) such as flash chromatography, DCCC, low-pressure liquid chromatography (LPLC), medium-pressure liquid chromatography (MPLC) and HPLC are available, but a large number of the separations (especially the preliminary fractionation work) reported in the literature are still carried out by conventional open-column chromatography. An idea of separation conditions, solvent systems, etc. can be obtained from the relevant tabular information presented here but for further experimental details the relevant references should be consulted. The best results are usually achieved by strategies which employ a combination of methods and some of these will be mentioned below.

As a number of saponins are acidic, salts can form and on completion of chromatography, treatment with an ion-exchange resin may be necessary to obtain the free saponin. Examples are Dowex 50Wx8 (H^+ form) (Kitagawa *et al.* 1988b; Yoshikawa *et al.* 1991b), Amberlite IRC 84 (Okabe *et al.* 1989; Nagao *et al.* 1990) and Amberlite MB-3 (Mizutani *et al.* 1984). However, if neutrality or careful control of pH are necessary to prevent decomposition, steps involving filtration on ion-exchange resins must be avoided (Massiot *et al.* 1992b).

In certain instances, crude saponin fractions have been *methylated* (assuming that free COOH groups are present) in order to achieve satisfactory separations of closely related products (Okabe *et al.* 1989; Nagao *et al.* 1989a, 1990).

Open-column chromatography

A few of the classical solvent systems employed for the silica gel column chromatography of saponins are listed in Table 3.3. Other sorbents and solvent systems are given elsewhere (Woitke *et al.* 1970; Adler and Hiller, 1985). Open-column chromatography is often used as a first fractionation step for a crude saponin mixture but in certain cases may yield pure products. In general, though, the resolution is not high and complex mixtures are only partially separated. Other problems are the loss of material because of irreversible adsorption and the length of time required to perform the separations.

Silica gel chromatography with chloroform–methanol–water eluents is the most popular method and is still used in the majority of separations. When a biphasic system is used, the water-saturated chloroform phase is the eluent. Thus, a gradient of chloroform–methanol–water (65:35:5 → 65:40:10) was employed for the initial separation of a methanol extract of *Swartzia simplex* (Leguminosae) leaves on silica gel. Further chromatography on low-pressure columns gave a monodesmosidic molluscicidal saponin, while a bidesmosidic saponin was obtained by silica gel column chromatography with the solvent system acetone–*n*-propanol–water (35:35:5) (Borel *et al.* 1987). Initial fractionation of a methanol extract of *Tetrapleura tetraptera* (Leguminosae) fruits was by means of open-column chromatography on 40–60 μm silica gel. As solvent, a gradient of chloroform–methanol–water (85:11:1 → 70:30:5 → 50:50:0) was used (Maillard *et al.* 1989).

A complex mixture of triterpene glycosides has been isolated from the corms of *Crocosmia crocosmiiflora* (Iridaceae). Three of these, 2,9,16-trihydroxypalmitic acid glycosides of polygalacic acid, were obtained by a strategy involving open-column chromatography of a crude saponin mixture on silica gel 60 (60–230 μm), employing *n*-butanol–ethanol–water (5:1:4, upper layer) and chloroform–methanol–water (60:29:6) as eluents. Final purification was by HPLC (Asada *et al.* 1989a).

Extensive use of silica gel chromatography also enabled the separation of the dammarane glycosides actinostemmosides A–D from *Actinostemma lobatum* (Cucurbitaceae). After an MCI (Mitsubishi Chemical Industries) polystyrene gel column, the relevant fractions were chromatographed with a variety of solvents: chloroform–methanol–water (7:3:0.5, 32:8:1), chloroform–methanol (9:1, 1:1), chloroform–ethanol (17:3), ethyl acetate–methanol (4:1), chloroform–methanol–ethyl acetate–water (3:3:4:1.5, lower layer). By this means, pure actinostemmoside C was obtained while actinostemmosides A and B required an additional low-pressure LC step and actinostemmoside D required a final

separation on a C-18 column eluted with 70% methanol (Iwamoto *et al.*
1987).

Certain ester saponins have been chromatographed on silica gel im-
pregnated with 2% boric acid (Srivastava and Kulshreshtha, 1986, 1988).

As an addition to normal silica gel, coarse RP sorbents are now
employed in the open-column chromatography of saponins (Table 3.3).
As long as the granulometry is not too fine and the columns not too long,
gravity-fed columns are quite suitable. RP chromatography is generally
introduced after an initial silica gel separation step and enables a change
in selectivity for the substances being separated. Another possibility is to
introduce the reversed-phase separation after a DCCC step (Higuchi *et
al.* 1988).

Open-column chromatography with polymeric sorbents

The use of dextran supports, as found in Sephadex column
packings, has been current practice for several years. Sephadex LH-20
finds the most frequent application but the 'G' series of polymers is not
without interest (Table 3.4).

In recent work on the isolation of saponins, a new generation of
polymers has been exploited, particularly in Japan. Diaion HP-20
(Mitsubishi Chemical Industries, Tokyo), for example, is a highly porous
polymer which is widely used for the initial purification steps (Table 3.4).

Typically, the polymeric supports are washed with water after loading
the sample in order to elute monosaccharides, small charged molecules,
such as amino acids, and other highly water-soluble substances. Elution
with a methanol–water gradient (or with methanol alone) is then
commenced to obtain the saponin fractions. Other chromatographic
techniques are employed for the isolation of pure saponins.

Elution of HP-20 gels with acetone–water mixtures has also been
reported. For example, in the isolation of dubiosides D–F, bidesmosidic
glycosides of quillaic acid, from the tuber of *Thladiantha dubia* (Cucurbitaceae), methanol extracts were passed through a column of Diaion
CHP-20P and washed with water. The crude saponins were eluted with
40% acetone. Further separation involved silica gel chromatography
(ethyl acetate–methanol–water 6:2:1) and HPLC (Nagao *et al.* 1990).

For the isolation of fibrinolytic saponins from the seeds of *Luffa
cylindrica* (Cucurbitaceae), a water extract was chromatographed on an
Amberlite XAD-2 column eluted with methanol, followed by a second
XAD-2 column eluted with 40–70% methanol. The active principles were
obtained in the pure state after silica gel column chromatography

with chloroform–methanol–water (65:35:10, lower layer → 65:40:10) (Yoshikawa *et al.* 1991b).

Centrifugal thin-layer chromatography (CTLC)

The CTLC technique is a planar method related to preparative thin-layer chromatography (TLC) but without the need to scrape bands off the TLC plate (Hostettmann *et al.* 1980). CTLC relies on the action of a centrifugal force to accelerate mobile phase flow across a circular TLC plate. The plate, coated with a suitable sorbent (1, 2 or 4 mm thickness), is rotated at 800 r.p.m. by an electric motor, sample introduction occurs at the centre and eluent is pumped across the sorbent. Solvent elution produces concentric bands across the plate. These are spun off at the edges and collected for TLC analysis. Separations of 50–500 mg of a mixture on a 2 mm sorbent layer are possible.

A limited number of applications in the separation of saponins have been reported. For example, a combination of CTLC with chloroform–methanol–water (100:30:3) and column chromatography was used for the isolation of ginsenosides (Hostettmann *et al.* 1980). Saponins from *Vigna angularis* (Leguminosae) (Kitagawa *et al.* 1983e), from *Astragalus membranaceus* (Leguminosae) (Kitagawa *et al.* 1983f) and from *Cucurbita foetidissima* (Cucurbitaceae) (Dubois *et al.* 1988b) have also been obtained with chloroform–methanol–water mixtures on silica gel plates. Two protoprimulagenin A glycosides from *Eleutherococcus senticosus* roots (Araliaceae) were purified by CTLC (chloroform–methanol–water 65:35:7) after column chromatography on silica gel and gel filtration on Sephadex LH-20 (Segiet-Kujawa and Kaloga, 1991). For the isolation of cycloartane glycosides from *Passiflora quadrangularis* (Passifloraceae), the solvent system ethyl acetate–ethanol–water (8:2:1 or 16:3:2) was used at a flow rate of either 1 ml/min (Orsini *et al.* 1987) or 1.5 ml/min (Orsini and Verotta, 1985).

A Hitachi centrifugal liquid chromatograph, model CLC-5, was employed for the separation of acetyl-soyasaponins from soybeans. Chromatography was carried out on silica gel plates with the eluent chloroform–methanol–water (7:3:1 (lower phase) → 65:35:10 (lower phase)). A total of 1 g of the semi-purified saponin fraction was chromatographed on the circular plate (Kitagawa *et al.* 1988b; Taniyama *et al.* 1988a).

Flash chromatography

Flash chromatography (Still *et al.* 1978) is a preparative pressure liquid chromatography method which enables a considerable time saving

Table 3.3. *Open-column chromatography of saponins*

Adsorbent	Eluent	Reference
Silica gel	CHCl$_3$–MeOH–H$_2$O (65:35:10 (lower layer), 60:20:10 and other proportions and gradients)	Kondo and Shoji, 1975
		Tschesche et al., 1969e
	CHCl$_3$–EtOH–H$_2$O (7:10:3)	Jansakul et al., 1987
	CHCl$_3$–EtOAc–MeOH–H$_2$O (3:4:3:1.5 (bottom layer))	Iwamoto et al., 1987
	EtOAc–MeOH–H$_2$O (70:15:15)	Wagner and Hoffmann, 1967
	(12:2:1)	Mizutani et al., 1984
	EtOAc–EtOH–H$_2$O (8:2:1)	Shoji, 1981
	(7:2:1)	Kasai et al., 1987a
	(16:2:1)	Nakayama et al., 1986a
	(65:10:1)	Nakayama et al., 1986a
	n-BuOH–EtOAc–H$_2$O (1:1:1)	Bukharov and Karneeva, 1970a; Suga et al., 1978
	(4:1:2 (upper layer))	Shoji et al., 1971
	n-BuOH–EtOH–H$_2$O (5:1:4 (upper layer))	Asada et al., 1989a
	n-BuOH–HOAc–H$_2$O (4:1:5 (upper layer))	Chirva et al., 1968
	n-BuOH–EtOH–NH$_4$OH 25% (7:2:3)[a]	Sun et al., 1991
	CHCl$_3$–MeOH gradient	Waltho et al., 1986
	CH$_2$Cl$_2$–EtOAc–MeOH (10:3:4)	Wagner et al., 1987
	Me$_2$CO–n-PrOH–H$_2$O (35:35:5)	Borel et al., 1987
	CHCl$_3$–n-BuOH (0 → 20% n-BuOH)	Kizu et al., 1985c
	CHCl$_3$–n-BuOH–H$_2$O (25:10:0.5)	Kizu et al., 1985c

(cont.)

	Solvent system	Reference
Cellulose	$CHCl_3$–MeOH–H_2O (7:2:1 (lower phase))	Konishi and Shoji, 1979
Alumina (neutral)	MeOH	Chemli et al., 1987
Hydroxyapatite	$CHCl_3$–MeOH (5:1, 1:1, MeOH)	Tsukamoto et al., 1954
	CH_3CN–H_2O (85:15)	Shao et al., 1988
	(80:20)	Kasai et al., 1988a
C-8	MeOH–H_2O (65:35, 58:42)	Kasai et al., 1988a
		Higuchi et al., 1988
		Ding et al., 1986
C-18	MeOH–H_2O	Higuchi et al., 1984
	MeOH–H_2O (70:30, 55:45, 40:60)	Kusumoto et al., 1989
	(20:80)	Kitagawa et al., 1988b
	(1:5 → 3:1)	Wang et al., 1989c
	CH_3CN–H_2O (15:85, 25:75, 35:65, 50:50)	Kusumoto et al., 1989
	PrOH–H_2O (15:85)	
C-2	CH_3CN–i-PrOH–H_2O (32.3:4.2:63.5)	Kurihara et al., 1988
	→ MeOH–H_2O (10:3)	Reznicek et al., 1989b, 1990a
	MeOH–H_2O (60:40)	

[a] Dry column chromatography.

Table 3.4. *Open-column chromatography of saponins on polymeric supports*

Adsorbent	Solvent system	Reference
Amberlite XAD-2	$H_2O \rightarrow MeOH$	Zollo et al., 1987
Sephadex G-25 or G-50	$MeOH-H_2O$ (40:60 $\rightarrow$ 70:30)	Yoshikawa et al., 1991c; Kakuno et al., 1991a,b
	H_2O	Kimura et al., 1968a
		Uvarova et al., 1963
		Hiller et al., 1987b
		Brieskorn and Kilbinger, 1975
Sephadex LH-20	MeOH	Konoshima et al., 1980
		Domon and Hostettmann, 1983
		Pizza et al., 1987
		Orsini and Verotta, 1985
		Nagao et al., 1989b
Sephadex LH-60	$MeOH-H_2O$ (2:1)	Riccio et al., 1988
Toyopearl HW-40F	$MeOH-H_2O$ (50:50)	Oshima et al., 1984a
Fractogel TSK-HW40	MeOH	Carpani et al., 1989
Diaion HP-20	$H_2O \rightarrow MeOH$	Asada et al., 1989a
	$H_2O \rightarrow MeOH \rightarrow Me_2CO$	Kasai et al., 1988a
	$H_2O \rightarrow MeOH-H_2O \rightarrow MeOH \rightarrow CHCl_3$	Kasai et al., 1987a
	$H_2O \rightarrow MeOH-H_2O$ mixtures	Abe and Yamauchi, 1987
	$MeOH-H_2O$ mixtures $\rightarrow MeOH \rightarrow CHCl_3$	Nakayama et al., 1986a
	$H_2O \rightarrow MeOH-H_2O$ (50:50 $\rightarrow$ 85:15)	
	$\rightarrow MeOH \rightarrow Me_2CO$	Mizui et al., 1988
MCI gel (CHP-20P)	$H_2O \rightarrow MeOH \rightarrow CHCl_3$	Morita et al., 1985
	Me_2CO-H_2O mixtures	Iwamoto et al., 1987
	$H_2O \rightarrow Me_2CO-H_2O$ (40:60)	Nagao et al., 1990
Kogel B-G4600 (RP)	$MeOH-H_2O$ (10:90) $\rightarrow MeOH \rightarrow CHCl_3$	Morita et al., 1986a
Beads 60–80 mesh (Shoko-Tsusho Co.)	$H_2O \rightarrow MeOH-H_2O$ (10:90) $\rightarrow MeOH \rightarrow CHCl_3$	Namba et al., 1986

when compared with conventional open-column chromatography. Ordinary glass columns are used but eluent is driven through a sorbent by compressed air or nitrogen, reaching a maximum pressure of about 2 bar at the top of the column. The granulometry of the sorbent is somewhat reduced because solvent is being delivered under pressure; resolution is consequently higher.

Flash chromatography can be employed as a fast alternative to open-column chromatographic methods of preliminary fractionation; separations of 10 mg to 10 g of sample can be achieved in as little as 10 min. An example, taken from Table 3.5, involves the isolation of molluscicidal and fungicidal hederagenin, bayogenin and medicagenin glucosides from the roots of *Dolichos kilimandscharicus* (Leguminosae). A methanol extract (3.3 g) was fractionated on silica gel (63–200 μm granulometry) in a 60 × 4 cm column with the solvent system chloroform–methanol–water (50:10:1) at a flow rate of 15 ml/min. This was sufficient to remove contaminating material and obtain two saponin-rich fractions. The pure triterpene glycosides were obtained by a combination of DCCC and LPLC on C-8 supports (Marston *et al.* 1988a).

Although most applications have involved silica gel sorbents, there is an increasing trend towards RP materials (Table 3.5). RP flash chromatography enables the easy separation of saponins from other, more polar, components such as oligosaccharides.

Low-pressure liquid chromatography (LPLC)

LPLC is fast becoming one of the most popular methods for the isolation of pure saponins because of the speed of separation and ease of manipulation. LPLC employs columns containing sorbents with a particle size of 40–60 μm. High flow rates at pressures of up to 10 bar are possible and columns are mostly made of glass. Commercially available pre-packed columns (the 'Lobar' range from E. Merck, for example) in different sizes are ideal for the preparative chromatography of saponins in the 50–500 mg sample range. A high and uniform packing density guarantees a good separation efficiency.

It is relatively easy to transpose analytical HPLC conditions onto an LPLC separation, given that the chemistry of the sorbents is similar (Marston and Hostettmann, 1991b).

Most applications have been performed on RP sorbents, eluted with methanol–water mixtures (Table 3.6). It is generally only pre-purified samples which are injected in this case. A good illustration of LPLC is provided by the separation of molluscicidal oleanolic acid and gypsogenin glycosides from *Swartzia madagascariensis* (Leguminosae). The dried,

Table 3.5. *Applications of flash chromatography in the separation of saponins*

Plant	Support	Granulometry (µm)	Column dimensions (mm)	Solvent	Flow rate (ml/min)	References
Talinum tenuissimum	Silica gel	63–200	600 × 40	CHCl$_3$–MeOH–H$_2$O (65:40:5)	15	Gafner *et al*, 1985
Dolichos kilimandscharicus	Silica gel	63–200	600 × 30	CHCl$_3$–MeOH–H$_2$O (50:10:1)	15	Marston *et al*, 1988a
Solidago virgaurea	Silica gel	<80	400 × 40, 100 × 40	CHCl$_3$–MeOH–H$_2$O (15:10:2)	5	Hiller *et al*, 1987a
Cucurbita foetidissima	Silica gel	<63		EtOAc–MeOH–H$_2$O (63:25:9)		Dubois *et al*, 1988b
Guaiacum officinale	Silica gel			CHCl$_3$–MeOH (90:10 → 85:15)		Ahmad *et al*, 1986a
Medicago sativa	Silica gel	63–200		EtOAc–MeOH (4:1 → 2:1 → 1:1)	10	Levy *et al*, 1986
Bellis perennis	Silica gel	<63		CHCl$_3$–MeOH–H$_2$O (15:10:2, 17:10:2)		Schöpke *et al*, 1991
Schefflera octophylla	Silica gel	63–200		CHCl$_3$–MeOH–H$_2$O (70:30:3 → 60:35:8)		Sung *et al*, 1991
Cylicodiscus gabunensis	Silica gel			CHCl$_3$–MeOH (3:1)		Pambou Tchivounda *et al*, 1991
Dracaena mannii (steroid saponins)	Silica gel	63–200	600 × 30	EtOAc–MeOH–CHCl$_3$ (5:5:3) (8:13:1)		Okunji *et al*, 1991
Glycine max	C-18			H$_2$O → MeOH		Burrows *et al*, 1987 Curl *et al*, 1988b
Pisum sativum	C-18	40		H$_2$O, MeOH, MeOH–H$_2$O		Price and Fenwick, 1984
Cucurbita foetidissima	C-8	40–63		MeOH–H$_2$O (65:35)		Dubois *et al*, 1988b
Ziziphus jujuba	C-2			CH$_3$CN–PrOH–H$_2$O (32.3:4.2:63.5)		Kurihara *et al*, 1988

Table 3.6. LPLC of saponins

Plant	Support	Granulometry (μm)	Solvent	Reference
Swartzia simplex	C-8	40–63	MeOH–H$_2$O (70:30, 80:20) (55:45, 60:40)	Borel et al., 1987
S. madagascariensis	C-8	40–63	MeOH–H$_2$O (75:25, 55:45)	Borel and Hostettmann, 1987
Talinum tenuissimum	C-8	40–63	MeOH–H$_2$O (55:45)	Gafner et al., 1985
Dolichos kilimandscharicus	C-8	40–63	MeOH–H$_2$O (75:25, 40:10)	Marston et al., 1988a
Aphloia theiformis	C-8	40–63	MeOH–H$_2$O (80:20)	Gopalsamy et al., 1988
	Silica gel	40–63	CHCl$_3$–MeOH (8:2)	Gopalsamy et al., 1988
Sesamum alatum	C-8	40–63	MeOH–H$_2$O (40:60, 60:40)	Potterat et al., 1992
	DIOL	40–63	CHCl$_3$–MeOH	Potterat et al., 1992
Agave cantala (spirostanol glycosides)	C-8		MeOH–H$_2$O (9:1)	Pant et al., 1986
Dracaena mannii (spirostanol glycoside)	C-8	40–63	MeOH–H$_2$O (72:28)	Okunji et al., 1990, 1991
Allium giganteum (spirostanol glycosides)	C-18 (100 mm × 20 mm)	20	MeOH–H$_2$O (7:3)	Kawashima et al., 1991
Actinostemma lobatum	C-18		MeOH–H$_2$O (65:35)	Fujioka et al., 1987
	Silica gel		EtOAc–PrOH–H$_2$O (20:3:0.3)	Iwamoto et al., 1987
Xanthoceras sorbifolia	C-18		MeOH–H$_2$O (70:30)	Chen et al., 1985
Chenopodium quinoa	C-18		MeOH–H$_2$O (70:30)	Meyer et al., 1990
Gypsophila paniculata	C-18	40–63	MeOH–H$_2$O (30:70 → 100% MeOH)	Frechet et al., 1991
Polypodium glycyrrhiza	Silica gel	40–63	CHCl$_3$–MeOH–H$_2$O (6:3:1 (lower layer))	Kim et al., 1988
Bupleurum kunmingense	Silica gel	10–40	CHCl$_3$–MeOH–H$_2$O (30:20:1)	Luo et al., 1987

ground fruit pods were extracted with water and this extract was partitioned between *n*-butanol and water. After open-column chromatography of the organic phase, saponins **189** and **287** were separated on a Lobar LiChroprep C-8 column (40–63 μm; 27 × 2.5 cm) with methanol–water (75:25) as eluent (Borel and Hostettmann, 1987).

$$Rha\overset{3}{-}GlcA$$

189 R = -CH$_3$
287 R = -CHO

A column packed with DIOL (eluent chloroform–methanol, 85:15) was used in conjunction with C-8 columns for the isolation of alatoside C (**693**), a seco-ursane saponin from *Sesamum alatum* (Pedaliaceae) (Potterat *et al.*, 1992).

Joining LPLC columns in series permits an increase in loading capacity and/or separating power. This approach was used during the separation of dammarane glycosides from *Actinostemma lobata* (Cucurbitaceae), when three Lobar 27 × 2.5 cm columns were connected. The eluent also contained a small amount of water (ethyl acetate–*n*-propanol–water 20:3:0.3) (Iwamoto *et al.*, 1987).

Medium-pressure liquid chromatography (MPLC)

When relatively large amounts of pure saponins are required, MPLC is very useful (Table 3.7). Unlike commercially available LPLC equipment, gram quantities of sample can be loaded onto the columns, while separations are run at pressures of up to 40 bar. The granulometry of the support normally lies in the 25–40 μm range and separations are rapid, requiring considerably less time than open-column chromatography. A direct transposition of separation conditions from analytical HPLC to MPLC can be achieved on reversed-phase supports, thus facilitating the choice of solvent (Hostettmann *et al.* 1986).

As an example, molluscicidal saponins from *Cussonia spicata* (Araliaceae) were obtained in sufficient quantities for biological testing by MPLC on a C-8 sorbent with methanol–water (2:1) (Gunzinger *et al.* 1986). In fact, this method required just two steps (one on a silica gel support and the second on RP material) for isolation of saponins from a butanol extract of the stem bark (Table 3.7).

The isolation of saponins from *Phytolacca dodecandra* (Phytolaccaceae) berries is shown in Fig. 3.7. The final purification steps of saponins **481**, **213**, **218**, **255**, **399** and **482**, after rotation locular countercurrent chromatography (RLCC), were performed by MPLC on LiChroprep RP-8 (25–40 µm; column 46 × 2.6 cm) with methanol–water mixtures (Dorsaz and Hostettmann, 1986).

One MPLC technique uses axially compressed (Jobin-Yvon) columns and this method has been applied to the separation of saponins from several plants such as *Polygala chamaebuxus* (Polygalaceae) (Hamburger and Hostettmann, 1986), *Talinum tenuissimum* (Portulacaceae) (Gafner *et al.* 1985), *Calendula arvensis* (Asteraceae) (Chemli *et al.*, 1987), *Passiflora quadrangularis* (Passifloraceae) (Orsini and Verotta, 1985) and *Hedera helix* (Araliaceae) (Elias *et al.* 1991).

Steroidal saponins (mono-, di- and trisaccharides) have been isolated from the bulbs of *Camassia cusickii* (Liliaceae) by chromatography on CIG (Kusano Kagakukikai) pre-packed 100 × 22 mm glass columns filled with 20 µm C-18 material. The solvent systems methanol–water (19:1 and 9:1) were employed (Mimaki *et al.*, 1991).

High-performance liquid chromatography (HPLC)

Chromatography by HPLC is a powerful technique for obtaining multi-milligram quantities of saponins from mixtures of closely related compounds and, in this respect, is very frequently employed as a final purification step. Whereas MPLC makes use of larger particles (25–100 µm), semi-preparative HPLC sorbents lie in the 5–30 µm granulometry range and consequently permit a higher separation efficiency.

The vast majority of separations (Table 3.8) have been performed on octadecyl silica packings with refractive index detection. Methanol–water or acetonitrile–water mixtures generally constitute the eluent.

Semi-preparative HPLC was employed to separate oleanolic acid triglycoside **219** from its partial hydrolysis products **161** and **1312**. This was necessary in order to determine whether the galactose moiety in **219** (from *Phytolacca dodecandra*, Phytolaccaceae) was attached at position C-3 or C-4 of the glucose residue. Isolation of isomeric saponins **161** and **1312** was performed on a 7 µm LiChrosorb RP-8 column (250 × 16 mm) with acetonitrile–water (38:62) at a flow rate of 10 ml/min. Detection was at 206 nm and from 50 mg of mixture, 5 mg of **219**, 17 mg of **1312** and 4 mg of **161** were obtained (Décosterd *et al.* 1987).

Table 3.7. *Applications of MPLC in the separation of saponins*

Plant	Support	Solvent	Reference
Triterpene saponins			
Cussonia spicata	Silica gel	$CHCl_3$–MeOH–H_2O (6:4:1)	Gunzinger *et al.* 1986
	C-8	MeOH–H_2O (2:1)	Gunzinger *et al.* 1986
Calendula arvensis	C-8	MeOH–H_2O (65:35, 73:27)	Chemli *et al.* 1987
C. officinalis	Silica gel	$CHCl_3$–MeOH–H_2O (61:32:5)	Vidal-Ollivier *et al.* 1989b
	C-18	MeOH–H_2O (60:40, 80:20)	Vidal-Ollivier *et al.* 1989b
Polygala chamaebuxus	Silica gel	CH_2Cl_2–MeOH–H_2O (80:20:2)	Hamburger and Hostettmann, 1986
	C-8	MeOH–H_2O (55:45)	Hamburger and Hostettmann, 1986
Swartzia madagascariensis	C-8	MeOH–H_2O (65:35)	Borel and Hostettmann, 1987
Talinum tenuissimum	C-8	MeOH–H_2O (60:40)	Gafner *et al.* 1985
Sesbania sesban	C-8	MeOH–H_2O (55:45, 60:40)	Dorsaz *et al.* 1988
Tetrapleura tetraptera	C-8	MeOH–H_2O (70:30)	Maillard *et al.* 1989
Albizzia lucida	C-8	MeOH–H_2O (6:4 → 9:1)	Orsini *et al.* 1991
	C-18	MeOH–H_2O (7:3)	Orsini *et al.* 1991
Passiflora quadrangularis	C-18	MeOH–H_2O (17:3)	Orsini and Verotta, 1985
Hedera helix	C-18	MeOH–H_2O gradient	Elias *et al.* 1991
Primula veris	C-18	MeOH–H_2O (5:5 → 7:3)	Calis *et al.* 1992
	Silica gel	$CHCl_3$–MeOH–H_2O (61:32:7)	Calis *et al.* 1992
Steroid saponins			
Balanites aegyptiaca	Silica gel	$CHCl_3$–MeOH–H_2O (80:20:1 → 25:25:2 and 70:30:3)	Hosny *et al.* 1992

A large-scale separation of saikosaponins a, c and d from *Bupleurum falcatum* (Umbelliferae) roots has been achieved on axially compressed columns, dimensions 100 × 11 cm I.D. Preliminary purification of a methanol extract was carried out by solvent partition and chromatography on HP-20 polymer. The preparative HPLC column was packed with C-18 silica gel (20 μm particle size; 5 kg) and eluted at a flow rate of 210 ml/min with an aqueous acetonitrile step gradient. A charge of 10 g

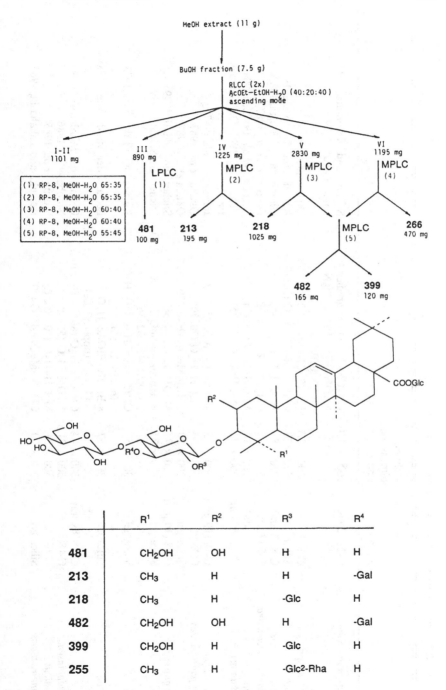

Fig. 3.7. Isolation of saponins from *Phytolacca dodecandra* berries.

Table 3.8. *Applications of HPLC in the separation of saponins*

Plant	Support	Column dimensions (mm)	Solvent	Reference
Bupleurum falcatum	C-18 (20 μm)	1000 × 110	CH_3CN–H_2O (step gradient)	Sakuma and Motomura, 1987
	C-18		$MeOH$–H_2O–$HOAc$–Et_3N (75 : 20 : 0.2 : 0.2)	Kimata et al., 1979
Bupleurum spp.	C-18	300 × 7.5	$MeOH$–H_2O (55 : 45)	Ding et al., 1986
B. falcatum	Silica gel	300 × 20	$CHCl_3$–$MeOH$–H_2O (30 : 10 : 1)	Kaizuka and Takahashi, 1983
Anagallis arvensis	C-8 (7 μm)	250 × 8	$MeOH$–H_2O (65 : 35)	Glombitza and Kurth, 1987a,b
Phytolacca dodecandra	C-8 (7 μm)	250 × 16	CH_3CN–H_2O (38 : 62)	Décosterd et al., 1987
Solidago gigantea	C-8 (7 μm)	250 × 16	$MeOH$–H_2O (59 : 41)	Reznicek et al., 1989b, 1990a
Calendula arvensis	C-18 (10 μm)	300 × 7.8	$MeOH$–H_2O (65 : 35)	Pizza et al., 1987; De Tommasi et al., 1991
Ardisia crispa	C-18 (10 μm)	500 × 15	$MeOH$–CH_3CN–H_2O (3 : 2 : 6)	Jansakul et al., 1987
Entada phaseoloides	C-18	300 × 50	$MeOH$–H_2O (70 : 30)	Okada et al., 1987
	Silica gel	300 × 20	$CHCl$–$MeOH$–H_2O (30 : 20 : 4)	Okada et al., 1987
Panax ginseng	C-18 (10 μm)	300 × 7.8	$MeOH$–H_2O (44 : 56)	Sticher and Soldati, 1979
	Silica gel	300 × 7.8	*n*-Heptane–*n*-BuOH–CH_3CN–H_2O (1000 : 446 : 132 : 36)	Sticher and Soldati, 1979
	Silica gel	300 × 57	$CHCl_3$–$MeOH$–$EtOH$–H_2O (62 : 16 : 16 : 66)	Kaizuka and Takahashi, 1983
P. trifolius	Silica gel	300 × 57	*n*-BuOH–$EtOAc$–H_2O (4 : 1 : 5) (upper phase)	Lee and der Marderosian, 1988
	Carbohydrate	300 × 7.8	CH_3CN–H_2O (86 : 14, 80 : 20)	Lee and der Marderosian, 1988
P. pseudoginseng	C-18	300 × 21.5	$MeOH$–H_2O (55 : 45)	Namba et al., 1986
P. japonicus	C-18	300 × 21.5	$MeOH$–H_2O (70 : 30 etc.)	Morita et al., 1985
Anemone rivularis	Silica gel	300 × 20	$CHCl_3$–$MeOH$–H_2O (30 : 17 : 2)	Kaizuka and Takahashi, 1983

	Silica gel	100 × 8	$CH_2Cl_2–EtOH–H_2O$ (9:5:1)	Mizutani et al., 1984
	C-18	600 × 7.5	$MeOH–H_2O$ (87:13)	Mizutani et al., 1984
Gleditsia japonica	C-18	250 × 10	$MeOH–H_2O$ (70:30)	Konoshima and Sawada, 1982
Androsace saxifragifolia	C-18		$MeOH–H_2O$ (70:30)	Pal and Mahato, 1987
	C-18	250 × 10	$MeOH–H_2O$ (60:40)	Waltho et al., 1986
Sanicula europaea	C-18 (7 μm)	250 × 8	$CH_3CN–HCOONH_4$	Kühner et al., 1985
Medicago sativa	C-18	500 × 10	$MeOH–H_2O$ (70:30 → 85:15)	Levy et al., 1986
Camellia japonica	C-18	250 × 10	$CH_3CN–H_2O–HOAc$ (37:62.9:0.1)	Nagata et al., 1985
Ziziphus jujuba	C-18 (5 μm)	250 × 7.8	$MeOH–i\text{-}PrOH–H_2O–HOAc$ (60.0:6.0:33.9:0.1)	Kurihara et al., 1988
Guaiacum officinale	C-18		$CH_3CN–H_2O$ (24:76 → 22:78)	Yoshikawa et al., 1991b
	C-18 (10 μm)		$MeOH–H_2O$ (70:30)	Ahmad et al., 1988a
Uncaria tomentosa	C-18 (10 μm)	300 × 7.8	$MeOH–H_2O$ (55:45, 65:35)	Cerri et al., 1988; Aquino et al., 1989b
Guettarda platypoda	C-18 (10 μm)	300 × 7.8	$MeOH–H_2O$ (13:12, 70:30)	Aquino et al., 1988, 1989a
Crossopteryx febrifuga	C-18 (10 μm)	250 × 22.7	$MeOH–H_2O$ (4:1, 7:3)	Babady-Bila et al., 1991
Sapindus delavayi	C-18		$MeOH–H_2O$ (60:40, 87:13)	Nakayama et al., 1986a
	Silica gel		$EtOAc–MeOH–H_2O$ (65:10:1)	Nakayama et al., 1986a
Chenopodium quinoa	C-18	300 × 21.5	$MeOH–H_2O$ (68:32, 73:27)	Mizui et al., 1988
Primula denticulata	C-18		$MeOH–H_2O$ (17:3, 4:1)	Ahmad et al., 1988b
Ilex cornuta	C-18	300 × 20	$MeOH–H_2O$ (65:35)	Qin et al., 1986
I. chinensis	C-18	100 × 22	$MeOH–H_2O$ (12:1)	Inada et al., 1987b
Patrinia scabiosaefolia	C-18 (10 μm)	300 × 7.8	$MeOH–H_2O$ (1:1)	Inada et al., 1988
Glycine max	C-18 (5 μm)	250 × 7.6	$CH_3CN–H_2O$ (4:6 → 1:1)	Kitagawa et al., 1988b
	C-18		$CH_3CN–n\text{-}PrOH–H_2O–HOAc$ (32.3:4.2:63.4:0.1)	Shiraiwa et al., 1991
Eleutherococcus senticosus (Acanthopanax senticosus)	C-18	300 × 21.5	$MeOH–H_2O$ (70:30)	Shao et al., 1988
Hedera helix	Silica gel	300 × 50	$CHCl_3–MeOH$ (17:3)	Hostettmann, 1980
	C-18	250 × 10	$CH_3CN–H_2O$ (32:68)	Wagner and Reger, 1986a
	C-18	300 × 7.8	$CH_3CN–H_2O$ (35:65)	Elias et al., 1991

(cont.)

Table 3.8. (*Cont.*)

Plant	Support	Column dimensions (mm)	Solvent	Reference
Stauntonia chinensis	C-18		CH_3CN-H_2O (34:66)	Wang et al., 1989c
Glycyrrhiza inflata	C-18		$CH_3CN-1\%$ HOAc	Kitagawa et al., 1989b
Aster tataricus	C-18	250×10	$MeOH-H_2O$ (65:35)	Nagao et al., 1989b
Periandra dulcis	C-18 (15 μm)	250×20	CH_3CN-H_2O (35:65) + 0.95% TFA	Ikeda et al., 1991
	Silica gel (15 μm)	250×20	$CH_2Cl_2-MeOH-HOAc-H_2O$ (80:10:8:2)	
Crocosmia crocosmiiflora	C-18 (7 μm)	300×10	Dioxane–H_2O (45:55)	Asada et al., 1989a
	Silica gel (5 μm)	300×10	$MeOH-H_2O$–dioxane (65:35:5) $CHCl_3-MeOH-H_2O$ (60:33:7)	Asada et al., 1989a
Gypsophila paniculata	C-18 (5 μm)	250×8	$CH_3CN-0.1\%$ H_3PO_4 (72:28 → 65:35)	Frechet et al., 1991
Holothuria pervicax (sea cucumber)	C-18		$CH_3CN-MeOH-H_2O$ (1:1:1)	Kitagawa et al., 1989c
Holothuria forskalii (sea cucumber)	C-18 (10 μm)	300×19	$MeOH-H_2O$ (7:3)	Rodriguez et al., 1991
Steroid saponins				
Balanites aegyptiaca	C-18	300×21.5	$MeOH-H_2O$ (67:33) + 0.05% TFA	Kamel et al., 1991
Allium vineale	C-18 (5 μm)	250×10	$MeOH-H_2O$ (93:7, 95:5, 90:10, 92:8)	Chen and Snyder, 1989
Rhodea japonica	C-18	250×4.6	$MeOH-H_2O$ (90:10)	Miyahara et al., 1983
Smilax sieboldii	C-18 (5 μm)	250×10	CH_3CN-H_2O–2-methoxyethanol (9:11:2)	Kubo et al., 1992
Halityle regularis (starfish)	C-18	300×7.8	$MeOH-H_2O$ (45:55)	Riccio et al., 1985a
Pisaster ochracens (starfish)	C-18	300×7.8	$MeOH-H_2O$ (9:11)	Zollo et al., 1989
Glycoalkaloids				
Potato glycoalkaloids	NH_2	250×9.4	$CH_3CN-THF-H_2O$ (25:55:20)	Bushway and Storch, 1982
Fritillaria persica	C-18	250×10	$MeOH-H_2O$–2-methoxyethanol (12:2:1, 14:6:1)	Ori et al., 1992

219 R = -Glc^4-Glc
|^3
Gal
1312 R = -Glc^4-Glc

161 R = -Glc^3-Gal

was sufficient to give 400 mg of saikosaponin c, 1200 mg of saikosaponin a and 1600 mg of saikosaponin d (Sakuma and Motomura, 1987).

Ginsenosides have been isolated from *Panax trifolius* (Araliaceae) by a two-step procedure, involving chromatography on a Waters Prep 500 system (radially compressed columns) with three silica gel cartridges (300 × 57 mm) arranged in series. The eluent was the upper phase of *n*-butanol–ethyl acetate–water (4:1:5) and charges of 4 g were injected. Semi-preparative HPLC on a carbohydrate column (Waters, 300 × 7.8 mm) with acetonitrile–water (86:14 or 80:20) at a flow rate of 2 ml/min was employed for final purification (Lee and der Marderosian, 1988).

The preparative separation of ginsenosides (notably Rb$_1$, Rb$_2$, Rc, Rd, Re and Rg$_1$) can conveniently be performed on ODS which has an optimal pore size of 55 nm and a narrow particle size distribution (20 μm). For example, 10 g of a methanol extract of *Panax ginseng* root (after preliminary sample clean-up on a cartridge) injected onto a 500 × 50 mm column and eluted with an acetonitrile–water step gradient gave directly 4 mg ginsenoside Rg$_1$, 7 mg ginsenoside Re and 7–10 mg of ginsenoside Rb$_1$. Separation was complete within 8 h at a flow rate of 35 ml/min and detection was at 203 nm (Kanazawa *et al.* 1990b). The only drawbacks, therefore, to the utility of this method are the high consumption of acetonitrile and the separation time. The same packing material has been applied to the separation of malonyl ginsenosides from white ginseng. Gradient elution was necessary with acetonitrile–50 mM KH$_2$PO$_4$ mixtures (flow rates 15-40 ml/min). Inorganic phosphate was removed from the separated fractions by a solid-phase extraction technique (Sep-Pak^R C-18 cartridges) (Kanazawa *et al.* 1991).

For the successful separation of rhodeasapogenin glycosides it was necessary to keep the column (TSK-GEL LS-410, C-18) at 0 °C. The sample was loaded as a 5% solution in dimethyl formamide and the mobile phase (methanol–water 90:10) eluted at 0.5 ml/min (Miyahara *et al.* 1983).

Countercurrent chromatography

Liquid–liquid partition methods have proved ideal for application to the field of saponins. Very polar saponins lend themselves especially well to countercurrent chromatographic separation, especially as there is no loss of material by irreversible adsorption to packing materials. This aspect has been of especial use for the direct fractionation of crude extracts.

Droplet countercurrent chromatography (DCCC)

DCCC relies on the continuous passage of droplets of a mobile phase through an immiscible liquid stationary phase contained in a large number of vertical glass tubes. The solute undergoes a continuous partition between the two phases. Depending on whether the mobile phase is introduced at the top or at the bottom of these tubes, chromatography is in the 'descending' or 'ascending' mode, respectively. The separation of closely related saponins by DCCC and even the isolation of pure products has been possible (Hostettmann *et al.* 1984). In fact, certain separations which have not been possible by liquid–solid chromatography have been achieved by this technique. DCCC was capable of separating isomeric saponins differing only in the positions of substitution of acetate groups on the sugar residues (from *Platycodon grandiflorum*, Campanulaceae) (Ishii *et al.* 1984).

A baseline separation of saikosaponins a (**547**), d (**548**) and c (**544**) (Appendix 2) from *Bupleurum falcatum* (Umbelliferae) was achieved with chloroform– methanol–water–benzene–ethyl acetate (45:60:40:2:3) as solvent in the acending mode. As expected with the more polar layer as the mobile phase, the trisaccharide **544** eluted first and was followed by the disaccharides **547** and **548**. The separation of the latter two saponins, which differ only in configuration at C-16, by column chromatography is very difficult (only one spot is observed on TLC). Even separation by HPLC is not easy (Otsuka *et al.*, 1978b).

A number of solvent systems have been employed for the DCCC separation of saponins (Table 3.9; see also Hostettmann *et al.* 1986) and, of these, the system chloroform–methanol–water (7:13:8) has been involved in the greatest number of applications. Chloroform–methanol–water systems can be used either in the ascending mode for very polar

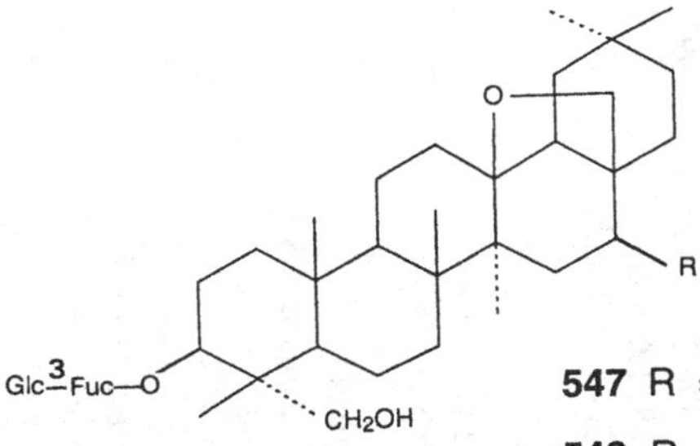

547 R = β-OH

548 R = α-OH

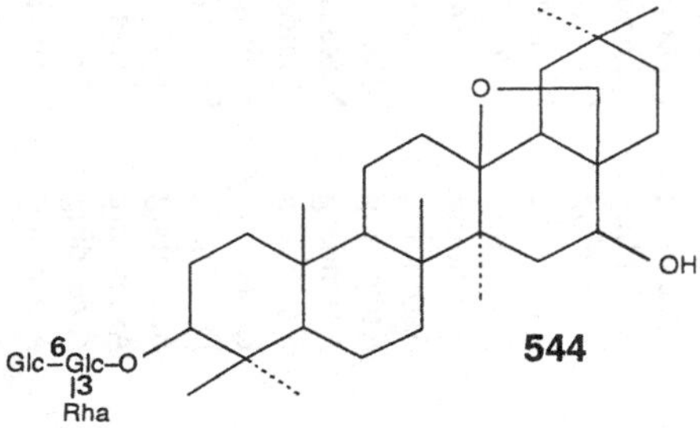

Glc³⁻Fuc—O · · CH₂OH

Glc⁶⁻Glc—O |3 Rha

544

saponins or in the descending mode for saponins possessing one or two sugars and few free hydroxyl groups.

A molluscicidal methanolic extract of *Cornus florida* (Cornaceae) bark was fractionated by Sephadex LH-20 gel filtration, but efficient isolation of the active saponins was not possible by subsequent open-column chromatography or TLC. However, DCCC with chloroform–methanol–water (7:13:8) (lower layer as mobile phase) allowed the separation of the sarsasapogenin glycosides **916**–**918**. Tubes of length 40 cm and I.D. 2 mm were used and injection of 45 mg of the active fraction gave 1 mg of **916**, 8 mg of **917** and 28 mg of **918** (Fig. 3.8) (Hostettmann *et al.* 1978).

A large-scale DCCC procedure for preliminary purification, using 18 columns (30 cm × 10 mm I.D.) with *n*-butanol-saturated water as the stationary phase and water-saturated *n*-butanol as the mobile phase, has been applied to the isolation of a cholest-9(11)-en-23-one glycoside sulphate from the starfish *Luidia maculata* (Asteroidea) (Komori *et al.* 1983).

In some cases, two (or more) DCCC separations are run to obtain the pure saponins, e.g. for the molluscicidal saponins of *Hedera helix* (Araliaceae) (Hostettmann *et al.*, 1979) and for the saponins from *Cussonia barteri* (Araliaceae) (Dubois *et al.* 1986).

Table 3.9. *DCCC of saponins*

Organism	Solvent system	Mode	Reference
Triterpene saponins			
Hedera helix	$CHCl_3$–$MeOH$–H_2O (7:13:8)	Descending	Hostettmann *et al.*, 1979
Dolichos kilimandscharicus	$CHCl_3$–$MeOH$–H_2O (7:13:8)	Descending	Marston *et al.*, 1988a
Phaseolus vulgaris	$CHCl_3$–$MeOH$–H_2O (7:13:8)	Descending	Jain *et al.*, 1988
Phytolacca thyrsiflora	$CHCl_3$–$MeOH$–H_2O (7:13:8)	Descending	Haraguchi *et al.*, 1988
Tetrapleura tetraptera	$CHCl_3$–$MeOH$–H_2O (7:13:8)	Ascending	Maillard *et al.*, 1989
Calendula arvensis	$CHCl_3$–$MeOH$–H_2O (7:13:8)	Ascending	Pizza *et al.*, 1987; De Tommasi *et al.*, 1991
Guettarda platypoda	$CHCl_3$–$MeOH$–H_2O (7:13:8)	Ascending	Aquino *et al.*, 1988, 1989a,b
Anagallis arvensis	$CHCl_3$–$MeOH$–H_2O (7:13:8)	Ascending	Glombitza and Kurth, 1987a,b
Xanthoceras sorbifolia	$CHCl_3$–$MeOH$–H_2O (7:13:8)	Ascending	Chen *et al.*, 1985
Bupleurum rotundifolium	$CHCl_3$–$MeOH$–H_2O (7:13:8)	Ascending	Akai *et al.*, 1985b
Celmisia petriei	$CHCl_3$–$MeOH$–H_2O (7:13:8)	Ascending	Rowan and Newman, 1984
Tetrapanax papyriferum	$CHCl_3$–$MeOH$–H_2O (7:13:8)	Ascending	Takabe *et al.*, 1985
Hovenia dulcis	$CHCl_3$–$MeOH$–H_2O (7:13:8)	Ascending	Inoue *et al.*, 1978
Ilex rotunda	$CHCl_3$–$MeOH$–H_2O (7:13:8)	Ascending	Nakatani *et al.*, 1989
Lonicera nigra	$CHCl_3$–$MeOH$–H_2O (25:45:30)	Descending	Domon and Hostettmann, 1983
Fagus sylvatica	$CHCl_3$–$MeOH$–H_2O (13:7:8)	Descending	Romussi *et al.*, 1987
Sanguisorba minor	$CHCl_3$–$MeOH$–H_2O (9:13:8)	Descending	Reher *et al.*, 1991
Passiflora quadrangularis	$CHCl_3$–$MeOH$–H_2O (13:7:2)	Ascending	Orsini *et al.*, 1987
Desfontainia spinosa	$CHCl_3$–$MeOH$–H_2O (5:5:3)	Ascending	Houghton and Lian, 1986a,b
Chenopodium quinoa	$CHCl_3$–$MeOH$–H_2O (5:5:3)	Ascending	Meyer *et al.*, 1990
Hovenia dulcis	$CHCl_3$–$MeOH$–H_2O (5:6:4)	Ascending	Inoue *et al.*, 1978
Chenopodium quinoa	$CHCl_3$–$MeOH$–H_2O (5:6:4)	Ascending	Meyer *et al.*, 1990
Bencomia caudata	$CHCl_3$–$MeOH$–n-$PrOH$–H_2O (5:6:1:4)	Ascending	Reher and Budesinsky, 1992
Albizzia anthelmintica	$CHCl_3$–$MeOH$–n-$PrOH$–H_2O (5:6:1:4)	Ascending	Carpani *et al.*, 1989

Passiflora quadrangularis	$CHCl_3$–MeOH–n-PrOH–H_2O (5:6:1:4)	Ascending	Orsini et al., 1986
	$CHCl_3$–MeOH–n-PrOH–H_2O (45:60:6:40)	Ascending	Orsini and Verotta, 1985
Platycodon grandiflorum	$CHCl_3$–MeOH–n-PrOH–H_2O (5:6:1.2:4)	Descending	Ishii et al., 1984
Cussonia barteri	$CHCl_3$–MeOH–n-PrOH–H_2O (9:12:1:8)	Ascending	Dubois et al., 1986
Colubrina asiatica	$CHCl_3$–MeOH–n-PrOH–H_2O (9:12:1:8)		Wagner et al., 1983
Dodonaea viscosa	$CHCl_3$–MeOH–n-PrOH–H_2O (9:12:2:8)	Ascending	Wagner et al., 1987
Sarcopoterium spinosum	$CHCl_3$–MeOH–n-PrOH–H_2O (375:400:75:30)	Ascending	Reher et al., 1991
Aphloia theiformis	$CHCl_3$–MeOH–i-PrOH–H_2O (5:6:1:4)	Descending	Gopalsamy et al., 1988
Gundelia tournefortii	$CHCl_3$–MeOH–H_2O–n-PrOH–EtOH (9:6:8:1:8)	Descending	Wagner et al., 1984b
Quillaja saponaria	$CHCl_3$–MeOH–H_2O–HOAc (5:6:1:4) (9:10:8:1:2)	Descending	Wagner et al., 1984b
Bupleurum falcatum	$CHCl_3$–MeOH–H_2O–C_6H_6–EtOAc (45:60:40:2:3)	Descending	Higuchi et al., 1988
		Ascending	Shimizu et al., 1983
Sargentodoxa cuneata	$CHCl_3$–MeOH–H_2O–C_6H_6–EtOAc (45:60:40:2:3)	Ascending	Rücker et al., 1991
Lonicera nigra	n-BuOH–Me_2CO–H_2O (35:10:55)	Descending	Domon and Hostettmann, 1983
Cussonia barteri	n-BuOH–Me_2CO–H_2O (35:10:55)	Ascending	Dubois et al., 1986
Holothuria forskalii (Sea cucumber)	$CHCl_3$–MeOH–H_2O (7:13:8)	Ascending	Rodriguez et al., 1991
Steroid saponins			
Cornus florida	$CHCl_3$–MeOH–H_2O (7:13:8)	Descending	Hostettmann et al., 1978
Allium vineale	$CHCl_3$–MeOH–H_2O (7:13:8)	Descending	Chen and Snyder, 1989
Balanites aegyptiaca	CH_2Cl_2–MeOH–H_2O (8:13:7)	Ascending	Liu and Nakanishi, 1982
			Hosny et al., 1992
Trigonella foenum-graecum	$CHCl_3$–MeOH–H_2O (65:35:10)	Ascending	Liu and Nakanishi, 1982
Agave cantala	$CHCl_3$–MeOH–H_2O (7:13:8)	Descending	Gupta et al., 1984, 1985, 1986
	$CHCl_3$–MeOH–H_2O (7:13:8)	Descending	Jain, 1987a

(cont.)

Table 3.9. (*Cont.*)

Organism	Solvent system	Mode	Reference
Dracaena mannii	$CHCl_3$–MeOH–H_2O (7:13:8)	Descending	Okunji et al., 1990
Polypodium glycyrrhiza	$CHCl_3$–MeOH–*i*-PrOH–H_2O (5:6:1:4)	Ascending	Kim et al., 1988
Oreaster reticulatus (starfish)	*n*-BuOH–Me_2CO–H_2O (3:1:5)	Ascending	Segura de Correa et al., 1985
Luidia maculata (starfish)	*n*-BuOH–H_2O	Ascending	Komori et al., 1983
Asterias amurensis (starfish)	*n*-BuOH–Me_2CO–H_2O (3:1:5)	Descending	Riccio et al., 1988
	$CHCl_3$–MeOH–H_2O (40:42:18)	Ascending	Riccio et al., 1988
	$CHCl_3$–MeOH–H_2O (7:13:8)	Ascending	Riccio et al., 1988
Pisaster ochracens (starfish)	*n*-BuOH–Me_2CO–H_2O (3:1:5)	Descending/ ascending	Zollo et al., 1989
Steroid alkaloid saponins			
Fritillaria thunbergii	$CHCl_3$–MeOH–1% NH_3 (7:12:8)	Ascending	Kitajima et al., 1982
Solanum incanum	$CHCl_3$–MeOH–PrOH–H_2O–NH_4OH (35:65:40:5:1)	Ascending	Fukuhara and Kubo, 1991

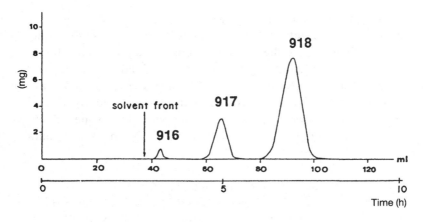

916 R = -Gal

917 R = -Gal²-Xyl

918 R = -Gal²-Glc

Fig. 3.8. DCCC separation of *Cornus florida* saponins.

Rotation locular countercurrent chromatography (RLCC)

RLCC has few applications in the separation of saponins (triterpene saponins: Domon and Hostettmann, 1984; Dorsaz and Hostettmann, 1986; steroid saponins: *i*-butanol–*n*-propanol–water (5:3:10), Chen and Snyder, 1989). In the RLCC apparatus, an assembly of 16 glass columns (each composed of 37 compartments or loculi) is arranged around a rotational axis. The speed of rotation of the columns can be varied but the optimum is around 70 r.p.m. The angle of inclination can also be adjusted, from 0° to 90° from the horizontal (Hostettmann *et al.* 1986). As in DCCC, one phase of a two-phase solvent system is introduced into the columns as the stationary phase and the other phase serves as the mobile phase. Ascending or descending modes of operation are also possible.

Solasonine (**1147**) and solamargine (**1148**) (Table 2.10, p. 99) have been obtained from the fruits of *Solanum incanum* (Solanaceae) by a strategy involving liquid–liquid partition techniques alone. Following simple solvent partition of a methanol extract, the remaining aqueous layer was loaded onto a rotation locular countercurrent chromatography instrument, with water as the stationary phase. Step-gradient elution ((i) water-saturated *n*-hexane, (ii) water-saturated ethyl acetate, (iii) ethyl acetate–butanol–water 4:1:1, 2:1:1) in the ascending mode gave a glycoalkaloid-rich fraction which was chromatographed by DCCC (chloroform–methanol–water–propanol–ammonium hydroxide 35:65: 40:5:1, ascending mode) to obtain the two pure substances (Fukuhara and Kubo, 1991).

Centrifugal partition chromatography (CPC)

The recently introduced technique of CPC holds great promise because of its speed and versatility (Marston *et al.* 1988b, 1990). CPC relies on a centrifugal field (produced by rotation at 800–2000 r.p.m. or faster) rather than a gravitational field for retention of the stationary phase. The principle of the method involves a continuous process of non-equilibrium partition of solute between two immiscible phases contained in rotating coils or cartridges. Instruments based on rotating coils can involve either planetary or non-planetary motion about a central axis. One of these, the high speed countercurrent chromatograph (HSCCC) consists of a Teflon tube of 1.6 or 2.6 mm I.D. wrapped as a coil around a spool. One, two or three spools constitute the heart of the instrument. In the case of cartridge instruments, the cartridges are located at the circumference of a centriguge rotor, with their longitudinal axes parallel to the direction of the centrifugal force. The number and volume of the cartridges can be varied, depending on the application to which the instrument is put. Compared with DCCC and RLCC, in which separations may take 2 days or longer, CPC can produce the same results in a matter of hours. Instruments based on rotating coils or cartridges have capacities up to the gram scale. A multilayer coil planet instrument has been used, for example, for the preliminary purification of cycloartane glycosides from *Abrus fruticulosus* (Leguminosae) (Fullas *et al.* 1990). Molluscicidal triterpene glycosides from *Hedera helix* (Araliaceae) have been separated on a different instrument, the Sanki LLN chromatograph (six cartridges; total volume 125 ml). A methanol extract of the fruit was partitioned between *n*-butanol and water. The butanol fraction was injected directly into the instrument in 100 mg amounts, using the lower layer of the solvent system chloroform–methanol–water (7:13:8) as mobile phase. Separation of the *Hedera helix* saponins **346**, **350**, **368** and **377** was complete within 2 h (Fig. 3.9). Changing the elution mode to upper phase as mobile phase after fraction 30 allowed the elution of two unknown more-polar saponins, together with a large amount of polysaccharide (Marston *et al.* 1988b).

The two main saponins asiaticoside and madecassoside from *Centella asiatica* (Umbelliferae) have been separated with the aid of an Ito multi-layer coil separator-extractor (P.C. Inc.) equipped with a 66 m × 2.6 mm I.D. column (350 ml capacity), turning at 800 r.p.m. A sample of 400 mg could be resolved with the solvent system chloroform–methanol–2-butanol–water (7:6:3:4; mobile phase was lower phase). Detection was by means of on-line TLC (Diallo *et al.* 1991). The same instrument was employed during the isolation of a triterpene disaccharide from *Sesamum alatum* (Pedaliaceae). The lower phase of the solvent chloroform–methanol–*i*-propanol–water (5:6:1:4) was chosen as the mobile phase and a charge of 1.25 g was injected (Potterat *et al.* 1992).

Other chromatographic techniques

Certain saponins have been isolated by different techniques from those already mentioned. Although not widely employed, one or two will briefly be presented here. Further details of these methods exist elsewhere (Hostettmann *et al.* 1986).

Vacuum liquid chromatography (VLC). This technique involves the use of reduced pressure to increase the flow rate of a mobile phase through a short bed of solvent. The chromatography column is packed with silica gel (generally 10–40 μm TLC grade) and the sample is eluted with appropriate solvent mixtures, starting with solvent of low polarity and gradually increasing the elution strength. Application of a vacuum to the bottom of the column pulls the solvent through the sorbent. This technique has been applied to the isolation of hederagenin saponins from *Schefflera impressa* (Araliaceae) (Srivastava and Jain, 1989). Three saponins were obtained from the roots of *Collinsonia canadensis* (Labiatae) by a procedure involving VLC on Merck 60H silica gel as the only chromatographic tool (Joshi *et al.* 1992).

Overpressured layer chromatography (OPLC). This is a modified planar chromatographic technique in which the vapour phase is eliminated by covering the sorbent layer with an elastic membrane under variable external pressure. The mobile phase is forced through the sorbent layer with a pump. The separation of the intensely sweet polypodoside A (**1070**) from other polar constituents (from *Polypodium glycyrrhiza* rhizomes) in a pre-purified sample was achieved by OPLC. Introduction of a 55 mg sample gave 16 mg of pure polypodoside A (Fullas *et al.* 1989).

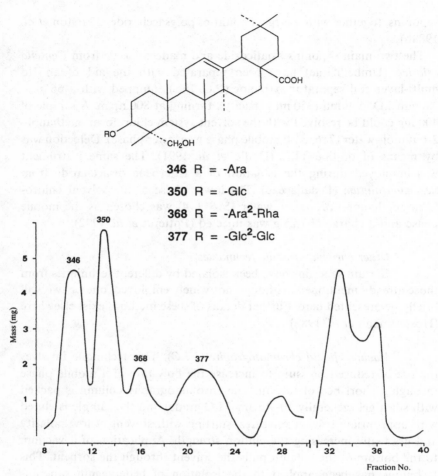

346 R = -Ara

350 R = -Glc

368 R = -Ara^2-Rha

377 R = -Glc^2-Glc

Fig. 3.9. CPC of a methanol extract of *Hedera helix* berries after *n*-butanol–water partition. Solvent system: acetonitrile–methanol–water (7:13:8), descending mode; flow rate: 1.5 ml/min; rotational speed: 700 r.p.m.

Combination of methods

It is very rare that a single chromatographic step is sufficient to isolate a pure saponin from an extract. As a general rule, several preparative techniques are required in series to obtain the necessary product. A combination of classical techniques (such as open-column chromatography) and modern high-resolution methods (such as HPLC) has proved suitable for the separation of many saponins, but the newer techniques of liquid–liquid chromatography and low- and medium-pressure systems are gaining in importance.

As the first of several examples, the glycosides **522** and **523** of

$3\beta,23,27,29$-tetrahydroxyolean-12-en-28-oic acid were obtained from the aerial parts of *Polygala chamaebuxus* (Polygalaceae) by a combination of MPLC on silica gel and RP material, LPLC and centrifugal TLC (Hamburger and Hostettmann, 1986).

522 $R^1 = $-H, $R^2 = $-Ara

523 $R^1 = $-Glc, $R^2 = $-H

Similarly, the isolation of five triterpene saponins from *Swartzia madagascariensis* (Leguminosae) required open-column chromatography, LPLC and MPLC (Borel and Hostettmann, 1987).

Steroid saponins from *Allium vineale* (Liliaceae) were obtained from methanol extracts of the bulbs by a combination of RLCC (*i*-butanol–*n*-propanol–water (5:3:10); ascending mode; 4 g injected per run), silica gel chromatography (chloroform–methanol–water 78:20:2, 70:30:10, 65:35:10) and semi-preparative HPLC (C-18 5 µm, 250 × 10 mm; methanol–water 93:7, 95:5; 0.5 ml/min and 1 ml/min; detection 210 nm) or by a combination of DCCC and RP HPLC (Chen and Snyder, 1989).

CPC has been used in conjunction with flash chromatography and OPLC for the isolation of triterpene glycosides from *Abrus fruticulosus* (Leguminosae). A multilayer coil instrument (solvent chloroform–methanol–water 7:13:8, lower phase as mobile phase) provided initial purification, while flash chromatography and OPLC were effective for obtaining the pure substances (Fullas *et al.* 1990).

The straightforward combination of flash chromatography on unmodified silica gel with either flash chromatography or open-column chromatography on RP material can sometimes be sufficient for the purification of saponins (Schöpke *et al.* 1991).

A strategy employed frequently in Japan is to pass extracts (after preliminary partition) over highly porous polymers and to follow this step by further fractionation of the crude saponin mixtures. This approach was used in the isolation of 3β-hydroxyolean-12-en-28,29-dioic acid glycosides from *Nothopanax delavayi* (Araliaceae). A methanol extract of the leaves and stems was partitioned between hexane and water. The aqueous layer was chromatographed on a Diaion HP-20 column and eluted with water, 10% methanol, 50% methanol, 80% methanol, methanol and chloroform. The glycosides were obtained by subsequent column chromatography of the 80% methanol eluate on silica gel with ethyl acetate–ethanol–water (7:2:1) (Kasai *et al.* 1987a). For the isolation of triterpene and nor-triterpene saponins from *Acanthopanax senticosus* (Araliaceae), the procedure began with a fractionation of the methanol extract of the leaves on Diaion HP-20 polymer. The fraction eluted with methanol was chromatographed on silica gel (chloroform–methanol–water 30:10:1) and all the resulting fractions were subjected to column chromatography on LiChroprep RP-8. Final purification was achieved by HPLC on TSK-GEL ODS-120T (300 × 21 mm; methanol–water 70:30; 6 ml/min; RI detection) or chromatography on a hydroxyapatite column (aceto-nitrile–water 85:15) (Shao *et al.* 1988). A similar scheme has been employed to obtain saponins from the bran of quinoa (*Chenopodium quinoa*, Chenopodiaceae) (Mizui *et al.* 1988, 1990).

A very successful strategy employed for the isolation of numerous steroid glycosides from starfish involves a preliminary Amberlite XAD-2 step, followed by Sephadex LH-60 fractionation and DCCC (often with *n*-butanol–acetic acid–water 3:1:5 in the ascending mode). Final purification is achieved by HPLC on C-18 semi-preparative columns with methanol–water mixtures (Riccio *et al.* 1988).

A related procedure, employing a combination of Sephadex LH-20 (methanol), DCCC (chloroform–methanol–water 7:13:8) and HPLC (C-18, methanol–water 65:35), has been applied to the separation of oleanolic acid glycosides from the aerial parts of *Calendula arvensis* (Asteraceae) (De Tommasi *et al.* 1991).

However, there are occasions when mixtures fail to be resolved, as is the case for *Dodonea viscosa* (Sapindaceae). Two saponin esters have been isolated from a methanol extract of the seeds but they have not yet been separated by existing chromatographic techniques (Wagner *et al.* 1987).

4

Structure determination

There are several basic problems to be solved in the structure elucidation of saponins:

- the structure of the genuine aglycone
- the composition and sequence of the component monosaccharides in the carbohydrate moiety
- how the monosaccharide units are linked to one another
- the anomeric configuration of each glycosidically linked monosaccharide unit
- the location of the carbohydrate moiety on the aglycone.

The necessary approach is to apply a combination of methods in order to arrive at a final conclusion for the structure. A simplified scheme for the structure elucidation of saponins, with a selection of the available techniques, is shown in Fig. 4.1. This is a stepwise process, in which the saponin is gradually broken down into smaller fragments which themselves are analysed spectroscopically. By a judicious handling of the data from the fragments, an idea of the composition of the saponin can gradually be built up.

With advances in the modern techniques available, preparative separations of increasing numbers of saponins are becoming conceivable. At the same time, however, the quantities of pure saponins isolated are often small (sometimes only several milligrams) and there is always a need for highly sensitive, high-resolution and, if possible, non-degradative methods in order to aid the structure determination of a saponin. Recourse to innovations in NMR spectroscopy and mass spectrometry (MS) has been essential for further advances in the investigations of complex saponins. Thus, FAB-MS gives information about the molecular weight and, in many cases, the sugar sequence, while 1-D and 2-D NMR techniques

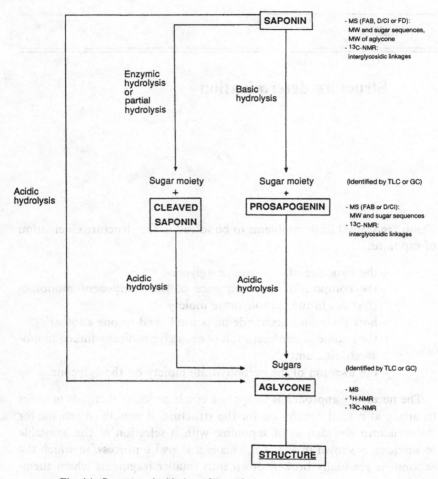

Fig. 4.1. Structure elucidation of saponins.

permit the localization of sugar linkages and contribute to the structure elucidation of the aglycone.

A number of the cleavage methods and the subsequent spectroscopic techniques required for the analyses of the fragments are described below.

Structure determination and chemical studies of the ginsenosides have been thoroughly discussed in a review by Tanaka and Kasai (1984) and a summary of the structure elucidation of steroid saponins has appeared (Mahato *et al.* 1982).

4.1   Cleavage reactions

Saponins are glycosides in which the hemiacetal hydroxyl groups of saccharides in their cyclic pyranose or furanose forms build acetals with a triterpene or steroid residue. The ether linkage between the hemiacetal hydroxyl and the triterpene or steroid is known as a glycosidic linkage. The monosaccharide constituents of the oligosaccharides are also bound by ether linkages (interglycosidic bonds).

On complete hydrolysis of a glycoside, the glycoside linkage is cleaved to liberate the component monosaccharides and the non-carbohydrate moiety (the aglycone or genin). The non-carbohydrate portion from the hydrolysis of saponins is termed a *sapogenol* or *sapogenin*. All known saponins are *O*-glycosides, with ether or ester linkages.

Numerous chemical reactions and methods have been employed for breaking down saponins into smaller units for more ready analysis (see, for example, Kitagawa, 1981).

4.1.1   Acidic hydrolysis

Acidic hydrolysis is carried out by refluxing the saponin in acid for a fixed length of time, typically with 2–4 M hydrochloric acid for 4 h. The aqueous solution remaining after hydrolysis is extracted with diethyl ether, chloroform or ethyl acetate to obtain the aglycone. Extraction of the sugars from the aqueous layer is performed with pyridine, after neutralizing the solution (with alkali or basic ion exchange resin) (Tschesche and Forstmann, 1957; Sandberg and Michel, 1962) and evaporation to dryness. The saponins are completely cleaved into their constituents by this method so information is obtained as to the identity of the aglycone and the number and nature of monosaccharides present. If a prosapogenin (obtained after cleavage of an ester linkage by basic hydrolysis) is acid hydrolysed, the nature of the sugar chains which are ether-linked to the aglycone can be established. An aqueous reaction medium can be replaced by alcohol or dioxane.

In addition to hydrochloric acid, sulphuric acid is also employed for the hydrolysis of saponins. With sulphuric acid there is less chance of degradation or rearrangement of the molecule (Kubota *et al.* 1969) but cleavage of ether linkages is not as efficient (Yamauchi, 1959). A convenient method of obtaining gypsogenic acid from *Dianthus* saponins, for example, involved hydrolysis with 1 M sulphuric acid in dioxane (Oshima *et al.* 1984a). A comparative study of hydrolytic conditions with hydrochloric acid and sulphuric acid in water and water–ethanol has shown that the best recoveries of saccharides are achieved by heating the saponin for 2 h with 5% sulphuric acid/water in a sealed vacuum ampoule (Kikuchi *et al.* 1987).

Somewhat milder hydrolyses can be achieved with trifluoroacetic acid. Thus, for the liberation of echinocystic acid from the corresponding saponins, refluxing for 3 h in 1 M trifluoroacetic acid was sufficient to obtain the non-degraded aglycone (Carpani *et al.* 1989).

Steroid saponins are generally hydroysed with hydrochloric or sulphuric acid (1–2 M) in 50% methanol or dioxane (2–5 h at 100 °C or 10 h at 80 °C). Artifacts formed from furostanol glycosides by this treatment include spirostanols, spirosta-3,5-dienes and C-25 epimers (Voigt and Hiller, 1987). For very sensitive saponins, such as the glycosides of nuatigenin, destruction of the aglycone is avoided by employing mild conditions. In one case, hydrolysis was performed with 2% hydrochloric acid in methanol for 8 days at 37 °C (Kesselmeier and Budzikiewicz, 1979).

An alternative to the hydrolysis of saponins in solution is to hydrolyse them directly on a TLC plate by treatment with hydrochloric acid vapours. Once the acid has been evaporated, normal elution with the TLC solvent is performed in order to identify the monosaccharides present (Kartnig and Wegschaider, 1971; He, 1987). By this means, the terminal sugars xylose and galactose were identified after partial hydrolysis of agaveside B (945). The TLC plate was developed with the solvent chloroform–methanol–water (8:5:1) and the detection was by means of aniline–diphenylamine–H_3PO_4–methanol (1:1:5:48) (Uniyal *et al.* 1990).

945

$Xyl\overset{3}{-}Glc\overset{3}{-}Glc-O$
$\quad |2 \quad\ |2$
$\quad Xyl \quad Gal$

Analysis of aglycones after hydrolysis

Once the hydrolysis is complete, the aglycones can be separated from the hydrolysate either by simple filtration or by a water–organic solvent partition and analysed against known triterpenes or steroids. The most common method is by TLC, using a solvent such as diisopropyl ether–acetone (75:30). Spray reagents are frequently those employed for the analysis of saponins (see Table 3.1).

Gas–liquid chromatography requires derivatization of the triterpenes, as indicated by the following examples: methyl esters of oleanolic and ursolic acids have been separated by GC on a glass column packed with 30% OV-17 or SE-30 (Fokina, 1979); triterpenes can be determined by GC after derivatization with *N,O*-bis(trimethylsilyl)acetamide and chloro-trimethylsilane, as is the case for soyasapogenols A–E and medicagenic acid in alfalfa (Jurzysta and Jurzysta, 1978).

The technique of GC–MS is also valuable for the characterization of sapogenins. The trimethylsilyl derivatives are normally prepared and then analysed in the spectrometer. An example is the application to the investigation of oleanane- and ursane-type triterpenes. Nine silylated triterpenes were separated by GC on OV-101 packing and their mass spectral patterns were investigated; those containing a 12-en double bond underwent a characteristic retro-Diels–Alder reaction (Burnouf-Radosevich *et al.* 1985). This technique has also been used for the determination of triterpenes from licorice (Bombardelli *et al.* 1979).

HPLC analysis does not require derivatization and gives excellent reproducibility and sensitivity for the analysis of triterpenes. Both normal-phase (analysis of quinoa sapogenins; Burnouf-Radosevich and Delfel, 1984) and RP-HPLC (Lin *et al.* 1981) can be employed, but a disadvantage of RP-HPLC is that the compounds tend to precipitate in the aqueous

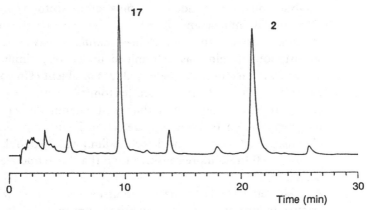

Fig. 4.2.. HPLC of a hydrolysed methanol extract of *Chenopodium quinoa* grains. Column: Nucleosil RP-8, 5 μm (125 × 4 mm); eluent: acetonitrile–water 50:50 → 70:30 over 30 min; flow-rate: 1 ml/min; detection: 206 nm.

mobile phases. However, in the case of the analysis of aglycones from a hydrolysate of *Chenopodium quinoa* (Chenopodiaceae) grain, the butanol extract could be dissolved in a dichloromethane–methanol (1 : 1) mixture for injection onto a C-8 column. Gradient elution with acetonitrile–water gave a clear separation of the two sapogenins oleanolic acid (**2**) and hederagenin (**17**, Table 2.1, p. 21) (Fig. 4.2). Standardization of different quinoa batches was also possible by quantitative analysis of these aglycones (Marston *et al.* 1991).

As well as the usual spectroscopic techniques for the characterization of an aglycone, classical chemical reactions and degradation techniques are also possible (Hiller *et al.* 1966).

Definitive structure determination of the aglycone is described in Section 4.2 and representative lists of the more common sapogenins are given in Chapter 2.

Analysis of sugars after hydrolysis

Analysis of the monosaccharides is carried out by TLC on, for example, silica gel plates with solvents such as ethyl acetate–methanol–water–acetic acid (65 : 25 : 15 : 20) and *n*-butanol–ethyl acetate–*i*-propanol–acetic acid–water (35 : 100 : 60 : 35 : 30) (Shiraiwa *et al.* 1991). Detection is with *p*-anisidine phthalate, naphthoresorcin, thymolsulphuric acid (Kartnig and Wegschaider, 1971) or triphenyltetrazolium chloride (Wallenfels, 1950; Kamel *et al.* 1991). Alternatively, a quantitative analysis of the monosaccharides is possible by GC or HPLC.

Several HPLC methods have been reported:

– Analysis on NH_2-bonded columns with acetonitrile–water (75 : 25) (Glombitza and Kurth, 1987a). The molar ratio of monosaccharides from an *Agave cantala* (Agavaceae) fruit spirostanol saponin was determined under very similar conditions, with acetonitrile–water (73 : 27) as eluent (Uniyal *et al.* 1990). Kikuchi *et al.* (1987) used acetonitrile–water (80 : 20) as eluent at 5 °C. They found that post-column derivatization with tetrazolium blue enabled detection in the visible region (546 nm). Furthermore, the detection limit was 20 µg/ml, which is about 100 times more sensitive than that for refractive-index detection.

– Analysis on C-18 columns (acetonitrile–water 4 : 1) with refractive index detection. For quantitative purposes, integration of the HPLC peaks was compared with standards (Adinolfi *et al.* 1987).

– Analysis on an Aminex ion exclusion HPX-87H column (Bio-Rad) with 0.005 M sulphuric acid as eluent (0.4 ml/min) (Adinolfi *et al.* 1990).

– Analysis of sugar *p*-bromobenzoates (formed by methanolysis of the saponin with 5% hydrochloric acid–methanol and subsequent *p*-bromobenzoylation of the methyl sugars) by HPLC and identification by comparison with authentic derivatives (Kawai *et al.* 1988; Sakamoto *et al.* 1992).

For GC, the persilylated sugars are used (Sweeley *et al.* 1963; Sawardeker and Sloneker, 1965; Wulff, 1965) or a GC–MS analysis of alditol acetate derivatives is carried out (see Section 4.2.6). GC-Fourier-transformed IR (FTIR) analysis of suitably derivatized monosaccharides is an alternative procedure (Chen and Snyder, 1989).

The most commonly found sugars are D-glucose, D-galactose, L-arabinose, D-xylose, D-fucose, L-rhamnose, D-quinovose, D-glucuronic acid and D-ribose.

Artifact formation

Acidic hydrolysis is not without risk because prolonged heating with an inorganic acid can give rise to complications involving artifact formation, low yields and low selectivity (Tschesche and Wulff, 1972; Kitagawa, 1981). This is true not only of triterpene saponins but also of steroid saponins and saponins from marine organisms (Kitagawa *et al.* 1985b), often making the job of structure elucidation very complicated. For instance, a study of the effects of various hydrolytic procedures on the sapogenin profile of soya saponins has shown that soyasapogenols B_1, C, D and E are probably formed as artifacts on aqueous hydrolysis. Thus, the olean-13(18)-en soyasapogenol B_1 is isolated after hydrolysis of soy flour with hydrochloric acid in ethanol (Ireland *et al.* 1987). The true aglycones, soyasapogenols A (**60**) and B (**33**) (Table 2.1, p. 21), are obtained by hydrolysis with sulphuric or hydrochloric acid in anhydrous methanol (Ireland and Dziedzic, 1986a). Another problem arises during the acidic hydrolysis of oleanolic acid and hederagenin glycosides in dioxan, when there is the possibility of lactone formation (Hiller *et al.* 1987a,c)

Sulphuric acid hydrolysis of hovenosides (glycosides of jujubogenin) similarly gives a lactone, ebelin lactone (**1313**) (Inoue *et al.* 1978).

On hydrolysis, an acid-catalysed double-bond migration in triterpenes can occur. For example, some olean-12-enes are isomerized to olean-13(18)-enes with hydrochloric acid in aqueous ethanol (Kubota *et al.*

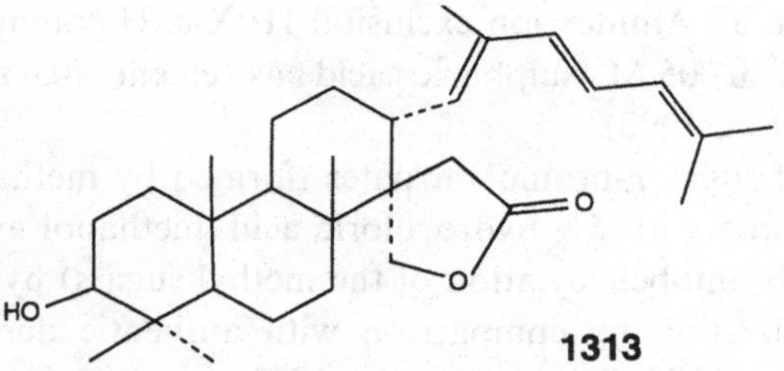

1313

1969). Quillaic and echinocystic acids are both isomerized to the corresponding olean-13(18)-enes under these conditions (Fig. 4.3) (Kubota *et al.* 1969).

Epimerization is possible during acid hydrolysis, as shown by the conversion of arjungenin (**56**) (from the corresponding 28-glycoside) to tomentosic acid (**55**). That the transformation proceeds via the pathway described (Fig. 4.4) is proved by the isolation of the intermediate lactone (**1314**) (Mahato *et al.* 1990).

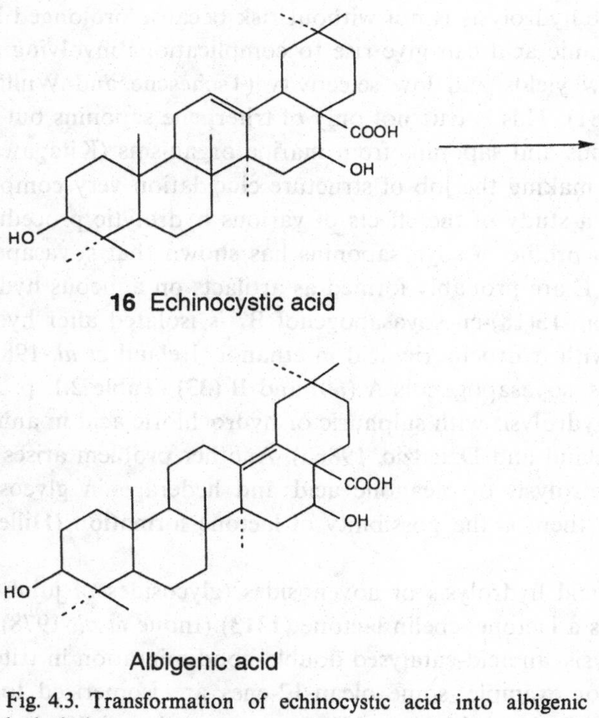

16 Echinocystic acid

Albigenic acid

Fig. 4.3. Transformation of echinocystic acid into albigenic acid on acidic hydrolysis.

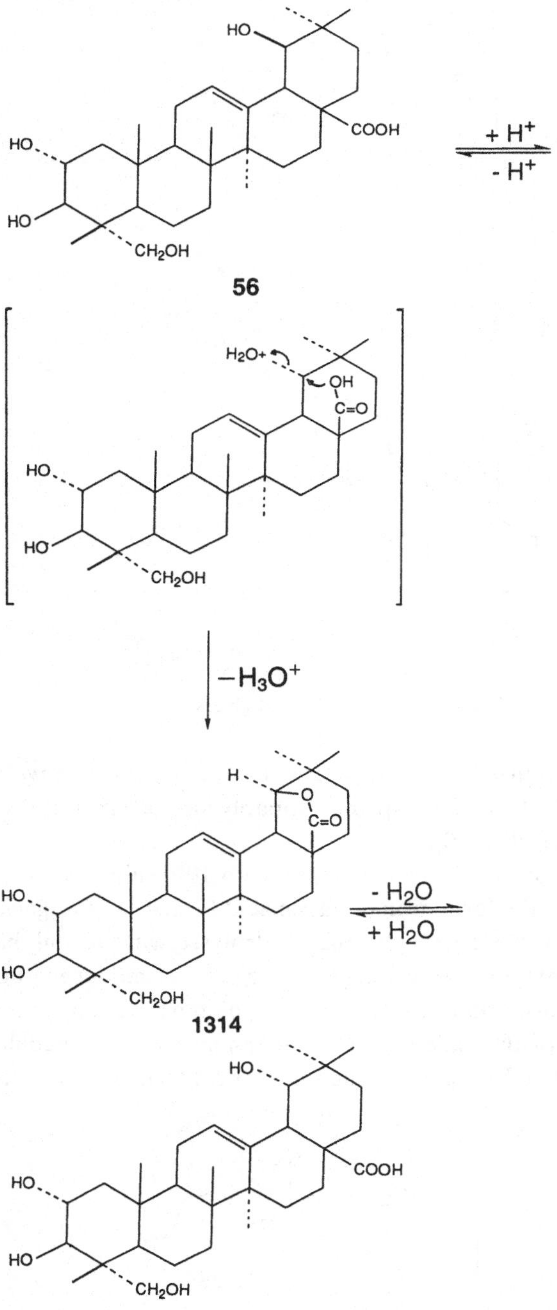

Fig. 4.4. Epimerization of arjungenin.

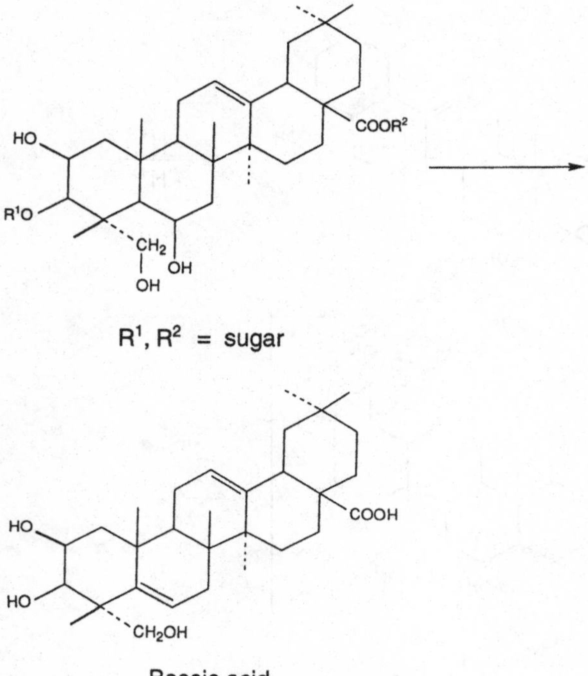

R¹, R² = sugar

Bassic acid

Fig. 4.5. Acid hydrolysis of protobassic acid glycosides.

The transformation of cochalic acid to echinocystic acid involves an epimerization at the 16-OH group and probably also occurs via a 28 → 16 lactone (Mahato *et al.* 1990).

Hydrolysis of protobassic acid saponins with 10% sulphuric acid leads to dehydration and the formation of bassic acid (Fig. 4.5) (Kitagawa *et al.* 1978b). To obtain the true aglycone, hydrolysis with a soil bacteria preparation was carried out (see Section 4.1.5). Other milder methods for avoiding artifact formation are described in the next section.

Acid hydrolysis of alatoside A (**691**) from *Sesamum alatum* (Pedaliaceae) similarly resulted in a dehydration reaction. The artifact, an 18,19-secoursa-

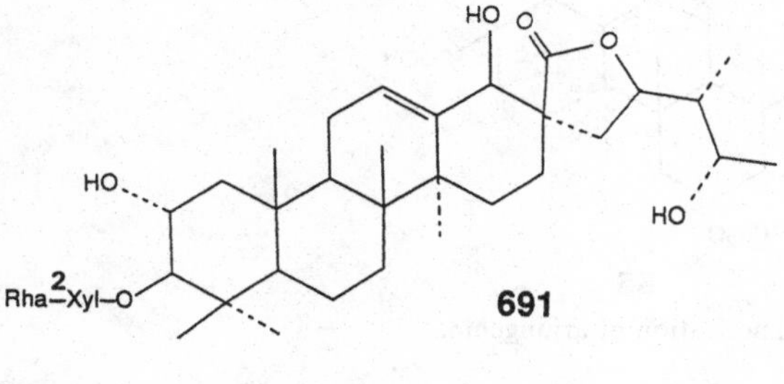

691

11,13(18)-dien-28,21-lactone (**691a**), exhibited strong UV absorption (λ_{max} 250 nm), resulting from the conjugated diene system formed. The true

691a

aglycone, alatogenin, was only obtained after enzymatic hydrolysis of alatoside A (**691**) (Potterat *et al.* 1992).

Furostanol bidesmosides are converted on refluxing with methanol into the less polar 22-OCH$_3$ derivatives, which regenerate the original 22-OH compounds on treatment with boiling water or dilute dioxane (Kawasaki, 1981).

Even the extraction procedure can influence the nature of pure compounds isolated from a plant. The monodesmosidic : bidesmosidic saponin ratio can vary considerably as a function of the extraction conditions (see Chapter 3) (Domon and Hostettmann, 1984; Kawamura *et al.* 1988).

4.1.2   *Basic hydrolysis*

Cleavage of *O*-acylglycosidic sugar chains is achieved under basic hydrolysis conditions, typically by refluxing with 0.5 M potassium hydroxide (Domon and Hostettmann, 1984; Kochetkov and Khorlin, 1966). Alternatively, 1–20% ethanolic or methanolic solutions of potassium hydroxide may be used but there is a risk of methylation, especially of the carboxyl group of triterpene acids. Ion exchangers such as Dowex 1 provide mildly basic hydrolysis conditions (Bukharov and Karlin, 1970a). Another method is to use lithium iodide in collidine (Kochetkov *et al.* 1964b).

By carefully controlling the reaction conditions, it is possible to selectively cleave different ester moieties. Hydrolysis of kizuta saponin K$_{11}$ (**822**, Table 2.7, p. 65) by refluxing in 0.5 M potassium hydroxide for 30 min removed the sugar at C-28 of the bidesmoside. However, stirring the saponin for 20 h in 0.1 M potassium hydroxide at room temperature selectively removed the acetate group on the C-28 ester glycoside chain (Kizu *et al.* 1985b).

4.1.3   *Partial hydrolysis*

In certain instances, when saponins have highly branched or long sugar chains, a procedure involving partial hydrolysis is necessary in order to obtain fragments more accessible to structure elucidation. This can be achieved with acid or, indeed, with enzymes. The oligosaccharide and/or the remaining saponin portions are isolated and then characterized.

For example, saponin **254** from *Phytolacca dodecandra* (Phytolaccaceae) was hydrolysed by 0.1 M hydrochloric acid for 45 min, to give a mixture of three products: **165**, **217**, and **1312** (see Fig. 4.6). These compounds were separated by RP-LPLC and their sugar sequences determined by MS, ^{13}C-NMR and GC–MS of alditol acetates. Putting all this information together enabled **254** to be assigned the formula 3-*O*-{*O*-α-L-rhamnopyranosyl-(1 → 2)-*O*-β-D-glucopyranosyl-(1 → 2)-*O*-[β-D-glucopyranosyl-(1 → 4)-β-D-glucopyranosyl} oleanolic acid (Dorsaz and Hostettmann, 1986).

Hydrolysis in dioxane gives milder conditions and, as shown in Fig. 4.7, partial hydrolysis is possible. In this example, the saponin was

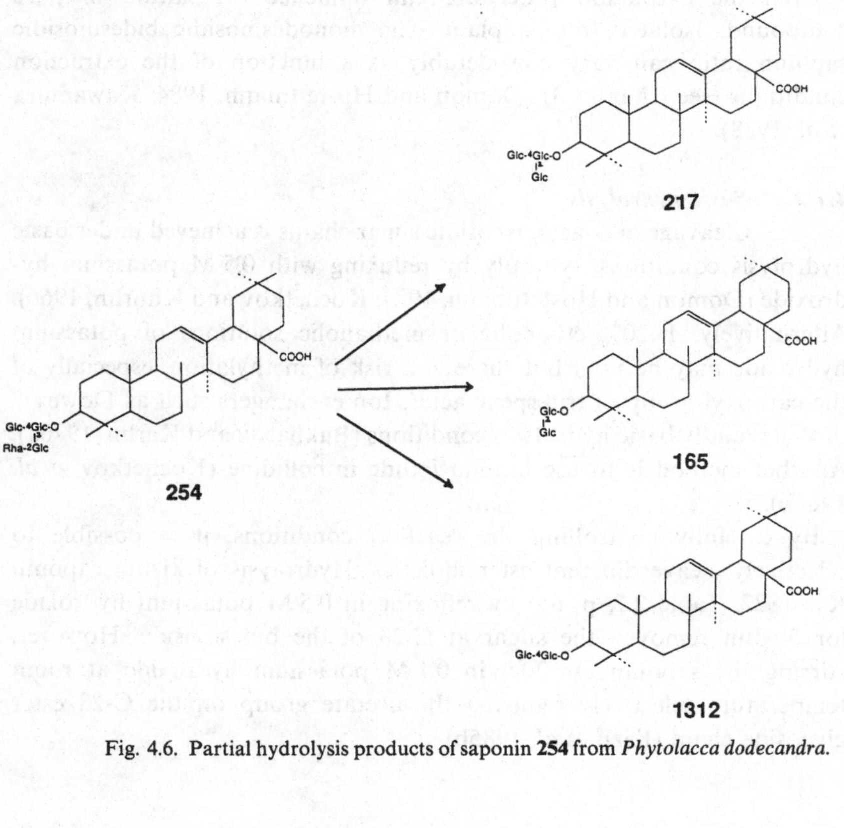

Glc-⁴Glc-O
|¹
Glc

217

Glc-⁴Glc-O
|¹
Rha-²Glc

254

Glc-O
|¹
Glc

165

Glc-⁴Glc-O

1312

Fig. 4.6. Partial hydrolysis products of saponin **254** from *Phytolacca dodecandra*.

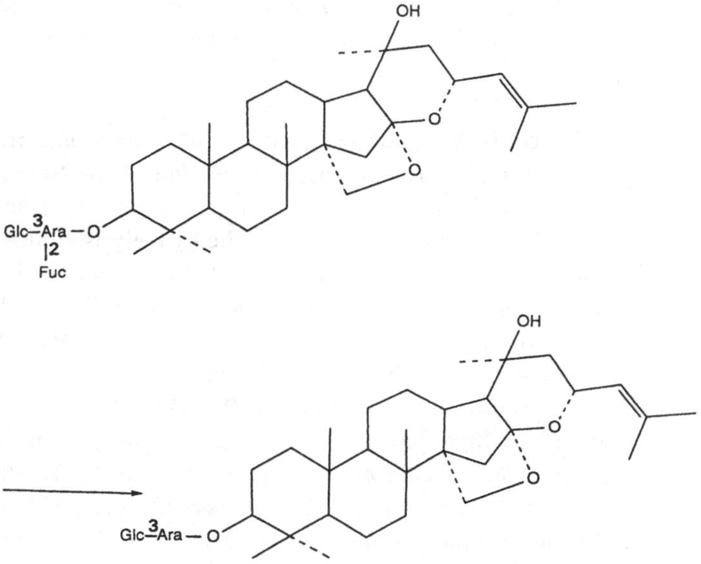

Fig. 4.7. Partial hydrolysis with dioxane-0.1 M hydrochloric acid.

refluxed for 6 h in dioxane–0.1 M hydrochloric acid (1:3) (Ikram *et al.* 1981). Another method for partially hydrolysing saponins is to treat a solution of the triterpene glycoside in alcohol with an alkali metal (sodium or potassium) and then add a trace of water (Ogihara and Nose, 1986).

4.1.4   *Hydrothermolysis*

Hydrothermolysis of triterpene and steroid glycosides leads to the formation of the corresponding aglycones and prosapogenins and thus can aid structure determination (Kim *et al.* 1992). The method involves heating the glycoside with water or water–dioxane at 100 °C to 140 °C for a period of 10 to 140 h, depending on the sample. Hydrothermolysis of triterpene 3,28-*O*-bisglycosides gives the corresponding 3-*O*-glycosides (Kim *et al.* 1992).

4.1.5   *Enzymatic hydrolysis*

A very efficient and mild method for the cleavage of sugar residues from saponins without artifact formation is enzymatic hydrolysis. Although the relevant hydrolases for all the sugars are not commercially available, cleavages of β-glucose residues by β-glucosidase (and emulsin; Konishi and Shoji, 1979; Son *et al.* 1990) etc. are perfectly straight-forward. A supplementary benefit of cleavage by specific enzymes is that the *anomeric configuration* of the sugar moiety is automatically proved.

To illustrate the possibilities available, certain enzyme preparations are described below:

- β-Galactosidase hydrolyses galactose moieties from furostanol glycosides of the bittersweet (*Solanum dulcamara*, Solanaceae).
- Cellulase has been used in the cleavage of glucose from kizuta saponin K_{7A} (Kizu *et al.* 1985a), in the hydrolysis of luperosides (Kusumoto *et al.* 1989; Okabe *et al.* 1989) and in the study of *Dodonaea* saponins (Wagner *et al.* 1987). It can lead to either complete (Kakuno *et al.* 1991b) or partial hydrolysis (Yoshikawa *et al.* 1991b) of saccharide moieties.
- Crude hesperidinase (Kohda and Tanaka, 1975; Mizui *et al.* 1988) gives high yields of the true aglycones from ginseng saponins, e.g. from ginsenosides Rb_1, Rb_2, Rc, Rg_1 (Kohda and Tanaka, 1975; Tanaka and Kasai, 1984). Hesperidinase cleaved a rhamnose moiety from the triterpene disaccharide alatoside A (**691**) (Potterat *et al.* 1992). In the structure elucidation of cauloside D (**372**, Appendix 2) from *Polyscias dichroostachya* (Araliaceae), enzymatic hydrolysis with hesperidinase removed the terminal rhamnose moiety from the C-28 sugar chain. Hydrolysis with a mixture of hesperidinase and β-glucosidase led to simultaneous cleavage of the terminal rhamnose moiety on the C-28 trisaccharide chain and the terminal glucose moiety on the C-3 disaccharide chain of cauloside G (**363**, Appendix 2) from the same plant (Gopalsamy, 1992). Hydrolysis of desulphated pervicoside B (**1222**, Table 2.13, p. 110), a lanostane-type triterpene tetrasaccharide from the sea cucumber *Holothuria pervica*, with crude hesperidinase enabled the cleavage of a glucose and a methylglucose moiety from the linear sugar chain (Kitagawa *et al.* 1989c).
- Pectinase from *Aspergillus niger* is employed for the cleavage of terminal rhamnose units (Kimata *et al.* 1983; Watanabe *et al.* 1983). Crude pectinase is also known to cleave glucose moieties (Ohtani *et al.* 1990).
- Naringinase D (200 units of naringinase, which is both a rhamnosidase and a β-glucosidase, together with 10 units of pectinase from *Aspergillus niger*) cleaves rhamnosyl moieties (Elujoba and Hardman, 1985) and has also found use in the cleavage of xylose from hovenosides (Inoue *et al.* 1978).
- Protease (type XIII from *Aspergillus satoi*) cleaves xylose

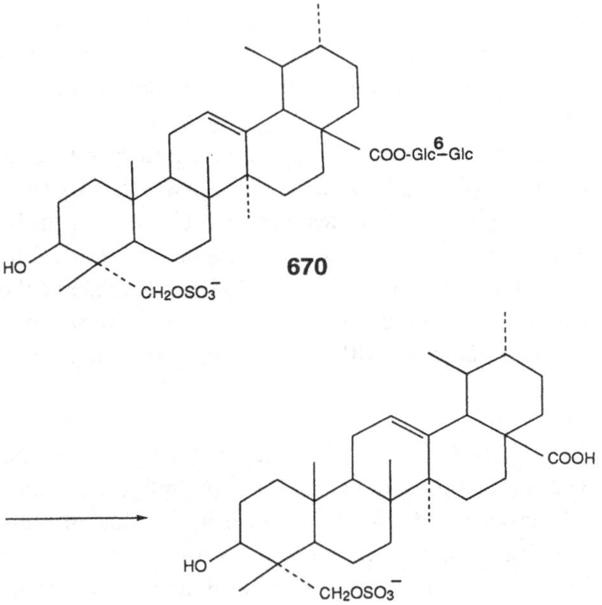

Fig. 4.8. Cleavage of gentiobioside by protease.

moieties (Inada *et al.* 1987b) and has also been used to cleave a gentiobioside unit from a sulphated saponin (Fig. 4.8) (Inada *et al.* 1988). Treatment with molsin removes the xylose unit from the cycloartane glycosides found in *Beesia* species (Sakurai *et al.* 1986).

– Xylosidase cleaves terminal xylose moieties and has been employed in the structure determination of saponins from *Sesamum alatum* (Pedaliaceae) (Potterat *et al.* 1992).

– Avenacosidase, a β-glucosidase from oats, splits off glucose units at C-26 of furostanol saponins. Hydrolysis of saponins from tobacco seeds with this enzyme enabled deglucosidation at C-26 *without* subsequent ring formation and spirostanol formation (Grünweller *et al.* 1990).

A systematic study involving crude preparations of hesperidinase, naringinase, pectinase, cellulase, amylase and emulsin has shown that hesperidinase, naringinase and pectinase were the most effective in hydrolysing ginsenosides (Kohda and Tanaka, 1975).

Enzyme preparations from *Helix pomatia* snails and other sources provide an alternative way of hydrolysing sugar moieties. For example, the fresh hepatopancreatic juice of *Helix pomatia* snails cleaves the gentiobioside residue of quadranguloside, a cycloartane glycoside from *Passiflora quadrangularis* (Passifloraceae) (Orsini *et al.* 1986). Another alternative is to hydrolyse the sugar chain with mouse intestinal flora (Shimizu *et al.* 1985).

Difficulties do arise when, for example, steric hindrance in the sugar chain (as in branched chains) prevents catalytic hydrolysis by the enzyme. In cases in which β-glucose residues are not cleaved or only slowly cleaved by β-glucosidase, more efficient reactions can be produced with β-glucuronidase (Dorsaz and Hostettmann, 1986). Cleavage of the 3-*O*-β-galactopyranosyl-(1 → 2)-β-glucuronopyranosyl disaccharide of thladio-side H_1 has also been achieved with β-glucuronidase (Nie *et al.* 1989).

4.1.6   *Microbial hydrolysis*

Microbial hydrolysis, in effect, is enzymatic hydrolysis by a combination of enzymes and is an extension of the method which employs crude glycosidase mixtures. It involves culturing a microorganism with a medium containing a saponin as a carbon source. Since pure cultures of microoganisms are used, extensive preliminary screening tests are necessary to find a suitable strain.

Diosgenin, for example, has been obtained on a semi-industrial scale by microbial hydrolysis of the saponin from *Dioscorea tokoro* (Dioscoreaceae) with *Aspergillus terreus* (Kitagawa, 1981). Hecogenin is liberated from cultures of the fungi *Corynespora* sp. or *Alternaria* sp. which contain *Agave sisalana* (Agavaceae) saponin as the sole carbon source (Hassall and Smith, 1957) and sapogenols of *Medicago sativa* (Leguminosae) can be obtained from cultures of *Aspergillus* sp., *Rhizopus* sp. or *Mucor* sp. (Lourens and O'Donovan, 1961).

Soil microorganisms

Soil microorganisms have also been successfully applied to the hydrolysis of saponins. A preliminary selection of a suitable microbial strain from soil samples is necessary and, following large-scale culture with the saponin in question, the culture broth is extracted with a suitable organic solvent (Kitagawa, 1981).

Among the genuine aglycones (sapogenols) obtained by this method is presenegenin (**47**, Table 2.1, p. 21) from senega root saponin (Yosioka *et al.* 1966). The effective microorganism in the soil sample used for hydrolysis was identified as a *Pseudomonas* species. The true sapogenols from *Panax japonicum, Aesculus turbinata, Sanguisorba officinalis, Styrax japonica* and *Metanarthecium luteo-viride* (steroid sapogenol) have all likewise been identified (Kitagawa, 1981). The establishment of protobassic acid (**52**, Table 2.1, p. 21) (and not bassic acid) as the genuine aglycone of the saponins from seed kernels of *Madhuca longifolia* (Sapotaceae) was by means of soil bacterial hydrolysis (Kitagawa *et al.* 1972a; Yosioka *et al.* 1974).

The method is valuable not only for obtaining aglycones but also for obtaining partial hydrolysis products, which are of great importance for the structure elucidation of the parent saponins (Kitagawa, 1981).

4.1.7 Mild and selective cleavage methods

As mentioned above, acid hydrolysis can result in a modification of the aglycone. This problem, encountered during the structure elucidation of saponins, has to be resolved by selecting other methods for obtaining the genuine aglycone. One way of achieving this is to employ enzymatic hydrolysis, but chemical methods, described below, are also available.

Acid hydrolysis in a two-phase medium

If a non-miscible organic solvent (benzene, ethyl acetate, etc.) is added to the aqueous medium during acid hydrolysis, the liberated sapogenol is transferred to the organic phase, so that secondary modification of the sapogenol is reduced.

This method has been used for the hydrolysis of cyclamin (**538**) from the roots of *Cyclamen europaeum* (Primulaceae) (Tschesche *et al.* 1964). Normal acid hydrolysis of cyclamin gives cyclamiretin D, while hydrolysis with ethanolic hydrochloric acid containing benzene gives the true aglycone cyclamiretin A as sapogenol (Fig. 4.9).

Two-phase systems with hydrochloric acid in carbon tetrachloride have also been recommended for the hydrolysis of steroid alkaloid glycosides (van Gelder, 1984).

Periodate degradation

Smith degradation. Periodate oxidation of vicinal hydroxyl groups (in rhamnose, for example) forms aldehydes. These are reduced and the modified sugar residue is cleaved by mild acid hydrolysis (Goldstein *et al.* 1965) (Fig. 4.10). Saccharide chains are thus broken up into shorter, more readily analysed units, with cleavage occurring wherever a monosaccharide with a glycol moiety is present (Kochetkov *et al.* 1963). Periodate degradation has been applied to the structure elucidation of several saponins, for example those of *Guaiacum officinale* (Zygophyllaceae) (Ahmad *et al.* 1986b). It is, furthermore, a means of obtaining the genuine sapogenins from ginsenosides (Tanaka and Kasai, 1984). The method also

538 Cyclamin

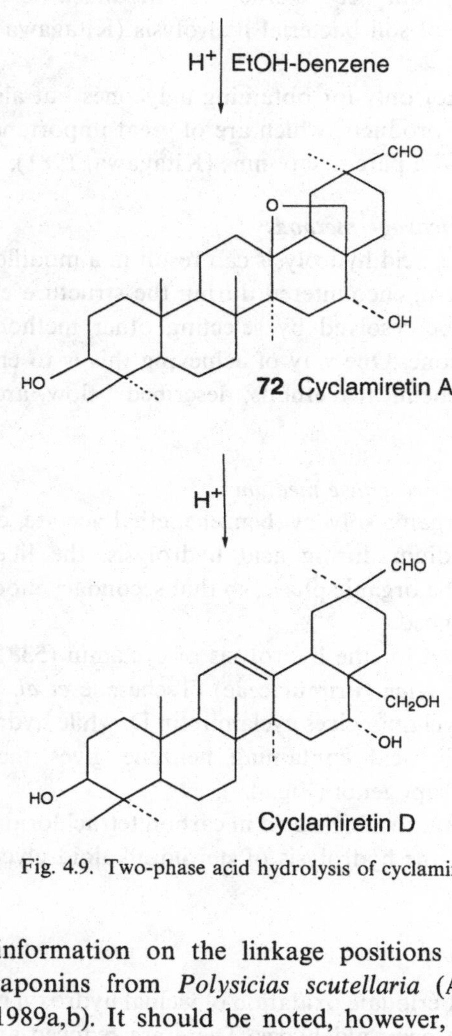

H$^+$ | EtOH-benzene

72 Cyclamiretin A

H$^+$

Cyclamiretin D

Fig. 4.9. Two-phase acid hydrolysis of cyclamin.

gives information on the linkage positions of sugars, as is the case with saponins from *Polysicias scutellaria* (Araliaceae) (Paphassarang *et al.* 1989a,b). It should be noted, however, that yields are rather low and that the periodate will attack vicinal hydroxyl groups on the aglycone.

Smith–de Mayo degradation. This is a modified Smith degradation method which has found application in the isolation of the true aglycone from *Polygala senega* (Polygalaceae) (Dugan and de Mayo, 1965) and in the cleavage of sugars from hovenosides (Inoue *et al.* 1978).

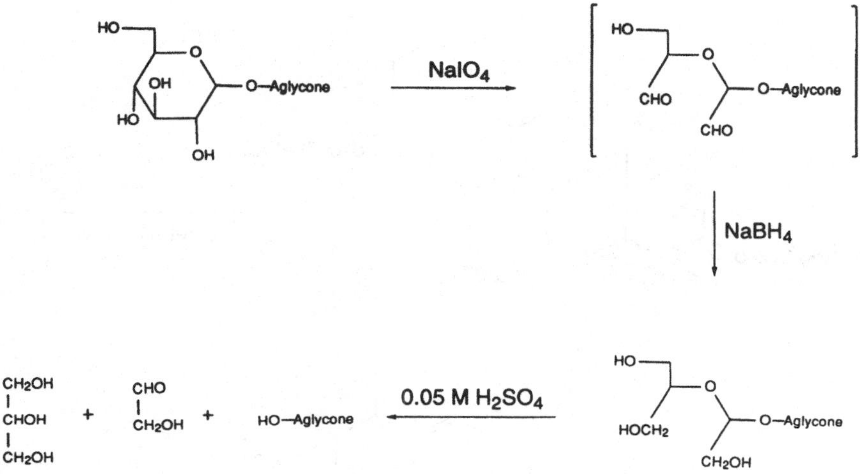

Fig. 4.10. Smith degradation.

Alkali metals

Cleavage of glycoside bonds is possible with an alcoholic solution of an alkali metal. To promote the reaction, a trace of water is necessary (Ogihara and Nose, 1986). By this means, saikogenin E and a mixture of prosapogenins were obtained from saikosaponin c (**544**, Appendix 2) (Ogihara and Nose, 1986).

Treatment of saponin D from *Hovenia dulcis* (Rhamnaceae) (**778**, Table 2.5, p. 58) with sodium metal in ethanol at 50 °C for 7 h gave a mixture of the aglycone (jujubogenin), the 3-glucosyl-20-rhamnosyl and 20-rhamnosyl derivatives, together with unchanged saponin D (Ogihara *et al.* 1987).

For the isolation of the genuine aglycones of saponins from *Anagallis arvensis* (Primulaceae), a solution of sodium in *n*-butanol was used. The aglycone anagalligenin B (**77**, Table 2.2, p. 24) was obtained after heating anagallisin A (**551**, Appendix 2) with the alkali metal solution for 40 h at 95 °C (Mahato *et al.* 1991).

Lithium iodide

A selective cleavage of ester glycoside linkages is achieved by refluxing the saponin with anhydrous lithium iodide and 2,6-lutidine (or collidine) in anhydrous methanol (Ohtani *et al.* 1984). This is generally a quantitative reaction and it provides the intact saccharide moiety after cleavage. The sugar sequence can then be determined by alditol acetate formation etc., as illustrated by the example in Fig. 4.11 (Mizutani *et al.* 1984).

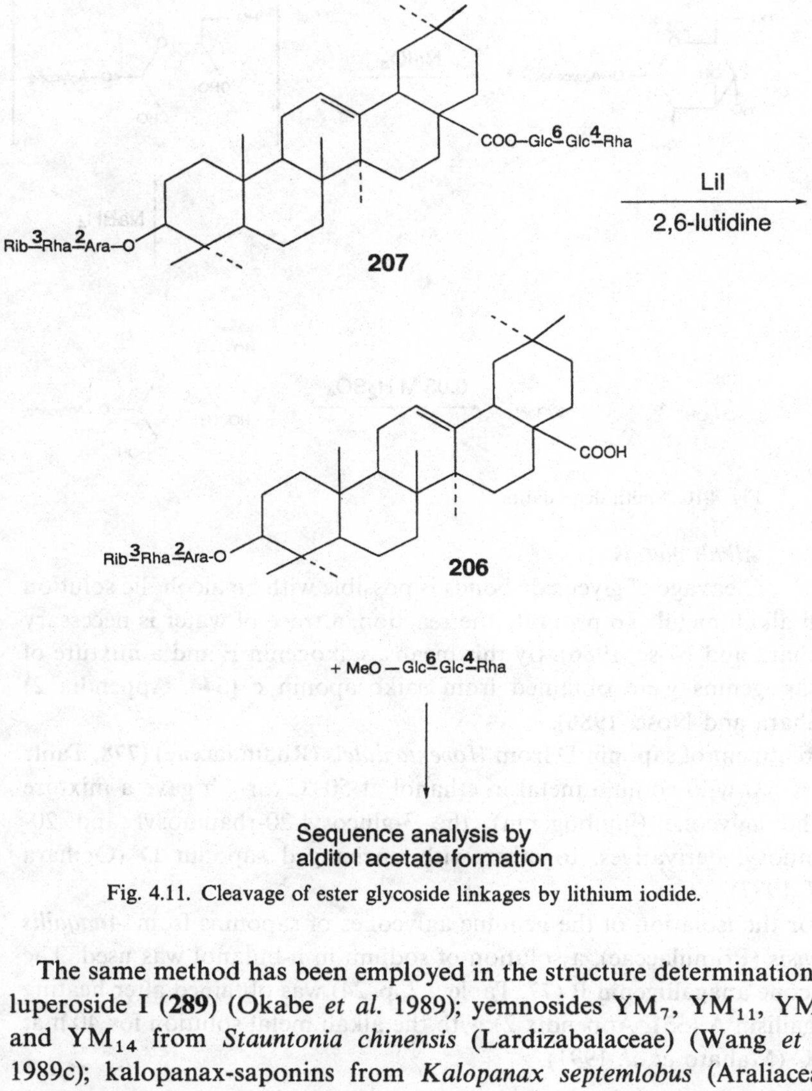

Fig. 4.11. Cleavage of ester glycoside linkages by lithium iodide.

The same method has been employed in the structure determination of luperoside I (**289**) (Okabe *et al.* 1989); yemnosides YM_7, YM_{11}, YM_{13} and YM_{14} from *Stauntonia chinensis* (Lardizabalaceae) (Wang *et al.* 1989c); kalopanax-saponins from *Kalopanax septemlobus* (Araliaceae) roots (Shao *et al.* 1989b); thladioside H1 (**282**) from *Thladiantha hookeri* (Cucurbitaceae) tubers (Nie *et al.* 1989); and aster saponins from *Aster tataricus* (Asteraceae) (Nagao *et al.* 1989b).

Lithium aluminium hydride

Lithium aluminium hydride is used for the reductive cleavage of saccharides from saponins (see also alditol acetates, section 4.2.6). For example, treatment of permethylated entada saponin-III with $LiAlH_4$ led to cleavage of the saccharide moiety from C-3 (Okada *et al.* 1987). In the case of *O*-acylglycosidic sugar chains (as found in most bidesmosidic saponins, for example), reduction of the permethylated saponin with LiAlH$_4$ leads to cleavage of the acylglycosidic linkage (Konoshima *et al.* 1980; Okada *et al.* 1980). This can be useful for elucidating the structure of bidesmosidic saponins because, following cleavage of the ester-linked saccharide moiety, determination of the nature and interglycosidic linkages of the sugars in the ether-linked saccharide moiety can then be carried out. Structural investigation of the reductively cleaved sugar chain is performed by NMR (Higuchi and Kawasaki, 1972; Tsukitani and Shoji, 1973; Tsukitani *et al.* 1973), by hydrolysis and subsequent chromatographic analysis of the partially methylated monosaccharides or by GC–MS analysis (Higuchi and Kawasaki, 1972) of the corresponding alditol acetates (see Section 4.2.6).

Lead tetraacetate oxidation

Oxidation with lead tetraacetate selectively cleaves glucuronide linkages in a saponin (Kitagawa *et al.* 1976c). Permethylated saponins with a glucuronic acid moiety (containing a free carboxyl function) connected to carbon C-3 of the aglycone are cleaved by treatment first with lead tetraacetate and then with alkali. In the case of monodesmosidic saponins, the methylated sapogenin is obtained, together with sugar-chain fragments.

Anodic oxidation

Another method for the selective cleavage of glucuronide linkages in saponins is anodic oxidation (Kitagawa *et al.* 1980a). Non-derivatized saponins are first subjected to anodic oxidation and then the labilized uronic acid moiety is split off by reaction with alkali.

Photochemical cleavage

Photodegradation also enables selective cleavage of a glycoside at a glucuronide residue (Kitagawa *et al.* 1973, 1974). Irradiation of a methanolic solution of the saponin in a quartz tube with a mercury lamp provides the necessary photolysis (Yoshikawa *et al.* 1985; Kitagawa *et al.* 1988b). By this means, protoprimulagenin A (**71**, Table 2.2, p. 24) was obtained from primula saponin (**531**, Appendix 2) (Tschesche *et al.* 1983) and a prosapogenol (**1315**) from sophoraflavoside I (**455**) (Fig. 4.12) (Yoshikawa *et al.* 1985). Sometimes, however, the yields are low and, in addition, the sapogenin must be photostable (Kitagawa *et al.* 1974).

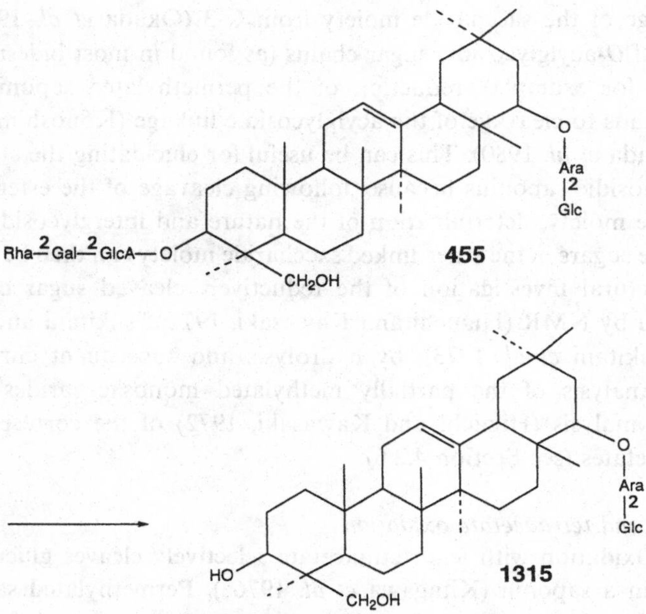

Fig. 4.12. Photochemical cleavage of sophoraflavoside.

Pyridine–acetic anhydride

When a glucuronic acid moiety is attached directly to the aglycone, the saccharide chain is removed by treating the saponin with pyridine–acetic anhydride (Kitagawa *et al.* 1977). This procedure has been used to identify the genuine aglycones of camellidins I and II (Nagata *et al.* 1985; Nishino *et al.* 1986; Numata *et al.* 1987) (Fig. 4.13). Normal hydrolysis with a 5 M sulphuric acid–dioxane (1:1) yielded an artifact, but cleavage of the glucuronic acid moiety at C-3 with pyridine–acetic anhydride gave the acetate of camellidin I aglycone. Deacetylation of the acetate with 1 M potassium hydroxide–ethanol (1:1) produced the genuine aglycone of camellidin II.

A similar procedure was used for the cleavage of the 3-*O*-β-D-glucopyranosyl(1 → 2)glucuronopyranosyl moiety of saniculosides A and B (Kühner *et al.* 1985).

Diazomethane degradation

Cleavage of a sugar chain at position C-3 of an aglycone containing an aldehyde moiety at C-23 has been achieved by adding ethereal diazomethane to a solution of the corresponding saponin in

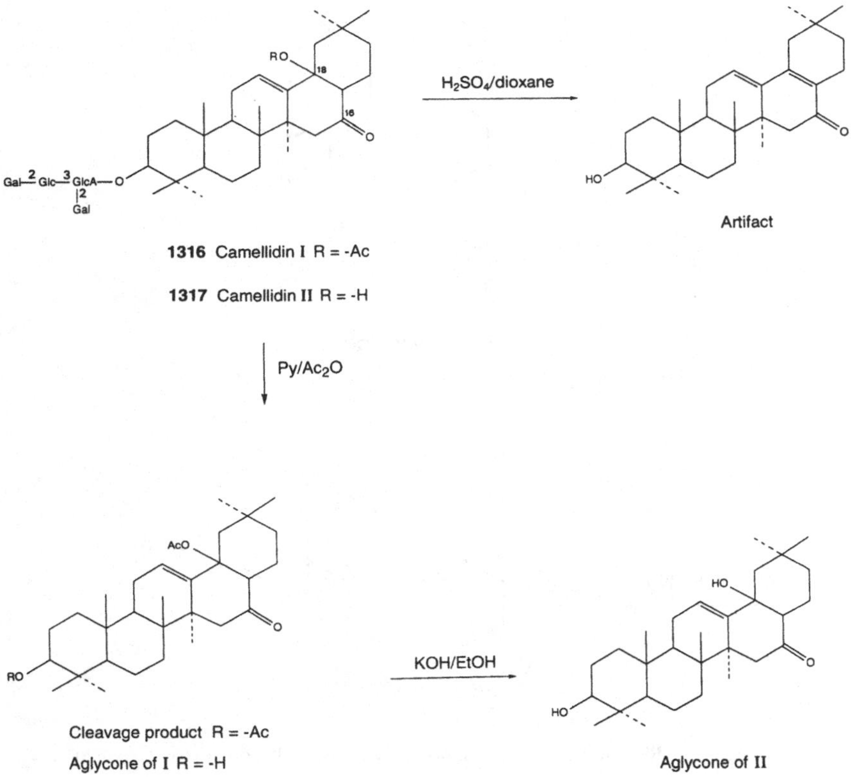

Fig. 4.13. Pyridine–acetic anhydride cleavage of saponins.

methanol (Higuchi *et al.* 1985). As is illustrated for quillaia saponin DS-1, cyclization occurs to form the corresponding 3-*O*,23-methyleneolean-12-en-28-oate (Higuchi *et al.* 1987) (Fig. 4.14). The advantage of this method is that the saccharide moiety at position C-28 is left intact.

4.2   Spectroscopic and other techniques for structure elucidation

The structure elucidation of saponins and the corresponding aglycones relies not only on chemical methods but also on spectro-scopic and related techniques – IR, UV, NMR, MS, optical rotary dispersion (ORD), circular dichroism (CD), X-ray analysis. Modern advances in some of these techniques, most notably in NMR spectro-scopy and MS, have facilitated enormously the difficult task of analysing saponins and their corresponding fragments from cleavage reactions,

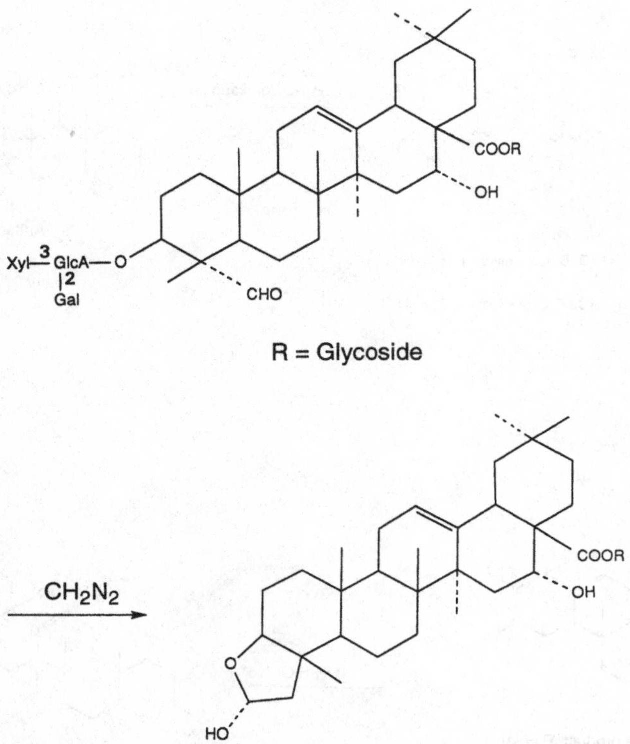

R = Glycoside

Fig. 4.14. Reaction of quillaia saponin DS-1 with diazomethane (Higuchi *et al.* 1987).

that the information can be collated and the relevant structures determined. Furthermore, NMR spectroscopy is a non-destructive technique and both NMR and MS allow examination of the *intact* saponin.

An integrated approach for solving saponin structures is necessary, with the different spectroscopic techniques each providing a certain contribution to the ensemble of data.

4.2.1 *Mass spectrometry (MS)*

The choice of ionization method in MS depends on the polarity, lability and molecular weight of the compound to be analysed. It is principally the so-called 'soft' ionization techniques such as FAB and desorption/chemical ionization (D/CI) which are employed to obtain molecular weight and sugar sequence information for naturally occurring glycosides (Wolfender *et al.* 1992). These permit the analysis of glycosides without derivatization. In certain cases, fragmentations of aglycones are observed, but electron impact mass spectra (EI-MS) are more useful for this purpose.

Fast atom bombardment MS (FAB-MS)

In FAB studies, an accelerated beam of atoms (or ions) is fired from a gun towards a target which has been preloaded with a viscous liquid (the 'matrix' – usually glycerol or 1-thioglycerol) containing the sample to be analysed (Barber *et al.* 1981, 1982). When the atom beam collides with the matrix, kinetic energy is transferred to the surface molecules, a large number of which are sputtered out of the liquid into the high vacuum of the ion source. Ionization of many of these molecules occurs during the sputtering, giving both positive and negative ions. Either can be recorded by an appropriate choice of instrumental parameters but negative ions have proved more useful, on the whole, for saponin work (Fraisse *et al.* 1986). The complementary nature of the information from both modes should not be underestimated, however. Positive ions are the result of protonation, $[M + H]^+$, or cationization, $[M + cation]^+$, where as negative ions are preponderantly $[M - H]^-$, but can also be formed by the addition of an anion, that is, $[M + anion]^-$.

As FAB-MS can easily be applied to the ionization of non-volatile, non-derivatized, polar, high-molecular-weight compounds, this 'soft' ionization method is particularly suited to the analysis of saponins. In addition to pseudomolecular ions, information concerning the fragmentation of monosaccharide components of saponins can be obtained.

A typical example of the application of FAB-MS to the structure elucidation of saponins is shown below (Fig. 4.15) (Borel and Hostettmann, 1987). The monodesmosidic molluscicidal saponin **189** from *Swartzia madagascariensis* (Leguminosae) gave a pseudomolecular ion $[M - H]^-$ at m/z 777. Cleavage of a rhamnosyl moiety gave a negative ion $[(M - H) - 146]^-$ at m/z 631, while subsequent cleavage of the glucuronic acid moiety resulted in an ion from the aglycone, oleanolic acid, $[(M - H) - 332]^-$ at m/z 455. Thus, information about the molecular weight of the saponin, the number, nature and sequence of sugars attached at position C-3 of the aglycone, and the molecular weight of the aglycone was obtained.

Further examples of FAB-MS spectra of oleanolic acid and hederagenin glycosides have been reported (Domon and Hostettmann, 1983) and the technique has also been applied to the analysis of glycoalkaloids (Mahato *et al.* 1980b; Price *et al.* 1985), dioscin and gracillin (Mahato *et al.* 1980a).

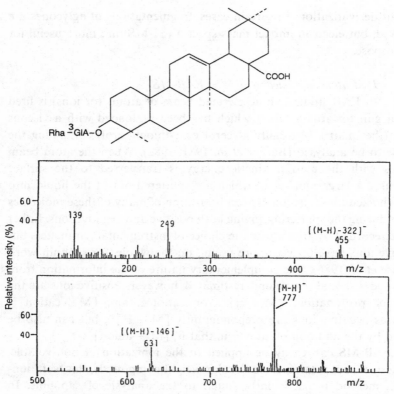

Fig. 4.15. FAB mass spectrum of saponin **189** (negative ion mode).

FAB-MS also furnishes molecular weight data for bidesmosidic saponins and, by considering the fragmentation pattern, the existence of branching in the sugar chain can be established. In the case of the oleanolic acid saponin **255** from *Phytolacca dodecandra* (Phytolaccaceae), pseudomolecular ions at m/z 1285 [M + Cl]⁻ and 1249 [M − H]⁻ were observed in the FAB mass spectrum, indicating a molecular weight of 1250. A signal at m/z 1087 [(M − H) − 162]⁻ resulted from the loss of a glucosyl unit at C-28, while the signals at m/z 941 and 925 corresponded to the simultaneous cleavage of rhamnosyl [(M − H) − 308]⁻ and glucosyl [(M − H) − 324]⁻ moieties, respectively. This proved the existence of a branched sugar. Other peaks at m/z 779 [(M − H) − 470]⁻, 617 [(M − H) − 632]⁻ and 455 [(M − H) − 794]⁻ represented the successive loss of three glucopyranosyl units (Fig. 4.16) (Dorsaz and Hostettmann, 1986).

The structure determination of saponins from marine organisms has been greatly aided by the introduction of FAB-MS, as demonstrated by

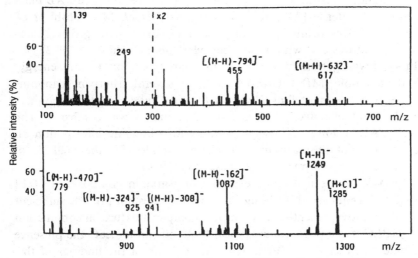

Fig. 4.16. FAB mass spectrum of saponin **255** (negative ion mode).

the asterosaponins (Minale *et al.* 1982). Negative-ion FAB-MS is also useful for molecular weight determinations of sulphated saponins because a strong molecular ion $[RSO_3]^-$ is observed. In addition, fragment ions are produced (Itakura and Komori, 1986a).

Sequence-specific ions, which are generally far less abundant than $[M + H]^+$ ions, can be enhanced by collisional activation (CA) mass spectra of the $[M + H]^+$ ion. This technique has the added advantage of eliminating interfering peaks from impurities or the matrix. FAB followed by CA has been applied, for example, to the structural analysis of a series of steroidal glycosides containing two to four sugar units from *Paris polyphylla* and *P. axialis* (Liliaceae). Intense protonated molecular ions were observed and differentiation between branched and linear saccharide chains was possible (Chen *et al.* 1987b). Another study of CA spectra from the mass selected ion $[M - H]^-$ of a series of saponins in the FAB mode gave a series of characteristic fragment ions which were useful for the determination of the sugar sequences. In the positive-ion mode, there was a strong signal for the aglycone in the CA spectra, whereas this was absent in the negative-ion mode (Chen *et al.* 1987a). In the normal FAB mass spectra, the negative-ion mass spectra exhibited more predominant quasi-molecular ions, as mentioned above. The matrix polyethylene glycol-200 gave better results than glycerol or thioglycerol (see also a report by Rose *et al.* (1983) on the FAB-MS of steroid glycosides). The nature of the matrix indeed plays a very important rôle in the quality of spectra obtained. Addition of KCl, for example, gives positive FAB mass spectra with intense $[M + K]^+$ ions (Mahato *et al.* 1991; Joshi *et al.* 1992). Successful results have also been obtained with glycerol/thioglycerol mixtures and with *m*-nitrobenzyl alcohol (Joshi *et al.* 1992).

In an investigation of the FAB mass-analysed ion kinetic energy/CA dissociation (MIKE/CAD) spectra of several triterpene saponins (ginsenosides, chrysantellins, *Hedera helix* and *Centella asiatica* saponins), the advantage of polyethylene glycol-200 as matrix was confirmed. The MIKE/CAD technique is useful not only for obtaining information on the structure of the carbohydrate chain but also for the analysis of saponins in complex mixtures (Fraisse *et al.* 1986).

FAB-MS has been used to examine the saponin composition of 13 varieties of legume seed. Although the method does not provide sufficient data to definitively identify saponins in a complex mixture, in conjunction with HPLC it is a powerful analytical tool and can indicate the presence of unidentified saponins (Price *et al.* 1988). One of the findings of this study, confirmed by other workers, is that in positive-ion FAB-MS none of the fragment ions appear to be cationized.

Californium plasma desorption MS (PD-MS)

Californium-252 plasma desorption MS uses the interaction of 100 MeV heavy ions from the spontaneous fission of ^{252}Cf with a solid matrix to induce desorption and ionization. The ^{252}Cf fission fragments sputter molecules deposited on thin nickel foil. The technique has low resolution but provides easily interpretable information, including molecular weight data. It has been successfully applied to the structure determination of steroid saponins from *Cornus florida* (Cornaceae) (Hostettmann *et al.* 1978).

Peracetylated saponins give quasimolecular ions $[M + Na]^+$ in PD-MS, accompanied by fragments which arise from sequential ruptures of the glycosidic bonds (Massiot *et al.* 1988b). Sequencing of saccharide moieties in saponins such as those from *Tridesmostemon claessenssi*

(Sapotaceae) has been greatly aided by this technique. Analysis by PD-MS of peracetylated tridesmosaponin A (which contains a gentiobiosyl moiety at C-3 and a pentasaccharide at C-28), for example, gave a quasi-molecular ion $[M + Na]^+$ at m/z 2411. Intense fragments at m/z 273 and 331 resulted from terminal rhamnose and glucose moieties. Disaccharide peaks were observed at m/z 489 (Xyl–Rha) and 619 (Glc–Glc), while an intense ion consistent with a tetrasaccharide composed of three rhamnosyl units and a xylose appeared at m/z 950. That this tetrasaccharide was situated on a xylose moiety was shown by a peak at m/z 1166 (Xyl–Rha(Rha)– Xyl–Rha), thus confirming the structure of the C-28 sugar chain, deduced from extensive NMR analysis (Massiot *et al.* 1990).

Analysis of the PD mass spectra of acetate derivatives also helped in the structure elucidation of 21-*O*-angeloyl- and 21-*O*-tigloyl-R_1-barrigenol glycosides from the leaves of *Steganotaenia araliacea* (Umbelliferae) (Lavaud *et al.* 1992).

Secondary ion mass spectrometry (SIMS)

This is another particle-induced desorption technique, in which keV ions impinging on the surface of a thin film of biomolecule induce the same desorption ionization as in PD-MS (Benninghoven and Sichtermann, 1978). The utility of this method in the structural investigation of three new bitter and astringent bidesmosides, acetyl-soyasaponins A_1, A_2 and A_3, isolated from American soybean seeds (*Glycine max*, Leguminosae) has been demonstrated. The significant fragment ion peaks provided information regarding the mode of acetylation in the monosaccharide units, as well as the sequence of these units (Kitagawa *et al.* 1988b).

Laser desorption (LD)

In LD it has been demonstrated that excitation by short duration laser pulses (< 10 ns) produces patterns of desorbed molecular ions similar to PD and SIMS.

Laser desorption/Fourier transform mass spectrometry (LD/FTMS), a technique which is also suitable for the analysis of complex glycosides, produces spectra which are different from and complementary to FAB-MS. Positive-ion mass spectra (with potassium chloride doping) of α-solanine and α-tomatine showed pronounced $[M + K]^+$ peaks, while negative-ion mass spectra gave $[M + Cl]^-$ and $[M - H]^-$ peaks. The positive-ion spectra showed little fragmentation, but in the negative-ion spectra some fragment peaks appeared, arising from cleavages at glycosidic linkages. The positive-ion spectrum of the pentaglycoside digitonin

(972, Appendix 3) gave an abundant $[M + K]^+$ peak at m/e 1276 and peaks corresponding to the loss of the chain-ending xylose, glucose and galactose moieties. In the negative-ion spectrum, the chloride attachment ion $[M + Cl]^-$ was observed at m/e 1263, the $[(M + Cl)-Xyl]^-$ ion at m/e 1131, the $[(M + Cl)-Glc]^-$ ion at m/e 1101, the $[(M + Cl)-Xyl-Glc]^-$ ion at m/e 969 and the $[(M + Cl)-Glc-Gal]^-$ ion at m/e 939. When comparing linear and branched glycosides, the fragmentation in the spectra appeared to be more closely related to the degree of branching in the sugar chain than to the number of sugars in the chain. However, chain length did seem to be involved since positive- and negative-ion spectra of digitonin contained more fragment ions than those of α-tomatine, which in turn contained more fragment ions than α-solanine (Coates and Wilkins, 1986).

Field desorption MS (FD-MS)

This technique is also practical for determining the molecular weights of saponins, together with the number, nature and sequence of sugar residues (Schulten and Games, 1974; Beckey and Schulten, 1975; Komori *et al.* 1985). However, the experimental complexity of FD-MS and the fact that FAB-MS produces longer-lasting spectra has meant that the FD-MS approach has declined in popularity recently. FD mass spectra have the added drawback that they are complicated by the presence of cationized fragments, making interpretations difficult. All the same, FD-MS has been very successfully applied to the structure elucidation of, for example, saponins from *Hedera helix* (Araliaceae) (Hostettmann, 1980), akebiasaponin G (Schulten *et al.* 1978) and gleditsiasaponin C from *Gleditsia japonica* (Leguminosae) (Schulten *et al.* 1982). The FD-MS of the hederagenin glycoside **377** (Appendix 2) from *Hedera helix* is shown in Fig. 4.17. As well as the pseudo-molecular ion m/z 797 $[M + H]^+$, the peak at m/z 635 corresponds to the loss of one glucose unit. No aglycone peak could be observed (Hostettmann, 1980).

The FD-MS of steroid saponins have been reported: for polyphyllin H (**1100**, Table 2.9, p. 90) (Singh *et al.* 1982a), and for saponins from *Cornus florida* (Cornaceae) (Hostettmann *et al.* 1978) and *Balanites aegyptiaca* (Balanitaceae) (Liu and Nakanishi, 1982). A comparison of FD and FAB mass spectra of spirostanol and furostanol saponins from *Paris polyphylla* (Liliaceae) has been carried out (Schulten *et al.* 1984; Komori *et al.* 1985), giving information about molecular weight and sugar sequences. In the FD mass spectra of furostanol saponins, the relative abundances of the peaks indicated the configuration and ring size of the sugar units.

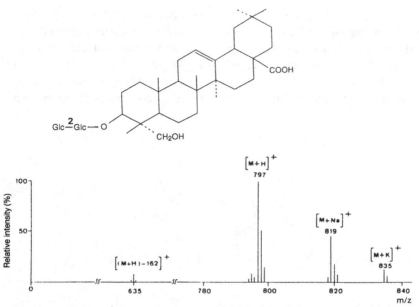

Fig. 4.17. FD-MS of compound **377** from *Hedera helix*.

Furthermore, the cleavage of sugars having a furanose structure was easier than those with a pyranose structure.

Within the steroid alkaloid saponins, FD-MS data of those from *Fritillaria thunbergii* (Liliaceae) (Kitajima *et al.* 1982) and *F. verticillata* (Komori *et al.* 1985) have been published.

Desorption–chemical ionization MS (D/CI-MS)

The mechanisms of ionization and fragmentation in D/CI-MS are analogous to those occurring in CI-MS and the same reactant gases (methane, ammonia or isobutane) are used. Ionization is by protonation $[M + H]^+$ or formation of a complex, for example $[M + NH_4]^+$ when ammonia is used as reactant gas. Whereas information about molecular ions is not always readily available by D/CI-MS, more fragmentation occurs than in FAB-MS, often including the very useful complementary ions of the glycosyl moieties.

Applications of D/CI-MS in the study of naturally occurring glycosides have been described by Hostettmann *et al.* (1981). An example of the use of the method in the structure elucidation of saponins is given for the bayogenin glycoside **481**. The D/CI-MS (reactant gas ammonia, negative-ion mode) showed a quasi-molecular ion $[M - H]^-$ at m/z 973 and signals at m/z 811 $[(M - H) \ominus 162]^-$ and m/z 649 $[(M - H) \ominus 324]^-$, corresponding to the loss of one and two glucose moieties, respectively. In the positive-ion mode the quasi-molecular ion was not obtained but the following signals were observed: m/z 830 $[(M + NH_4) - 162]^+$, m/z 668 $[(M + NH_4) - 324]^+$ and m/z 506 $[(M + NH_4) - 486]^+$, corresponding to the loss of one, two and three glucose units (Domon and Hostettmann, 1984).

481

D/CI-MS has also been applied to the structure elucidation of saponins from *Lonicera nigra* (Caprifoliaceae) (Domon and Hostettmann, 1983), *Aphloia theiformis* (Flacourtiaceae) (Gopalsamy *et al.* 1988), *Polyscias dichroostachya* (Araliaceae) (Gopalsamy *et al.* 1990) and *Tetrapleura tetraptera* (Leguminosae) (Maillard *et al.* 1989).

Electron impact MS (EI-MS)

For EI-MS, the most common MS method, samples need to be volatilized and, of course, saponins cannot be analysed by this technique unless they are converted to permethyl or peracetyl derivatives; such derivatization is not applicable to saponins possessing more than four sugar residues (Komori *et al.* 1975). However, mention should be made here of the application of EI-MS to the structure elucidation of the aglycones obtained from saponins. It is possible to observe the molecular peak of the aglycone and also to arrive at conclusions about the structure of the terpenoid skeleton from the fragmentation pattern. One of the diagnostically important features of the EI-MS of aglycones possessing 12(13)-double bonds is the retro-Diels– Alder rearrangement which cleaves the molecule at ring C and enables information to be furnished about substitution in the ring systems (Fig. 4.18) (Djerassi *et al.* 1962; Budzikiewicz *et al.* 1963; Karliner and Djerassi, 1966). An especially good indication of the distribution of hydroxyl groups between rings A and B, and rings D and E can be obtained.

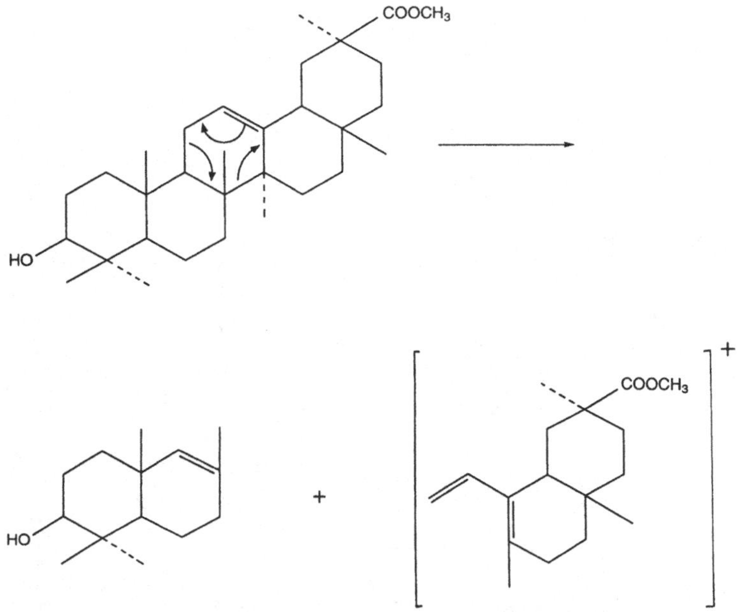

Fig. 4.18. Retro-Diels–Alder reaction of 12(13)-unsaturated oleananes.

Mass spectral information for many pentacyclic triterpenes is available in a review by Agarwal and Rastogi (1974). Ageta and co-workers (Shiojima *et al.* 1992) have documented EI-MS of triterpenoids belonging to the hopane, swertane, lupane, taraxerane, glutinane, friedelane, taraxastane, ursane and other classes. Fragmentations of steroids are found in papers by Coll *et al.* 1983; Fukuda *et al.* 1981; Faul and Djerassi, 1970; Evans, 1974 and Tomowa *et al.* 1974. Some characteristic mass spectral fragments formed by cleavage of the E and F rings of a steroid sapogenin are shown in Fig. 4.19 (Takeda, 1972).

The EI-MS of peracetylated saponins gives useful information as to the construction of the sugar chain. For example, fragment ions corresponding to terminal glucose, $(Glc)Ac_4$ (m/z 331), to terminal rhamnose, $(Rha)Ac_3$ (m/z 273) and to $(Rha–Glc)Ac_6$ (m/z 561) aided in the structure elucidation of a cyclamiretin A tetrasaccharide from *Ardisia crenata* (Myrsinaceae) (Wang *et al.* 1992).

4.2.2 *Nuclear magnetic resonance (NMR)*

Of all the modern methods for the structure elucidation of oligosaccharides and glycosides, NMR spectroscopy provides the most complete information, with or without prior structural knowledge

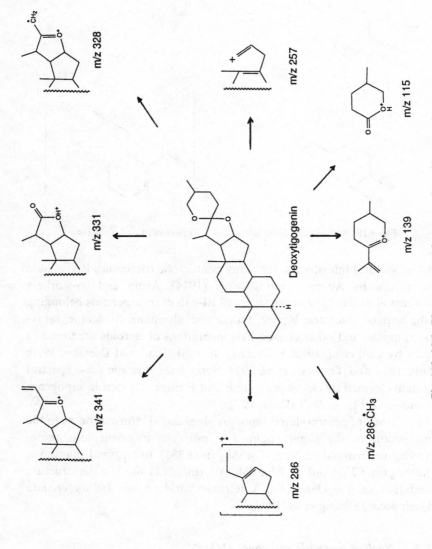

Fig. 4.19. Main EI mass spectral fragments of steroid sapogenins.

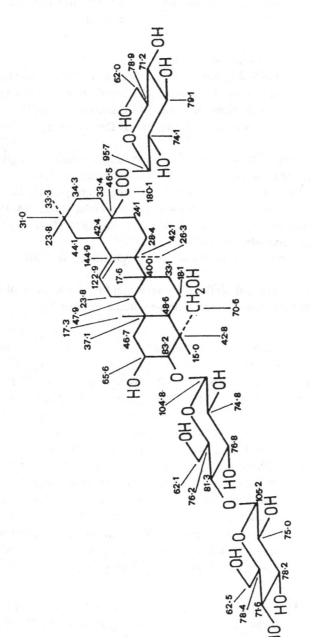

Fig. 4.20. ¹³C-NMR data for a typical bidesmosidic saponin.

(Agrawal, 1992). It is the only approach which can, in principle, give a complete structure without resort to any other method.

^{13}C-Nuclear magnetic resonance

Carbon-13 NMR spectroscopy, now widely used for the structure determination of saponins, is a fast and *non-destructive* method but requires quite large quantities of sample (mg amounts). Analysis of the spectra allows conclusions to be drawn about the following aspects:

- positions of attachment of the glycosidic chains to the aglycone
- sequence, nature and number of monosaccharides
- configuration and conformation of the interglycosidic linkages
- presence of acylglycosides in the chains (Ishii *et al.* 1978, 1984)
- nature of the aglycone (Doddrell *et al.* 1974; Tori *et al.* 1974; Blunt and Munro, 1980)
- structures of attached ester acids (Okada *et al.* 1980).

For assigning chemical shifts, it is very helpful to compare observed data with data reported for model and related compounds.

Triterpene saponins. As a guide to some of the typical chemical shifts in the ^{13}C-NMR spectrum of a triterpene saponin, the values obtained for the bayogenin glycoside **481** are shown in Fig. 4.20 (Domon and Hostettmann, 1984), It is of interest that sugar carbon resonances occur largely in a region distinct from that of the sapogenin moiety. Compilations of assignments of ^{13}C-NMR signals for oleanane (Patra *et al.* 1981; Agrawal and Jain, 1992), ursane, lupane (Wenkert *et al.* 1978; Sholichin *et al.* 1980), hopane (Wenkert *et al.* 1978; Wilkins *et al.* 1987) and lanostane (Parrilli *et al.* 1979) triterpenes have been made (Nakanishi *et al.* 1983). The relevant data for dammarane glycosides have been summarized in a review (Tanaka and Kasai, 1984), while ^{13}C-NMR spectroscopy of saikogenins (Tori *et al.* 1976a) and of saikosaponins (Tori *et al.* 1976b) has been described. Ginseng sapogenins and related dammarane triterpenes have also been studied (Asakawa *et al.* 1977).

Steroid saponins. The use of ^{13}C-NMR for the structural analysis of steroid saponins has been reviewed by Agrawal *et al.* (1985). This article also contains compilations of NMR data for sapogenins and saponins. In addition, ^{13}C-NMR data have been published for 5β-steroid sapogenins (Tori *et al.* 1981), furostanes (Hirai *et al.* 1982), hecogenin

(Eggert and Djerassi, 1975), gitogenin (van Antwerp *et al.* 1977), sarsasa-pogenin (Eggert and Djerassi, 1975) and diosgenin (Puri *et al.* 1993). Carbon-13 NMR spectral analysis of steroid saponins serves among other things to distinguish between spirostane and furostane skeletons and between (25*R*)- and (25*S*)-isomers.

Glycoalkaloids. Full assignments of the ^{13}C-NMR spectra of tomatine (in d_6-dimethyl sulphoxide; Weston *et al.* 1977); solasodine (in d-chloroform; Puri *et al.* 1993); solasonine, solamargine and khasianine (all in d_5-pyridine; Mahato *et al.* 1980a); and demissidine, solanidine, tomatidine and soladulcidine (Radeglia *et al.* 1977; Weston *et al.* 1977; Willker and Leibfritz, 1992) have been published.

Discrimination of the configuration at C-22 (*R*: solanidine type; *S*: tomatidine type) can be decided by the chemical shift of C-26 in the ^{13}C-NMR spectrum of glycoalkaloids (Radeglia *et al.* 1977).

When hydroxyl groups are derivatized, i.e. glycosylated, methylated (or acetylated), the α- and β-carbons of both the sugar and aglycone moieties undergo characteristic shifts: the α-CH signals are shifted downfield, while the β-C signals are shifted upfield (shift resulting from the general γ-upfield shift). Thus, glycosylation of an aglycone causes a downfield shift of the α-carbon and an upfield shift (glycosidation shifts) of the adjacent carbon atoms (Tori *et al.* 1976b; Kasai *et al.* 1977). In oleananes, glycosidation of the 3β-OH group causes C-3 to shift downfield by *c.* 8.0–11.5 p.p.m., C-2 and C-4 to shift by $+0.9$ or -0.9 to -1.9 p.p.m., C-23 to shift upfield by 0.5–5.1 p.p.m. and C-24 to shift by -0.2 to $+1.6$ p.p.m. Glycosidation of the 28-COOH group causes the carboxylic carbon resonance to move upfield (2.5–5.0 p.p.m.) and the C-17 signal to move downfield (1.0–2.5 p.p.m.) (Agrawal and Jain, 1992). A comparison of the ^{13}C-NMR data of the aglycone and saponin, therefore, gives the site of the sugar linkage (Seo *et al.* 1978; Tanaka, 1985).

In a similar fashion, ^{13}C-NMR will give an indication (in simpler saponins) of interglycosidic linkages by considering displacements of chemical shifts when compared with model compounds (Konishi *et al.* 1978). For this purpose, a table (Table 4.1) of chemical shifts of methyl-glycosides has been included here (Gorin and Mazurek, 1975; Seo *et al.* 1978; Mahato *et al.* 1980b; Agrawal *et al.* 1985). Carbon-13 NMR data for methyl β-D-fucopyranoside have been tabulated by Seo *et al.* (1978), while ^{13}C-NMR signals for the more complex sugars are listed by Gorin and Mazurek (1975) and Dorman and Roberts (1970). Apiose gives characteristic ^{13}C-NMR signals and these have been documented (Sakuma and Shoji, 1982; Adinolfi *et al.* 1987; Reznicek *et al.* 1990a).

Table 4.1. ^{13}C-*NMR chemical shifts of methylglycopyranoside/furanoside pairs*

Methyl glycoside	C-1	C-2	C-3	C-4	C-5	C-6
β-D-Glucopyranoside	103.7	73.7	75.5	70.3	75.5	61.7
β-D-Glucopyranoside tetraacetate	101.1	70.9	71.4	68.1	72.5	61.6
α-D-Glucopyranoside	99.9	72.2	73.9	70.4	71.9	61.5
α-D-Glucopyranoside tetraacetate	96.3	68.2	70.4	66.8	69.7	61.6
β-D-Glucofuranoside	110.0	80.6	75.8	82.2	70.7	64.7
α-D-Glucofuranoside	104.7			80.4		62.6
β-D-Galactopyranoside	104.1	71.2	73.3	69.1	75.3	61.4
β-D-Galactopyranoside tetraacetate	101.5	68.5	70.2	66.8	70.6	61.0
α-D-Galactopyranoside	99.8	69.9	70.2	68.9	71.2	61.8
α-D-Galactopyranoside tetraacetate	96.5	67.0	67.6	65.7	67.6	61.2
β-L-Rhamnopyranoside	102.7	72.2	75.4	73.8	73.5	18.5
α-L-Rhamnopyranoside	102.4	71.9	72.5	73.6	69.4	18.4
β-D-Fucopyranoiside	105.9	72.0	75.2	72.6	71.3	17.2
α-D-Fucopyranoside	101.6	70.0	71.5	73.1	66.9	17.1
α-D-Xylopyranoside	100.3	72.3	74.2	70.3	61.9	
α-D-Xylopyranoside triacetate	96.4	70.5	69.1	68.8	57.7	
β-D-Xylopyranoside	104.8	73.9	76.7	70.1	66.0	
β-D-Xylopyranoside triacetate	100.9	70.2	71.0	68.3	61.3	
α-D-Arabinopyranoside	104.7	71.6	73.3	69.1	66.9	
α-D-Arabinopyranoside triacetate	101.9	69.3	70.4	67.9	63.2	
β-D-Arabinopyranoside	100.7	69.8	69.8	69.1	63.4	
β-D-Arabinopyranoside triacetate	97.6	68.4	69.3	67.2	60.2	
α-D-Arabinofuranoside	109.1	81.5	77.2	84.7	62.0	
β-D-Arabinofuranoside	103.0	77.2	75.3	82.8	63.9	
β-L-Arabinofuranoside	103.3	77.9	76.3	83.1	64.3	
α-L-Arabinofuranoside	109.5	82.0	77.9	84.8	62.5	

Methylglycosides were measured in D_2O and peracetates in $CDCl_3$.
Agrawal *et al.* 1985.

Experience indicates that the chemical shifts of monosaccharide units within polysaccharide chains are similar to those of monosaccharides except for substituent effects: glycosylation causes a downfield shift of 4–10 p.p.m. of the α-carbon, while neighbouring β-carbon atoms are shifted upfield by 0.5–4 p.p.m. (Bock and Thorgensen, 1982; Bock and Pedersen, 1983). Substitution at C-28 (i.e. esterification) causes an upfield shift of *c.* 3 p.p.m. for this carbon atom. However, caution should be exercised with this procedure and a second method (formation of alditol acetates etc.) is recommended to confirm the conclusions drawn. Especially in complex cases, such as tubeimoside I (**645**, 30) (Kong *et al.* 1988), the predictions are not always reliable when employing empirical glycosylation and acetylation shift rules for the sequencing of sugar chains by ^{13}C-NMR.

Consideration of ^{13}C-NMR signals is of value for determining the positions of acetyl (or other ester) groups in acylated glycosides. Konishi *et al.* (1978) were able to establish the structures of the monoacetylated saponins platycodins A and C by this means. Similarly, ^{13}C-acylation shifts were diagnostic for the acetylation position of pseudoginsenoside F_8 (Tanaka and Kasai, 1984; p. 49).

Distortionless enhancement by polarisation transfer (DEPT) (and attached proton test (APT) and insensitive nuclei EPT(INEPT)). These are used for the determination of carbon multiplicities in the ^{13}C-NMR spectrum.

Relaxation data. Sugar sequences and linkages can also be ascertained by considering ^{13}C-NMR relaxation data. Different subunits in the saccharide chain have different relaxation times (T_1) and the terminal member has the largest mean T_1 value (Neszmelyi *et al.* 1977). The T_1 value for a particular carbon atom multiplied by the number (N) of directly bonded hydrogen atoms, gives a measure (NT_1) of the flexibility of the molecule at that point. In the case of saponins, the values of NT_1 are smallest for the triterpene and the inner sugar and greatest for the terminal sugar. This method has been applied to the structure elucidation of colubrinoside (**790**) from *Colubrina asiatica* (Rhamnaceae) (Wagner *et al.* 1983); to the assignment of the Ara–Glc–Ara–sequence of a noroleanane saponin (**686**) from *Celmisia petriei* (Asteraceae) (Rowan and Newman, 1984); and to the Rha–^{6}Glc–^{4}Glc–sequence of a lupane glycoside from *Schefflera octophylla* (Araliaceae) (Kitajima and Tanaka, 1989).

Partially relaxed Fourier transform spectra (PRFT). Another ^{13}C-NMR method for the confirmation of sugar linkages is to run PRFT spectra. By this means, the sequence of sugar linkages in the prosapogenin of entada saponin III from *Entada phaseoloides* (Leguminosae) was confirmed (Okada *et al.* 1987). Other examples of PRFT have been reported by Ishii *et al.* (1981), Hirai *et al.* (1982) and Chen and Snyder (1987).

Anomeric carbons. The number of anomeric carbons in a saponin is readily determined by ^{13}C-NMR, thus defining the number of individual sugar residues present. By comparing the chemical shifts with appropriate model sugars (Bock and Pedersen, 1974; Gorin and Mazurek, 1975; Seo *et al.* 1978; Agrawal *et al.* 1985), the ring size (furanose or pyranose form)

and type of each monosaccharide can be established. Specific shifts for
C-1, C-2 and C-3 (all appearing 4–14 p.p.m. downfield in the furanose
form) and C-5 (shifted 5–7 p.p.m. upfield in the furanose form) are the
markers which allow sugars of different ring sizes to be identified. The
anomeric carbon atoms in pyranoses and their derivatives resonate at
90–110 p.p.m., while carbon atoms bearing secondary hydroxyl groups in
pyranoses give signals at 65–85 p.p.m.; carbon atoms carrying primary
hydroxyl groups are found at 60–64 p.p.m. (Bock and Thorgensen,
1982; Bock and Pedersen, 1983). A characteristic feature of saponins
glycosylated at the carboxyl group is the appearance of the anomeric
carbon of the carboxylic group-bound sugar residue at remarkably upfield
positions (93–97 p.p.m.) (Agrawal, 1992) (see Fig. 4.20). The chemical
shifts of anomeric carbons provide an easy means of determining anomeric
carbon configurations, i.e. either α- or β-glycosides (Seo *et al.* 1978; Kasai
et al. 1979; Tanaka and Kasai, 1984). For example, smilagenin-3-*O*-α-D-
glucopyranoside has an anomeric carbon signal at 98.7 p.p.m., which in
smilagenin-3-*O*-β-D-glucopyranoside is at 103.1 p.p.m. (Agrawal *et al.*
1985). The α-carbon of the aglycone is commonly displaced downfield
by about 7 p.p.m. on β-D-glucosylation, β-L-rhamnosylation and
α-L-arabinosylation, while the downfield shift for α-D-glucosylation,
α-L-rhamnosylation and β-L-arabinosylation is somewhat smaller (*c.*
5–6 p.p.m.) (Kasai *et al.* 1979; Tanaka and Kasai, 1984).

[1]H-Nuclear magnetic resonance

Although ^{13}C-NMR spectral analysis and signal assignment has
become a routine procedure in the structure determination of saponins,
the complete assignment of their ^{1}H-NMR spectra has only seldom
been reported. The ^{1}H-NMR spectra have characteristically proved
complex and tedious to analyse. The vast majority of proton resonances
of the carbohydrate moiety appear in a very small spectral width of
3.0–4.2 p.p.m., with subsequent problems of overlapping. These derive
from the bulk of non-anomeric sugar methine and methylene protons
which have very similar chemical shifts in different monosaccharide
residues.

However, the methyl peaks of triterpenes are readily discernible and
most proton resonance positions in oleanene, ursene and related skeletons
have been assigned since the 1960s (Kojima and Ogura, 1989) by a variety
of techniques. For example, the complete ^{1}H- and ^{13}C-NMR spectral
assignments of soyasapogenol B (**33**) and the configuration of the C-4
hydroxymethyl substituent have been established by a combination of
[13]C-DEPT, ^{13}C-APT, 2-D correlation spectroscopy (COSY) ([1]H–[13]C-

COSY, ^{1}H–^{1}H COSY) and ^{1}H–^{1}H ROESY (2-D nuclear Overhauser enhancement (NOE) in a rotating frame) techniques (Baxter *et al.* 1990). The assignments of quaternary carbon resonances in this sapogenin have been confirmed by ^{1}H-detected heteronuclar multiple-bond (HMBC) and one-bond (HMQC) spectroscopy (Massiot *et al.* 1991b). A full interpretation of the ^{1}H-NMR spectra of diosgenin and solasodine has also been achieved (Puri *et al.* 1993).

Some useful data can be obtained from ^{1}H-NMR spectra for the anomeric configurations and linkages of the sugar chain. For example, the coupling constant of the C-1 proton of α-linked glucose units is approximately 3 Hz, while β-linked units have a coupling constant of 6–7 Hz. More details on the coupling constants of anomeric sugar protons can be found elsewhere (Lemieux *et al.* 1958; Capon and Thacker, 1964; Kizu and Tomimori, 1982).

When difficulties arise in determining configurations of hydroxyl groups at C-2, C-3 and C-23, C-24 of oleanene and ursene triterpenes, analysis of the ^{1}H-NMR signal peaks of the protons on oxygen-bearing carbon atoms gives valuable information (Kojima and Ogura, 1989).

The chemical shifts of H-16 and H_2-26 in glycoalkaloids can be applied to the determination of configuration at C-22 and C-25, i.e. whether they are *R* or *S* (Boll and von Philipsborn, 1965).

The ^{1}H-NMR spectra of (25*S*)-spirostanols have been tabulated by Kutney (1963), while spectral data for about 150 steroidal sapogenins and their derivatives have been summarized by Tori and Aono (1964).

Different 1-D and 2-D NMR techniques

With the advent of high-field-strength spectrometers (400, 500 or 600 MHz), modern NMR techniques (notably 2-D techniques) and pulse sequences are increasingly used for non-degradative glycoside sequence and configuration studies (Schöpke *et al.* 1991). By a judicious choice of techniques, complete structural assignments are possible with only a few milligrams of material. Some examples are described below.

COSY. There are two fundamental types of 2-D NMR spectroscopy: J-resolved spectroscopy in which one frequency axis contains spin coupling (J) and the other chemical shift information, and correlated spectroscopy in which both frequency axes contain chemical shift (δ) information (Agrawal, 1992). One of the major benefits of 2-D analysis is that it provides a method of overcoming the problem of spectral crowding. In high-field ^{1}H-COSY this is especially true of the 2.5–4.0 p.p.m. region, thus simplifying the assignment of saccharide protons. Under favourable conditions, all the protons present in a given sugar residue can be identified. Several general conclusions may be drawn from COSY spectra: substitution positions of monosaccharide units can be determined by the presence or absence of a corresponding hydroxyl proton; ring sizes of the monosaccharides can be determined directly; the nature of the cross-peaks reveals the multiplicity of overlapping peaks and an estimate of coupling constants. The technique has been applied to structure elucidation of triterpene tetrasaccharides from *Androsace saxifragifolia* (Primulaceae) (Waltho *et al.* 1986); luperoside I (**289**, Appendix 2) (Okabe *et al.* 1989); camellidins I (**1316**) and II (**1317**) from *Camellia japonica* (Theaceae) (Fig. 4.13) (Nishino *et al.* 1986); and the cyclamiretin A glycosides ardisiacrispins A (**537**) and B (**542**) (Appendix 2) (Jansakul *et al.* 1987).

In certain cases, structure elucidation of a saponin, together with its sugar sequence, has been achieved by ^{1}H-NMR 1-D and 2-D spectroscopy alone (Massiot *et al.* 1986, 1988b). The saponin is first peracetylated and if the field strength is sufficient (> 300 MHz), the sugar proton resonances split into two zones: one between 4.75 and 5.40 p.p.m. assigned to CHOAc and the other between 3.0 and 4.3 p.p.m. assigned to CH_2OAc, CHOR and CH_2OR. Anomeric protons are located between these two zones in the case of ether linkages or at a higher frequency than 5.5 p.p.m. for ester linkages. Peracetylation also gives derivatives which are soluble in chloroform, benzene or acetone. In the equivalent perdeuterated solvents, the mobility of the molecules is such that signals are observed more clearly and coupling constants can be measured with high accuracy. For the acetylated alfalfa root saponin **1318** (this is a derivative of medicoside I, **383**, Appendix 2), COSY and long-range COSY experiments were sufficient to identify the structure as Ara–^{2}Glc–^{2}Ara–^{3}hederagenin^{28}–Glc (Fig. 4.21) (Massiot *et al.* 1986).

The structures of further peracetylated saponins from the leaves of alfalfa (*Medicago sativa*, Leguminosae) and from *Tridesmostemon claessenssi* (Sapotaceae) have been elucidated by similar techniques to those outlined above. Confirmation of assignments and sugar sequences was obtained from HMQC (for ^{1}J couplings) and HMBC (for ^{2}J and ^{3}J couplings) and homonuclear Hartmann–Hahn (HOHAHA) triply relayed COSY and ROESY experiments (Massiot *et al.* 1990, 1991b). The ester sugar chains of the saponins from *T. claessenssi* contain a β-D-xylose moiety in the unusual ${}^1C_4$ configuration (all the substituents are axial).

At 600 MHz, the ^{1}H-NMR spectrum may be sufficiently well resolved

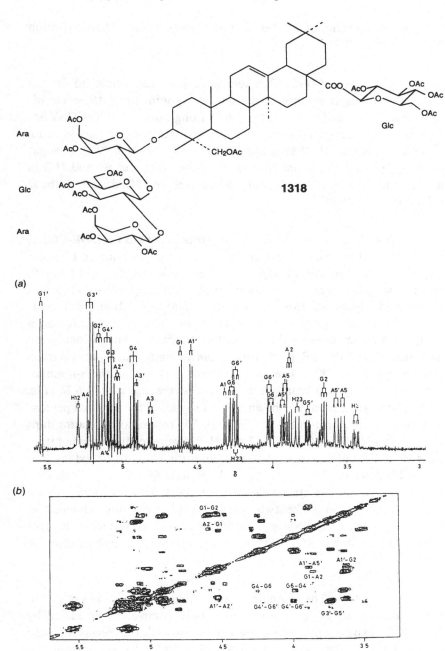

Fig. 4.21. (a) Part of the 500 MHz ^{1}H-NMR spectrum of a peracetylated alfalfa saponin (**1318**) in d-chloroform. (b) Part of the long-range COSY contour plot. For further details, see Massiot *et al.* 1986.

to allow assignment of all ¹H chemical shifts without peracetylation (Schöpke *et al.* 1991).

Long-range COSY. This technique has been employed for the assignment of sugar protons in the steroid saponins from *Allium vineale* (Liliaceae) (Chen and Snyder, 1987, 1989). Long-range ¹H–¹³C COSY has also been used for aglycone structure determination in a cycloastragenol saponin (Wang *et al.* 1989b) and for the location of a ⁴J inter-sugar coupling between the anomeric proton of the inner glucose and H-2 of the inner arabinose of the *Medicago sativa* saponin **1318** described above (Massiot *et al.* 1986).

Double quantum filtered, phase-sensitive COSY (DQF-COSY, DQ-COSY). This technique was applied to the assignment of sugar protons in *Allium* steroid saponins (Chen and Snyder, 1987, 1989) and to the assignments of ¹H chemical shifts in the 16α-hydroxyprotobassic acid glycosides from *Crossopteryx febrifuga* (Rubiaceae) roots (Gariboldi *et al.* 1990). The same technique was used to provide a complete assignment of saccharide protons in the acetylated hederagenin derivative **826** (Table 2.7, p. 66) from *Sapindus rarak* (Sapindaceae) fruits (Hamburger *et al.* 1992). The anomeric proton H − 1″ of the α-rhamnopyranosyl moiety was readily attributable to the signal at δ 6.28 (Fig. 4.22). The starting point for the analysis of signals of the terminal pentose was the anomeric proton H − 1‴ (δ 5.42). The *trans* diaxial relationships of H-2‴ and H-3‴, H-3‴ and H-4‴ and H-5a‴ were readily apparent from their vicinal coupling constants, confirming the sugar to be xylose. The low field shifts of the H-3‴ and H-4‴ signals (δ 5.70 and 5.24) were clearly caused by the strong deshielding effect of O-acetyl groups on the geminal protons. The two acetyl groups were thus attached at C-3‴ and C-4‴ of the xylopyranosyl moiety. Interglycosidic linkages were established by NOE difference spectroscopy (Hamburger *et al.* 1992).

HOHAHA. The proton coupling networks of aglycones of gypsogenin and quillaic acid glycosides have been completely elucidated by HOHAHA experiments. These are similar to COSY experiments (and related to total correlation spectroscopy − TOCSY) except that the observed correlation cross peaks are in phase, thereby preventing accidental nulling of overlapping peaks. For the carbohydrate chains, vicinal coupling constants extracted from HOHAHA experiments allowed the

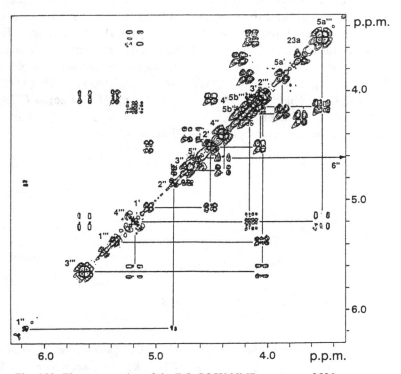

826 R¹ = R² = -OAc, R³ = -H

Fig. 4.22. The sugar region of the DQ COSY NMR spectrum of **826**.

determination of the relative stereochemistry of each asymmetric centre, thus enabling identification of the monosaccharides. Heteronuclear H–C relay experiments were used for assignment of ^{13}C resonances in the saccharide moieties and sugar linkages were determined from HMBC spectra (Frechet *et al.* 1991).

HMBC, HMQC. The use of HMQC and HMBC ^{13}C multiple-quantum coherence spectra is valuable not only for aglycone assignments but also for sugar sequence details. These experiments are analogous to ^{13}C–^{1}H heteronuclear correlated spectroscopy (HETCOR) but instead of observing ^{13}C, the more abundant ^{1}H is detected. In the case of bellissaponins BA$_1$ (**835**) and BA$_2$ (**836**) (Table 2.7, p. 66) from *Bellis perennis* (Asteraceae), it was possible to assign all the chemical shifts in the ^{1}H-NMR spectrum by considering ^{13}C-NMR data in conjunction with 2-D ^{1}H-detected HMQC and HMBC spectra. Cross peaks corresponding to two and three bond couplings were observed for nearly all possible correlations in the molecule. Similarly, long-range ^{1}H–^{13}C correlations in the HMQC and HMBC spectra enabled the determination of the sequence and positions of attachment of the sugar moieties. In **835**, cross peaks were observed for: H-1 (Rha2) → C-3 (Xyl), H-1 (Xyl) → C-4 (Rha1), H-1 (Rha1) → C-2 (Fuc), H-1 (Fuc) → C-28, H-1 (Rha3) → C-3. The attachment of the butenoic acid to C-4 of fucose was confirmed by the cross peak H-4 (Fuc) → C-1 (ester) (Schöpke *et al.* 1991).

835

Long-range heteronuclear correlation spectroscopy incorporating bilinear rotation decoupling pulses (FLOCK). This sequence has found application in the observation of ^{1}H–^{13}C long distance couplings in alatoside A (**691**) from *Sesamum alatum* (Pedaliaceae). Thus, interactions between the proton at C-18 and the carbon atoms C-13, C-17 and C-28 were observed. In conjunction with long-range heteronuclear $^{13}C-^{1}H$ correlation (XHCORR), much information was gathered about the structure of the novel seco-ursene aglycone (Potterat *et al.* 1992).

691

COLOC. An example of $^{13}C-^{1}H$ 2-D correlation spectroscopy optimized for long-range couplings ($^{2}J_{CH}$ and $^{3}J_{CH}$) is to be found in the structure elucidation of saponins from *Crossopteryx febrifuga* (Rubiaceae) (Gariboldi *et al.* 1990).

NOE. This technique has found extensive use in the structure determination of saponins – for example, in the assignment of saccharide protons and sugar sequence of luperoside I (**289**, Appendix 2) (Okabe *et al.* 1989) and camellidins I (**1316**) and II (**1317**) (Fig. 4.13) (Nishino *et al.* 1986). A NOE between the H-2 of arabinose and the anomeric proton of rhamnose helped to confirm the Rha–^{2}Ara– disaccharide linkage in ziziphin (**788**, Table 2.5, p. 58) (Yoshikawa *et al.* 1991b). The method has wide applications since connectivities are often observed between the anomeric proton and the aglycone proton at the linkage position. Negative NOE have been observed between the proton at the C-3 position and the anomeric proton of the 3-*O*-glycoside residue in cycloastragenol and other saponins (Wang *et al.* 1989b).

2-D-NOESY. This was applied to the determination of the sugar sequence of cyclamiretin A glycosides (ardisiacrispins A and B) (Jansakul *et al.* 1987) and the monosaccharide sequence of saxifragifolin A from *Androsace saxifragifolia* (Primulaceae) (Waltho *et al.* 1986). The location of rhamnosyl and glucosyl linkages on the arabinose moiety of kalopanax saponin C (**390**, Appendix 2) were confirmed by NOESY (after sugar sequence analysis of the permethylated saponin). Cross peaks were observed between H-1 of the rhamnosyl moiety and H-2 of the arabinosyl moiety, as well as between H-1 of the glucosyl moiety and H-3 of the arabinosyl moiety (Shao *et al.* 1989b). The structures of the sugar moieties of furostanol saponins from *Balanites aegyptiaca* (Balanitaceae) have been elucidated by means of 2-D NOESY on a 400 MHz NMR instrument (Kamel *et al.* 1991).

ROESY. In cases where attempts to obtain reliable cross peaks are unsuccessful, a ROESY spectrum can show all NOE cross peaks defining interglycosidic linkages (Agrawal, 1992).

Semi-selective excitation. The main disadvantage of inter-residue NOE or long-range scalar coupling across glycosidic bonds for elucidation of sugar chain primary structure is the low sensitivity of the methods and the heavy signal overlap. One solution is to perform 1-D versions of 2-D experiments. These are more sensitive than their 2-D counterparts and give a much higher digital resolution. They are obtained by applying Gaussian-shaped pulses for selective COSY experiments, combined with one- or two-step relayed coherence transfer (RCT), and have been used to determine the structure of a bidesmosidic saponin containing 10 sugar residues (see below) (Reznicek *et al.* 1989a,b). Similarly, semi-selective excitation of one carbohydrate proton, combined with multistep RCT and a terminal NOE transfer, can be used for the determination of the sequence of complex oligosaccharides (Haslinger *et al.* 1990).

RCT. This has been used to assign saccharide structures of steroid saponins from *Allium vineale* (Liliaceae) (Chen and Snyder, 1989).

2D-J. One application of this technique is to be found in studies of the *Camellia* saponins camellidins I (**1316**) and II (**1317**) (Nishino *et al.* 1986).

SINEPT. Selective INEPT NMR spectroscopy, which establishes connectivity between an anomeric proton and a carbon atom three bonds away, has been used for the determination of linkages of saccharide units to the aglycone in polypodoside B. By this means, the glucosyl moiety was shown to be attached at position C-3 and the rhamnosyl moiety at C-26 (Kim and Kinghorn, 1989).

A summary of some of the applications of NMR to the structure elucidation of saccharide moieties found in saponins is shown in Fig. 4.23.

The concerted use of 2-D NMR techniques led to complete ^{13}C and ^{1}H assignments for the oligosaccharide segment of the sarsasapogenin glycoside **921**, 3-*O*-[{α-L-rhamnopyranosyl(1 → 4)}{β-D-glucopyranosyl (1 → 2)}-β-D-glucopyranosyl]-(25*S*)-5β-spirostan-3β-ol. A combination of DEPT, HETCOR, long-range HETCOR, different homonuclear techniques, NOESY and INEPT were applied to the structure elucidation in order to resolve problems caused by overcrowding of the proton spectrum (Pant *et al.* 1988d).

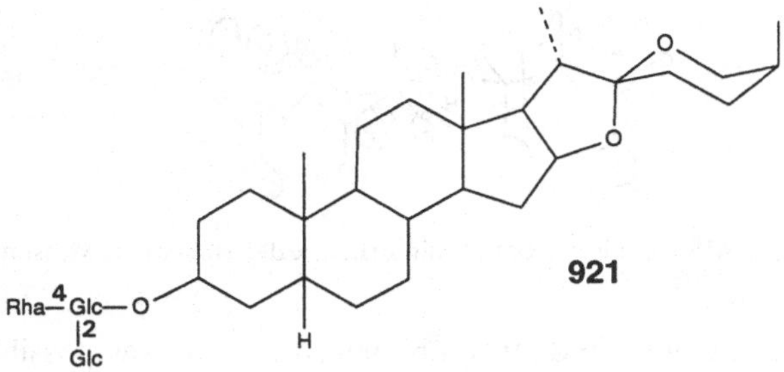

921

Rha—⁴Glc—O
    |2
    Glc

The identification and sequencing of sugars in the pentasaccharide saponin 3-*O*-[β-D-xylopyranosyl(1 → 3)-α-L-arabinopyranosyl(1 → 4)-β-D-glucopyranosyl(1 → 3)-α-L-rhamnopyranosyl(1 → 2)-α-L-arabinopyranosyl]-hederagenin (**404**, Appendix 2) from *Blighia welwitschii* (Sapindaceae) was possible by NMR techniques alone, using a 500 MHz instrument. The saponin was first acetylated; subsequent analysis of the DQF-COSY, NOESY and ROESY spectra allowed assignment of structure. Information obtained from NOE data was most helpful for establishing the sugar sequence (Penders *et al.* 1989).

A saponin containing ten sugar residues from *Solidago gigantea* (Asteraceae) was identified by NMR, based on multi-step RCT experiments. This involved COSY, heteronuclear COSY, COSY-type H–H–C coherence transfer and 2-D NOESY experiments. Extensive degradation

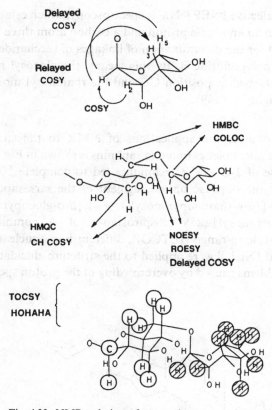

Fig. 4.23. NMR techniques for saponin structure determination (Massiot, 1991).

studies were thus avoided and structure determination was possible with 30 mg of the product (Reznicek *et al.* 1989a,b). Similar techniques were employed for the structure determination of another four glycosides, giganteasaponins 1–4 (**489–492**, Appendix 2) (bidesmosides of bayogenin containing nine or ten sugar units), from the same plant (Reznicek *et al.* 1990a).

A combination of 2-D COSY, HMBC and ROESY NMR experiments was sufficient to give the sequence and linkage positions of the hexasaccharide in mimonoside A (**266**, Appendix 2), an oleanolic acid saponin from *Mimosa tenuiflora* (Leguminosae) (Jiang *et al.*, 1991).

The 2-D NMR of peracetylated and underivatized chrysantellin A (**326**, Appendix 2) has allowed the assignment of protons and the sequencing of sugars. The esterified xylose moiety was shown to exist in the β-form and have a $^1C_4$ conformation. Among the techniques employed were HMQC, HMBC and ROESY (or, more precisely, CAMELSPIN (cross-relaxation appropriate for minimolecules emulated by locked spins)) on the peracetylated derivative and HOHAHA, TOCSY on the native saponin. The ROESY experiment was particularly useful for determining the sugar sequence (Massiot *et al.* 1991a).

326

The presence of an ester-linked β-xylose moiety with a $^4C_1$ conformation in aster saponin C from *Aster tataricus* (Asteraceae) has also been proved by the 1-D HOHAHA approach (Nagao *et al.* 1989b).

The sequences of sugar and interglycosidic linkages of triterpene glycosides from marine organisms have been established from NT_1 data and NOESY experiments (Miyamoto *et al.* 1990) but this methodology is limited by the complexity of the ^{1}H-NMR spectra in the 3–5 p.p.m. region, which usually precludes the measurement of NOE for a large number of protons. However, a combination of COSY, NOESY and direct and XHCORR NMR spectroscopy has allowed complete signal assignment and structural analysis of pentasaccharide triterpene saponins from the sea cucumber *Holothuria forskalii* (Rodriguez *et al.* 1991).

In the structure determination of santiagoside, an asterosaponin from the Antarctic starfish *Neosmilaster georgianus*, the techniques of COSY, TOCSY, HMQC and ROESY NMR spectroscopy were extensively applied, with ROESY experiments being used to resolve the exact sequence of sugars, their points of attachment and the stereochemistry (Vazquez *et al.* 1992).

As for steroid saponins, the structure determination of acylated alliogenin derivatives from *Allium giganteum* (Liliaceae) was successfully completed with the aid of DEPT, ^{1}H–^{1}H COSY, ^{1}H–^{13}C COSY, 2-D NOESY and 2-D INADEQUATE (incredible natural abundance double quantum) techniques (Sashida *et al.* 1991a).

The 1-D TOCSY and ROESY experiments used in studies of tomatine (**1158**, Table 2.10, p. 99) were helpful for the assignment of sugar moieties and for establishing the connectivity and conformation of the carbohydrate chain (Willker and Leibfritz, 1992).

Other structure determinations of glycoalkaloids by modern NMR techniques have been performed by Puri *et al.* (1994).

4.2.3   Infra-red spectroscopy (IR)

Apart from the usual applications of IR, there are one or two features which are of relevance to the structure elucidation of saponins. IR is useful for the characterization of steroid sapogenins because several strong bands between 1350 and 875 cm^{-1} are diagnostic for the spiroketal side chain (Wall *et al.* 1952; Rothman *et al.* 1952a; Eddy *et al.* 1953; Jones *et al.* 1953; Hayden *et al.* 1954; Jain, 1987b). Four bands, 980 (A band), 920 (B band), 900 (C band) and 860 cm^{-1} (D band) have been assigned as characteristic of the E and F rings. With 25*S*-sapogenins the B band has a stronger absorbance than the C band, while in the 25*R*-series this relationship is reversed. In sapogenins having oxygen substituents in the E and F rings or at position 27, the four bands are considerably changed (Takeda, 1972).

The presence of ionized carboxyl groups in saponins can be ascertained by bands in the IR spectrum at 1610 and 1390 cm^{-1} (Numata *et al.* 1987). This information is useful during the isolation procedure, when it is important to know whether carboxyl groups in the molecule are ionized.

4.2.4   X-ray crystallography

Because of the difficulty in obtaining suitable single crystals for X-ray crystallographic analysis, this technique has found virtually no application in the structure determination of saponins. One of the very rare exceptions is the trisaccharide triterpene asiaticoside (**597**, Appendix 2) from *Centella asiatica* (Umbelliferae), the molecular geometry of which was determined by single crystal X-ray analysis. Crystallization was from dioxane (Mahato *et al.* 1987). X-ray diffraction analysis was also successful for confirmation of the structure of mollic acid 3-β-D-glucoside (**626**, Appendix 2) (Pegel and Rogers, 1985).

X-ray crystallography has, however, been an important means of solving structural problems of aglycones. Useful information for the determination of the structure of the aglycone of alatoside A (**691**, p. 39) (from *Sesamum alatum*, Pedaliaceae) was obtained by an X-ray diffraction analysis of the crystalline triacetate of the artifact produced after acid hydrolysis (Potterat *et al.* 1992). An X-ray crystallographic study of medicagenic acid (**31**, Table 2.1, p. 21), the parent aglycone of medicagenic acid 3-*O*-glucoside (**434**, Appendix 2) from the tubers of *Dolichos kilimandscharicus* (Leguminosae), showed the molecule to have *cis*-fused D and E rings. Ring C had a slightly distorted sofa conformation, while rings A, B, D and E had chair conformations (Stoeckli-Evans, 1989).

4.2.5 Molecular optical rotation

A correlation exists between molecular rotation difference (ΔM) and the structure and stereochemistry of triterpenes. Thus it has been possible to classify different stereoskeletal types on the basis of their ΔM values (Ogunkoya, 1977).

Klyne's rule applies molecular rotation [M]$_D$ values to the determination of the configuration of sugar linkages (Klyne, 1950). Its use has not been frequently reported; examples are the structure elucidation of rotundioside A (**666**) from *Bupleurum rotundifolium* (Umbelliferae) (Akai *et al.* 1985a,b) and the saponins from *Patrinia scabiosaefolia* (Valerianaceae) (Choi and Woo, 1987).

Nakanishi and Liu have described a method based on split CD curves of pyranose polybenzoates for determining glycosidic linkages at the branching points. In the case of balanitin 1, a tetrasaccharide of yamogenin from *Balanites aegyptiaca* (Balanitaceae), the saponin was first permethylated and methanolysed. After *p*-bromobenzoylation, two UV-absorbing products were isolated. These were shown by CI-MS to be a mono- and a dibenzoate of glucose, thus proving that one of the glucose moieties was monosubstituted and one disubstituted. No other branched monosaccharides could be present, since all terminal units are permethylated methyl glycosides and are not UV active. Measurement of the amplitudes of the CD curve from the dibenzoate, for example, established the presence of a 1,3-*ee* dibenzoate and thus the branching points could only be at C-2 and C-4 (Liu and Nakanishi, 1981). Although the method has not yet been widely used, it has the advantage of using only microgram quantities of material and no reference samples are required.

Incidentally, CD spectra are also valuable for furnishing details of configuration, such as that at C-22 of steroid saponins (Jennings *et al.*, 1964; Yamashita *et al.* 1991).

4.2.6 Permethylation and alditol acetate formation

A combination of ^{13}C-NMR and FAB-MS gives useful information about the structures of saponins but is not confirmatory in all cases. For example, in the structure elucidation of saponins from *Swartzia madagascariensis* (Leguminosae) (Borel and Hostettmann, 1987) and *Anagallis arvensis* (Primulaceae) (Glombitza and Kurth, 1987b), recourse to other methods was essential.

One of the important procedures for the determination of inter-glycosidic linkages in the saccharide moieties of saponins is permethyl-ation, followed by cleavage and identification of the modified sugar residues. Among the numerous methods for permethylation (Wallenfels *et al.* 1963), the Hakomori method has found the most widespread application (Hakomori, 1964). Reaction is carried out with methyl iodide in dimethyl sulphoxide, in the presence of sodium hydride. Verification of the degree of methylation is necessary, by MS, ^{1}H-NMR, IR (dis-appearance of –OH bands) and TLC. Complete methylation is sometimes difficult and during the procedure some cleavage of acylglycosidic linkages may occur. Improved results can frequently be obtained by alternative methods – for example, the use of sodium hydroxide/sodium *t*-butoxide mixtures instead of sodium hydride (Ciucanu and Kerek, 1984; Gunzinger *et al.* 1986; Sung *et al.* 1991). The methylated product can be hydrolysed into its component methyl sugars (Kuhn *et al.* 1955a), which after chromatographic purification are identified by comparison with authentic samples (Tschesche and Wulff, 1972). It has been claimed, however, that methyl 3,4,6-tri-*O*-methyl-D-glucopyranoside and methyl 2,3,4-tri-*O*-methyl-D-glucopyranoside are indistinguishable if their GC retention times are compared. This implies that sophoroside and gentiobioside moieties cannot be discriminated by this method (Oobayashi *et al.* 1992).

Another method has been described in which methylation and *bromo-benzoylation* were used to deduce the interglycosidic linkages of sugars in *Guaiacum officinale* (Zygophyllaceae) saponins (Ahmad *et al.* 1988a): permethylation of the saponin was followed by methanolysis and sub-sequent *p*-bromobenzoylation. The mixture of *p*-bromobenzoyl deriv-atives was separated by HPLC to give axial anomers as major and equatorial anomers as minor products. Their substitution patterns were determined by NMR spectroscopy.

A better way of determining interglycosidic linkages is to carry out a GC–MS analysis of the *alditol acetates* of the sugars from the permethyl-ated saponins. The permethylated product is hydrolysed, reduced and subsequently acetylated, giving the corresponding monosaccharide deriv-atives, which are analysed and compared with the data from authentic samples (Björndal *et al.* 1967, 1970; Lindberg, 1972; Jansson *et al.* 1976). This approach was used to sequence the trisaccharide moiety of an oleanolic acid saponin isolated from *Anemone rivularis* (Ranunculaceae). The saponin was first permethylated and then hydrolysed to cleave the sugars. Reduction and acetylation led to formation of the alditol acetates. When reduction was carried out with sodium borohydride, it was impossible to distinguish 2-linked xylose from 4-linked xylose because of formation of the same di-*O*-methylalditol diacetate from both compounds. The problem was resolved by the use of deuterated sodium borohydride: the characteristic MS fragments at m/z 117 and 190 associated with 2-linked xylose were distinguished from those at m/z 118 and 189 associated with a 4-linked xylose (Fig. 4.24) (Mizutani *et al.* 1984). Further examples involving the alditol acetate procedure are numerous in the literature and only a selection will be mentioned here: Becchi *et al.* (1979), Wagner *et al.* (1984b), Dubois *et al.* (1986) and Glombitza and Kurth (1987b).

On a cautionary note, comparison of GC retention times for alditol acetates is not an infallible means of establishing interglycosidic linkages. The fact that 1,4,5-tri-*O*-acetyl-2,3-di-*O*-methylpentitol and 1,2,5-tri-*O*-acetyl-3,4-di-*O*-methylpentitol have identical retention times most probably led to the erroneous assignment of 4-substitution to arabinose in ziziphin (**788**, Table 2.5, p. 59) (Yoshikawa *et al.* 1991b).

While analysis of alditol acetates gives information about the linkage positions of monosaccharides, the actual distribution of substituents on the inner sugars cannot be determined. It is then necessary to obtain supplementary data from methods such as partial hydrolysis (see Section 4.1.3) and methylation–ethylation analysis (Valent *et al.* 1980; Glombitza and Kurth, 1987b).

For bidesmosides it is necessary to first cleave the acylglycosidic sugar chain of the permethylated saponin with lithium aluminium hydride. This is then hydrolysed and the component monosaccharides are reduced, acetylated and analysed by GC–MS. If reduction of the cleaved sugar chain is performed with sodium borodeuteride, the sugar attached directly to the aglycone can be determined, as this is not deuterated.

1. Permethylation/reduction/acetylation

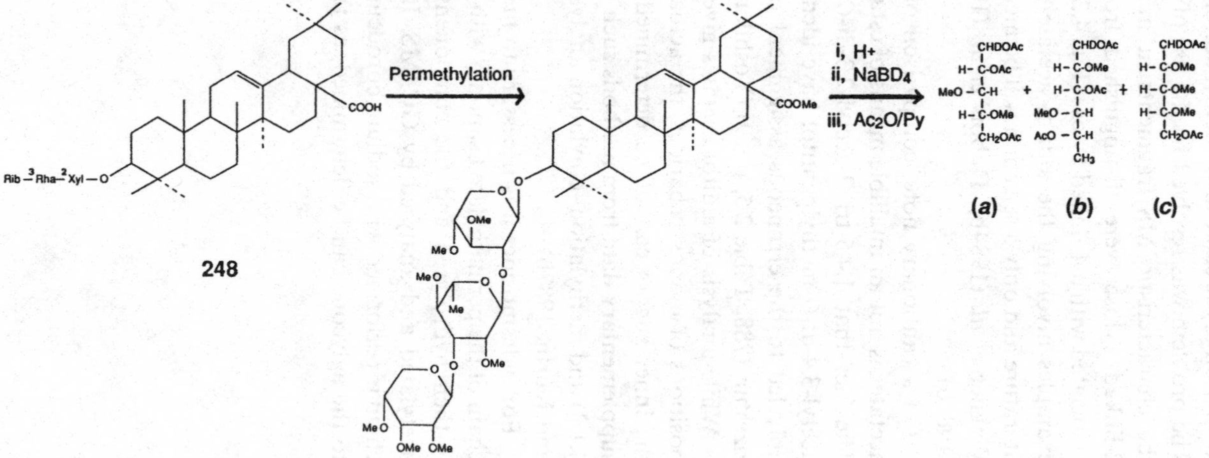

Rha–³Rha–²Xyl–O

COOH

Permethylation →

248

COOMe

OMe OMe OMe Me

i, H+
ii, NaBD₄
iii, Ac₂O/Py
→

(a)

CHDOAc
H–C–OAc
MeO–C–H
H–C–OMe
MeO–C–H
CH₂OAc

+ (b)

CHDOAc
H–C–OMe
H–C–OAc
MeO–C–H
AcO–C–H
CH₃

+ (c)

CHDOAc
H–C–OMe
H–C–OMe
H–C–OMe
CH₂OAc

2. GC-MS analysis

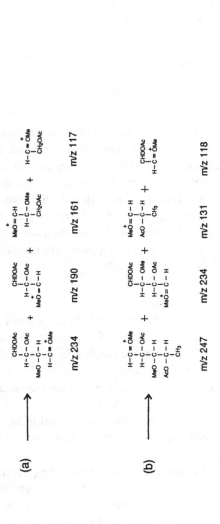
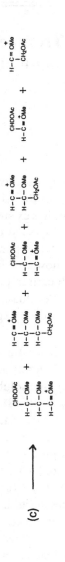

Fig. 4.24. Analysis of alditol acetates of a saponin from *Anemone rivularis* (Mizutani *et al.* 1984).

5

Triterpene saponins – pharmacological and biological properties

The widespread occurrence of saponins has meant that an awareness of these glycosidic plant constituents and their detergent and piscicidal properties has existed for centuries. Early investigations on their biological activities involved crude saponin mixtures but, with the introduction of sophisticated separation and structure elucidation techniques, the tendency is now to work with pure saponins or mixtures of pure saponins. Although certain properties of saponins are well defined, some effects are difficult to evaluate and, in other instances, responses are only manifested at high doses in experimental animals.

5.1 General properties of saponins

Generalizations about the solubilities of saponins are hazardous but many are soluble in water (particularly if the water contains small amounts of alkali) and alcohol; some are soluble in ether, chloroform, benzene, ethyl acetate or glacial acetic acid.

As already mentioned in Chapter 1, it is impossible to characterize saponins purely by their physical or biological properties. It is true that many share haemolytic, piscicidal, cholesterol complexation and foaming properties (McIlroy, 1951) but the exceptions are too numerous for the use of these criteria in a global manner.

However, on the basis of historical considerations, the purely classical 'general' properties of saponins will be described. These rely mainly on their tensioactive nature. Of great importance is the interaction between saponins and cell membranes, with effects on the hydrophobic–lipophilic balance (HLB) and on membrane permeability. The precise nature of this phenomenon is unfortunately still unclear.

One of the most striking features of saponins is the enormous difference between mono- and bidesmosidic saponins, with bidesmosides either

232

lacking or exhibiting an attenuation of the properties of the corresponding monodesmosides. An exception to this phenomenon is surface activity. This is more pronounced in the bidesmosides and increases with length and branching of the saccharide moiety (Wulff, 1968). Despite the large differences, however, the inactive bidesmosidic compounds can often be readily converted into the corresponding biologically active mono-desmosidic saponins by simple basic hydrolysis.

The amphiphilic nature of saponins dominates their physical properties in solution. They are strongly surface active, form stable foams, act as emulsifying agents and form micelles in much the same way as detergents.

5.1.1   Formation of stable foams

This is the most characteristic of saponin properties and the phenomenon has been exploited as a test for the presence of saponins (Steiner and Holtzem, 1955). Caution should be exercised, however, because not all saponins foam in aqueous solution and, furthermore, other natural products are capable of producing foams. The mechanism of foam formation is not clear; it is uncertain whether the hydrophilic–hydrophobic (amphiphilic) character of saponins is responsible.

The emulsifying properties of saponins are well known, the advantage over soaps and many other surfactants being that their salt-free nature makes them less likely to be affected by alkaline or acid conditions. Foam-forming activity can be measured by the method of O'Dell *et al.* (1959) which involves shaking 5 ml 0.01–1% saponin in $\frac{1}{15}$ M K_2PO_4 for 1 min in a 25 ml graduated cylinder and observing the amount of formed foam after another minute. This simple method gives a good correlation between the amount of foam and the concentration of saponin for various saponins. Other methods of measuring foaming characteristics have been described by Huffman *et al.* (1975) and Mangan (1958).

5.1.2   Formation of complexes with cholesterol

The property of complex formation between saponins and cholesterol was discovered by Windaus (1909). Systematic studies on the ability of different classes of saponin to complex with cholesterol have been carried out (Tschesche and Wulff, 1964; Wulff, 1968). Complexes or loose compounds can also be formed with lecithin, ergosterol, amyl alcohol, terpene alcohols, phenols and thiophene. High-molecular-weight proteins and gallotannins are likewise known to interact with saponins.

5.1.3   *Haemolysis*

The ability of saponins to cause haemolysis of blood *in vitro* was first reported by Kobert in 1887. Low concentrations of saponins are capable of destroying (lysing) erythrocyte membranes, causing a release of haemoglobin. The phenomenon involves a reduction of surface tension between the aqueous and lipid phases of the erythrocyte membrane, causing emulsion of the lipids and their subsequent departure from the membrane. Through these holes Na^+ and water are allowed to enter the cell, while K^+ leaves. This flux occurs until the membrane ruptures and haemoglobin is shed into the plasma. The sensitivity of red blood cells to saponins has led to the widespread use of haemolysis as a quantitative determination method (see Chapter 3), with the haemolytic index defined as $30\,000 \times a/b$ (where a is the quantity of standard saponin (in grams) required for complete haemolysis of blood and b is the quantity of test saponin (in grams) required for complete haemolysis). In the *Pharmacopoea Helvetica VII*, the haemolytic potential of saponin-containing preparations is expressed in Ph. Helv. units. One Ph. Helv. unit represents the haemolysis caused by 10 mg of standard saponin.

Haemolytic activity varies considerably with the structure of the glycoside (Table 5.1). Monodesmosidic steroid and triterpene saponins (except acylglycosides and glycyrrhizin) are strongly haemolytic but bidesmosidic furostanol saponins and triterpene bidesmosides are virtually inactive, with certain exceptions, such as gypsoside A (**292**, Appendix 2). Ester saponins, however, often have high haemolytic activity. Polar groups, as represented by saccharide moieties or numerous free hydroxyl groups, on rings D and E are probably responsible for a reduction in the haemolytic activity (Table 5.2). Concerning the carbohydrate moiety, haemolysis decreases with length and increases with branching of the sugar chain, although branching can also lead to a diminution of activity (Romussi *et al.* 1980).

Methyl oleanolate is considerably more haemolytic than oleanolic acid and, in a series of glycosides of the former, the haemolytic activities of methyl oleanolate diglycosides were stronger than those of the monoglycosides (Takechi and Tanaka, 1992).

The relationships between chemical structure and haemolytic activity of 12 triterpene saponins with oleanolic acid and hederagenin as aglycones from the aerial parts of *Lonicera japonica* (Caprifoliaceae) have been studied. Monodesmosidic saponins possessed strong haemolytic activity; in particular, saponins with a glucosylarabinosyl moiety at C-3 showed very strong activity. No difference between the activity of oleanolic acid

Table 5.1. *Haemolytic index of selected saponins*

Saponin	Type	Source	Haemolytic index
Gypsoside A (**292**)	Bidesmosidic triterpene saponin	*Gypsophila* spp.	29 300[a]
α-Hederin (**368**)	Monodesmosidic triterpene saponin	*Hedera helix*	150 000
Primula saponin (mixture)	Monodesmosidic triterpene saponin	*Primula elatior*	50 000
Aescine (mixture)	Triterpene ester saponin (monodesmosidic)	*Aesculus hippocastanum*	98 000
Glycyrrhizin (**269**)	Monodesmosidic triterpene saponin	*Glycyrrhiza* spp.	<2 000
Cyclamin (**538**)	Monodesmosidic triterpene saponin	*Cyclamen europaeum*	390 000
Sarsaparilloside (**1075**)	Bidesmosidic steroid saponin	*Smilax* spp.	<2 000
Digitonin (**972**)	Monodesmosidic steroid saponin	*Digitalis purpurea*	88 000
Tomatine (**1158**)	Glycoalkaloid	*Lycopersicon lycopersicum*	170 000

[a] Measured according to the *Pharmacopoea Helvetica V*, relative to Swiss standard saponin (HI = 25 000), at a blood dilution of 1 : 200.
Modified from Wulff, 1968.

Table 5.2.. *Haemolytic activity of saponins*

Saponins with polar groups in rings D and E	Haemolytic activity[a]	Saponins without polar groups in rings D and E	Haemolytic activity[a]
Sarsaparilloside (**1075**)	>250[a]	Parillin (**923**)	4.0
Convallamaroside (**1129**)	>250	Convallamarogenin triglycoside (**1039**)	2.0
Lanatigoside (**1091**)	>250	Lanatigonin (**948**)	0.6
Methyl-protogracillin (**1103**)	>200	Gracillin (**995**)	2.0
Hederasaponin C (**372**)	>250	α-Hederin (**368**)	3.0
Hederasaponin B (**156**)	>250	β-Hederin (**154**)	0.9
Clematoside C (**265**)	>250		
Gypsoside A (**292**)	30		
Asiaticoside (**597**)	>250		
		Aescine	2.5
		Theasaponin	20

[a] Concentration in μg/ml required for complete haemolysis with washed erythrocytes.
Tschesche and Wulff, 1972.

and hederagenin saponins was observed, while bidesmosidic saponins were only very weakly active (Kawai *et al.* 1988).

The oleanolic acid saponin calenduloside B (**164**, Appendix 2) from the roots of *Calendula officinalis* (Asteraceae) has a haemolytic index of 15 000 (Kofler method) (Yatsyno *et al.* 1978), while saponins from the flowers have haemolytic indices of up to 48 600 (measured by the *French Pharmacopoeia IX* method) (Isaac, 1992).

A study of saikosaponin a, saikosaponin d and their intestinal metabolites has shown the order of haemolytic activity to be monoglycoside > diglycoside > aglycone. In fact, the genins were more or less inactive. The strongest haemolytic activity was shown by prosaikogenin G (**1319**) and saikosaponin d (**548**, Appendix 2), which both possess an α-OH function at C-16.

1319 R = -Fuc
548 R = -Fuc-Glc

Saikosaponin a (**547**, Appendix 2) has a β-OH group at C-16 and has weaker haemolytic properties. The corticosterone-secreting potencies of these compounds correlated to a certain extent with haemolytic activity. For adsorbability on cell membranes, the following relationship was observed: monoglycoside = diglycoside > aglycone, closely paralleling haemolytic activity. One of the most important factors in the haemolytic activity of saikosaponins seems to be the interplay between the sugar moiety (polar) and the aglycone (non-polar) (Nose *et al.* 1989b).

Ginsenosides only show weak haemolytic activity. Saponins from marine organisms, however, possess considerable haemolytic potency. Haemolysis of mouse, rabbit and human erythrocytes by holothurin A (**1241**, Table 2.13, p. 111) was greater than that by digitonin or quillaia saponin (Burnell and ApSimon, 1983). Fusetani *et al.* (1984) have documented the haemolytic activities of starfish steroid saponins.

Sapogenins of haemolytic saponins are themselves supposed to cause haemolysis of red blood cells (Segal *et al.* 1974), but their water-insolubility makes measurement difficult. For aglycones to have activity, they require a polar grouping in ring A and a moderately polar grouping in rings D or E (Schlösser and Wulff, 1969). Although it has been proposed that the haemolytic activity of saponins depends on the structure of the aglycone (Segal *et al.* 1974) and only to a small extent on the sugar moiety, this assertion has been challenged (Anisimov and Chirva, 1980; Roddick and Rijnenberg, 1986 – see Chapter 6). In any case, the mechanism of haemolysis by saponins is not clear. One possibility is that the erythrocyte membranes are perturbed by complex formation of the glycoside with cholesterol and other constituents of the lipid bilayer (Anisimov and Chirva, 1980). At the same time, there is no direct parallel between haemolytic activity and cholesterol complex formation; this can be illustrated by parillin, cyclamin and aescine, which have high haemolytic indices and weak complexing ability with cholesterol (Wulff, 1968). Similarly, there is no relationship between haemolysis and the capability to reduce surface activity (Segal *et al.* 1974). It is possible, however, that hydrolysis of the glycosidic bond precludes haemolysis and that lack of haemolytic activity is the result of either the non-adsorbability of the saponin to the red blood cell or the lack of a suitable membrane glycosidase necessary for the hydrolysis. Evidence for this hypothesis comes from the finding that, although saponins are rapidly adsorbed by red blood cells, no glycosides could ever be extracted from ghost cells obtained from saponin-haemolysed erythrocytes (Segal *et al.* 1974).

In an investigation of the time course of haemolysis, it was found that the haemolytic activity of the steroid saponins tested reached a maximum value well before the triterpene saponins. This is likely to be because of the higher affinity of the steroid aglycone for the erythrocytes. Incubation of the saponins with a small amount of cholesterol caused a dramatic inhibition of haemolysis, consistent with the hypothesis that cholesterol is a common target of haemolytic saponins on erythrocyte membranes (Takechi *et al.* 1992).

5.1.4   Bitterness

Many saponins have a reputation for being bitter. The seed-coat of quinoa (*Chenopodium quinoa*, Chenopodiaceae), for example, contains bitter saponins which have to be removed before cooking (Mizui *et al.* 1988). Similarly, partially acetylated soyasaponins are responsible for the bitter and astringent taste of soybeans (Kitagawa *et al.* 1988b; Shiraiwa *et al.* 1991). Saponins impart a bitter taste to asparagus (*Asparagus officinalis*, Liliaceae) and the 'bottom cut' has to be discarded during processing. A tridesmoside of 16α-hydroxymedicagenic acid (zahnic acid) was found to be the most bitter and throat-irritating compound of all the pure saponins isolated from the aerial parts of alfalfa (*Medicago sativa*, Leguminosae) (Oleszek *et al.* 1992). By comparison there are exceptions: glycyrrhizin, the principal saponin of licorice, for example, is some 50 times sweeter than sucrose.

A simple test for bitterness has been described (Ma *et al.* 1989) and this has been used in the bioactivity-guided fractionation of saponins from seeds of quinoa (which are also toxic to brine shrimp) (Meyer *et al.* 1990). Bitterness of vernoniosides, steroid glycosides from *Vernonia amygdalina* (Asteraceae), a medicinal plant used by both humans and wild chimpanzees, has been determined by a filter paper tasting method. Bitter activities (BA) of vernonioside A_4 (BA 30 µg) and the corresponding aglycone (BA 7 µg) were comparable to that of quinine sulphate (BA 17 µg) (Jisaka *et al.* 1993).

5.2   Solubilizing properties

It has been shown that saponins (especially bidesmosides) have the ability to increase solubilities of water-insoluble compounds. The monodesmosidic saponins Ara–^{3}Rha–^{2}Ara–^{3}hederagenin (**386**) and Ara$f$–^{3}Rha–^{2}Ara–^{3}hederagenin (**388**) (Appendix 2) from *Sapindus mukurossi* (Sapindaceae) are normally only soluble at 0.017 and 0.0076 mg/ml in water, but on addition of a mixture of three bidesmosidic saponins (0.1% solution) from the same plant, the solubilities increase to 5.7 and 5.0 mg/ml, respectively (Kimata *et al.* 1983). The bidesmosides Xyl–^{3}Rha–^{2}Ara–^{3}hederagenin–Glc^2–Glc (**394**) and Ara–^{3}Rha–^{2}Ara–^{3}hederagenin–Glc^2–Glc (**387**) (Appendix 2) begin to exhibit their solubilizing effects on **386** and **388** near their critical micelle concentrations, in a similar manner to the synthetic surfactants (Nakayama *et al.* 1986b).

The saponins of *Bupleurum falcatum* (Umbelliferae) are also solubilized by ginseng saponins and, furthermore, both drugs are found together in certain oriental herbal formulations. Thus, saikosaponin a (**547**, Appendix 2) is normally soluble at 0.1 mg/ml in water, but on addition of ginsenoside Ro (chikusetsusaponin V) (**184**, Appendix 2) this goes up to 3.4 mg/ml (Watanabe *et al.* 1988). The solubility of saikosaponin d (**548**) is similarly dramatically increased. The neutral dammarane saponins of ginseng have themselves no solubilizing effect so it would appear that the glucuronide moiety of **184** is important for the increased solubility of saikosaponin a (Watanabe *et al.* 1988). However, it has been observed that the solubilizing effect of **184** is potentiated by the presence of the neutral dammarane saponins.

In a similar fashion, the glucuronyl saponins **228** and **233** (Appendix

2) from *Hemsleya macrosperma* (Cucurbitaceae) have the power to solubilize saikosaponin a. It was found that 1 ml of a 0.1% aqueous solution of **228** or **233** could dissolve 8.7 mg and 5.0 mg of saikosaponin a, respectively (Morita *et al.* 1986b). These two saponins are closely related structurally to chikusetsusaponin V (**184**).

The strongest solubilising effect on saikosaponin a found so far is produced by tubeimoside I (**645**), a cyclic bidesmoside from *Bulbostemma paniculatum* (Cucurbitaceae) (Kasai *et al.* 1986a). Tubeimosides I–III all increase the solubility of ($\pm$)-α-tocopherol, and with 1-anilino-8-naphthalene-sulphonate it is thought that an inclusion complex is formed by tubeimoside I (Miyakoshi *et al.* 1990).

Solubility enhancement has also been observed with saponins from licorice roots (Sasaki *et al.* 1988).

5.3   Biological and pharmacological activities of triterpene saponins

The list of biological activities associated with saponins is very long. Certain attributes of saponins, such as their fungicidal and piscicidal effects, have been known for many years, while new activities are continually being discovered. Antisweetness and utero-contraction are just two of these.

5.3.1   *Antimicrobial activity*

The function of saponins in plants has often been questioned and there is not always a satisfactory explanation for their very high content (up to 30%) in some species. One theory is that they protect plants against fungal attack (Défago, 1977). As there is often an increase in saponin content of the plant part undergoing microbial attack, this supposition would seem to be quite reasonable. It has been proposed that bidesmosidic saponins, which frequently lack the typical properties of saponins and yet are soluble in water, exist as a transport form from the organs not at risk (e.g. leaves) to those parts of the plant (e.g. roots, bark, seeds) under attack by various microorganisms (Tschesche and Wulff, 1972). When plant tissue is damaged, the released enzymes act on the bidesmosidic saponins. Once transformed into their monodesmosidic derivatives, the saponins can provide a defence against microbial invasion at the threatened area. An example is the production of α-hederin (**368**) from hederasaponin C (**372**) (Appendix 2) in *Hedera helix* (Araliaceae). The main saponin of *H. helix* is hederasaponin C, which is virtually inactive as an antimicrobial agent. Ivy leaves contain an enzyme which cleaves the ester-linked sugar moiety of this bidesmosidic saponin to give the highly antibiotic α-hederin (Tschesche, 1971; Schlösser, 1973).

Similarly, while leaves of spinach *Spinacia oleracea* (Chenopodiaceae) are practically free of saponins, the roots contain the monodesmosidic saponins spinasaponins A (**185**) and B (**379**) (Appendix 2), the aglycones of which are oleanolic acid and hederagenin (both antimicrobial) (Tschesche, 1971).

Pioneering work by Tschesche and Wulff (Tschesche and Wulff, 1965; Wulff, 1968) and Wolters (1966a) has shown that antimicrobial activity is fairly widespread in both steroid and triterpene glycosides. To a large extent, the antimicrobial activity parallels the haemolytic activity.

Fungicidal activity

Numerous early examples of antifungal saponins exist in the literature (Wolters, 1966b, 1968a,b). Wulff (1968) has documented results with a range of different saponins and several fungal strains, both plant pathogenic (*Fusarium solani* etc.) and pathogenic to humans (*Trichophyton* spp., *Microsperma* spp., etc.).

The strongest activities are exhibited by the monodesmosidic saponins. Maximum activity is shown by monodesmosides with four or five monosaccharides. Shorter carbohydrate chains lead to lower water solubility and weaker antifungal activity (Anisimov and Chirva, 1980). A study of the *in vitro* activity of 49 pentacyclic triterpenes against *Saccharomyces carlsbergensis* confirms these observations, since the glycosides of oleanolic acid and hederagenin and a free carboxyl group were the best inhibitors (Anisimov *et al.* 1979).

Certain ester saponins, e.g. aescine, also inhibit fungal growth, while bidesmosides used to be considered inactive (Hiller, 1987). However, a small number of bidesmosides do indeed possess antifungal properties, as evidenced by hederasaponin C (**372**), which is active *in vivo* against *Candida albicans* (mouse) but not *in vitro* (Timon-David *et al.* 1980). During this *in vivo* testing of triterpene glycosides from *Hedera helix* (Araliaceae) in mice infected with *Candida albicans*, test samples were administered orally by means of a tube to the stomach. Daily doses of 100 mg/kg of α-hederin over 10 days and 100 mg/kg of hederasaponin C over 25 days were sufficient to completely cure the mice. This compared with daily doses of 2.5 mg/kg of amphotericin B over 6 days and 200 000 units of nystatin over 4 days to achieve the same effect. Minimum inhibitory concentrations (MIC) of α-hederin *in vitro* were 0.5 mg/ml against *Candida albicans* and 0.05 mg/ml against the dermatophyte *Microsporum canis* (20 times less than that of amphotericin B) (Timon-David *et al.* 1980). Some simple comparisons of haemolytic and antifungal activities for α-hederin (**368**) and two derivatives (**350** and **1320**) have been drawn up. In one analogue, (**350**), the terminal rhamnose was absent and in the other, (**1320**), the carboxyl group was methylated. The haemolytic activities of the analogues were 57% and 25%, respectively, of α-hederin values, and the antifungal activities (against *Trichophyton mentagrophytes*) were 13% and 60%, respectively, of α-hederin. The aglycone, hederagenin, was inactive. Therefore, no parallel could be drawn between haemolytic and antifungal activities as modification of the structure of α-hederin gave differing quantitative effects (Takechi and Tanaka, 1990).

Synthetic methyl oleanolate glycosides, diglycosides and a triglycoside were almost completely inactive against *Trichophyton mentagrophytes* (Takechi and Tanaka, 1992).

368 R^1 = -Rha, R^2 = -H

350 R^1 = R^2 = -H

1320 R^1 = -Rha, R^2 = -CH$_3$

The fungicidal activity of *Dolichos kilimandscharicus* roots (Leguminosae) has been attributed to the 3-*O*-glucosides of hederagenin (**350**), bayogenin (**479**) and medicagenic acid (**434**) (Appendix 2) (Marston *et al.* 1988a). These saponins inhibited growth of the plant-pathogenic fungus *Cladosporum cucumerinum* in a bioautographic assay (Homans and Fuchs, 1970). This bioassay involved first migrating pure compounds or extracts on silica gel TLC plates with a suitable solvent and then spraying with spores of the fungus and incubating for 3 days in the dark to find zones of inhibition of growth of the fungus.

Medicagenic acid 3-*O*-β-D-glucopyranoside is also effective against *Cryptococcus neoformans* (Polacheck *et al.* 1986) and the plant pathogenic fungi *Trichoderma viride*, *Sclerotium rolfsii*, *Rhizopus mucco*, *Aspergillus niger*, *Phytophthora cinamommi* and *Fusarium oxysporum* (Levy *et al.* 1986); another alfalfa root saponin, medicagenic acid 3-*O*-β-D-glucopyranosyl-glucopyranoside (436), inhibits the growth of a similar range of fungi (Levy *et al.* 1989).

Noroleanane saponins active against *Cladosporium cladosporoides* have been found in *Celmisia petriei* (Asteraceae) extracts (Rowan and Newman, 1984).

Oat roots contain a series of triterpene trisaccharides avenacins A-1 (851), A-2 (852), B-1 (853) and B-2 (854) (Section 2.1.6) which effectively protect the plants against 'take-all' disease, caused by the fungus *Gaeumannomyces graminis*. Those esterified with *N*-methylanthranilic acid (A-1, B-1) are the more toxic (Crombie *et al.* 1987).

Two saponins, camellidins I (1316) and II (1317), from *Camellia japonica* (Theaceae), have activity against the fungi *Botrytis cinerea*, *Pyricularia oryzae*, *Cochliobolus miyabeanus* and *Pestalotia longiseta* (Nagata *et al.* 1985). Fungitoxicity towards *Pyricularia oryzae* is very important, especially in view of the damage this organism inflicts in rice (Nishino *et al.* 1986). Further details on the isolation and structure determination of antifungal saponins from *Camellia* species have been published (Nagata, 1986; Nishino *et al.* 1986). The ester saponin mixture, theasaponin, from *C. sinensis* is also fungistatic for *Pyricularia oryzae* (Tschesche *et al.* 1969a).

Whereas extracts of *Solidago virgaurea* (Asteraceae) are active against plant-pathogenic fungi, the genuine (ester) saponins are inactive against human-pathogenic fungi. However, partial alkaline hydrolysis of the saponin mixture gave three major products which were inhibitory to a number of human pathogenic yeasts such as *Candida albicans*, *Candida tropicalis*, etc. (Hiller *et al.* 1987a). Two of these (characterized as bidesmosidic saponins of polygalacic acid), virgaureasaponins 1 and 2, together with bellissaponin 1 and the three corresponding mono-desmosides have now been systematically tested against *Candida* and *Cryptococcus* strains in agar diffusion and liquid cell culture tests. Interestingly, the three bidesmosidic saponins were more active than the monodesmosides. Glycosides with one monosaccharide moiety in position C-3 of the aglycone were more strongly antimycotic than glycosides with two monosaccharide moieties at C-3 (Bader *et al.* 1990; Hiller *et al.* 1990a).

Saponins from *Primula acaulis* (Primulaceae) have shown fungicidal activity against several strains, including *Candida*. Their efficacy was comparable to the antibiotics nystatin and stamicin (Margineau *et al.* 1976).

A protoprimulagenin pentasaccharide which exhibits inhibition of *Cladosporium cucumerinum* growth and molluscicidal activity has been isolated from the leaves of *Rapanea melanophloeos* (Myrsinaceae) (Ohtani *et al.* 1993). Other plants which contain antifungal triterpene glycosides include *Clerodendrum wildii* (Verbenaceae) (Toyota *et al.* 1990) and *Sapindus mukurossi* (Sapindaceae) (a hydrophilic ointment containing the purified extract is effective against athlete's foot and ringworm) (Tamura *et al.* 1990). The bidesmosidic Mi-saponin A (**502**) from the roots of the former plant was found to be responsible for activity against *Cladosporium cucumerinum* (spore formation was inhibited by 1.5 µg of the saponin in the TLC bioassay) and for the bitter taste of the plant. The aglycone of Mi-saponin A, protobassic acid (**52**), was also active against the fungus (3.3 µg on the TLC plate was sufficient to inhibit spore formulation) (Toyota *et al.* 1990).

52 Protobassic acid $R^1 = R^2 = $ -H

502 Mi-saponin A $R^1 = $ -Glc, $R^2 = $ -Ara^2-Rha^4-Xyl^3-Rha

The sarsapogenin triglycoside As-1 (**922**, Appendix 3) from *Asparagus officinalis* (Liliaceae) has MIC values varying from 0.5 to 30 µg/ml against *Candida albicans, Cryptococcus albidus, Epidermophyton floccosum, Microsporum gypseum* and *Trichophyton* spp. (Shimoyamada *et al.* 1990).

The antifungal activity of holostane triterpene glycosides from marine organisms has also been noted (Shimada, 1969). The growth of a number of fungal strains is considerably inhibited by these saponins and their derivatives, with MIC values as low as 0.78 µg/ml (Kitagawa, 1988). Thus, holothurins A (**1241**) and B (**1240**), echinosides A (**1219**) and B (**1218**), and the stichlorosides are all antifungal (Habermehl and Krebs, 1990); bivittoside D (**1212**) exhibited significant activity against *Penicillium chrysogenum, Trichophyton rubrum, Candida utilis*, etc. (Kitagawa *et al.* 1989a). The presence of the 12α-OH function seems to be essential for the activity and, furthermore, there are indications that the linear sequence of the carbohydrate chain is important (Kitagawa *et al.* 1989a). Further structure–activity relationships have been investigated by Kuznetsova *et al.* (1982) and Maltsev *et al.* (1985). The latter authors found that substitution of a glucose moiety by quinovose in the monodesmosidic side chain (stichoposides C and D, thelenotosides A and B) gave a four-fold higher activity against *Candida albicans*. The optimal number of monosaccharides in the monodesmosidic sugar chain was established to be between 4 and 6 residues. Desulphated triterpene tetrasaccharides from the sea cucumber *Holothuria pervicax* are active against a number of microorganisms, including *Aspergillus niger*, *Penicillium citrinum* and *Trichophyton rubrum* (Kitagawa *et al.* 1989c).

The major saponin, laeviuscoloside A (**1297**), from the starfish *Henrica laeviuscola* is a sulphated diglycoside and inhibits growth of the plant pathogenic fungus *Cladosporium cucumerinum* at a level of less than 1 µg in a TLC bioassay (D'Auria *et al.* 1990).

The mechanism of action of saponins against fungi would appear to involve the formation of complexes with sterols in the plasma membrane, thus destroying the cellular semi-permeability and leading to death of the cell (Wolters, 1968a; Défago, 1977). One theory is that the saponins themselves are inactive and comprise only water-soluble transport forms. In the presence of cell membrane glycosidases, there is formation of the aglycone which is the active membranolytic component (Segal and Schlösser, 1975). However, it must be emphasized that other mechanisms must also be involved since there are exceptions to the parallel between cholesterol-binding and fungicidal properties, e.g. parillin.

Antibacterial activity

Although strongly antifungal, triterpene glycosides have virtually no antibacterial activity. Some are weakly active against Gram-positive bacteria but show no effect on Gram-negative bacteria (Anisimov and Chirva, 1980). It has been reported, however, that saponins of *Hydrocotyle asiatica* (Umbelliferae) are active against diarrhoea-causing bacteria (*Salmonella, Shigella, Escherichia coli*) (Chowdhury *et al.* 1987).

Antiviral activity

The *in vitro* antiviral activity of several glycosides of acylated β-amyrin aglycones against influenza virus has been reported. Various gymnemic acids from *Gymnema sylvestre* (Asclepiadaceae) are inhibitory, as are aescine, cyclamin A, primula saponin and theasaponin (Rao *et al.* 1974). Gymnemic acid A is also active *in vivo* and may be involved in early viral infection events, involving inhibition of virus–host cell attachment (Rao *et al.* 1974).

When tested against herpes simplex type I, vaccinia, vesicular stomatitis viruses, adenovirus type 6 and poliovirus, a whole-plant extract of *Anagallis arvensis* (Primulaceae) was active against herpes simplex virus and poliovirus (Amoros *et al.* 1987, 1988). The saponins responsible for this activity have been isolated (Amoros and Girre, 1987) but there is some doubt as to their correct structure (Glombitza and Kurth, 1987b). The *in vitro* antiviral effect manifests itself by an inhibition of the cytopathic action and a reduction of the virus population. However, the saponins are not virucidal and the activity is thought to involve inhibition of virus–host cell attachment (Amoros *et al.* 1988). Single-cycle experiments have shown that interference with a step of viral replication occurs (Amoros *et al.* 1987). For *in vivo* experiments, herpes simplex keratitis induced in rabbit eyes was used as a model. It was found that a saponin ointment preparation was as effective at treating the keratitis as adenine arabinoside, more effective than idoxuridine and slightly less effective than acycloguanosine. However, the therapeutic index was low and small increases of saponin concentration in the ointment gave toxic effects (Amoros *et al.* 1988).

Glycyrrhizin (**269**, Appendix 2) inhibits the growth of several DNA and RNA viruses but has no effect on protein synthesis and replication of uninfected cells. Herpes simplex virus particles are also irreversibly inactivated (Pompei *et al.* 1979) and *in vitro* antiviral action against varicella-zoster virus has been documented (Baba and Shigeta, 1987).

Saikosaponin d (**548**, Appendix 2), from the roots of *Bupleurum falcatum* (Umbelliferae), causes inactivation of measles and herpes viruses at concentrations greater than 5 μM *in vitro*. However, at these concentrations it is also cytotoxic to the Vero host cells. Viral inactivation by the saponin was suggested to involve interaction with membrane glycoproteins since there was no effect on poliovirus, which lacks an envelope (Ushio and Abe, 1992). Furthermore, saikosaponin d is known to decrease membrane fluidity (Abe *et al.* 1978) and behaves as a surface-active agent on peritoneal macrophages (Ushio and Abe, 1991).

A series of quinovic acid glycosides from *Uncaria tomentosa* (Rubiaceae) and *Guettarda platypoda* (Rubiaceae) has been tested against RNA viruses. An *in vitro* inhibitory effect against vesicular stomatitis virus (a minus-strand RNA virus) was evident for all nine compounds, although at relatively high concentrations with respect to the toxic dose. The MIC of quinovic acid 28-*O*-diglucoside required to reduce plaque number by 50% (IC_{50}) compared with a control of infected cells was 31 µg/ml. This compared with an IC_{50} value of 100 µg/ml for cytotoxicity in chicken embryo-related (CER) cells. The presence of a free C-27 carboxyl group was important and activity was also affected by the nature of the sugar moiety. Two of the glycosides, both bidesmosidic and possessing a free C-27 carboxyl group, were active against rhinovirus type 1B infection in HeLa cells (a plus-strand RNA virus) (Aquino *et al.* 1989b).

Five oleanolic acid glycosides from *Calendula arvensis* (Asteraceae), including both monodesmosidic and bidesmosidic saponins, were found to reduce vesicular stomatitis virus (VSV) plaque formation in CER monolayer cell cultures at non-toxic concentrations. However, only **213** reduced rhinovirus type 1B (HRV 1B) infection. Thus, the enveloped virus VSV was more sensitive than the naked virus HRV. Action by the saponins on one or more steps of the viral replication cycle seems probable. *Calendula arvensis* saponins appear to have an antiviral potential higher than glycyrrhizin but are likely to function by the same mechanism (De Tommasi *et al.* 1991).

Giganteaside D and flaccidin B from *Anemone flaccida* (Ranunculaceae) inhibit reverse transcriptase from an RNA tumour virus at concentrations from 10^{-5} to 10^{-4} M (Shen *et al.* 1988).

A more general study of triterpene saponins has been performed with the following DNA and RNA virus groups: herpes simplex types 1 and 2, adenovirus type 2, poliovirus type 2 and vesicular stomatitis virus. By

Table 5.3. *Antitumour activity of certain saponins*

Substance	ED_{50} $(mg/kg)^a$
Aescine	20
Cyclamin (**538**)	10
Senegin	1.5
Primulasaponin	40
α-Hederin (**368**)	4
Quillaioside	–
Hederasaponin C (**372**)	–

[a] Dose necessary to reduce growth of Walker carcinomas in rats by 50% (compared with controls).
Tschesche and Wulff, 1972.

evaluating cytotoxicity to Vero cells, a measure of the therapeutic index was also obtained. The most active saponins were those containing ursolic acid, 18β-glycyrrhetinic acid, hederagenin and oleanolic acid; the glycosides were more potent than the respective aglycones. Preliminary structure–activity relationships were delineated but no simple correlation between number, position or nature of the sugar moieties and antiviral activity was possible, even when the aglycone was the same (Simoes *et al.* 1990).

Some saponins have been investigated for their effects on human immunodeficiency virus (HIV); glycyrrhizin is one of these (Ito *et al.* 1987) (see Chapter 7). It has also been reported that soybean saponins have an inhibitory effect on the infectivity and cytopathic activity of HIV (Nakashima *et al.* 1989). A mixture of soyasaponins Ba and Bb at concentrations greater than 0.5 mg/ml completely inhibited HIV-induced cytopathic effects and virus-specific antigen expression 6 days after infection (Nakashima *et al.* 1989).

5.3.2   *Cytotoxic and antitumour activity*

A cytotoxic activity for many saponins has been documented (Tschesche and Wulff, 1972). *In vivo* experiments with these saponins in rats carrying Walker carcinomas confirmed the *in vitro* data (Table 5.3) but the high toxicity of the saponins precludes any practical applications (Tschesche and Wulff, 1972).

Saponins from *Saponaria officinalis*, *Myrsine africana*, *Hedera helix* and *Entada phaseoloides* have all been claimed to have antitumour activity but the picture is complicated again by the phenomenon of general toxicity (Tschesche and Wulff, 1972). One exception appears to be a saponin from *Acer negundo* (Aceraceae) which is active and yet has a high therapeutic index, giving good survival rates at doses of around 1.5 mg/kg in rats with Walker carcinosarcoma 256 tumours (Kupchan *et al.* 1967). The IC_{50} values of Entada saponins II, III and IV (from *Entada phaseoloides*, Leguminosae) against L-5178Y tumour cells are 3.10, 0.83 and 0.96 μg/ml, respectively (Okada *et al.* 1987).

The cyclic bidesmoside tubeimoside I (**645**) from the Chinese medicinal plant *Bulbostemma paniculatum* (Cucurbitaceae) (itself used in the treatment of tumours) is reported to have *in vivo* antitumour activity (Kong *et al.* 1986).

The nortriterpene glucuronides pfaffoside A, pfaffoside C, pfaffoside F and their aglycone, pfaffic acid, inhibit the growth of a B-16 melanoma cell line, albeit at high concentrations, from 30 to 100 g/ml (Nakai *et al.* 1984; Nishimoto *et al.* 1984).

Monoglycosides of 14,18-cycloapoeuphane triterpenes exhibit potent selective cytotoxicity against MOLT-4 human leukaemia cells, with ED_{50} values as low as 0.00625 μg/ml (Kashiwada *et al.* 1992).

The glycoside 4″,6″-di-*O*-acetylsaikosaponin d from *Bupleurum kunmingense* (Umbelliferae) is cytotoxic to Ehrlich tumour cells (Jap. Pat. 62 240 696), while 2″-*O*-, 3″-*O*- and 4″-*O*-acetylsaikosaponin a inhibit the growth of HeLa and L-1210 leukaemia cells (Jap. Pat. 61 282 395).

Antitumour formulations containing saikosaponins isolated from *Bupleurum falcatum*, *B. longeradiatum*, *B. nipponicum* and *B. triradiatum* (Umbelliferae) are the subject of a Japanese patent (Kono and Odajima, 1988).

A saponin preparation from *Dolichos falcatus* (Leguminosae) is effective against sarcoma-37 cells both *in vitro* and *in vivo* (mouse) (Huang *et al.* 1982a).

Certain glycosides of medicagenic acid and polygalacic acid from *Crocosmia crocosmiiflora* (Iridaceae) are purported to have antitumour activities (Nagamoto *et al.* 1988). Fractionation of a water extract of the bulbs gave a purified saponin fraction which increased the lifespan of mice implanted with Ehrlich ascites carcinoma cells and inhibited the growth of Ehrlich solid carcinomas implanted in Jc1-ICR mice (Nagamoto *et al.* 1988).

Antitumour activity has been shown by foetoside C and cyclofoetoside B from *Thalictrum foetidum* (Ranunculaceae) and thalicoside A from *Thalictrum minus* at doses from 30 to 50 mg/kg (i.p.) per day in rats with implanted tumours. Foetoside C was the most effective of the three glycosides (Rakhimov *et al.* 1987).

There are plans for clinical applications of a small number of saponins. Ginsenosides Rg_1 (**758**) and Rb_1 (**749**) (Table 2.4, p. 52) are effective against stomach cancer (Hayashi and Kubo, 1980), while soyasaponins from *Glycine max* (Leguminosae) are effective against a number of tumours (Jap. Pat. 58 72,523). Gypenosides from *Gynostemma pentaphyllum* (Cucurbitaceae) have been proposed for the treatment of carcinomas of the skin, uterus and liver (Jap. Pat. 58 59,921).

A small number of triterpene aglycones have been shown to possess cytotoxic or cytoinhibitory properties. Betulic acid (betulinic acid **106**, Table 2.2, p. 24), for example, from the stem bark of *Crossopteryx febrifuga* (Rubiaceae), was cytoinhibitory in the Co-115 human colon carcinoma cell line (ID_{50} 0.375 µg/ml) but inactive in the human epidermal carcinoma of the nasopharynx (KB) and P-388 lymphocytic leukaemia cell lines (Tomas-Barberan and Hostettmann, 1988).

Holothurins from marine animals such as *Actinopyga agassizi* are antitumour in mice carrying sarcoma 180 and Krebs-2 ascites tumours (Scheuer, 1987). They also possess *in vitro* cytotoxicity and are cytotoxic in the sea urchin egg assay (holothurin A (**1241**, Table 2.13, p. 109) is active at 0.78 µg/ml) (Anisimov *et al.* 1980). The holothurinosides A (**1242**), C (**1239**) and D (**1238**) (Table 2.13, p. 109) from *Holothuria forskalii* are toxic to P-388, A-549, HeLa and B-16 cells *in vitro* (Rodriguez *et al.* 1991). Stichoposides A (**1233**) and C (**1235**), holothurins A (**1241**) and B (**1240**) (Table 2.13, p. 109), and cucumarioside G show higher cytotoxicity to sea urchin eggs than saponins of starfish and of plant origin (Anisimov *et al.* 1980). Stichoposide A blocks fertilized egg division and causes anomalies in development of sea urchin embryos; the cytotoxic effects appear to be the result of an alteration in the selective membrane permeability (Anisimov *et al.* 1978).

Antineoplastic (P-388 leukaemia) activity has been reported for stichostatin 1 from *Stichopus chloronotus* and thelenostatin 1 from *Thelenota ananas* (Pettit *et al.* 1976).

In a short-term *in vitro* assay for *tumour-promoting* activity in human lymphoblastoid Raji cells infected with Epstein–Barr virus (EBV), saponins from *Gleditsia japonica* and *Gymnocladus chinensis* exhibited inhibitory effects. A total of three saponins reduced EBV activation induced by the tumour promoters 12-O-tetradecanoylphorbol-13-acetate (TPA) and teleocidin B. It was shown that a sugar chain at C-3 of the aglycone and/or an acylglycosidic sugar chain were essential for activity (Tokuda *et al.* 1988). In a similar fashion, wistaria saponin D and its parent aglycone, soyasapogenol E, from *Wistaria brachybotrys* (Leguminosae) were able to inhibit TPA-induced EBV early antigen

activation. Soyasapogenol E (**23**) exhibited 100% inhibition at a
5×10^2 mole ratio to TPA (Konoshima *et al.* 1991). Soyasaponin I from
W. brachybotrys exhibited strong inhibition of mouse skin tumour
promotion. It was tested in a two-stage carcinogenesis model using
DMBA as initiator and TPA as a promoter (Konoshima *et al.* 1992).

   Antimutagenic activity. Antimutagenic activity of some saponins
has been investigated using the Ames test and either the known
promutagen benzo[a]pyrene (BaP) or a mutagenic urine concentrate from
a smoker. While saponins from both *Calendula arvensis* (Asteraceae) and
Hedera helix (Araliaceae) inhibited the mutagenic activity of BaP, only
the saponins **154**, **346** and **368** from *Hedera helix* leaves decreased the
mutagenicity of smokers' urine. The results were compared with those of
a known antimutagen, chlorophyllin, the sodium-copper salt of chloro-
phyll (Elias *et al.* 1990). Doses of 200 µg for active saponins corresponded
well with the number of revertants in the *Salmonella* microsomal assay
obtained from a 100 µg dose of chlorophyllin.

5.3.3   *Piscicidal activity*

   The piscicidal properties of plants have been exploited for
hundreds of years as a means of harvesting fish from streams and ponds.
There is even a mention by Aristotle of the ichthyotoxic properties of
Verbascum (Kofler, 1927). Tabulations of these plants show that saponin-

containing species, such as those from the Sapindaceae (particularly the genera *Paullinia*, *Serjania* and *Sapindus*), figure prominently (Howes, 1930; Marston and Hostettmann, 1991a). Among the plants with a high saponin content, those which have the longest history of use as fish poisons include:

- – *Swartzia madagascariensis* (Leguminosae)
- – *Sesbania sesban* (Leguminosae)
- – *Neorautanenia pseudopachyrhiza* (Leguminosae)
- – *Sapindus saponaria* (Sapindaceae)
- – *Securidaca longepedunculata* (Polygalaceae)
- – *Securinega virosa* (Euphorbiaceae)
- – *Strychnos innocua* (Loganiaceae)
- – *Xeromphis spinosa* (Rubiaceae)
- – *Balanites aegyptiaca* (Balanitaceae)
- – *Cyclamen* species (Primulaceae).

Brazilian Indians use stems of *Serjania lethalis* (Sapindaceae) for immobilizing fish, and it has been found that the plant contains the piscicidal saponins serjanosides A, B and C (Texeira *et al.* 1984). Similarly, the bark of *Schima mertensiana* (Theaceae), employed by fishermen of the Bonin Islands (Japan), has been shown to contain a saponin mixture (boninsaponin) consisting of glycosides of primulagenin A, dihydroprivero-genin A, A_1-barrigenol, barringtogenol C and R_1-barrigenol. This mixture is toxic to killie-fish (LC_{50} 1.5 µg/ml) (Kitagawa *et al.* 1976d).

The saponins isolated from *Madhuca butyracea* (Sapotaceae) seeds are piscicidal to guppy fish (LC_{50} 11 µg/ml, LC_{90} 14 µg/ml) (Lalitha *et al.* 1987).

A crude saponin fraction from the stem of a South American liana *Thinouia coriacea* (Sapindaceae) is toxic at 1 p.p.m. (1 mg/l) to the fish *Phallocerus caudi maculatus* and *Geophagus brasiliensis*. This mixture is known to contain at least eight oleanolic acid monodesmosides (Schenkel *et al.* 1991).

Death of fish results from damage to the gill capillaries, which are not only the respiratory organs but function also in the regulation of ion balance and osmotic pressure.

Piscicidal saponins are also found in marine organisms and most probably function as predator repellants. The poisonous properties of sea cucumbers, for example, have been known for centuries in parts of the Pacific, where mashed or chopped sea cucumbers have traditionally been used to poison fishes in tide pools (Burnell and ApSimon, 1983). Recent investigations of these organisms has shown the presence of triterpene saponins in large quantities (see Chapter 2).

5.3.4   *Molluscicidal activity*

Saponins have a pronounced action on molluscs and organisms which use gills for breathing, such as frogs and fish (Kofler, 1927). The mechanism of this toxicity most likely involves the binding of the saponin to the gill membranes, resulting in an increase in their permeability and a subsequent loss of important physiological electrolytes. Studies on the mode of action, however, have produced conflicting results. Even the assumption that there is a correlation between haemolysis, cholesterol complexation and molluscicidal activity does not hold; for example, digitonin is highly haemolytic and forms stable cholesterol complexes but is only weakly molluscicidal. It has been postulated that membrane channel formation or modification may actually be involved (Brain *et al.* 1990).

The toxicity to snails (molluscicidal activity) is stimulating great interest for the local control of schistosomiasis, a tropical parasitic disease affecting approximately 250 million people worldwide. Snails (of the genera *Biomphalaria, Bulinus* and *Oncomelania*) are directly implicated in the transmission of this disease since they act as intermediate hosts for the miracidial state of the life cycle of the parasite (Mott, 1987). Certain plants are very effective at killing these water-borne snails and several have already reached the stage of field trials, for use in areas of high incidence of the disease where synthetic chemicals would be too expensive (Hostettmann, 1989). Those plants which contain saponins (Hostettmann and Marston, 1987) are among the most promising for large-scale application since:

- many saponins have high molluscicidal activity
- saponins are water soluble
- saponins frequently occur in large amounts in plants
- saponins are widely distributed in the plant kingdom
- oral toxicity is low.

In fact, the first plant proposed for use as a molluscicide was *Balanites aegyptiaca* (Balanitaceae), which contains steroid saponins (Liu and Nakanishi, 1982; Marston and Hostettmann, 1985).

Phytolacca dodecandra (Phytolaccaceae) is a bush known in Ethiopia as 'endod' and elsewhere is referred to as 'soapberry'. It has small berries which, when dried, powdered and mixed with water, yield a foaming detergent solution. The berries have a long history of use in Ethiopia as a soap substitute. During a study of the distribution and ecology of schistosomiasis-transmitting snails in the town of Adwa in northern Ethiopia, Lemma noticed that in places along rivers or streams where people washed clothes with powered endod berries there were more dead snails than in adjacent areas. Subsequently it was shown that a preparation of *P. dodecandra* possessed molluscicidal properties (Lemma, 1970), comparable to those of the synthetic molluscicide niclosamide. A 5-year pilot snail control programme was undertaken in Ethiopia with extracts of the plant, leading to a significant reduction of *Schistosoma mansoni* infection rates (Goll *et al.* 1983). Dried endod berries contain up to 25% saponins. The structures of the three molluscicidal saponins have been determined as oleanoglycotoxin-A (**217**), lemmatoxin (**219**) and lemmatoxin C (**220**) (Table 5.4). They are glycosides of oleanolic acid and are active in the 1.5–3.0 mg/l (p.p.m.) range (Parkhurst *et al.* 1974).

Other characterized triterpene glycosides from plants with molluscicidal activity are shown in Table 5.4. Activities of certain saponins, most notably the rhamnosyl glucuronide of oleanolic acid (**189**) from *Swartzia madagascariensis*, approach those of synthetic molluscicides (Borel and Hostettmann, 1987).

Certain saponin-containing plants have undergone field trials. Two of the most important are *Swartzia madagascariensis* (Suter *et al.* 1986) and, as mentioned above, *Phytolacca dodecandra* (Hostettmann, 1989). The saponins responsible for molluscicidal activity have been characterized and when toxicological testing is complete, the way will be open for the large-scale use of these plant-derived molluscicides in focal areas.

The African species *S. madagascariensis* appears to be one of the most promising plants for the control of schistosomiasis. Each tree can carry up to 30–40 kg of pods and there is subsequently no shortage of vegetable material. Water extracts exerted significant molluscicidal activity against *Biomphalaria glabrata* and *Bulinus globosus* snails at dilutions down to 100 mg ground pods/l. Field trials were undertaken in Tanzania at the end of the dry season when water levels were very low. Mature pods were ground and then extracted with water. Application of the water extract to ponds at an initial molluscicide concentration of not less than 100 mg/l was performed. Densities of *B. globosus* snails dropped to zero 1 week after a single application of the vegetable extract. Following application of the extract, saponin content was monitored by TLC and a simple haemolytic test. The saponins had short half-lives of 12–24 h under field

Table 5.4. *Molluscicidal activities of triterpene glycosides*

No.	Name	Structure	Molluscicidal activity (mg/l)	Plant	Reference
Oleanolic acid aglycone (2)					
140		3-GlcA	2 (LC_{100})	*Lonicera nigra* (Caprifoliaceae)	Domon and Hostettmann, 1983
649	Aridanin	3-Glc²-NHAc	0.88 (LC_{50})	*Tetrapleura tetraptera* (Leguminosae)	Adewunmi et al. 1988
			20 (LC_{100})		Maillard et al. 1989
650		3-Glc²-NHAc |⁴ Gal	2.5 (LC_{100})	*T. tetraptera* (Leguminosae)	Maillard et al. 1989
651		3-Glc²-NHAc |⁶ Glc	5 (LC_{100})	*T. tetraptera* (Leguminosae)	Maillard et al. 1989
144		3-Ara²-Glc	2 (LC_{100})	*Lonicera nigra* (Caprifoliaceae)	Domon and Hostettmann, 1983
145		3-Ara²-Glc 28-Glc⁶-Glc	n.a.	*L. nigra* (Caprifoliaceae)	Domon and Hostettmann, 1983
154		3-Ara²-Rha	80 (LC_{100})	*Phytolacca dodecandra* (Phytolaccaceae)	Hostettmann et al. 1982
220	Lemmatoxin-C	3-Glc²-Glc²-Rha	3 (LC_{90})		Parkhurst et al. 1973b
222		3-Glc⁶-Rha²-Glc	16 (LC_{100})	*Xeromphis spinosa* (Rubiaceae)	Sati et al. 1986
208		3-Ara²-Rha³-Xyl	16 (LC_{100})		Hostettmann et al. 1982
206		3-Ara²-Rha³-Rib	32 (LC_{100})		Hostettmann et al. 1982
1321		3-Ara²-Rha³-Xyl⁴-Glc	32 (LC_{100})		Hostettmann et al. 1982
1322		3-Ara²-Rha³-Rib⁴-Glc	16 (LC_{100})		Hostettmann et al. 1982

217	Oleanoglycotoxin-A	3-Glc^4-Glc $\quad\vert^2$ $\quad$Glc	*Phytolacca dodecandra* (Phytolaccaceae)	6 (LC$_{100}$) 3 (LC$_{90}$)	Domon and Hostettmann, 1984 Parkhurst *et al.* 1973a
218		3-Glc^4-Glc $\quad\vert^2$ $\quad$Glc 28-Glc	*P. dodecandra* (Phytolaccaceae)	n.a.	Domon and Hostettmann, 1984
219	Lemmatoxin	3-Glc^4-Glc $\quad\vert^3$ $\quad$Gal	*P. dodecandra* (Phytolaccaceae)	1.5 (LC$_{90}$)	Parkhurst *et al.* 1974
194		3-GlcA^3-Xyl	*Talinum tenuissimum* (Portulacaceae)	1.5 (LC$_{100}$)	Gafner *et al.* 1985
195		3-GlcA^3-Xyl 28-Glc	*T. tenuissimum* (Portulacaceae)	n.a.	Gafner *et al.* 1985
189		3-GlcA^3-Rha	*Swartzia madagascariensis* (Leguminosae)	3 (LC$_{100}$)	Borel and Hostettmann, 1987
190		3-GlcA^3-Rha 28-Glc	*S. simplex* (Leguminosae)	n.a.	Borel *et al.* 1987
176		3-GlcA^4-Ara*f*	*Cussonia spicata* (Araliaceae)	12.5 (LC$_{100}$)	Gunzinger *et al.* 1986
225		3-GlcA^4-Ara*f* $\quad\vert^2$ $\quad$Gal	*C. spicata* (Araliaceae)	100 (LC$_{100}$)	Gunzinger *et al.* 1986
184		3-GlcA^2-Glc 28-Glc	*Panax japonicus* (Araliaceae)	n.a.	Hostettmann *et al.* 1982
159		3-Gal^3-Glc	*Xeromphis spinosa* (Rubiaceae)	15 (LC$_{100}$)	Sati *et al.* 1987
210		3-Gal^3-Glc^2-Glc	*X. spinosa* (Rubiaceae)	20 (LC$_{100}$)	Sati *et al.* 1987
256		3-Glc^6-Rha^2-Glc $\quad\vert^2$ $\quad$Glc	*X. spinosa* (Rubiaceae)	15 (LC$_{100}$)	Sati *et al.* 1986

(*cont.*)

Table 5.4. (*Cont.*)

No.	Name	Structure	Molluscicidal activity (mg/l)	Plant	Reference
262		$3\text{-Xyl}^2\text{-Glc}^6\text{-Glc}^6\text{-Glc}$ $\quad\mid^3$ $\quad\text{Ara}$	$5\ (LC_{100})$	*Gundelia tournefortii* (Asteraceae)	Wagner *et al.* 1984b
264		$3\text{-Xyl}^2\text{-Glc}^6\text{-Glc}^6\text{-Glc}$ $\quad\mid^3$ $\quad\text{Ara}$	$5\ (LC_{100})$	*G. tournefortii* (Asteraceae)	Wagner *et al.* 1984b
Hederagenin aglycone (17)					
350		3-Glc	$15\ (LC_{100})$	*Hedera helix* (Araliaceae)	Hostettmann, 1980
346		3-Ara	$3\ (LC_{100})$	*H. helix* (Araliaceae) *Lonicera nigra* (Caprifoliaceae) *Polyscias dichroostachya* (Araliaceae)	Hostettmann, 1980 Domon and Hostettmann, 1983 Gopalsamy *et al.* 1990
348		3-Ara $28\text{-Glc}^6\text{-Glc}$	n.a.	*Lonicera nigra* (Caprifoliaceae)	Domon and Hostettmann, 1983
349		3-Ara $28\text{-Glc}^6\text{-Glc}^4\text{-Rha}$	n.a.	*Hedera helix* (Araliaceae)	Hostettmann *et al.* 1982
355		3-GlcA	$16\ (LC_{100})$	*Lonicera nigra* (Caprifoliaceae)	Domon and Hostettmann, 1983
377		$3\text{-Glc}^2\text{-Glc}$	$12\ (LC_{100})$	*Hedera helix* (Araliaceae)	Hostettmann, 1980
368		$3\text{-Ara}^2\text{-Rha}$	$8\ (LC_{100})$	*H. helix* (Araliaceae) *Polyscias dichroostachya* (Araliaceae)	Hostettmann, 1980 Gopalsamy *et al.* 1990

No.	Structure	Plant (Family)	Activity	Reference
369	3-Ara^2-Rha 28-Glc	P. dichroostachya (Araliaceae)	n.a.	Gopalsamy et al. 1990
372	3-Ara^2-Rha 28-Glc^6-Glc^4-Rha	Hedera helix (Araliaceae)	n.a.	Hostettmann et al. 1982
360	3-Ara^2-Glc	Lonicera nigra (Caprifoliaceae)	8 (LC$_{100}$)	Domon and Hostettmann, 1983
362	3-Ara^2-Glc 28-Glc^6-Glc	Polyscias dichroostachya (Araliaceae)	n.a.	Gopalsamy et al. 1990
363	3-Ara^2-Glc 28-Glc^6-Glc^4-Rha	Lonicera nigra (Caprifoliaceae)	n.a.	Domon and Hostettmann, 1983
393	3-Ara^2-Rha^3-Xyl	Hedera helix (Araliaceae)	8 (LC$_{100}$)	Hostettmann et al. 1982
391	3-Ara^2-Rha^3-Rib	Sapindus mukurossi (Sapindaceae)	16 (LC$_{100}$)	Hostettmann et al. 1982
403	3-Ara^2-Rha^3-Xyl^4-Glc	Anemone rivularis (Ranunculaceae)	8 (LC$_{100}$)	Hostettmann et al. 1982
1323	3-Ara^2-Rha^3-Rib^4-Glc		8 (LC$_{100}$)	Hostettmann et al. 1982

398	3-Glc^4-Glc
	  |^2
	  Glc
		Phytolacca dodecandra (Phytolaccaceae)	12 (LC$_{100}$)	Domon and Hostettmann, 1984

399	3-Glc^4-Glc
	  |^2
	  Glc
	28-Glc
		P. dodecandra (Phytolaccaceae)	n.a.	Domon and Hostettmann, 1984

826	3-Ara^2-Rha^3-Xyl^4-Ac
	  |^3
	  Ac
		Sapindus rarak (Sapindaceae)	6.25 (LC$_{100}$)	Hamburger et al. 1992

(cont.)

Table 5.4. (*Cont.*)

No.	Name	Structure	Plant	Molluscicidal activity (mg/l)	Reference
Bayogenin aglycone (45)					
479		3-Glc	*Dolichos kilimandscharicus* (Leguminosae)	7.5 (LC_{100})	Marston et al. 1988a
480		3-Glc^4-Glc	*Phytolacca dodecandra* (Phytolaccaceae)	12 (LC_{100})	Domon and Hostettmann, 1984
481		3-Glc^4-Glc 28-Glc	*P. dodecandra* (Phytolaccaceae)	n.a.	Domon and Hostettmann, 1984
Phytolaccagenin aglycone (49)					
496		3-Xyl	*P. americana* (Phytolaccaceae)	60 (LC_{90})	Johnson and Shimizu, 1974
497		3-Xyl^2-Glc	*P. americana* (Phytolaccaceae)	80 (LC_{90})	Johnson and Shimizu, 1974
Barringtogenol C aglycone (64)					
1324		3-GlcA^2-Glc 21-Angelate 22-Angelate	*Aesculus indica* (Hippocastanaceae)	10 (LC_{100})	Sati and Rana, 1987
R_1-Barrigenol aglycone (70)					
805	Dodonoside A	3-GlcA-Gal | Ara*f* 21,22-(2-methylbutyrate, 2,3-dimethoxiran-2-carboxylate)	*Dodonaea viscosa* (Sapindaceae)	25 (LC_{100})	Wagner et al. 1987

| 806 | Dodonoside B | 3-GlcA-Gal
|
Ara*f*
21,22-(angelate,
2,3-dimethoxiran-
2-carboxylate) | *D. viscosa* (Sapindaceae) | 25 (LC$_{100}$) | Wagner *et al.* 1987 |

Medicagenic acid aglycone (31)

| 434 | | 3-Glc | *Dolichos kilimandscharicus*
(Leguminosae) | 25 (LC$_{100}$) | Marston *et al.* 1988a |

Echinocystic acid aglycone (16)

| 654 | | 3-Glc^6-Ara^2-Ara
|^2
NHAc | *Albizzia anthelmintica*
(Leguminosae) | 25 (LC$_{100}$) | Carpani *et al.* 1989 |
| 649 | | 3-Glc^2-NHAc | *Tetrapleura tetraptera*
(Leguminosae) | 20 (LC$_{100}$) | Maillard *et al.* 1989 |

Protobassic acid aglycone (52)

| 502 | | 3-Glc
28-Ara^2-Rha^4-Xyl^3-Rha | *Clerodendrum wildii*
(Verbenaceae) | 25 (LC$_{100}$) | Toyota *et al.* 1990 |

Protoprimulagenin A aglycone (71)

| 533 | Lysikokianoside 1 | 3-Arap^4-Glc^2-Xyl
|^2
Glc | *Lysimachia sikokiana*
(Primulaceae) | 2 (LC$_{100}$) | Kohda *et al.* 1989 |

Saikogenin E aglycone (74)

| 544 | Saikosaponin c | 3-Glc^6-Glc
|^4
Glc | | >20 (LC$_{100}$) | Kohda *et al.* 1989 |

(cont.)

Table 5.4. (*Cont.*)

No.	Name	Structure	Plant	Molluscicidal activity (mg/l)	Reference
Saikogenin F aglycone (76)					
547	Saikosaponin a	3-Fuc^3-Glc	*Bupleurum falcatum* (Umbelliferae)	20 (LC$_{100}$)	Kohda *et al.* 1989
Saikogenin G aglycone (77)					
548	Saikosaponin d	3-Fuc^3-Glc	*B. falcatum* (Umbelliferae)	20 (LC$_{100}$)	Kohda *et al.* 1989

Lethal concentration to *Biomphalaria glabrata* snails (after 24 h): LC$_{100}$, 100%; LC$_{90}$, 90%; LC$_{50}$, 50%; n.a. inactive.

conditions; this rapid biodegradability is important since secondary effects are kept to a minimum (Suter *et al.* 1986).

Tetrapleura tetraptera (Leguminosae) is a well-known Nigerian tree, known locally as 'aridan', with a multitude of medicinal uses. Extracts of the fruits possess molluscicidal properties comparable to *Phytolacca dodecandra* (Adesina *et al.* 1980). A water extract gave 100% mortality of *Bulinus globosus* snails at a concentration of 50–100 mg/l over a period of 24 h in the field (Adewunmi, 1984). The fruits of *T. tetraptera* contain rare *N*-acetylglycosides of oleanolic acid (**649–651**) and echinocystic acid (**652**) and are molluscicidal at between 2.5 and 20 mg/l (Adesina and Reisch, 1985; Maillard *et al.* 1989). At position C-3 of the aglycone, they all have a 2-acetamido-2-deoxy-β-D-glucopyranose moiety.

In field trials (Ile-Ife, Nigeria) over a two-year period, the water supply of four villages was treated every 3 months with aqueous extracts of aridan to give a known concentration of about 20 mg/l. By this method, the density of snails was reduced by a factor of 30 during the weeks after application. Very importantly, the transmission sites were also kept free from schistosome cercariae for a minimum of 1 month after application of molluscicide (Adewunmi, 1984).

Molluscicidal activity has so far only been observed for *monodesmosidic* saponins (Domon and Hostettmann, 1984), with bidesmosidic saponins requiring basic or enzymatic hydrolysis to the corresponding monodesmosides before the induction of activity. The ratio of monodesmosidic to bidesmosidic saponins depends on the extraction procedure. In certain cases, as with the berries of *Phytolacca dodecandra*, extraction with water provides larger amounts of active monodesmosidic saponins, while extraction with methanol provides mainly the inactive bidesmosides (Domon and Hostettmann, 1984). This phenomenon is also observed, for example, in the extraction of *Talinum tenuissimum* (Portulacaceae) tubers with methanol, when mainly the inactive bidesmosidic saponin (**195**) is obtained. In contrast, the aqueous extract contained predominantly the monodesmosidic molluscicidal saponin (**194**) and only traces of (**195**) (Gafner *et al.* 1985). It could be that methanol inactivates hydrolytic enzymes which cleave the ester-linked sugar chains during extraction with water.

Saponins of oleanolic acid or hederagenin are the most active, while the corresponding aglycones are inactive (Marston and Hostettmann, 1985). Other factors which are important for the activity are: the nature of the sugar chains, the sequence of the monosaccharides and the interglycosidic linkages.

5.3.5   *Spermicidal and contraceptive activities*

In an effort to find new sources of contraceptive agents, numerous medicinal plants have been investigated for their antifertility activity. The use of plants as antifertility agents has a long history in ethnomedicine (Farnsworth and Waller, 1982). Among the substances with important activities are the saponins (Bhargava, 1988a). They have various effects on sperm, the most important being their spermicidal activity. The mechanism of action probably involves a disruption of the spermatozoid plasma cell membrane, in much the same way as the commercially available spermicide nonoxynol-9.

A list of plants whose spermicidal activity results from their triterpene saponins is shown in Table 5.5. In addition, the spermicidal activity of certain pure triterpene glycosides is known but it is sometimes difficult to compare results because the determination methods employed are different. One method, recommended by the WHO, involves the observation of spermatozoid motility after a 3 min incubation period with the substance under test (Waller *et al.* 1980).

Lemmatoxin (**219**), lemmatoxin C (**254**) and oleanoglycotoxin-A (**217**) (Appendix 2) from *Phytolacca dodecandra* give 100% inhibition of human spermatozoid motility after 3 min at concentrations of 80, 100 and 500 μg/ml, respectively (Stolzenberg and Parkhurst, 1974). This compares with a figure of 40 μg/mg for nonoxynol-9. The oleanolic acid saponins (**176**) and (**225**) from *Cussonia spicata* (Araliaceae) showed 100% inhibition of spermatozoid motility at 100 μg/ml and 3000 μg/ml, respectively (Gunzinger *et al.* 1986), while the molluscicidal oleanolic acid rhamnosyl glucuronide (**189**, Appendix 2) from *Sesbania sesban* (Leguminosae) was active at 1000 μg/ml (Dorsaz *et al.* 1988). Spermicidal activity of nepalins 1 and 2, hederagenin glycosides from *Hedera nepalensis* (Araliaceae), was measured by another method, giving 100% immobilization at 0.5 and 0.25% levels, respectively (Pant *et al.* 1988b). Inhibition is less marked when the aglycone is acylated at position C-28.

A cream formulated with the total saponins of *Sapindus mukurossi* (Sapindaceae) is presently undergoing clinical trials as a vaginal contraceptive (Bhargava, 1988b). The total saponins of this plant are equipotent to nonoxynol-9 and have no reported toxic effects (Setty *et al.* 1976). A saponin fraction from the fruit pulp of *Sapindus trifoliatus* is reported to have both antispermatogenic activity (in dogs) and postcoital antifertility effects in female rats (Bhargava, 1988b).

Elbary and Nour (1979) have correlated the spermicidal activity of certain saponin-enriched plant extracts with their haemolytic activity (Table 5.6). In general, strong inhibition of spermatozoid motility was associated with a high haemolytic index, with the saponins from *Gypsophila paniculata* (Caryophyllaceae) giving the highest activities. Only *Terminalia horrida* (Combretaceae) had a low haemolytic index.

5.3.6   Insecticidal and antifeedant activity

Resistance of wood to termite attack has in certain cases been attributed to the saponin constituents. Mono- and bidesmosidic saponins repel termites, probably because of the presence of enzymes in the termite's digestive tract which convert the bidesmosides into monodesmosides (Tschesche *et al.* 1970; Tschesche, 1971). Both *Ternstroemia japonica* (Theaceae) (Watanabe *et al.* 1966) and *Kalopanax septemlobus* (Araliaceae) (Kondo *et al.* 1963) give protection against termites, as do certain legumes. A number of legumes which are used as food sources have been investigated for their resistance to other insects and their larvae (Applebaum *et al.* 1965, 1969).

Legume seeds are prone to infestation with beetles (e.g. the Azuki bean beetles, *Callosobruchus chinensis*) during storage. However, soybeans inhibit the development of *C. chinensis* and it has been proposed that the presence of saponins in these beans and certain other legume seeds causes resistance to the attack of insects (Applebaum *et al.* 1969).

It has been reported that, in addition to its antifungal properties, camellidin II (**1317**, p. 197) from *Camellia japonica* (Theaceae) has antifeedant activity against 5th instar larvae of the yellow butterfly *Eurema hecabe mandarina* (Numata *et al.* 1987). Furthermore, the saponin mixture from the seeds of *Camellia sinensis* (*Thea sinensis*) has insecticidal properties (Tschesche *et al.* 1969a).

Peduncloside (**592**) and rotungenoside (**593**) (Appendix 2) from the fruits of *Ilex rotunda* (Aquifoliaceae) showed weak antifeedant activity at 2000 p.p.m. against larvae of *Spodoptera litura* in a leaf disk assay (Nakatani *et al.* 1989). Young leaves of *Ilex opaca*, which contain high levels of saponins, deter the southern red mite *Oligonychus ilicis*. This foliage is also unpalatable to the fall webworm *Hyphantria cunea* and the eastern tent caterpillar *Malacosoma americanum* (Potter and Kimmerer, 1989).

There is a correlation between growth of the yellow mealworm *Tenebrio molitor* and the saponin content of lucerne leaves (*Medicago sativa*, Leguminosae) fed to the insects. Glycosides of medicagenic acid appear to exhibit the highest degree of feeding inhibition (Pracros, 1988).

Table 5.5. *Plants with reported spermicidal activity resulting from the presence of triterpene glycosides*

Species	Family	Active principle(s)	Reference
Cussonia spicata	Araliaceae	Oleanolic acid saponins	Gunzinger *et al.* 1986
Hedera nepalensis	Araliaceae		Pant *et al.* 1988b
Panax japonicus	Araliaceae	Chikusetsusaponin IV	Stolzenberg and Parkhurst, 1974
Pentapanax leschenaultii	Araliaceae		Pant *et al.* 1988c
Schefflera capitata	Araliaceae	Scheffleroside	Farnsworth and Waller, 1982
Calendula officinalis	Asteraceae	Oleanolic acid saponins	Stolzenberg and Parkhurst, 1974
Solidago virgaurea	Asteraceae		Setty *et al.* 1976
Gypsophila cerastioides	Caryophyllaceae		Setty *et al.* 1976
G. paniculata	Caryophyllaceae		Setty *et al.* 1976
Saponaria officinalis	Caryophyllaceae		Elbary and Nour, 1979
Vaccaria segetalis	Caryophyllaceae	Vacsegoside	Farnsworth and Waller, 1982
Terminalia belerica	Combretaceae		Setty *et al.* 1976
T. horrida	Combretaceae		Setty *et al.* 1976
Bulbostemma paniculatum	Cucurbitaceae		Su and Guo, 1986
Dimeria gracilis	Gramineae		Setty *et al.* 1976
Acacia auriculiformis	Leguminosae	Acaciasides A and B	Pakrashi *et al.* 1991
A. caesia	Leguminosae		Banerji and Nigam, 1980
A. concinna	Leguminosae	Acacic acid saponin	Farnsworth and Waller, 1982
			Banerji *et al.* 1979a
Aeschynomene indica	Leguminosae	Echinocystic acid saponins	Setty *et al.* 1976
Albizzia chinensis	Leguminosae	Echinocystic acid and oleanolic acid saponins	Rawat *et al.* 1989
A. lebbek	Leguminosae		Farnsworth and Waller, 1982
A. procera	Leguminosae		Farnsworth and Waller, 1982
Anisomeles malabarica	Leguminosae	Oleanolic acid saponin	Setty *et al.* 1976
Enterolobium cyclocarpum	Leguminosae		Elbary and Nour, 1979
Melilotus siculus	Leguminosae		Elbary and Nour, 1979

Pithecellobium dulce	Leguminosae	Oleanolic and echinocystic acid saponins	Farnsworth and Waller, 1982
Pterolobium indicum	Leguminosae		Setty *et al.* 1976
Samanea saman	Leguminosae	Samanin D	Farnsworth and Waller, 1982
(*Pithecellobium saman*)			
Sesbania sesban	Leguminosae	Oleanolic acid saponins	Dorsaz *et al.* 1988
Ardisia neriifolia	Myrsinaceae	Primulagenin A saponins	Malaviya and Khanna, 1989
Phytolacca acinosa	Phytolaccaceae	Oleanolic acid saponins	Farnsworth and Waller, 1982
P. americana	Phytolaccaceae	Oleanolic acid saponins	Stolzenberg and Parkhurst, 1974
P. dodecandra	Phytolaccaceae	Oleanolic acid and bayogenin saponins	Stolzenberg and Parkhurst, 1974
			Dorsaz and Hostettmann, 1986
Pittosporum nilghrense	Pittosporaceae	Pittosides A and B	Farnsworth and Waller, 1982
Anagallis arvensis	Primulaceae	Anagalligenone saponin	Farnsworth and Waller, 1982
Cyclamen persicum	Primulaceae		Primorac *et al.* 1985
Primula vulgaris	Primulaceae		Primorac *et al.* 1985
Caltha palustris	Ranunculaceae		Setty *et al.* 1976
Blighia sapida	Sapindaceae	Oleanolic acid and hederagenin saponins	Farnsworth and Waller, 1982
Sapindus mukurossi	Sapindaceae	Hederagenin saponins	Farnsworth and Waller, 1982
Madhuca butyracea	Sapotaceae	Bassic acid saponin	Farnsworth and Waller, 1982
M. latifolia	Sapotaceae	Bassic acid saponin	Farnsworth and Waller, 1982
Mimusops elengi	Sapotaceae	Bassic acid saponin	Farnsworth and Waller, 1982
M. hexandra	Sapotaceae	Bassic acid saponin	Farnsworth and Waller, 1982
M. littoralis	Sapotaceae	Bassic acid saponin	Farnsworth and Waller, 1982
Herpestis monniera	Scrophulariaceae	Bacoside	Farnsworth and Waller, 1982
Clerodendrum serratum	Verbenaceae		Setty *et al.* 1976

Table 5.6. *Spermicidal and haemolytic activities of selected plant saponin fractions*

Plant	Haemolytic index	Spermicidal activity[a] (%)
Gypsophila paniculata	9524	1
Saponaria officinalis	3846	10
Enterolobium cyclocarpum	1250	>40
Terminalia horrida	222	40
Melilotus siculus	2857	35

[a] Concentration of saponin fraction in isotonic sodium chloride required for total inhibition of spermatozoidal motility after 3 min.
From Elbary and Nour, 1979.

5.3.7   Anthelmintic activity

Some saponin preparations have a practical use as anthelmintics in the traditional medicine of certain countries. In fact, a large number of saponins exhibit an anthelmintic activity but many find no practical application because of the parallel irritation of mucous membranes (Jentsch *et al.* 1961). The evaluation of *Filix*-type vermicidal activity of *Albizzia anthelmintica* (Leguminosae) and one of its saponins, musennin, has confirmed the use of the bark of this tree as a nematocide in Ethiopia (Jentsch *et al.* 1961). Musennin (**345**) is a linear tetrasaccharide of echinocystic acid (Tschesche and Kämmerer, 1969).

A triterpene glycoside from the bark of *Schima argentea* (Theaceae) was found to be effective as an oral anthelmintic (Chen and Wu, 1978).

Both a saponin preparation and α-hederin (**368**, Appendix 2) from the leaves of *Hedera helix* (Araliaceae) killed the liver flukes *Fasciola hepatica* and *Dicrocoelium* spp. at concentrations of around 0.005 mg/ml *in vitro*. The saponin preparation also eliminated the trematodes of *Dicrocoelium* spp. from sheep (Julien *et al.* 1985).

5.3.8   Expectorant and antitussive activity

Saponins possess a general and non-specific ability to produce local irritation, especially of mucous membranes (Vogel, 1963). This can take place in the nasal cavity, the throat, the bronchi, the lungs and in kidney epithelia. One effect is the sneezing provoked by saponins in powder form. The expectorant activity of saponin drugs (*Senegae radix*, *Primulae radix*, *Liquiritiae radix*, *Gypsophila* spp., ivy leaves (*Hederae folium*)) is attributed at least in part to local irritation of mucous membranes, although their mechanism of action is not certain. The irritation of the throat and respiratory tract probably increases respiratory fluid volume by drawing more water into the bronchial secretions, hence diluting the mucus and reducing its viscosity. Alternatively, the surface activity of the saponins may render the sputum less viscid, making it more mobile and easier to eject. Another possibility is that the amphiphilic nature of saponins causes them to spread out as a monomolecular film at the back of the throat and subsequently aid elimination of mucus. More extensive details of individual expectorants are to be found in Chapter 7.

Recently, a series of *antitussive* glycosides has been isolated from *Luffa cylindrica* (Cucurbitaceae), including lucyoside A (**469**), a bidesmoside of 21β-hydroxyhederagenin (Jap. Pat. 60 48,995).

5.3.9 Diuretic activity

The diuretic action of many saponins can also be traced back to a local irritation of kidney epithelia. Plants with diuretic activity caused at least in part by saponins include: *Astragalus membranaceus* (Jap. Pat. 57 165,400), *Herniara glabra*, *H. hirsuta*, *Solidago virgaurea*, *S. gigantea*, *S. canadensis* (Schilcher, 1990), *Agropyron repens*, *Betula pendula*, *B. pubescens*, *Equisetum arvense*, *Ononis spinosa*, *Viola tricolor*, *Vigna sinensis*, *V. radiata* and *V. mungo* (Sood *et al.* 1985; Schilcher and Emmrich, 1992).

Diuresis caused by certain saponin-containing plants such as *Ononis*, *Herniana*, *Betula* and *Solidago* species is relatively mild, and the effect may just as well originate from the accompanying flavonoids and essential oils (Hänsel and Haas, 1983). An alternative theory is that the potassium content of these plants is in effect the diuretic agent (Steinegger and Hänsel, 1988).

The flowering tips of the branches of *Solidago virgaurea*, *S. serotina* and *S. canadensis* are all used in the treatment of inflammatory complaints of the urinary tract and urinary calculi but they vary considerably in their saponin content: *S. serotina* and *S. canadensis* contain considerably more saponins than *S. virgaurea* (Anonymous, 1991).

5.3.10 Cholesterol metabolism

Saponins reach the human body in relatively large amounts as a result of the intake of food rich in these glycosides and in the form of pharmaceuticals. Their actual biological action is still not well understood but there are certain effects which have been investigated.

Work by Cheeke (1971) has shown that triterpene glycosides in feed given to hens decrease the amount of cholesterol in both blood and tissues. This may be a result of formation of a cholesterol–triterpene glycoside complex in the digestive tract which is not resorbed or there is the possibility of a direct effect on cholesterol metabolism. The former theory is supported by the fact that, in the blood of laboratory animals, saponins have been shown to form complexes with plasma lipids (Oakenfull and Topping, 1983). An increased faecal elimination of bile acids, followed by increased formation of bile acids from endogenous cholesterol might also contribute to the lowering of cholesterol (Rao and Kendall, 1986). Ivanov *et al.* (1987) studied hypolipidaemic properties of semi-synthetic betulin glycosides. Both mono- and bidesmosidic glycosides were prepared. Liposomes containing the glycosides led to *in vitro* decrease in cholesterol content by 187–211%. In a study in cats with experimentally induced hypercholesterolaemia, the glycosides also decreased cholesterol levels. The highest activity was produced by monodesmosidic betulin glycosides. As a consequence of these observations, the suggestion has been made that foods rich in saponins may reduce the risk of heart disease (Oakenfull, 1981). In view of the current frenetic search for means of decreasing cholesterol levels in the human body, the results obtained with saponins may provide further impetus for more detailed studies with this class of substance.

Oleanolic acid saponins from *Aralia mandshurica* (Araliaceae), *Calendula officinalis* (Asteraceae) and *Beta vulgaris* (Chenopodiaceae) fed by mouth reduce total lipid content, triglyceride content and cholesterol content of rats by up to 37%, 30% and 25%, respectively. The rats were fed on a fat-rich diet containing coconut oil and cholesterol. This led to an increase in the lipid content measured in the blood serum and liver homogenates (total lipid, triglyceride, total cholesterol, free fatty acids were all determined by established methods) (Lutomski, 1983).

Alfalfa saponins are reported to depress concentrations of lipids and cholesterol in the livers of mice. It is thought that the medicagenic acid glycosides are responsible for these effects (Reshef *et al.* 1976). Adzuki saponins from *Vigna angularis* reduce adriamycin-induced hyperlipaemia in rats (Yuchi and Uchida, 1988), while soybean saponins reduce intestinal uptake of cholesterol in the rat (Sidhu *et al.* 1987).

Ilexsaponin B_3 (**582**, Appendix 2) from the roots of *Ilex pubescens* (Aquifoliaceae) (used for the treatment of cardiovascular diseases and hypercholesterolaemia) inhibits hypercholesterolaemia in mice by 68% at an oral dose of 300 mg/kg (Hidaka *et al.* 1987a).

Ginseng extracts reduce total cholesterol and triglyceride concentrations but do not have an effect on HDL-bound cholesterol (Moon *et al.* 1984); ginseng saponins, moreover, reduce the spread of arteriosclerosis in a rabbit model (Koo, 1983).

Other effects on metabolism have been studied in the case of the ginsenosides: stimulation of DNA, RNA and lipid synthesis (see Chapter 7).

5.3.11   *Cardiovascular activities*

Saponins are able both to increase the permeability of cell plasma membranes and to cause *positive inotropic effects* in isolated cardiac muscles. In one study on saponins where seven compounds were tested, only those saponins with haemolytic activity showed a positive inotropic action on isolated cardiac muscles of the guinea pig. The mode of action of the most potent saponin (holothurin-A) resembled ouabain rather than isoprenalin, the effect being caused by a modification of the calcium channels, as in the case of isoprenalin (Yamasaki *et al.* 1987; Enomoto *et al.* 1986).

Antiarrhythmic action. Antiarrhythmic activity in rats and rabbits has been demonstrated by panaxatriol saponins from *Panax notoginseng* (Araliaceae) (Li and Zhang, 1988). The total saponins of *Panax quinquefolius* at 80 mg/kg i.v. inhibited cardiac arrhythmias induced by chloroform in mice, by barium chloride in rats and by ouabain in guinea pigs (Zhang *et al.* 1985).

Seed saponins of *Ziziphus jujuba* are *hypotensive* in normal rats (i.v. and i.p.) and cats (i.p.). They inhibit the central pressor reflex and the spontaneous discharge of preganglionic fibres of the greater splanchnic nerve, while decreasing plasma renin activity (Gu *et al.* 1987). The antihypertensive activity of ginseng saponins is attributed to a vascular relaxant effect caused by blocking Ca^{2+} channels without affecting voltage-operated Ca^{2+} channels or Ca^{2+} release from the intracellular store (Guan *et al.* 1988). Since ginseng saponins act as antagonists to presynaptic α_2-receptors, their hypotensive action can also result from a decrease in transmitter release from sympathetic nerves (Zhang and Chen, 1987).

Vasodilatory action. Vasodilatory effects have been shown by crude saponins of *Panax notoginseng*. Calcium channel blockers of the verapamil type exhibit similar action. Ginsenosides Rg_1 and Re inhibited intracellular Ca^{2+} release; ginsenoside Rb_1 and total saponins of *P. notoginseng* inhibited the extracellular Ca^{2+} influx (Guan *et al.* 1985; Wu and Chen, 1988). Ginsenosides Rb_1 and Ro (i.v.) induced vasodilatation in normal rabbits and in rabbits with noradrenaline-induced vasoconstriction. These effects were not observed in the case of ginsenoside Rg_1 and nifedipine; thus the vasodilatory effects of ginsenosides Rb_1 and Ro are non-selective, whereas those of ginsenoside Rg_1 are selective for Ca^{2+}-induced vasoconstriction (Kaneko *et al.* 1985).

Table 5.7. *Corticosterone release in rats*

Saponin or saponin preparation[a]	ED_{50} (mg/kg)[b]
Ginsenoside Rd	7
Ginsenoside Rb_2	23
Ginsenoside Rc	49
Ginsenoside Rb_1	128
Ginsenoside Re	107
Ginsenoside Rg_1	>131
Saikosaponin a	0.89
Saikosaponin d	0.33
Panax japonicus	54
Gypsophila paniculata	3.2
Sapindus mukurossi	2.4
Polygala senega var. *latifolia*	2.3
Thea sinensis	4.9
Bupleurum chinense	2.7
Aesculus hippocastanum	2.0
Styrax japonicus	0.7
Platycodon grandiflorum	1.0

[a] Administered i.p.
[b] Evaluated from log dose–response curve obtained by
 a competitive protein-binding method.
From Hiai *et al.* 1983; Yokoyani *et al.* 1982a.

Other actions. Certain astrasieversianins (cycloartane saponins) from *Astragalus sieversianus* (Leguminosae) have *hypotensive* activity (Gan *et al.* 1986a,b). *Astragalus membranaceus* saponins have *antihypertensive* and *cardiotonic* properties (Jap. Pat. 57 165,400). The *neurotropic, cardiotonic, hypotensive* and *tonic* effects of certain triterpene saponins such as calenduloside B, aralosides A, B and C, patriside A, patrinoside D and clematoside C have been discussed (Sokolov, 1986). These have, in general, low toxicity.

It should be emphasized, however, that caution must be exercised when extrapolating the results of such studies with saponins on animals to the human situation. The saponins are often parenterally applied in animals whereas in humans it is almost exclusively oral administration which is practicable.

5.3.12   *Effects on the adrenocorticotrophic system*
Many saponins are capable of inducing the release of adrenocorticotrophic hormone (ACTH) (Hiai, 1986) and corticosterone in rats (Table 5.7) (Yokoyama *et al.* 1982a,b, 1984; Hiai *et al.* 1983) but this

does not appear, all the same, to be a general phenomenon of triterpene saponins. The mode of action, involving the hypothalamus and/or the pituitary, has been discussed (Hiai, 1986).

Ginsenoside Rb_1 (**749**, Table 2.4, p. 52) not only induces corticosterone release but also is capable of stimulating ACTH secretion in primary cultures of pituitary cells. Since ginsenosides Rc, Rd, Re and Rg_1 do not stimulate ACTH secretion, increased plasma corticosterone levels in rats treated with these ginsenosides are not the direct result of ginsenoside-induced ACTH secretion by the hypophysis (Odani *et al.* 1986). The increase in plasma corticosterone level after administration of total saponins of *Panax notoginseng* (Araliaceae) roots was inhibited by dexamethasone, thus implicating a central mechanism of saponin-induced adrenal corticosterone release (Chen *et al.* 1987c). Intraperitoneal administration of total saikosaponins or saikosaponin d (**548**, Appendix 2) caused an increase in the weight of the adrenals and a decrease in the weight of the thymus of rats (Hiai and Yokoyama, 1986; Hiai *et al.* 1987). Saikosaponin c (**544**, Appendix 2) influenced neither the adrenal nor thymus weight (Hiai *et al.* 1987). There was no weight gain of the adrenal or thymus after administration of saikosaponin d to hypophysectomized rats, showing that effects on the adrenal or thymus by saikosaponins are caused by stimulation of ACTH secretion by the hypophysis (Hiai and Yokoyama, 1986). The site of action of saikosaponins a (**547**, Appendix 2) and d is closely related to that of dexamethasone (Yokoyama *et al.* 1984).

Certain metabolities of saikosaponins a and d formed in the murine alimentary tract show corticosterone sectretion-inducing activity when administered intraperitoneally. On oral administration, secretion induction activities were similar to those obtained after i.p. administration (Nose *et al.* 1989a).

The saponins $3\beta,16\beta,28$-trihydroxyolean-9(11),12-dien 3-*O*-β-D-glucopyranoside (**1325**) and $3\beta,16\beta,28$-trihydroxyolean-11,13(18)-dien3-*O*-β-D-glucopyranosyl-(1 → 4)-α-L-rhamnopyranoside (**1326**) increase serum corticosterone levels in mice and have been used in the preparation of anti-inflammatory agents (Nose *et al.* 1988a,b).

5.3.13   *Anti-inflammatory, antiexudative and antioedematous activities*

The initial stages of an inflammatory response are characterized by increased blood vessel permeability and release (exudation) of histamine, serotonin and basic polypeptides and proteins. This is accompanied by hyperaemia and oedema formation. Subsequently, there is cellular infiltration and formation of new conjunctive tissue. Substances with antiexudative and antioedematous properties, therefore, form a subgroup of anti-inflammatory (or antiphlogistic) agents, in that they influence the earlier stages of inflammation. In some cases the anti-inflammatory action seems to be independent of the pituitary–adrenal axis, whilst in other cases the drugs are said to have a direct action on the adrenal cortex (e.g. glycyrrhetinic acid – see Chapter 7).

The reduction of rat paw oedema by the saponin mixture aescine from the horse chestnut (*Aesculus hippocastanum*; Hippocastanaceae) is 600 times more effective than rutin on parenteral application. No effect is observed, however, on oral application (Lorenz and Marek, 1960). Ester groupings are important for activity, since the deacylated derivative, aescinol, is not anti-inflammatory.

Studies on the anti-inflammatory activities of hederagenin (**17**, Table 2.1, p. 21) and the crude saponins from *Sapindus mukurossi* (Sapindaceae) fruits have been performed with the carrageenin-induced oedema, granuloma pouch and adjuvant arthritis models in mice. The crude saponin exhibited anti-inflammatory activity both i.p. and p.o. at doses of around 100–200 mg/kg but hederagenin was only effective against carrageenin-induced oedema by the i.p. route. Similar results were obtained with other triterpene aglycones. The crude saponins also produced a decrease in vascular permeability (Takagi *et al.* 1980). The monodesmoside α-hederin (**368**, Appendix 2) has antiexudative properties, whereas the corresponding bidesmoside hederasaponin C (**372**) lacks this activity in animal models.

Saponins from *Eryngium planum* (Umbelliferae) and *Sanicula europaea* (Umbelliferae) have a beneficial effect in oedema in the rat and reduce swelling considerably. The time required for evidence of activity is approximately 1 h, whereas aescine is effective after 17 h (Jacker and Hiller, 1976). *Hedera helix*, *Thea sinensis*, *Hydrocotyle vulgaris*, *Polemonium caeruleum* and *Solidago virgaurea* have also been investigated for the antiexudative activity of their constituent saponins. Of these, the saponins from *Solidago virgaurea* (Asteraceae), *Polemonium caeruleum* (Polemoniaceae) and *Hydrocotyle vulgaris* (Umbelliferae) were the most effective (Jacker *et al.* 1982), with activity approximating to that of aescine. The antiexudative activity of a *Thea sinensis* (Theaceae) saponin mixture ('theasaponin') has also been reported by Vogel *et al.* (1968b) and Tschesche *et al.* (1969a). Because of the longer sugar chain, the water solubility is better and the toxicity less than aescine. Tea seed saponin normalizes the capillary permeability that is disturbed in the initial stages of inflammation.

Since both aescine and theasaponin are mixtures of ester saponins, the structurally similar ester saponins from *Dodonea viscosa* (Sapindaceae) (**805** and **806**, Table 2.6, p. 62) were investigated for their effects on the viscarin– carrgeenin rat paw oedema test. At a concentration of 0.7 mg/kg, an oedema inhibition of 33% was observed (Wagner *et al.* 1987), a similar result to that obtained with aescine.

As all the above-mentioned saponins are injected i.v. a limiting factor is always their toxicity. The most useful are, therefore, those with the largest therapeutic index, such as aescine.

Ester saponins of olean-12-en-3β,16α,22α,28-tetraol from *Maesia chisia* var. *angustifolia* (Myrsinaceae) reduced rat paw oedema by 43% after 4 h at 25 mg/kg but had no effect on histamine- and serotonin-induced paw oedema. Positive responses were also shown against cotton pellet granuloma, formaldehyde-induced arthritis and Freund's adjuvant arthritis. The mechanism of action is thought to involve an inhibition of cyclo-oxygenase (Gomes *et al.* 1987).

The sailosaponins from *Bupleurum chinense* (Umbelliferae) which reduce carrageenin-induced oedema at high does (150–300 mg/kg i.p. in the rat; LD_{50} 1.53 g/kg) inhibit histamine release but have no effect on 5-hydroxytryptamine (5-HT) or prostaglandin E_1. They counteract a 5-HT- or histamine-induced increase in capillary permeability and are active in the cotton pellet granuloma test. Their application in the treatment of arthritis has been proposed.

The constituents of *Tetrapanax papyriferum* (Araliaceae) are antiphlogistic in both the carrageenin oedema and the cotton pellet granuloma tests (mouse). Papyriogenins are the most active (papyriogenin C has the same activity as prednisolone), followed by the propapyriogenins and the papyriosides. Papyriogenin A, with a keto group at C-3, is less active than papyriogenin C. The C-21 oxygen function and the C-28-COOH are important for the activities of the propapyriogenins and papyriosides but the C-11 oxygen is not essential. Triterpenes with a *trans*-conformation of rings D/E have more powerful antiphlogistic activity than those of a *cis*-conformation; papyriogenin C, 18α-glycyrrhetinic acid (D/E *trans*) and papyriogenin A are more active than, for example, 18β-glycyrrhetinic acid (D/E *cis*). It can logically be assumed, therefore, that the 11(12), 13(18)-diene system of papyriogenins A and C is also important for the activity (Sugishita *et al.* 1982).

The saponins asiaticoside (**597**) from *Hydrocotyle asiatica* (Umbelliferae) (see Chapter 7), saikosaponin a (**547**) from *Bupleurum falcatum* (Umbelliferae) and akeboside from *Akebia quinata* (Lardizabalaceae) are also reported to show anti-inflammatory activity (Shibata, 1977); others with anti-inflammatory properties include arvenoside A (**162**); calendulosides C (**214**) and D (**215**) from *Calendula arvensis* (Asteraceae) (Chemli *et al.* 1987; Vidal-Ollivier *et al.* 1989c); calenduloside B (**164**) from *Calendula officinalis* (Asteraceae) (Yatsyno *et al.* 1978); swericinctoside (**319**) from *Swertia cincta* (Gentianaceae) (Zhang and Mao, 1984); and a quinovic acid glycoside from *Uncaria tomentosa* (Rubiaceae) (Aquino *et al.* 1991). The 28-*O*-glucoside of 23-hydroxytormentic acid (**603**) from the leaves of *Quercus suber* (Fagaceae) has topical antiphlogistic activity (Romussi *et al.* 1991). (See Appendix 2 for all structures.)

The anti-inflammatory activities of the Mi-saponin mixture from *Madhuca longifolia* (Sapotaceae) are similar to β-aescine but weaker than phenylbutazone in mice (Yamahara *et al.* 1979).

Two saponins from the roots and bark of *Crossopteryx febrifuga* (Rubiaceae), crossoptine A (**1327**) and crossoptine B (**1328**) have anti-inflammatory, mucolytic and antioedemic activities (Foresta *et al.* 1986; Gariboldi *et al.* 1990). They are more effective than aspirin, phenylbutazone and aescine in inhibiting carrageenin-induced oedema of rat paw (Foresta *et al.* 1986).

1327 R = -H
1328 R = -Api

Glycyrrhetinic acid glycosides are found in the West African anti-inflammatory plants *Lonchocarpus cyanescens* (Leguminosae) and *Terminalia ivorensis* (Combretaceae), while oleanolic acid heterosides are present in *Boerhavia diffusa* (Nyctaginaceae), *Gymnema sylvestre* Asclepiadaceae), *Securidaca longepedunculata* (Polygalaceae), *Tetrapleura tetraptera* (Leguminosae) and *Phytolacca dodecandra* (Phytolaccaceae), all of which are used in traditional medicine in the treatment of inflammation (Bep Oliver-Bever, 1986).

5.3.14   *Effects on capillary fragility and venous stasis/venous insufficiency*

This section is closely tied up with the previous section in that agents which decrease capillary fragility also have a beneficial effect on capillary permeability. Weakness of the blood vessels (and, most importantly, the veins) can lead to several complications:

- hypertension (resulting from enlarged veins) (see cardiovascular activities)
- oedema (resulting from increased blood vessel permeability)
- reduced blood flow, with the danger of thrombosis.

Related disorders include haemorrhoids and varicose veins.

Substances which affect the veins ('Venenmittel' in German) can be divided into three categories (Felix, 1985):

- those which strengthen the veins (reduction of venous cross-sectional area)
- those which are diuretic
- those which protect against oedema (decrease capillary permeability).

Certain saponin-containing plants have capillaro-protective properties and it has been known for a long time, for example, that seed extracts of horse chestnut (*Aesculus hippocastanum*, Hippocastanaceae) have beneficial effects on capillary fragility (Lorenz and Marek, 1960). This was proved by the observation of positive results for the crude saponin aescine (see Chapter 7) in the petechial test for capillary resistance, which involves applying a sucking device to the shaved back skin of rats and measuring the tendency of the capillaries to bleed (Lorenz and Marek, 1960).

5.3.15   *Antiulcer activity*

Perhaps the best-known of the saponin antiulcerogenic drugs is glycyrrhiza (see Chapter 7), although it is now known that in addition to glycyrrhizin, other active factors are present in the roots and rhizomes of *Glycyrrhiza glabra*: these are flavonoids such as licoflavonol, licoricone and kumatekenin (Saitoh *et al.* 1976). Other saponins which have been studied include quillaia saponins I and II (Muto *et al.* 1987); gypenoside from *Gynostemma pentaphyllum* (Cucurbitaceae) (Takemoto, 1983); chikusetsusaponin III (**756**, Table 2.4, p. 52) from *Panax japonicum* (Araliaceae) (Yamahara *et al.* 1987b); the sodium salt of 3-(α-L-arabino-pyranosyl)-16,23-dihydroxyolean-12-en-28-oic acid (racemoside) (**466**, Appendix 2) from the roots of *Cimicifuga racemosa* (Ranunculaceae) (Jap. Pat. 59 20,298); sericoside from *Terminalia sericea* (Combretaceae) (Mustich, 1975); and calenduloside B (**164**, Appendix 2) from *Calendula officinalis* (Asteraceae) (Yatsyno *et al.* 1978).

The antiulcer action is not the result of an inhibition of gastric acid secretion but may be ascribed to one of the following:

- promotion of mucus formation
- direct activation of gastric membrane protective factors
- activation of natural body protection mechanisms via irritation of the gastric mucosa.

Protective effects in various experimentally induced gastric ulcer models in rats have been shown by a saponin-containing ethanol extract of *Pyrenacantha staudtii* (Icacinaceae) leaves. The observed effects were dose dependent in indomethacin-induced ulcers and comparable to those of the histamine blocker cimetidine – 28.5 mg/kg extract orally was more effective in reducing the incidence of ulcers than 100 mg/kg cimetidine (Aguwa and Okunji, 1986).

Ginseng flower saponins were effective in treating gastric ulcers (Zhang and Hu, 1985), while a patent has appeared for antiulcer tablets containing quillaiasaponins (Muto *et al.* 1987).

5.3.16   *Analgesic activity*
The root extract of *Platycodon grandiflorum* (Campanulaceae) (2 g/kg s.c. representing 160 mg platycodin/kg) has a stronger analgesic effect than a 100–200 mg dose of acetylsalicylic acid (aspirin) (Racz-Kotilla *et al.* 1982).

Barbatosides A and B, glycosides of quillaic acid, from *Dianthus barbatus* (Caryophyllaceae) are both analgesic and anti-inflammatory. In the acetic acid writhing test, barbatoside A (ED_{50} 95 mg/kg) and barbatoside B (ED_{50} 50 mg/kg) were more active than acetylsalicylic acid (ED_{50} 125 mg/kg) (Cordell *et al.* 1977).

Dianosides A (**298**) and B (**300**) (Appendix 2), glycosides of gypsogenic acid from *Dianthus superbus* var. *longicalycinus* herbs (Caryophyllaceae), have been found to possess analgesic activity. At 10 and 30 mg/kg s.c. respectively, they exhibited significant suppressive activity on acetic acid-induced writhing (Oshima *et al.* 1984a).

Analgesic effects are also shown by saponins from *Panax notoginseng* (Araliaceae) (100–250 mg/kg i.p. in the mouse gives the same effect as 150 mg/kg of aminophenazone) (Lei *et al.* 1984), from *Crossopteryx febrifuga* (Rubiaceae) (Gariboldi *et al.* 1990), from *Dolichos falcatus* (Leguminosae) (5 mg/kg i.p. in the mouse) (Huang *et al.* 1982b) and from the bark of *Zizyphus rugosa* (Rhamnaceae) (25–100 mg/kg i.p. in the mouse). Saponins of *Maesa chisia* var. *angustifolia* (Myrsinaceae) demonstrate analgesic activity in the mice writhing test (oral administration of 25 mg/kg caused 33% inhibition of writhing produced by *p*-phenylquinone and 52% inhibition of writhing produced by acetic acid). Ineffectiveness in the tail flick and hot plate models differentiated this activity from narcotic analgesics (Gomes *et al.* 1987).

5.3.17   Antipyretic activity

Root saponins of *Bupleurum chinense* (Umbelliferae) produce antipyretic effects on rabbits (Gan and Chen, 1982).

Prior administration of the saponin fraction of *Maesa chisia* (Myrsinaceae) at a dose of 50 mg/kg checked the rise in temperature caused by 2,4-dinitrophenol (DNP) in rats. A similar effect was noticed on DNP-induced pyrexia in rabbits. In rats treated with Brewer's yeast, the reduction in rectal temperature was faster on administration of 50 mg/kg p.o. of the saponin drug than in controls (Gomes *et al.* 1987; Indian Pat. 165,214).

5.3.18   Immunomodulation

Saponins belong to the classes of substance which have an effect on the immune system of the human body. Other important immunomodulators are lipids, lectins (or phytohaemagglutinins), polysaccharides and fungal metabolites.

The term 'immunomodulators' includes immunostimulants, which boost the immune system of an organism, and immunosuppressive factors, which block the immune system. Immunostimulants are employed during chemotherapy, in the treatment of infectious diseases and to stimulate the immune response after vaccination (adjuvants) (Richou *et al.* 1964). Saponins, for example, have been used as adjuvants against infection with *Schistosoma mansoni* (James and Pearce, 1988). Immunosuppressive agents are of especial use in the avoidance of transplant rejections and in the treatment of autoimmune diseases and allergies.

Immunomodulators may be divided into two categories, the specific and the non-specific. Specific immunomodulators include the adjuvants (e.g. quillaia saponins – see Chapter 7), while non-specific immunostimulants are given on their own in order to elicit a generalized state of resistance to pathogens or tumours.

Saponins from *Quillaja saponaria* (Rosaceae), when administered orally, markedly potentiated the humoral response of mice fed inactivated rabies vaccine, especially when given in advance of the oral vaccine (Maharaj *et al.* 1986; Chavali and Campbell, 1987). The increased antibody titre was maintained for at least 6 months after administration of saponin (Maharaj *et al.* 1986). Meanwhile, quillaiasaponin potentiated the antibody response of mice to the T-cell-dependent antigen keyhole limpet haemocyanin (KLH) (Scott *et al.* 1985). These saponins enhance both humoral and cellular immune responses; the immunostimulatory effects of saponins used as adjuvants do not depend solely on an increased uptake of the antigen.

The mode of action of immunostimulants involves an increased phagocytosis by granulocytes and macrophages, an activation of T-helper cells and a stimulation of cell division and transformation in lymphocytes. In immunosuppression, suppressor cells are activated and cytotoxic effects are observed on other cells. Some of the *in vitro* immuno-modulatory effects of saponins have been reported by Chavali *et al.* (1987).

Not only is the immune system important for general resistance to infection but it is also important in *geriatry*, since elderly persons are more prone to disease, to tumour formation and to autoimmune complications. In this respect, there is some overlap between immunomodulatory activity and adaptogenic activity of saponins.

Saponins can form ordered particulate structures of around 35 nm diameter (about the size of a virus particle) with the surface protein from enveloped viruses. These have been named immunostimulating complexes (ISCOMs) (Morein, 1988). ISCOM vaccines against a number of viruses, e.g. influenza, measles, EBV, HIV-1, have been prepared (Bomford, 1988; Kensil *et al.* 1991).

Ginseng is generally accepted to have immunomodulatory properties and this aspect is developed in Chapter 7.

The saponin fraction of *Aralia mandshurica* (Araliaceae) has been found to show immunoregulation, with the following consequences (Lutomski, 1983):

- stimulation of granulocytes
- stimulation of phagocytes
- inhibition of Ehrlich ascites tumour formation in mice
- increased lifespan of animals treated with sub-lethal doses of X-rays.

As the LD_{50} in animals is around 210 mg/kg, no general toxic effects were observed. Saponins from *Dodonaea viscosa* (Sapindaceae) have also been investigated for their effects on *in vitro* granulocyte systems: in the Brandt test, a dose-dependent enhancement of phagocytosis up to 25% was observed; in the chemoluminescence test, a phagocytosis-independent increase of luminescence of approximately 65% was found (Wagner *et al.* 1987).

Some of the non-specific effects of saponin preparations operating via a stimulation of an immune response might result from their irritant nature ('Reizkörpertherapie') (Hänsel, 1984). This so-called 'induction of paramunity' can be provoked via the oral route, through the mucous membranes of the respiratory, digestive and urogenital systems. Saponin-containing plants which can be considered as candidates for such a mode of action are: ginseng, eleutherococcus, *Albizzia lebbeck* (Leguminosae), *Aralia mandshurica* (Araliaceae), *Quillaja saponaria* (Rosaceae) and *Smilax regelii* (Liliaceae) (Hänsel, 1984).

Since orally fed saponins only enter the circulation to a very limited extent, it is not clear how they stimulate the immune system (enhanced cell proliferation, induction of helper T-cell activation and B-cell activation). One theory is that an initial effect on the mucosal immune system leads to secretion of mediators into the circulation. These then initiate a multitude of cellular events (Chavali *et al.* 1987).

In some instances there is controversy as to whether saponins are themselves the immunomodulatory agents. For example, there are claims (Yatsyno *et al.* 1978) and counterclaims (Lutomski, 1983) as to the effectiveness of calenduloside B (**164**) from *Calendula officinalis* (Asteraceae) as an immunostimulant.

5.3.19 Adaptogenic activity

According to Brekhman and Dardymov (1969b), an adaptogen should meet the following criteria:

- it should be innocuous and cause minimal disorder in the physiological functions of an organism
- its action should be non-specific
- it should have a normalizing action irrespective of the direction of the foregoing pathological change.

Plants important for their tonic and so-called 'adaptogenic' properties are *Panax ginseng*, *Eleutherococcus senticosus* (see Chapter 7) and *Aralia mandshurica* (Lutomski, 1983), all from the family Araliaceae.

5.3.20 Sedative activity

Certain Chinese herbal medicines are reputed to have sedative activity (Shibata, 1977):

- *Polygala tenuifolia* roots (Polygalaceae)
- *Ziziphus jujuba* seeds (Rhamnaceae)
- *Panax japonicum* rhizomes (Araliaceae).

Alcoholic extracts of the roots of *Patrinia scabiosaefolia* (Valerianaceae), on repeated pretreatment of mice, cause a significant prolongation of hexobarbital-induced sleeping time (Choi and Woo, 1987).

Wagner *et al.* (1983) have discovered a sedative action of the saponins colubrin (**789**) and colubrinoside (**790**) (Table 2.5, p. 58) from *Colubrina asiatica* (Rhamnaceae). They decrease spontaneous motility in mice and prolong barbiturate narcosis. With a dose of 2 mg/kg of colubrinoside, a prolongation time of 8 min was observed for hexobarbital narcosis. Similar effects were also obtained with the jujubogenin glycosides hovenoside G (**775**) and jujuboside B (**781**, Table 2.5, p. 58) from *Ziziphus jujuba*, thus confirming the ancient use of *Ziziphus jujuba* seeds as a remedy for insomnia (Wagner *et al.* 1983). A saponin (**400**, Appendix 2) from *Cussonia barteri* (Araliaceae) stem bark also showed a significant decrease of mobility at 1 mg/kg (oral) in the same test (Dubois *et al.* 1986).

A saponin mixture from *Panax notoginseng* (Araliaceae) had sedative effects and reduced stimulation caused by coffee. Thiopental-induced sleep was increased and the ED_{50} of pentobarbital was diminished. A synergistic effect was observed with chlorpromazin (Lei *et al.* 1984). Astragalussaponin 1 also prolongs thiopental-induced sleeping time in mice (Zhang *et al.* 1984).

5.3.21   Miscellaneous activities

Uterine contraction. *Ardisia crispa* (Myrsinaceae) is a plant that is used in Asian folk medicine for the treatment of menstrual disorders. Two utero-contracting saponins, ardisiacrispin A (**537**) and B (**542**) (Appendix 2), have been isolated from the roots. At a concentration of 8 µg/ml, both ardisiacrispins gave contractive responses of the isolated rat uterus corresponding to 84% of the contraction caused by a standard dose of acetylcholine (0.2 µg/ml) (Jansakul *et al.* 1987). The mixture of saponins from the seeds of *Thea sinensis* (Theaceae) has also been reported to possess utero-contracting activity (Tschesche *et al.* 1969a).

The rhizome of black cohosh (*Cimicifuga racemosa*, Ranunculaceae) is effective in the treatment of menopausal symptoms such as hot flushes, insomnia and depression. It has been shown that an ethanolic extract of the rhizomes, (Remifemin[R]), which contains the cycloartane glycosides actin (**1329**) and cimicifugoside (**1330**), has an oestrogenic effect and suppresses luteinizing hormone secretion in menopausal women (Düker *et al.* 1991). *Cimicifuga* extracts are also known to have hypotensive, anti-inflammatory, diuretic and antitussive activities (Leung, 1980).

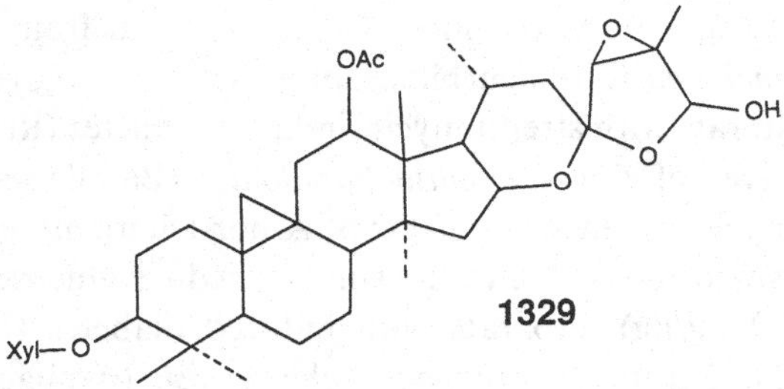

1329

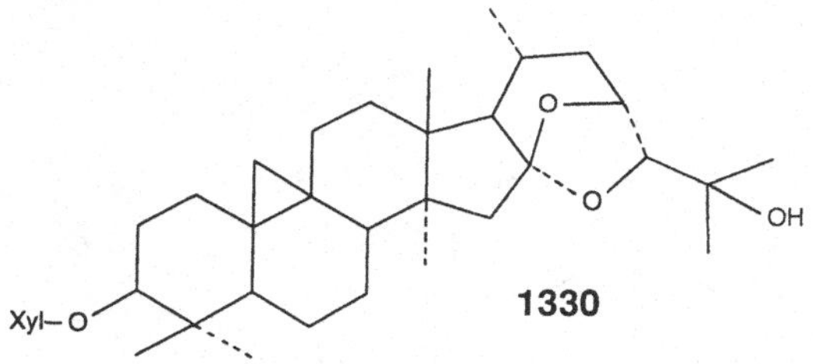

1330

Sweeteners. The sweet-tasting properties of glycyrrhizin (**269**, Appendix 2) have already been noted (Section 5.1.4). In addition, four sweet cycloartane glycosides are known to occur in the leaves of *Abrus precatorius* and *A. fruticulosus* (Leguminosae). They are non-toxic to rodents, non-mutagenic and 30–100 times sweeter than a 2% sucrose solution (Choi *et al.* 1989; Fullas *et al.* 1990). Other natural sweeteners periandrins I (**676**, Section 2.1.1), II and IV have been isolated from the roots of the Brazilian legume *Periandra dulcis* (Hashimoto *et al.* 1983).

Sweetness inhibition. While sweet saponins are known, there are certain examples which are sweetness inhibiting. They have been isolated from the leaves of *Gymnema sylvestre* (Asclepiadaceae) (Maeda *et al.* 1989; Yoshikawa *et al.* 1989a,b, 1991a) and from the leaves of *Ziziphus jujuba* (Rhamnaceae) (Kurihara *et al.* 1988; Yoshikawa *et al.* 1991b). While it was previously thought that acyl groups were essential for this activity (Kurihara *et al.* 1988), it is now known that acylation affects only the magnitude of sweetness inhibition (Yoshikawa *et al.* 1991a). The non-acylated glycosides gymnema saponins III (**514**), IV (**515**) and V (**516**) (Appendix 2) from *G. sylvestre* completely inhibit sweetness perception induced by 0.1 M sucrose; this corresponds to half the activity exhibited by the acylated saponins gymnemic acids I–VI (Yoshikawa *et al.* 1991a).

Hypoglycaemic action. Glycosides of oleanolic acid from tubers of *Momordica cochinchinensis* (Cucurbitaceae) possess hypoglycaemic activity at 25 mg/kg in rats with streptomycin-induced diabetes (Rezu-Ul-Jalil *et al.* 1986). Leaves of *Boussingaultia baselloides* (Basellaceae) are used in Columbia for the treatment of diabetes. A nor-triterpene saponin from the plant, boussingoside A (**1331**), does in fact reduce glucose levels when injected i.p. (20 mg/kg) into rats with induced diabetes (Espada *et al.* 1990). Chiisanoside from *Acanthopanax chiisanensis* (Araliaceae) is also antidiabetic (Hahn *et al.* 1986).

1331

Cyclic nucleotide phosphodiesterase (PDE) inhibition. A potent inhibitor has been found in the roots of *Periandra dulcis* (Leguminosae). This olean-18(19)-en ester saponin periandradulcin (**863**, Section 2.1.6) is more active than the known PDE inhibitors papaverine, vinpocetine and milurinone. Of the three diesterases (PDE-I, PDE-II, PDE-III) isolated from bovine hearts, PDE-I (activated by Ca^{2+} and calmodulin) is most effectively inhibited by the saponin (Ikeda *et al.* 1990, 1991).

Antiallergic action. Ilexosides A–F (**694–699**, Section 2.1.1) and G–I from *Ilex crenata* (Aquifoliaceae) exhibit antiallergic activity. They are approximately 10 times as potent as glycyrrhizin in inhibiting histamine release from mast cells (Kakuno *et al.* 1991a,b).

Fibrinolytic action. Lucyoside N (**428**), a glycoside of quillaic acid, and lucyoside P (**281**) (Appendix 2), a glycoside of gypsogenin from *Luffa cylindrica* (Cucurbitaceae), were virtually equipotent with ginsenoside Rg_1 in a test for fibrinolytic activity (Yoshikawa *et al.* 1991c).

Antileishmanial action. Antileishmanial activity has been reported for saponins of ivy, *Hedera helix* (Araliaceae). The monodesmosides α-, β- and δ-hederin were as effective *in vitro* as the reference antileishmanial drug pentamidine against promastigote (extracellular flagellated) forms of two strains of *Leishmania infantum* isolated from dog and a strain of *L. tropica* isolated from man. Only the aglycone hederagenin was active against amastigote (non-flagellated intracellular) forms (Majester-Savornin *et al.* 1991).

5.4   Pharmacokinetics and metabolism

On oral application, the saponins are poorly absorbed in animals. They are either excreted unchanged or metabolized in the gut. Consequently it is difficult to envisage how these saponins act and there is no direct evidence for many of the activities associated with the saponins. However, the following observations on the absorption of saponins have been made:

 - In order to achieve the same antiexudative effect as for intramuscular injection, saikosaponins have to be taken orally in 10-fold higher doses (Yamamoto *et al.* 1975).
 - α-Aescine is absorbed from rat duodenum to the extent of 10–20%. Maximum blood levels are reached after 1 h but elimination is very rapid (Henschler *et al.* 1971).
 - The pharmacokinetics of ginseng saponins are incompletely understood but it is known that only 0.1% of ginseng saponins with 3 sugar molecules (e.g. ginsenoside Rb_1) are absorbed from the alimentary canal of the rat. Ginsenosides with only 2 molecules of sugar (e.g. ginsenoside Rg_1) are better absorbed: 2–20% (Odani *et al.* 1983). The unabsorbed ginsenosides are rapidly metabolized in rats, with cleavage of the sugar moieties and formation of highly lipophilic products (Odani *et al.* 1983).

Detailed information on the fate of saponins in the animal gut is generally lacking but any breakdown of saponins in the alimentary canal is most likely caused by gut microorganisms, intestinal enzymes or gastric juice. The decomposition of ginsenosides Rb_1, Rb_2 and Rg_1 (Chapter 7) in the rat large intestine is supposed, for example, to involve enteric bacteria and/or enteric enzymes (Karikura *et al.* 1990).

Asiaticoside (**597**, Appendix 2) can be partially cleaved into sugar and aglycone portions in the digestive tract. The pharmacologically active aglycone can then be absorbed (Chausseaud *et al.* 1971). It is quite possible that other saponins can be similarly affected. For example, the major metabolite of glycyrrhizin (**269**, Appendix 2) is glycyrrhetinic acid (**6**, Table 2.1, p. 21), which itself has many of the properties of the parent saponin – anti-inflammatory activity, antiulcerogenic activity, etc.

Hydrolysis of glycyrrhizin in humans and rats and the pharmacokinetics of glycyrrhetinic acid have been described by Ishida *et al.* (1989). Ogihara and co-workers have reported that saikosaponin a (**547**), saikosaponin c (**544**) and saikosaponin d (**548**) (Appendix 2) are transformed into more than 27 metabolites in the alimentary tract; a mixture of at least 30 compounds is thus available for absorption into the bloodstream when the pure saikosaponins are orally administered (Nose *et al.* 1989a). Saikosaponins a and d and certain of these intestinal metabolites strongly induce corticosterone secretion but the aglycones and metabolites in which the ether ring of the aglycone is cleaved are inactive (Nose *et al.* 1989a).

It has been suggested that saponins may have a quasi-synergistic function in affecting penetration and transport of biologically active molecules (Freeland *et al.* 1985). In this connection, they have the capacity to increase the absorption of certain substances. Work by Fischer *et al.* (1959) has shown that the absorption of cantharidin is dramatically increased by the co-administration of gypsophila saponin, leading to significantly higher levels in the liver, kidneys, gall-bladder and bladder. Absorption of strophanthin, digitoxin, curare alkaloids, magnesium sulphate, calcium and iron salts, glucose and acetylsalicylic acid is aided by saponins (Lower, 1984) and there may also be a similar effect on flavones and phytosterols. This is, in fact, a phenomenon which has been known for many years; Jacobi in 1906 observed an increased uptake of adrenaline through frog skin after treatment by various saponin solutions.

A remarkable enhancement of absorption of the antibiotic sodium ampicillin and other β-lactam antibiotics from rat intestine or rectum has been obtained with the crude saponin fraction of *Sapindus mukurossi* (Sapindaceae), the pericarps of which are used as an expectorant in Japanese folk medicine (Kimata *et al.* 1983). It was shown that a solution of monodesmosides solubilized by the bidesmosides from the plant were responsible for this effect (Yata *et al.* 1985).

As to whether saponins are responsible for increasing the absorption of components of whole plant preparations (e.g. extracts), no clear evidence is yet available.

The mechanism of absorption enhancement is not known but it may involve an increased diffusion of substances into the circulatory system after irritation of the gut wall.

5.5   Toxicity

The toxicity of saponins is an extremely important issue as a result of their widespread occurrence in foods (beans, peas, soya beans, peanuts, lentils, spinach and oats, just to mention a few examples). Mean daily intakes vary according to diet and it has been calculated that while the average family in the United Kingdom consumes 15 mg per person per day, a vegetarian has a daily intake of 110 mg per person. Caucasian males in the United Kingdom have a daily intake of 10 mg per person and, at the other extreme, vegetarian Asians in the same country consume 214 mg per person per day (Ridout *et al.* 1988). Luckily, the oral toxicity of saponins to warm-blooded animals is relatively low (George, 1965) and LD_{50} values are in the range 50–1000 mg/kg (Oakenfull, 1981). The reason for this low-risk phenomenon is the feeble absorption which saponins undergo in the body. The question as to whether consumption of saponins over the long term by humans leads to contraindications would seem to be answered by the fact that few negative effects are observed after continued intake of saponins from edible plants, or from ginseng, for example. Over-consumption, however, poses some risk, as illustrated by the examples of licorice. Prolonged exposure to excessive amounts has been known to produce hypertension, flaccid quadriplegia, hypokalaemia, fulminant congestive heart failure and hyperprolactinaemia with amenorrhoea (Spinks and Fenwick, 1990). As a consequence, patients with hypertension or circulatory disorders should avoid licorice (Epstein *et al.* 1977).

The safety of alfalfa saponins for human consumption has been extensively investigated by Malinow and co-workers. Rats fed alfalfa saponins at a level of 1% in the diet for up to 6 months showed no ill effects, although a potentially beneficial reduction in serum cholesterol and triglycerides was observed (Malinow *et al.* 1981). Triterpene saponins from *Quillaja* did not show significant toxic effects in short-term feeding studies in rats (Gaunt *et al.* 1974) and long-term toxicity studies in mice (Phillips *et al.* 1979).

Signs of intoxication by saponins include abundant salivation, vomiting, diarrhoea, loss of appetite and manifestations of paralysis.

There is the occasional instance when oral intake of saponin-containing plants can be lethal – cases of mortality have been recorded after consumption of *Agrostemma githago* (Caryophyllaceae) (Gessner, 1974) – but these instances are extremely rare.

On the contrary, parenteral (and especially intravenous) injection can have much more dramatic effects (Vogel and Marek, 1962). The LD_{50} values vary considerably, from 0.67 mg/kg for quillaioside from *Quillaja saponaria* (Rosaceae) to 50 mg/kg for hederasaponin C (**372**, Appendix 2) (Tschesche and Wulff, 1972). Holothurin A (**1241**, Table 2.13, p. 109) from sea cucumbers has been reported to have an LD_{50} as low as 0.75 mg/kg intravenously in mice (Habermehl and Krebs, 1990). Once in the bloodstream, liver damage, haemolysis of red blood cells, respiratory failure, convulsions and coma can all result (Martindale, 1982). Lethal doses of *Agrostemma* saponins cause liver necroses and bleeding in the alveoli, intestinal walls and other vessel walls – similar phenomena to those observed with *Senega* saponins (Vogel, 1963). However, it is interesting to note that there is no parallel between the toxicity and the haemolytic capacity of a saponin (Vogel, 1963).

Experiments with humans are, of course, rare but Keppler in 1878 injected 100 mg of saponin dissolved in water into his thigh and suffered from terrible pain, accompanied by an anaesthetizing effect around the area of injection (Boiteau *et al.* 1964).

As has previously been mentioned, saponins are highly toxic to fish, molluscs, frogs and other gill-breathing organisms; death can occur at concentrations down to $1:200\,000$. Permeabilization of respiratory membranes and rapid loss of physiological function is responsible for this toxicity.

In conclusion, the impact of saponins on human health has not been well studied. The most important overall consideration when weighing up the risks and evaluating the efficacy is the therapeutic index of a saponin. In this regard, aescine, primula saponin and α-hederin (**368**, Appendix 2) possess very favourable values.

The legal status and toxicity of saponins have been documented for questions pertinent to their industrial and medical use (Lower, 1984).

6

Steroid saponins and steroid alkaloid saponins: pharmacological and biological properties

A large number of steroid saponins and glycoalkaloids exhibit the foaming and haemolytic properties typical of saponins. They also share a large number of the biological activities of triterpene saponins, such as expectorant and diuretic effects. Furostanol glycosides are responsible for the bitter taste of asparagus (Kawano *et al.* 1975), tomato seed (Sato and Sakamura, 1973) and bittersweet (*Solanum dulcamara*, Solanaceae) (Willuhn and Köthe, 1983). Vernonioside A_1 (**869**) from *Vernonia amygdalina* leaves (Asteraceae) is bitter at 7 µg in a paper disc sensory test – an activity similar to that of quinine sulphate (Ohigashi *et al.* 1991).

Some spirostanol saponins are formally bidesmosidic, i.e. they have two sugar chains. In the case of *Convallaria* species (Liliaceae), both chains are attached in the ring A and ring B regions, unlike the situation for most bidesmosidic saponins. Not surprisingly, therefore, they have the characteristic properties of monodesmosidic saponins rather than those of bidesmosidic saponins. On the contrary, the bidesmosidic furostanol saponins lack haemolytic activity, have no bacteriostatic and fungicidal effects and do not complex with cholesterol. Their only typical saponin property is the ability to foam when shaken in aqueous solution (Tschesche *et al.* 1969b). Bidesmosidic furostanol saponins are found mainly in the leaves and metabolically active organs; they can be considered as transport forms since plant wounding leads to their transformation into biologically active spirostanol glycosides. Monodesmosidic spirostanol saponins themselves are often found stored in the seeds and roots (Voigt and Hiller, 1987).

As far as saponins from *marine organisms* are concerned, the toxicity of starfishes can be explained by the saponins which constantly exude from the epidermis and tube feet (Burnell and ApSimon, 1983). Because of their general toxicity, it is possible that saponins act primarily as chemical defence agents and discourage predators. Interestingly, pavoninins, steroid aminoglycosides from the sole *Pardachirus pavoninus* (Tachibana *et al.* 1984, 1985b), and mosesins, steroid monoglycosides from the Moses sole *Pardachirus marmoratus* (Tachibana *et al.* 1985b; Tachibana and Gruber, 1988), have a repellant activity against sharks and are considered to be the factors responsible for the predator-repelling property of these fish.

Starfish extracts and the corresponding purified saponins have a wide spectrum of physiological and pharmacological activities, some of which will be briefly mentioned in this chapter. The asterosaponins are highly surface active and most are haemolytic (Hashimoto, 1979).

6.1   Biological and pharmacological activities

6.1.1   *Antimicrobial activity*
Both steroid saponins (Tschesche, 1971) and steroid alkaloid saponins (Skinner, 1955; Wolters, 1965) possess antibiotic activity. As mentioned above, furostanol saponins (bidesmosidic) show no antibiotic activity. It is only on hydrolysis to the corresponding spirostanol saponins that they exert their biological effects.

Fungicidal activity
Monodesmosidic spirostanol saponins (and also spirosolanol glycosides, such as α-tomatine (**1158**, Table 2.10, p. 99)) show strong fungicidal activity and yet are only weakly effective against bacteria. Their fungitoxicity is greater than that of the triterpene glycosides but they are active against a more restricted range of organisms (Wulff, 1968). The glycoalkaloid α-tomatine (**1158**) was shown in 1948 to be the active principle in tomato that inhibits the growth of *Fusarium oxysporum*, the organism responsible for tomato wilt (Fontaine *et al.* 1948). Further studies indicated that α-tomatine was active against a whole range of microorganisms, including *Aspergillus*, *Candida albicans* and *Trichophyton* (with MIC values from 1.0 to 0.0001 M), saprophytic fungi, organisms of the ringworm group and even Gram-positive bacteria and protozoans (Jadhav *et al.* 1981; Arneson and Durbin, 1968). It was suggested that the tomatidine moiety is responsible for the antibiotic effects. Notable is also the synthesis of α-tomatine in response to a pathological stress caused by fungal infection.

The disease resistance of potato cultivars has been attributed to the presence of glycoalkaloids. α-Solanine (**1174**, Table 2.11, p. 102) is, in this respect, active against *Trichoderma viride*, *Helminthosporium carbonum*,

Table 6.1. *Haemolytic and antifungal activities of
synthetic diosgenin glycosides*

R	HD_{50} $(\mu M)^a$	ID_{50} $(\mu M)^b$
H (Diosgenin)	138	>97
Glc	235	7.5
Gal	111	8.1
Xyl	129	>73
Ara	180	>73
Rha	>357	>71
β-L-Fuc	87	>71
β-D-Fuc	>357	>71

[a] Concentration causing 50% haemolysis.
[b] Concentration giving 50% inhibition of fungal
growth.
From Takechi and Tanaka, 1991.

Fusarium caeruleum, Cladosporium fulvum and other fungal strains (Jadhav
et al. 1981).

In order to examine some structure–activity relationships of steroid
saponins, several monoglycosides of diosgenin have been synthesized. The
glycosides had similar haemolytic activity to the aglycone, while the
glucoside and galactoside were more active against the fungus *Trichophyton
mentagrophytes* (Table 6.1). Since the 3-*O*-glucoside of cholesterol was
virtually inactive when compared with diosgenin glucoside, the E and F
rings of diosgenin are important for the biological effects (Takechi and
Tanaka, 1991).

A similar study was performed on dioscin, dioscinin and their partial
acid-hydrolysis products. Activities were proportional to the number of
sugar residues, and those derivatives having branched sugar chains
showed higher activities than those with straight chains. The C-17
hydroxyl group of dioscinin derivatives caused a reduction of both
haemolytic and fungicidal activities (Takechi *et al.* 1991).

Digitonin (**972**, Appendix 3) (Wulff, 1968; Assa *et al.* 1972), lanatonin
(Wulff, 1968) and parillin (**923**) (Wulff, 1968) are all fungistatic; deltonin

and deltoside from *Dioscorea deltoides* (Dioscoreaceae) are active against *Fusarium solani* and *Phytophthora infestans* (Vasyukova *et al.* 1977). A sisalanin complex, composed of spirostan saponins, from the juice of sisal leaves (*Agave sisalana*, Agavaceae) possessed antifungal activity in tests with *Aspergillus flavus*, *A. oryzae*, *A. parasiticus*, *Candida albicans*, *Rhodotorula glutinis* and *Saccharomyces cerevisiae* (Ujikawa and Purchio, 1989).

Two spirostan saponins from *Allium ampeloprasum* (Liliaceae) have weak fungicidal activity against *Candida albicans* (Morita *et al.* 1988).

Following the discovery that extracts of the rhizome and aerial parts of *Trillium grandiflorum* (Liliaceae) showed significant *in vitro* activity against *Candida albicans*, fractionation led to the isolation of two saponins, dioscin (**1002**) and pennogenin rhamnosyl chacotrioside (**1031**) (Appendix 3). Both gave MIC values comparable to amphotericin B in an agar well-diffusion assay with *Candida albicans*. They also exhibited some *in vitro* activity against the yeasts *Cryptococcus neoformans* and *Saccharomyces cerevisiae*, and the filamentous fungi *Aspergillus flavus*, *A. fumigatus* and *Trichophyton mentagrophytes* (Hufford *et al.* 1988). Another pennogenin glycoside (**1030**, Appendix 3), from *Dracaena mannii* (Agavaceae), has been tested against 17 species of fungi (dermatophytes, pathogenic dermatiaceous fungi and yeasts) and showed antifungal activity against all strains. The dermatophyte *Trichophyton soudanense* was the most sensitive to the saponin and gave an MIC value of 6.25 µg/ml (Okunji *et al.* 1990).

Rhizome extracts of *Ruscus aculeatus* (Liliaceae) inhibit spore germination of both *Aspergillus fumigatus* and *Trichophyton mentagrophytes* (Guérin and Réveillère, 1984).

A sarsasapogenin triglycoside As-1 (**922**, Appendix 3) from *Asparagus officinalis* (Liliaceae) was found to be active against *Candida*, *Cryptococcus*, *Trichophyton*, *Microsporum* and *Epidermophyton*, with MIC values ranging from 0.5 µg/ml to >8 µg/ml (Shimoyamada *et al.* 1990).

As spirostanol glycosides are often found in the seeds of plants, it is reasonable to suppose that they exert a protective function against fungal attack. A monodesmosidic 26-desglucofurostanol derivative of a saponin from tobacco seeds (*Nicotiana tabacum*, Solanaceae) was active against *Cladosporium cucumerinum* in a TLC/agar bioassay. This steroid glycoside also caused a 60% growth reduction of *Puccinia recondita* spores sprayed onto wheat and acted as a plant protective agent. The corresponding bidesmosidic furostanol saponin was inactive against *C. cucumerinum* but protected against *Puccinia recondita* infection. This may have resulted from the degradation of the bidesmoside to a fungicidal monodesmoside by fungal glycosidases (Grünweller *et al.* 1990).

Antibacterial activity

The spirostanol saponin **1030** (Appendix 3) from *Dracaena mannii* (Agavaceae), which has strong antifungal activity (see above), was also tested against bacteria. It was inactive against Gram-negative bacteria but showed weak inhibition of the Gram-positive bacteria *Staphylococcus aureus* and *Streptococcus pyogenes* (Okunji *et al.* 1990).

Asterosaponins from starfish have antibacterial activities (Minale *et al.* 1982) and polyhydroxylated steroid glycosides from the same sources are active against the Gram-positive bacterium *Staphylococcus aureus* (Andersson, 1987). In an examination of 15 starfish saponins, the eight polyhydroxylated steroid glycosides tested were effective against *S. aureus* but none of the compounds were active against the Gram-negative bacterium *Escherichia coli* (Andersson *et al.* 1989).

Antiviral activity

Asterosaponins from *Asterias forbesi*, *Acanthaster planci* and *Asterina pectinifera* are reported to inhibit influenza virus multiplication (Shimizu, 1971), while crossasterosides B and D from *Crossaster papposus* inhibit plaque formation of pseudorabies virus (Suid herpes virus 1, SHV-1) by 43% and 27%, respectively, at a dose of 1 µg/ml (Andersson *et al.* 1989).

6.1.2   *Piscicidal activity*

Numerous saponin-containing plants are toxic to fish; those whose activity results from their steroid saponin content include *Balanites aegyptiaca* (Balanitaceae) and *Cyclamen* species (Primulaceae). Furthermore, ichthyotoxicity is not confined to plants but is also a feature of saponins from marine organisms. Defence secretions of the sole (*Pardachirus* spp.), for example, are effective at repelling sharks and are also toxic to fish. Contained in these secretions are steroid glycosides: the pavonins (Tachibana *et al.* 1985b) and mosesins (Tachibana and Gruber, 1988).

Norlanostane triterpene oligoglycosides, sarasinosides A_1, A_2, A_3, B_2, B_3, C_1, C_2 and C_3, from the Palauan marine sponge *Asteropus sarasinosum* are also ichthyotoxic; sarasinoside A_1 (**1303**) has an LD_{50} of 0.39 µg/ml against the killifish *Poecilia reticulata* and sarasinoside B_1 has an LD_{50} of 0.71 µg/ml (Kitagawa *et al.* 1987a).
?
6.1.3   *Molluscicidal activity*

In addition to the triterpene saponins, a number of steroid saponins are known to possess molluscicidal activity (Table 6.2). Here

Table 6.2. *Molluscicidal activities of spirostanol glycosides*

No.	Name	Structure	Plant	Molluscicidal activity, LC$_{100}$ (mg/l)	Reference
Sarsasapogenin aglycone (870)					
917		3-Gal²-Xyl	*Cornus florida* (Cornaceae)	6	Hostettmann *et al.* 1978
918		3-Gal²-Glc	*C. florida* (Cornaceae)	12	Hostettmann *et al.* 1978
			Anemarrhena asphoderoides (Liliaceae)	10	Takeda *et al.* 1989
921		3-Glc⁴-Rha |² Glc	*Aspergillus curillus* (Liliaceae)	20	Sati *et al.* 1984
1332		3-Glc⁴-Ara |² Glc Ara |⁶	*A. curillus* (Liliaceae)	5	Sati *et al.* 1984
924		3-Glc⁴-Rha |² Glc	*A. curillus* (Liliaceae)	5	Sati *et al.* 1984
Tigogenin aglycone (882)					
1333	Cantalasaponin-7	3-(Gal,2Glc,Xyl,Rha)	*Agave cantala* (Agavaceae)	50	Pant *et al.* 1987
947		3-Glc³-Gal⁴-Xyl |³ Xyl	*A. cantala* (Agavaceae)	9	Kishor, 1990

Gitogenin aglycone (884)

No.	Name	Structure	Source (Family)		Reference
962		3-Glc³-Xyl³-Ara³-Rha	*Yucca aloifolia* (Agavaceae)	10	Kishor and Sati, 1990
967		3-Glc³-Glc³-Gal³-Xyl |² Xyl³-Rha³-Xyl	*Y. aloifolia* (Agavaceae)	10	Kishor *et al.* 1991

Digitogenin aglycone (886)

972	Digitonin	3-Gal⁴-Glc³-Xyl |² Gal³-Glc		10	Alzerreca and Hart, 1982

Yamogenin aglycone (893)

990	Balanitin-1	3-Glc⁴-Glc²-Rha |² Rha	*Balanites aegyptiaca* (Balanitaceae)	5–10	Liu and Nakanishi, 1982
988	Balanitin-2	3-Glc³-Glc⁶-Xyl |² Rha	*B. aegyptiaca* (Balanitaceae)	5–10	Liu and Nakanishi, 1982
989	Balanitin-3	3-Glc⁴-Glc²-Rha	*B. aegyptiaca* (Balanitaceae)	5–10	Liu and Nakanishi, 1982
1334		3-Glc²-Rha	*Asparagus plumosus* (Liliaceae)	25	Sati *et al.* 1984
1335		3-Glc³-Rha |² Rha	*A. plumosus* (Liliaceae)	20	Sati *et al.* 1984

Diosgenin aglycone (894)

1005		3-Glc⁴-Rha³-Glc |² Rha	*Allium vineale* (Liliaceae)	50	Chen and Snyder, 1989

(cont.)

Table 6.2. (*Cont.*)

No.	Name	Structure	Plant	Molluscicidal activity, LC_{100} (mg/l)	Reference
1013		$3\text{-Glc}^4\text{-Rha}^4\text{-Glc}^6\text{-Glc}$ $\quad\ \mid^2 \qquad\ \ \mid^4$ $\quad\ \text{Rha} \qquad \text{Glc}$	*Allium vineale* (Liliaceae)	25	Chen and Snyder, 1989
Pennogenin aglycone (900)					
1030		$3\text{-Glc}^3\text{-Rha}$ $\quad\ \mid^2$ $\quad\ \text{Rha}$	*Dracaena mannii* (Agavaceae)	6	Okunji *et al.* 1991
Hecogenin aglycone (908)					
1044	Cantalasaponin-2	$3\text{-Gal}^4\text{-Xyl}$ $\quad\ \mid^2$ $\quad\ \text{Glc}^3\text{-Glc}$	*Agave cantala* (Agavaceae)	7	Pant *et al.* 1986
Nuatigenin aglycone (912)					
1053		$3\text{-Glc}^2\text{-Rha}$	*Allium vineale* (Liliaceae)	50	Chen and Snyder, 1989

LC_{100}, 100% lethal concentration to snails (after 24 h); n.a., inactive.

also the bidesmosidic glycosides are not noticeably active against *Biomphalaria glabrata* snails.

The first report on the use of a plant for the control of schistosomiasis, made by Archibald in 1933, concerned the fruits of *Balanites aegyptiaca* (Balanitaceae). He noted that the fruits were employed as a fish poison in Sudan and that they also killed *Bulinus* snails and the cercariae of schistosomes. It was subsequently discovered that saponins, tri- and tetraglycosides of yamogenin, were responsible for this activity (Liu and Nakanishi, 1982).

The furostanol saponin **1087**, a glycoside of (25*S*)-5β-furostan-3β,22,26-triol, from *Asparagus curillus* (Liliaceae) was found to be inactive against snails (Sati *et al.* 1984).

1087

Activities have been reported for certain steroid glycoalkaloids (Table 6.3). Commercially available tomatine (**1158**), which also occurs in the tomato, is molluscicidal at 4 mg/l against *Biomphalaria glabrata* (Hostettmann *et al.* 1982). A mixture of solasonine (**1147**) and solamargine (**1148**) (Table 2.10, p. 99) from *Solanum mammosum* killed all *Lymnaea cubensis* snails at 10 mg/l and all *Biomphalaria glabrata* snails at 25 mg/l (Alzerreca and Hart, 1982). The aglycones solasodine and tomatidine were completely inactive (Alzerreca and Hart, 1982), as was the glycoside α-solanine (**1174**, Table 2.11, p. 103) (Hostettmann *et al.* 1982).

6.1.4   *Insecticidal and antifeedant activity*

The steroid alkaloid saponins are especially well known as insect antifeedants; Roddick (1974) has reported on the insecticidal and insect-repellant properties of tomatine. Demissine (**1188**, Table 2.11, p. 102) and α-tomatine (**1158**, Table 2.10, p. 99) are effective against the larvae of potato beetles and provide a useful protection when sprayed on leaves (Schreiber, 1968). α-Tomatine and, to a certain extent, α-solanine (**1174**, Table 2.11, p. 102) and α-chaconine (**1177**) are active against the potato leafhopper (Jadhav *et al.* 1981), while the leptines from *Solanum*

Table 6.3. *Molluscicidal activities of steroid glycoalkaloids*

No.	Name	Structure	Plant	Molluscicidal activity, LD_{100} (mg/l)	Reference
Tomatidine aglycone (1143)					
1158	Tomatine	3-Gal^4-Glc^3-Xyl $\overset{\mid^2}{\text{Glc}}$	*Solanum japonense* (Solanaceae)	4	Hostettmann *et al.* 1982
Solanidine aglycone (1165)					
1174	Solanine	3-Gal^3-Glc $\overset{\mid^2}{\text{Rha}}$	*Solanum* spp. (Solanaceae)	n.a.	Hostettmann *et al.* 1982

LC_{100}, 100% lethal concentration to snails (after 24 h); n.a., inactive.

chacoense convey resistance against the Colorado beetle *Leptinotarsa decemlineata* (Kuhn and Löw, 1961; Sinden *et al.* 1986). The acetate group of the leptines is responsible for the high activity of the glycosides – a higher activity than the co-occurring glycoalkaloids solanine and chaconine.

Both steroid alkaloids and their glycosides act as strong termite repellants. Laboratory tests with *Reticulitermes flavipes* have shown that concentrations of 0.5% saponin kill the termites but the effect in nature manifests itself as an antifeedant activity (Tschesche *et al.* 1970).

A review has been published on the rôle of glycoalkaloids in the resistance of potato plants towards insect pests (Tingey, 1984).

A tetrasaccharide glycoside mixture of diosgenin and yamogenin from *Balanites roxburghii* (Balanitaceae) has antifeedant activity against *Diacresia obliqua* (Jain, 1987b).

The diosgenin tetrasaccharide **1009** (Appendix 3) from *Balanites roxburghii* (Balanitaceae) exhibited some weak feeding-deterrent activity against larvae of the caterpillar *Spilosoma obliqua*, a pest of crops in northern India (Jain and Tripathi, 1991).

Certain spirosolan and spirostan glycosides (solamargine, solasonine, tomatine) show larval growth inhibition of the spiny bollworm *Earias insulana*, a pest of cotton in the Middle East. Since the aglycones are sparingly soluble and inactive in the bioassay, there is a correlation between biological activity and hydrophilic/lipophilic character of the compounds, apparently reflected in a polarity or solubility effect. At the same time, the inhibitory effect does not depend on the composition of the carbohydrate unit of the glycosides (Weissenberg *et al.* 1986).

Aginoside (**971**, Appendix 3), an agigenin-type spirostanol saponin from the flowers of leek (*Allium porrum*, Liliaceae) and from *Allium giganteum*, is toxic to leek-moth larvae (*Acrolepiopsis assectuella*). The saponin causes symptoms of intoxication, manifested by digestive dysfunctions, reduced or arrested growth and by inability of the small larvae to ecdyse. The potency of aginoside is reduced by addition of cholesterol or sitosterol to the diet (Harmatha *et al.* 1987).

There is evidence that the action of glycoalkaloids as resistance factors against insects or as antifungal agents may be attributable to their membrane-destabilizing properties (Roddick *et al.* 1988). In this connection, the potato glycoalkaloids α-solanine (**1174**) and α-chaconine (**1177**) (Table 2.11, p. 102) are capable of disrupting animal (rabbit erythrocytes), plant (red beet cells) and fungal (*Penicillium notatum* protoplasts) cells. While α-chaconine is the more membrane-disruptive compound of the two, a 1:1 mixture of the glycoalkaloids produces pronounced synergistic effects in all three test systems. Synergistic interactions also occur with regard to *in vitro* cholesterol binding (Roddick *et al.* 1988).

It is also known that α-chaconine is able to inhibit insect acetyl-cholinesterase (of German cockroaches, *Aedes* mosquitoes, houseflies and Cottonwood leaf beetles) and this may be one of the factors in the host plant resistance to insect pests (Wierenga and Hollingworth, 1992).

6.1.5   *Plant growth inhibition*

A methanolic fruit extract of the shrub *Solanum incanum* (Solanaceae; Sodom-apple) exhibits plant growth inhibition, preventing the growth of lettuce seedlings (*Lactuca sativa*) at 3000 p.p.m. Fractionation of this extract has resulted in the isolation of solamargine (**1148**) and solasonine (**1147**) (Table 2.10, p. 99), both of which inhibit lettuce root growth at concentrations greater than 100 p.p.m. (Fukuhara and Kubo, 1991).

6.1.6   *Antifertility activity*

Saponins from *Costus speciosus* (Zingiberaceae) are abortifacient to pregnant goats, rats and cows (Chou *et al.* 1971); they prevent pregnancy in rats when fed at 5–500 μg/100 g body weight for 15 days (Chou *et al.* 1971).

The furostanol glycoside shatavarin-I (**1085**, Table 2.9, p. 90) from the Ayurvedic drug *Shatavari* (*Asparagus racemosus*, Liliaceae) possesses anti-oxytocin activity (Joshi and Dev, 1988).

Solutions (1.5% w/v) of ruscogenin 1-*O*-[α-L-arabinopyranosyl(1→2)]-β-D-glycopyranoside (**1015**) from *Ophiopogon intermedius* (Haemodoraceae) (Rawat *et al.* 1988) and a spirostanol glycoside (**921**) (Appendix 3) from *Asparagus officinalis* (Liliaceae) (Pant *et al.* 1988a) caused total im-mobilization of human spermatozoa.

The spermicidal activity of *Balanites roxburghii* (Balanitaceae) fruit has been reported to involve a diosgenin saponin (Farnsworth and Waller, 1982), although steroid saponins are less frequently encountered as spermicides than triterpene glycosides. Other, steroid saponin-containing plants with activity include *Trigonella foenum-graecum* (Leguminosae) (Setty *et al.* 1976), *Ruscus hypoglossum* (Liliaceae) (Elbary and Nour, 1979) and *Agave cantala* (Agavaceae) (Pant, 1988).

In contrast to the above findings, steroid glycosides from the underground parts of *Tribulus terrestris* (Zygophyllaceae) stimulate spermatogenesis in male rats and cause increased fertility in female rats (Tomowa *et al.* 1981; Naplotonova, 1984).

6.1.7   Cytotoxic and antitumour activity

Afromontoside (**1099**, Table 2.9, p. 90), a furostanol glycoside from *Dracaena afromontana* (Agavaceae), and trillin (diosgenin 3β-glucoside, **994**, Appendix 3) are both cytotoxic to cultured KB cells at 100 µg/ml, while the aglycone diosgenin and its derivative dihydrodiosgenin are significantly more active and kill the cells at 10 µg/ml (Reddy *et al.* 1984). Two diosgenin glycosides (**1004** and **1007**, see Appendix 3) from *Paris polyphylla* (Liliaceae) have *in vitro* cytotoxic activities (ED$_{50}$ of **1004**: P-388 0.44 µg/ml, L-1210 0.14 µg/ml, KB 0.16 µg/ml; ED$_{50}$ of **1007**: P-388 0.22 µg/ml, L-1210 0.43 µg/ml, KB 0.29 µg/ml) (Zhou, 1989).

The yamogenin glycosides **991–993** (Appendix 3) from the seeds of *Balanites aegyptiaca* (Balanitaceae) exhibited cytostatic activity against the P-388 lymphocytic leukaemia cell line, with ED$_{50}$ values of 0.41, 2.40 and 0.21 µg/ml, respectively, while the diosgenin tetrasaccharide **1014** (Appendix 3) had ED$_{50}$ 0.22 µg/ml. This compared with an ED$_{50}$ value of 0.08 µg/ml for the cancer chemotherapeutic agent 5-fluorouracil (Pettit *et al.* 1991).

Solamargine (**1148**) and solasonine (**1147**) (Table 2.10, p. 99) possess inhibitory effects against JTC-26 cells *in vitro*, while solamargine and khasianine (**1152**, Table 2.10, p. 99) exhibit cytotoxicity against human hepatoma PLC/PRF/5 cells (ED$_{50}$ 1.53 µg/ml and 8.60 µg/ml, respectively) (Lin *et al.* 1990).

It has been shown that β-solanine (**1175**, Table 2.11, p. 102) from *Solanum dulcamara* (Solanaceae) is effective against sarcoma 180 in mice at concentrations ranging from 15 to 30 mg/kg (Kupchan *et al.* 1965). Similarly, the glycoalkaloids extracted from *S. sodomaeum* show antitumour activity against sarcoma 180 in mice at 8 mg/kg (Cham *et al.* 1987). This mixture is composed of approximately 33% solasonine and 33% solamargine, together with mono- and diglycosides of solasodine. Furthermore, a cream formulation containing the glycoalkaloids from this plant is claimed to have remarkable effectiveness against malignant human skin tumours such as basal cell carcinomas and squamous cell carcinomas (Cham and Meares, 1987).

Solaplumbine (**1153**, Table 2.10, p. 99) from *Nicotiana plombaginifolia* (Solanaceae) has antitumour activity in rat model tumour systems (Singh *et al.* 1974).

Like the holothurins, cytotoxicity has been reported for asterosaponins. Ruggieri and Nigrelli (1974) reported considerable cytotoxicity of some asterosaponins towards human KB carcinoma cells *in vitro*. A systematic study of the effects of asterosaponins from *Acanthaster planci*, *Luidia maculata* and *Asterias amurensis* on fertilized sea urchin eggs and starfish

991 Balanitin 4 R = -Glc^4-Glc^3-Glc
                                  |2
                                  Rha

992 Balanitin 5 R = -Glc^4-Glc^3-Rha
                                  |2
                                  Rha

993 Balanitin 6 R = -Glc^4-Glc
                                  |2
                                  Rha

Xyl$\xrightarrow{3}$Glc$\xrightarrow{4}$Glc—O
                      |2
                      Rha

1014 Balanitin 7

eggs has been carried out because information from these bioassays can give some idea of the mode of action of drugs that inhibit cell division (Fusetani *et al.* 1984).

Sarasinoside A$_1$ (**1303**, Section 2.4), a saponin from the sponge *Asteropus sarasinosum*, exhibits cytotoxic activity (ED$_{50}$ 2.8 µg/ml) in a P-388 *in vitro* assay (Schmitz *et al.* 1988).

The pectiniosides from *Asterina pectinifera* have borderline activity against murine lymphoma L1210 and human epidermoid carcinoma KB cell lines (Dubois *et al.* 1988a). Despite the *in vitro* activities, however, these saponins have no useful *in vivo* activity.

The 24-*O*- and 28-*O*-glycosylated steroid polyols from starfish show only mild cytotoxicity (Stonik, 1986) and while certain starfish saponins are cytotoxic to bovine fetal cells at high doses of 100 µg/ml, the cells were unaffected at 1 µg/ml (Andersson *et al.* 1989). Activities at such levels imply that these saponins will probably find no further practical application.

6.1.8 Pharmacology

Saponins (1%) in the diet of rats decrease plasma cholesterol levels and increase bile production (Topping *et al.* 1978). More specifically, ruscoside A and ruscoside B from *Ruscus hyrcanus* (Liliaceae) decrease the cholesterol content of the blood, lipid deposition in the aorta and liver arterial tension. They also slow down the cardiac rhythm and respiration of humans and rabbits suffering from arteriosclerosis and have antisclerotic and hypotensive activities (Guseinov and Iskenderov, 1972).

Fenugreek (*Trigonella foenum graecum*, Leguminosae) seed saponins are responsible for lowering cholesterol levels in dogs. During passage through the digestive tract, these steroid saponins are transformed into diosgenin and gitogenin and these aglycones may be implicated, together with the parent saponins, in the inhibition of cholesterol absorption, the decrease of liver cholesterol concentration and the increased conversion of cholesterol into bile acids by the liver (Sauvaire *et al.* 1991). Garlic and onion (*Allium*) preparations have cholesterol-lowering and anti-arteriosclerotic effects, most probably as a result of their saponin content (Koch, 1993).

Digitonin (**972**, Appendix 3) prevents hypercholesterolaemia in monkeys (Malinow *et al.* 1978).

Solasodine (**1141**, Table 2.10, p. 99) was pharmacologically and clinically tested in the Soviet Union (Steinegger and Hänsel, 1988). It has a cortisone-like action in animals: blood vessel permeability is lowered and experimental lung oedema is reduced. In mice, there is some hypertrophy of the adrenals and a positive inotropic activity is observed on the cat heart. Solasodine and cortisone have the same antiphlogistic activity in rats, both at the same dose. In the clinic, trials with solasodine citrate on humans have shown cardiotonic, antiphlogistic and desensibilizing activities.

A comprehensive trial of the pharmacological and toxicological properties of tomatine (**1158**, Table 2.10, p. 99) and its derivatives has been performed (Wilson *et al.* 1961).

Anti-inflammatory activity

Among plants with anti-inflammatory activity, it has been demonstrated that a saponin-containing fraction of the leaves of *Yucca schottii* (Liliaceae) inhibits carrageenin-induced oedema in rats (Backer *et al.* 1972).

Extracts of *Ruscus aculeatus* (Liliaceae) rhizomes have important pharmacological properties, the most notable being anti-inflammatory activity (Bombardelli *et al.* 1971) and the treatment of venous insufficiency (Vanhoutte, 1986; Rauwald and Janssen, 1988). Investigations have shown that the saponins from the rhizomes of this ancient medicinal plant (described by Theophrastos, 327–288 BC) demonstrate anti-inflammatory effects on carageenin-induced rat paw oedema after i.p. administration (Capra, 1972). Extracts of butcher's broom are also active in the experimental arthritis model of the rat paw, provoked by kaolin (Chevillard *et al.* 1965). Both i.p. and rectal administration can be employed.

Paris polyphylla var. *chinensis* (Liliaceae), also contains anti-inflammatory saponins – all glycosides of diosgenin (Xu and Zhong, 1988b).

Solasodine (**1141**, Table 2.10, p. 99) from the unripe fruits of *Solanum aviculare* (Solanaceae) is anti-inflammatory. α-Tomatine (**1158**, Table 2.10, p. 99) inhibits carrageenin-induced paw oedema when administered to rats intramuscularly (1 to 10 mg/kg) or orally (15 to 30 mg/kg) (Filderman and Kovacs, 1969).

In West Africa, a number of anti-inflammatory plants are known which contain steroid glycosides: *Commiphora indica* (Burseraceae), *Costus afer* (Zingiberaceae), *Cyperus rotundus* (Cyperaceae), *Leptadenia pyrotechnica*, *Solanum torvum* (Solanaceae) (and other *Solanum* species) and *Withania somnifera* (Solanaceae) (Bep Oliver-Bever, 1986).

Saponins from the starfish *Asterias forbesi* are reported to have anti-inflammatory activity (Minale *et al.* 1982).

Capillary fragility and permeability

As in the case of triterpene saponins such as aescine or *Hedera helix* saponins, it has been postulated that certain steroid saponins have the ability to strengthen veins and decrease blood vessel permeability. Extracts of the rhizomes of *Ruscus aculeatus* (Liliaceae), a bush found around the Mediterranean, are used in tablet form as a general vein tonic (Mandor[R], Tissan[R]-Veno) and ruscogenin/neoruscogenin saponin preparations are employed in the treatment of haemorrhoids (Ruscorectal[R]). Another preparation of *R. aculeatus* available in France, Cyclo 3[R], is sold for the treatment of venous insufficiency and promotes a better erythrocyte circulation and improved function of the capillaries (Coget, 1983). Ointments containing extracts of *R. aculeatus* are known to increase capillary resistance in rats (Aubert and Anthoine, 1985).

Extracts of *Ruscus aculeatus* (Liliaceae) also cause contraction of isolated canine cutaneous veins. This is the result of an α-adrenergic activation of venous smooth muscle (Vanhoutte, 1986). The active principle has yet to be identified.

Antihepatotoxic activity

The spirosolan glycosides solasonine (**1147**) and solamargine (**1148**) (Table 2.10, p. 99) and the solanidan glycoside capsicastrine (**1190**, Table 2.11, p. 102) possess strong activity against carbon tetrachloride-induced hepatotoxicity in mice at doses of 0.1, 1.0 and 3.0 mg/kg, respectively (Lin *et al.* 1988; Lin and Gan, 1989). Solasonine and solamargine have been found in the berries of *Solanum incanum* (Solanaceae), which are reputed in Taiwan to cure hepatitis (Lin *et al.* 1988).

6.1.9   Other activities

Paris polyphylla var. *yunnanensis* (Liliaceae) has been used for over 500 years as a *haemostatic* drug in China. The two major saponins responsible for this activity are a trisaccharide (**995**, Appendix 3) of diosgenin and a trisaccharide of pennogenin (Zhou, 1989). Other haemostatic diosgenin glycosides have been isolated from *Paris polyphylla* var. *chinensis* (Ma and Lau, 1985; Xu and Zhong, 1988b).

Two pennogenin saponins (including **1031**, Appendix 3) from *Paris polyphylla* also possess strong *uterine-contracting* activity. Doses of 2 mg/kg in *in vivo* experiments on rats were sufficient to demonstrate contraction (Zhou, 1989).

α-Tomatine (**1158**, Table 2.10, p. 99) has an *anti-histamine* activity (Calam and Callow, 1964).

Steroid saponins from *Paris vietnamensis* (Liliaceae) have chronotropic effects on the spontaneous beating of mouse myocardial cells. At 0.016 mg/ml, a triglycoside of diosgenin stimulated the cell beating rate in calcium-depleted cells by more than 80%. The same compound also stimulated calcium uptake by the myocardial cells (Namba *et al.* 1989).

Sarsasapogenin glycosides from the Chinese antipyretic plant *Anemarrhena asphodeloides* (Liliaceae) inhibit Na/K-ATPase activity (Chen *et al.* 1982).

Anemarsaponin B (26-*O*-β-D-glucopyranosylfurost-20(22)-en-3β,26-diol-3-*O*-β-D-glucopyranosyl-(1→2)-β-D-galactopyranoside, **1077**, Table 2.9, p. 90) from the rhizomes of *Anemarrhena asphodeloides* inhibits PAF-induced rabbit platelet aggregation *in vitro* (IC_{50} 25 μM) (Dong and Han, 1991).

An aqueous extract of mesocarps of the fruits of *Balanites aegyptiaca* (Balanitaceae) exhibited a prominent *antidiabetic* activity by oral administration in streptozotocin-induced diabetic mice. The individual saponins did not show activity, while the recombination of these saponins resulted in a significant antidiabetic effect (Kamel *et al.* 1991).

6.2   Toxicity
    The toxicity of glycoalkaloids has largely been overestimated and their dangers are probably not as high as history would imply. Cases of poisoning do, however, periodically appear in humans (e.g. McMillan and Thompson, 1979). Steroid alkaloid glycosides are only sparingly resorbed on peroral application. They are then hydrolysed in the digestive tract to relatively non-toxic alkamines (the aglycones). Parenteral application, however, gives similar effects to those observed with k-strophanthoside and other cardiac glycosides; paralysis of the central nervous system results eventually in death (Nishie *et al.* 1971). The intraperitoneal LD_{50} of solanine (**1174**, Table 2.11, p. 102) in mice is 42 mg/kg (chaconine (**1177**) 28 mg/kg and tomatine (**1158**) 34 mg/kg), whereas oral application gives no reaction, even at 1000 mg/kg (Nishie *et al.* 1971). The toxic effects observed in six species of animal treated with glycoalkaloids and their aglycones were comparable to one another; α-chaconine was the most lethal and solasodine the least toxic (Morris and Lee, 1984). Chaconine is more active to a wide range of organisms than solanine, even though both share the same aglycone.

    In animals, adverse physiological effects of glycoalkaloids are manifested in a number of ways (reduced respiratory activity or blood pressure, bradykardia, haemolysis, etc.) which are thought to stem mainly from membrane disruption, inhibition of acetylcholinesterase or interference with sterol/steroid metabolism, or from combinations of these. As α-solanine and α-tomatine are serum cholinesterase inhibitors (Jadhav *et al.* 1981) and α-solanine and α-chaconine have been shown to significantly inhibit bovine and human acetylcholinesterase at a concentration of 100 μM (Roddick, 1989), at least part of their toxic effects are probably thus accounted for. Solamargine has little anticholinesterase activity but binds readily to cholesterol *in vitro* and has significant membrane-disrupting properties (Roddick *et al.* 1990).

    A more recent view, after observations on hamsters administered crude potato alkaloid preparations by gavage, is that the toxic effects of glycoalkaloids are predominantly a result of severe gastric and intestinal mucosal necrosis. This is accompanied by secondary fluid loss, septicaemia and bacteraemia (Baker *et al.* 1988). Similar effects in hamsters are observed with other members of the Solanaceae: *S. tuberosum, S.*

elaeagnifolium, S. dulcamara and *Lycopersicon esculentum* (Baker *et al.* 1989).

There is some uncertainty as to whether the aglycones or the glycosides themselves are responsible for the membranolytic properties of glycoalkaloids, but evidence that the intact carbohydrate moiety is important for bioactivity comes from the study of liposome membrane permeability. The potato glycoalkaloid α-chaconine (**1177**, Table 2.11, p. 102) caused release of entrapped peroxidase from phosphatidylcholine liposomes containing different free sterols but was ineffective against sterol-free liposomes. On the other hand, α-solanine (**1174**, Table 2.11, p. 102) had no effects on sterol-containing liposomes under these conditions. Another potato glycoalkaloid β₂-chaconine (**1179**, Table 2.11, p. 102), which has one monosaccharide less than α-chaconine, was also unable to disrupt sterol-free liposomes (Roddick and Rijnenberg, 1986). Even though these glycosides differ considerably in their activity spectrum, they all have the same aglycone.

1174 α-Solanine R = -Gal^3-Glc
                                        |^2
                                        Rha

1177 α-Chaconine R = -Glc^4-Rha
                                         |^2
                                         Rha

1179 β₂-Chaconine R = -Glc^2-Rha

All parts of the potato (*Solanum tuberosum*) contain toxic glycoalkaloids (Jones and Fenwick, 1981; Morris and Lee, 1984). Commercial cultivars commonly contain 2–15 mg of glycoalkaloids/100 g of unpeeled tuber (Slanina, 1990). The potato tuber usually only contains around 7 mg solanine per 100 g but, under certain conditions, the content can rise above the critical concentration of 35 mg/100 g. Factors influencing the formation of glycoalkaloids are light exposure or mechanical damage. The glycoalkaloids are not destroyed by boiling, baking, frying or drying at high temperatures. The parts with the greatest risk, if consumed, are the green tubers or the sprouts. These, together with the flowers and leaves, contain much higher amounts of glycoalkaloids. When screening new potato varieties for human consumption, according to one report, those which contain over 20 mg total glycoalkaloid per 100 g fresh tuber weight are considered toxic (Sapeika, 1969), while a more recent article recommends a maximum level for solanidine glycosides of 60 mg/kg fresh unpeeled potatoes (Parnell *et al.* 1984). By adhering to these limits, a sufficiently large safety margin is taken into account.

Several *Solanum* species have proven to be teratogenic in laboratory animal assays. The active agents have not always been identified with certainty but they appear to be solanidan and spirosolan steroid alkaloids, probably as their glycosides. The mechanism of action is not known and dose levels have not been established (Keeler *et al.* 1991). Congenital craniofacial malformations are induced in hamsters administered high doses of *Solanum elaeagnifolium* and *S. dulcamara* fruits and *S. tuberosum* sprouts by gavage, but *S. sarrachoides* and *S. melongena* give statistically insignificant results (Keeler *et al.* 1990). Although concern about ingestion of potato sprouts by pregnant women has been expressed, the real risk to humans has yet to be rigorously evaluated (Keeler *et al.* 1991).

In view of the association of certain glycoalkaloids with some very widely consumed foodstuffs, the importance of other steroid glycosides, with respect to their toxicity, has been largely overshadowed. It should not be forgotten, however, that other classes may have undesirable or potentially damaging properties. Asterosaponins, for example, are toxic to fish, molluscs, arthropods and vertebrates in very low concentrations (Patterson *et al.* 1978).

7

Commercially important preparations and products

From the very earliest times, saponin-containing plants have found widespread application in the treatment of various ailments – coughs, syphilis, rheumatism, gout, just to mention a few. Their use as panaceas has also been recorded. In Europe, certain saponin preparations have been employed for centuries. Of these, the most important include: *Saponariae radix, Hippocastani semen* (Weil, 1901), *Primulae radix, Hederae helicis herba* (or *folium*), *Senegae radix* and *Liquiritiae radix*. The *Rote Liste* has over 200 preparations which contain saponins. The most frequently employed are *Primulae radix, Senegae radix, Liquiritiae radix, Hippocastani semen, Hederae helicis folium, Ginseng radix, Hydrocotylis herba* and *Rusci radix*. The pharmacologically active saponins from these sources have now largely been characterized and their activities confirm the medicinal properties of the respective plant drugs. Thus, aescine, primula saponin and α-hederin exhibit antiexudative effects; primula saponin, senegin, glycyrrhizin and gypsophila saponin exhibit expectorant effects; saponins from ginseng and *Eleutherococcus* are adaptogenic.

Synthetic pharmaceuticals have, to a certain extent, replaced many of the earlier saponin-containing preparations, but, even so, a number of saponin drugs still assume a high degree of importance (Gerlach, 1966; Schantz, 1966). It must be stressed that most of the saponin preparations used in phytotherapy are either extracts or mixtures. In many of the studies on these preparations, the exact saponin composition is not known and extrapolation of the activities of monosubstances to the saponin-containing product may be risky. In view of this problem, there is a tendency to move towards the use of pure substances because they can be directly analysed and standardized.

Because of the large quantities of bidesmosidic saponins in plants used traditionally as soap substitutes (*Saponaria officinalis, Gypsophila* species,

Table 7.1. *Origin of raw materials used in the production of steroids*

Country	Source	Estimated production for 1978 (in tons diosgenin)
Mexico	Diosgenin (Barbasco)	600
	Sarsasapogenin (Palia China)	15
	Hecogenin (Agave Henequen)	5
USA	Stigmasterol (Soya beans)	350
	Total synthesis	60
	Sitosterol (Soya beans)	250
	Hecogenin (*Yucca brevifolia*)	20
Guatemala	Diosgenin (Barbasco)	20
Puerto Rico	Diosgenin (Barbasco)	5

From Onken and Onken, 1980.

Quillaja saponaria, etc.), these have, until recently, been valuable sources of industrially important triterpene glycosides, used as soaps, detergents, foam producers in fire extinguishers and also in the production of film. The following saponin-containing plants have been exploited for the industrial production of saponins: horse chestnuts (*Aesculus*), climbing ivy (*Hedera*), peas (*Pisum*), cowslip (*Primula*), soapwort (*Saponaria*) and sugar beet (*Beta*).

Since saponins have *emollient* properties, cosmetic preparations from *Saponaria officinalis*, *Viola tricolor*, *Sanicula europaea* and ginseng have been investigated. The sulphate salts of certain marine organism saponins have been employed to stimulate skin regeneration and they have been used with sterols and phosphatides to produce salve-like cosmetic emulsion bases. Saponins at a concentration of about 5% are frequently employed in soap, shampoo and bath salt formulations. Other commercial applications include their use in dentifrices, as emulsifiers in cod liver oils and to combat lice infestation in humans.

According to Onken and Onken (1980), more than 6% of all prescriptions in human medicine are steroid hormones. Of greatest importance are the corticosteroids and, especially with the introduction of oral contraceptives, the sex hormones. The major proportion of steroids used in therapy and as contraceptives are obtained by semi-synthesis from natural products: saponins, phytosterols, cholesterol and bile acids. Some statistics on the sources of raw materials for steroid production are shown in Table 7.1. The immense commercial significance of spirostanol saponins stems from the discovery by Marker (Marker *et al.* 1947; Fieser and Fieser, 1959) that their aglycones can be converted easily into pregnane derivatives, starting materials for the synthesis of steroid hormones. The first milestone was the isolation by Marker of diosgenin from Japanese *Dioscorea* (Dioscoreaceae), which he used for the synthesis of the female sex hormone progesterone in 1940. Then, in 1949, he discovered that the Mexican plant *Dioscorea barbasco* provided large quantities of diosgenin (**894**, Table 2.8, p. 79). The same year saw the attribution of anti-inflammatory and antirheumatic properties to cortisone by Hench and Kendall, thus triggering the subsequent explosion in steroid research and production (Onken and Onken, 1980). Diosgenin is the most important sapogenin from an economic standpoint and, although it occurs in several plant families, it is extracted almost entirely from *Dioscorea* species. *Dioscorea mexicana*, the Mexican yam, and *D. composita* (Barbasco) are two of the most important sources of diosgenin, while *D. floribunda*, also from Mexico, contains up to 10% diosgenin by dry weight. *Dioscorea deltoidea* (India) contains 2.7% and the diosgenin content of *D. spiculiflora* can be as much as 15% of the dry weight (Takeda, 1972).

Hecogenin (**908**, Table 2.8, p. 79), obtained from East African sisal (*Agave sisalana*, Agavaceae) juice, accounts for about 6% of the steroid precursors used by the pharmaceutical industry (Blunden *et al.* 1975). Mexican *Agave* species are rich sources of hecogenin but also contain tigogenin (ratio of hecogenin to tigogenin 6:4) which is difficult to separate. African sisal is free of tigogenin (Onken and Onken, 1980).

The seeds of *Yucca* (Agavaceae) species contain steroid sapogenins, reaching levels of 8–12% in *Y. brevifolia* and *Y. arizonica* (Wall and Serota, 1959). Seeds of *Y. filifera*, which grows in Mexico and the USA, contain 8% sarsasapogenin (**870**, Table 2.8, p. 79), which can be easily converted into 16-dehydropregnenolone (see below).

Attempts at cultivation of certain *Dioscorea* species (*D. floribunda*, *D. speculifora* and *D. composita*) have been made, but getting the right growing conditions has not been an easy task. However, it has been estimated that after about four years of growth, *D. composita* (Barbasco) tubers with a diosgenin content of 3.5–5% can be obtained, giving a total yield of 100–200 kg diosgenin per hectare (Onken and Onken, 1980).

The excessive collection of wild *Dioscorea* species (*D. floribunda*, *D. composita*, *D. spiculifora*, etc.) and the complexities of cultivating other species (*D. deltoidea* etc.) has led to a shortage of raw materials for the extraction of steroid glycosides. As a consequence, solasodine (**1141**, Table 2.10, p. 99) from various species of *Solanum* is assuming increasing importance as a source of pregnane derivatives in the commercial synthesis of steroids (Franz and Jatisatienr, 1983). The most promising producers of solasodine and its glycosides for this purpose seem to be

Solanum marginatum (the green fruits contain 5% solasodine), *S. khasianum* and *S. laciniatum* (Franz and Jatisatienr, 1983). Other possible commercial sources of steroid sapogenins are *Costus speciosus* (Zingiberaceae), *Trigonella foenum-graecum* (Leguminosae; the seeds contain 0.9% diosgenin), *Balanites aegyptiaca* (Zygophyllaceae) and *Allium* species (Liliaceae) (Kravets *et al.* 1990).

The raw materials for the production of corticoids and the essential intermediate partial synthetic steps are shown in Fig. 7.1. The key compound is 16-dehydropregnenolone, which can be transformed into corticosteroids, pregnanes, androstanes and 19-*nor*-steroids. On an industrial scale, either the saponin mixture is extracted from dried, finely-cut pieces of rhizome or tuber with methanol or ethanol, then hydrolysed with acid, or acid hydrolysis of the dried plants is followed by extraction with petroleum ether or another non-polar solvent (Takeda, 1972). The most important advance in the synthesis was the microbiological introduction of the oxygen function at C-11 by Upjohn (USA) in 1952. Previously many steps were required to achieve this transformation but with *Rhizopus nigricans* 85–95% yields of 11α-hydroxyprogesterone are achieved (Peterson *et al.* 1952). The conversion is also produced with another microorganism, *Rhizopus arrhizus*. Once the 11α-OH functionality has been introduced, the synthesis of cortisone (Mancera *et al.* 1953) and other steroids is relatively straightforward. By the end of the 1950s, 80–90% of the total world production of steroids used diosgenin as the starting material but this figure had dropped to 40–45% by the early 1970s.

7.1   *Sarsaparillae radix* (sarsaparilla root)

Sarsaparilla consists of the dried roots (and sometimes rhizomes) of *Smilax* species (Liliaceae). Approximately 200 species are presently known but the important varieties of sarsaparilla are as follows:

- American species
    Smilax aristolochiaefolia MILL. (Vera Cruz or Grey)
    S. regelii KILL. et MORTON (Honduras or Brown)
    S. febrifuga KUNTH (Guayaquil)
    S. papyracea DUHAM. (Brazilian)
    Undetermined species (Costa Rica or 'Jamaican')
- Chinese species
    S. sieboldi MIQ.
    S. stans MAXIM
    S. scobinicaulis C.H. WRIGHT
    S. glabra ROXB.

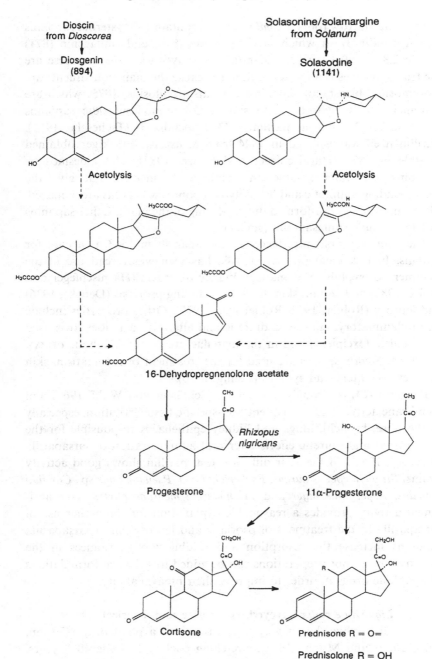

Fig. 7.1. Production of corticoids from diosgenin and solasodin.

The long, narrow roots of *Smilax* species contain 1–3% steroid saponins (see Appendix 3), in which sarsasapogenin (**870**) and smilagenin (**871**) (Table 2.8, p. 79) have been identified as aglycones. Whether these are the true aglycones is a matter of doubt because the main constituents are bidesmosidic furostanol saponins (e.g. sarsaparilloside, **1075**) which are extremely easily hydrolysed to spirostanol monodesmosidic saponins (see Chapter 2), such as parillin (**923**, Appendix 3) (Tschesche, 1971). Parillin itself was isolated in 1824 from *S. aspera*; Flückiger obtained crystals in 1877 (Harnischfeger and Stolze, 1983). The bidesmosidic saponins are water soluble and biologically inactive but give the corresponding antibiotic and haemolytic saponins when cells are damaged. They can even be transformed during drying of the plant. Other saponins isolated include smilasaponin (smilacin).

The root was first introduced by the Spanish in 1563 as a cure for syphilis. In folk medicine, sarsaparilla has seen widespread use in the treatment of syphilis, rheumatism, kidney disorders (Harnischfeger and Stolze, 1983) and certain skin diseases, including psoriasis (Deneke, 1936) and leprosy (Rollier, 1951; Rollier *et al.* 1979). Other properties include anti-inflammatory, antiseptic, diuretic and alterative activities. It is used as a vehicle (excipient) and in the manufacture of non-alcoholic drinks. In China, *Smilax* species are used for the treatment of rheumatism, skin diseases, dysentery and syphilis (Leung, 1980).

Parillin (**923**) is strongly haemolytic (Schlösser and Wulff, 1969) and membrane active. The consequent non-specific tissue irritation, especially of the nose, bronchi, lungs and kidney epithelia, is responsible for the expectorant and diuretic effects after oral administration of sarsaparilla (Jaretzky, 1951). In the agar diffusion test, parillin shows good activity against *Staphylococcus aureus*, *Escherichia coli*, *Pseudomonas* sp., *Candida albicans*, *Aspergillus niger* and *Trichoderma mentagrophytes*. This anti-fungal activity provides a reasonable explanation for the earlier use of sarsaparilla in the treatment of psoriasis and leprosy. Since sarsaparilla saponins increase the resorption and solubility of substances in the digestive tract, root preparations can be added to galenical formulations of lipophilic drugs in order to improve their bioavailability.

7.2   *Liquiritiae radix* (glycyrrhiza, licorice root, licorice)

Glycyrrhiza has a very long history as a plant drug (Gibson, 1978; Davis and Morris, 1991), stretching back over some 4000 years. Various therapeutic uses have been described by Hippocrates, Dioscorides, Pliny, Gerard and others. The physiological and pharmacological effects of licorice root have been described in great detail by Boiteau *et al.* (1964).

The best-known constituent is glycyrrhizin (glycyrrhizic acid or glycyrrhinic acid) (**269**). Licorice root contains up to 14% of this sweet-tasting saponin, in the form of potassium and calcium salts. The

aglycone 18β-glycyrrhetinic acid (or glycyrrhetic acid) (**6**) has no sweet taste. In its commercial form, a mixture of 18α- and 18β-glycyrrhetinic acid is present because of the isomerization of the aglycone during hydrolysis of glycyrrhizin. A number of minor sapogenins have also been identified (Price *et al.* 1987).

The highest concentrations of glycyrrhizin are found in the main roots, while small quantities are found in the lateral roots and none in the green parts (Fuggersberger-Heinz and Franz, 1984). Chinese licorice (*Glycyrrhiza uralensis*) and various *Abrus* species such as *A. melanospermus, A. precatorius* and *A. pulchellus* also contain glycyrrhizin.

Liquiritiae radix comes from the yellow-coloured varieties of *Glycyrrhiza glabra* L. (Leguminosae), a widely distributed shrub. Commercial sources are:

- Spanish glycyrrhiza (var. *typica*)
- Russian glycyrrhiza (var. *glandulifera*). This has a bitter after-taste
- Chinese glycyrrhiza, which may also contain *G. pallida* and *G. uralensis*
- Turkish glycyrrhiza (var. *glandulifera*)
- Iranian glycyrrhiza (var. *violacea*).

The drug has several uses:

- In the preparation of teas.
- In the preparation of extracts. These are subdivided into dry extracts which contain 4–12% glycyrrhizin (produced by spray drying), gummy extracts (*Succus Liquiritiae*) which also contain 9–12% glycyrrhizin and liquid extracts (4–6% glycyrrhizin).

- In the preparation of licorice (liquorice) by decoction. Freshly collected roots are cut into small pieces and boiled in water. After removal of most of the water by evaporation under reduced pressure, the resulting syrup is poured into moulds, to give *Succus Liquiritiae*. This can then be converted into the form of licorice required.

Licorice root is reputed as a condiment and flavouring; large quantities are used by the tobacco industry as a curing agent. Industrial applications have been outlined by Houseman (1944). In 1988, licorice-containing products represented 5% of the UK confectionery market, representing a value of £49 million, with a total market volume of 18 000 tonnes in 1987 (Spinks and Fenwick, 1990).

Apart from their use as sweeteners and taste modifiers, glycyrrhizin and *Glycyrrhiza glabra* extracts are useful *expectorants* and demulcents. The cough reflex is reduced with licorice sweets because saliva production is stimulated, with a consequent increase in the need to swallow.

Perhaps the most important action of glycyrrhizin is on the gastrointestinal tract. As early as the seventeenth century, apothecaries were using licorice extracts to treat 'inflamed stomachs', just as the Chinese had done centuries before. The first reports describing the treatment of *ulcus ventriculi* date from 1946 and concern the application of licorice root extracts in Holland (Revers, 1946). A phytochemical investigation of the roots and rhizomes of *Glycyrrhiza glabra* resulted in the discovery that the active factor responsible for curing peptic ulcers (gastric and duodenal ulcers) was glycyrrhizin (Doll *et al.* 1962). The action might be the result of its anti-inflammatory activity but other factors, such as the inhibition of gastric secretion, are important (Gibson, 1978). Glycyrrhizin has, at the same time, a stimulating effect on the adrenal cortex, inducing an increase of production of corticoids and adrenal androgens. It also inhibits the inactivation of corticoids in the liver and kidney and increases sodium and water retention in the adrenal cortex. At high doses, therefore, side effects (induction of oedema and hypertension) may be observed, as a result of action on the renin–angiotensin–aldosterone system and on the electrolyte balance. These symptoms resemble those observed in the Conn Syndrome and hyperaldosteronism. It is likely that inhibition of renal 11β-dehydrogenase (responsible for the conversion of corticosterone to 11-dehydrocorticosterone and cortisol to cortisone) and the subsequent effect of increased cortisol concentrations on mineralocorticoid receptors is the primary mechanism of these complications (Monder *et al.* 1989; Bielenberg, 1992).

As glycyrrhetinic acid is produced from glycyrrhizin by hydrolysis after administration of the latter to humans, rats and rabbits, it is thought that the aglycone contributes to the side effects. The pharmacokinetics of glycyrrhetinic acid have, therefore, been studied, with the aim of increasing the safety of glycyrrhizin therapy (Ishida *et al.* 1989).

In view of these adverse pharmacological effects of glycyrrhizin, proposals have been made by the German and Swiss health authorities to introduce a maximum daily dose. Although this limit is expected to lie between 100 and 300 mg per day, HPLC analysis of glycyrrhizin-containing confectionery has shown levels of up to nearly 2000 mg glycyrrhizin in 100 g of preparation (Rüegg and Waser, 1993). Therefore, in just 10 g of these sweets it is possible to find the maximum recommended daily dose of the saponin.

Glycyrrhizin, on boiling, forms glycyrrhetinic acid, which can be converted into the hemisuccinate carbenoxolone (**1336**). Carbenoxolone sodium (Biogastrone[R], Bioral[R], Duogastrone[R], Neogel[R], Sanodin[R]) is the water-soluble disodium salt. Starting in the 1960s, carbenoxolone sodium was the market leader in the treatment of gastric ulcers for about 10 years (Peskar, 1980). It functions by maintaining prostaglandin levels in mucosa via inhibition of the prostaglandin metabolizing enzyme 15-hydroxy-prostaglandin dehydrogenase. Bicarbonate secretion is consequently stimulated and there is an increase in the quantity and viscosity of gastric mucus, an increase in the lifespan of gastric epithelial cells and possibly

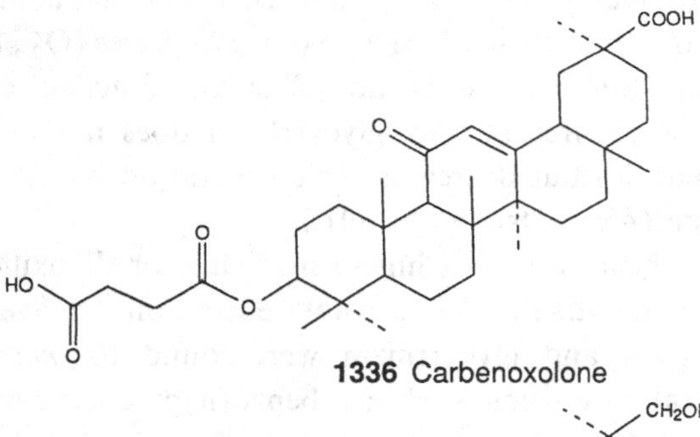

1336 Carbenoxolone

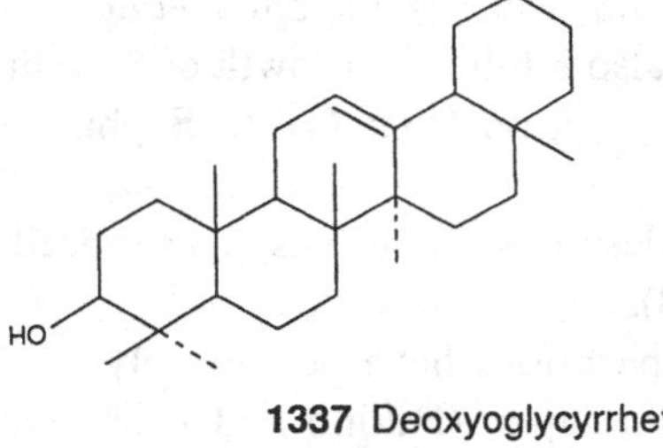

1337 Deoxyoglycyrrhetol

an inhibition of peptic activity (Lewis, 1974). However, its structural similarities to the steroid hormone aldosterone is probably responsible for side effects such as weight gain, oedema, hypokalaemia (potassium loss) and hypertension found in some patients.

The synthetic derivative deoxyoglycyrrhetol (**1337**) lacks the $\alpha\beta$-unsaturated C=O system in ring C of glycyrrhetinic acid, thus avoiding pseudoaldosteronism, which is sometimes a side effect of treatment with carbenoxolone. In addition, **1337** has a hydroxymethyl group attached to the terminal ring, a functional group which has an enhancing effect on the pharmacological activities of several saponins (Shibata *et al.* 1987).

Deglycyrrhizinized licorice (Caved-S[R], Rabro[R]) contains some residual antiulcer compounds which are polyphenolic in nature. These include the chalcone isoliquiritigenin and its glycoside, the flavonols kumatakenin and licoflavonol, the isoflavone licoricone, the chalcone licuzid and the coumestan glycyrrol (Saitoh *et al.* 1976).

Since glycyrrhizin and glycyrrhetinic acid have *anti-inflammatory* activities (Aleshinskaya *et al.* 1964), glycyrrhizin has been exploited in the preparation of skin cosmetics (Gibson, 1978) and in the treatment of dermatoses and pruritis (Boiteau *et al.* 1964). Furthermore, glycyrrhizin is used in Egypt as a substitute for cortisone. It reduces 12-*O*-tetradecanoylphorbol-13-acetate (TPA)-induced inflammation on mouse ear (Inoue *et al.* 1989) and suppresses the promoting effect of TPA on skin-tumour formation (Yasukawa *et al.* 1988). Regarding anti-inflammatory activity, glycyrrhizin both inhibits cortisone degradation in the liver and inhibits the generation of reactive oxygen species (O_2^-, H_2O_2, $^{\cdot}OH$) by neutrophils. Since there is no effect on reactive oxygen species generated in a cell-free system, glycyrrhizin does not act as a scavenger for these entities but decreases their generation by inhibiting neutrophil metabolism (Akamatsu *et al.* 1991).

As glycyrrhizin has been used in Chinese medicine for detoxification purposes, it was also investigated for possible effects on mutagenicity. Both glycyrrhiza extract and glycyrrhizin were found to *inhibit the mutagencity* of several mutagens such as benzo[*a*]pyrene, dimethyl-aminoazobenzene and Trp-P-1 (Tanaka *et al.* 1987). Glycyrrhizin and glycyrrhetinic acid also inhibit the growth of B-16 melanoma cells of mice, by inhibiting the transition from G1 to S phase of mitosis (Abe *et al.* 1987).

Glycyrrhizin, as described in herbals, does indeed have a *wound-healing* effect (Cunitz, 1968).

Glycyrrhizin preparations have been widely used for chronic *hepatitis* and liver cirrhosis in Japan (Hikino and Kiso, 1988). The antioxidative actions of glycyrrhizin might participate in the antihepatotoxic activity in carbon tetrachloride-induced liver lesions and the saponin may also protect liver cells from immunologically induced liver injuries (Hikino and Kiso, 1988). Another possibility is that glycyrrhizin enhances antibody formation through the production of interleukin I (Mizoguchi *et al.* 1984).

Glycyrrhizin demonstrates high accumulation in the liver and, when given intravenously to rats, about 80% of the injected dose is excreted into bile (Ichikawa *et al.* 1986). This phenomenon has been exploited in the preparation of liposomes that are surface-modified with glycyrrhizin and which then show four-fold more accumulation in rat liver than control liposomes. Consequently, targeting to the liver is enhanced with this ligand (Tsuji *et al.* 1991).

A recent discovery is that glycyrrhizin has *antiviral* activity and inhibits *in vitro* growth and cytopathogenicity of several DNA and RNA viruses, including herpes simplex virus type 1 (HSV-1), Newcastle disease virus, vesicular stomatitis virus, poliovirus type 1 (Pompei *et al.* 1979) and varicella-zoster virus (VZV) (Baba and Shigeta, 1987). Following reports that this saponin also inhibited replication of human immunodeficiency virus type 1 (HIV-1), the etiologic agent of acquired immune deficiency syndrome (AIDS) (Ito *et al.* 1987), a glycyrrhizin preparation, Stronger Neominophagen C (SNMC), was developed for clinical use (Hattori *et al.* 1989). Chemically modified derivatives of glycyrrhizin have been synthesized and evaluated for their inhibitory effects on the replication of HIV-1 and HSV-1. The 11-deoxo derivative **1338** is equipotent with glycyrrhizin against HIV-1 in MT-4 and MOLT-4 cells. It completely inhibits HIV-1-induced cytopathogenicity in both cell lines at a concentration of 0.16 mM (Hirabayashi *et al.* 1991).

Glycyrrhizin and glycyrrhetinic acid are effective antiallergic agents (Kumagai, 1967), while glycyrrhetinic acid is both antipyretic (Saxena and Bhalla, 1968) and antimicrobial.

A most unusual property of licorice root is its supposed ability to quench thirst, but as yet no scientific study has been performed to confirm this observation (Bielenberg, 1992).

7.3   *Hippocastani semen* (aescine)

Since aescine is one of the most frequently encountered saponin preparations, it has been pharmacologically investigated to a much greater extent than many other saponins and saponin mixtures (Boiteau *et al.* 1964).

Aescine (3–6%) is obtained commercially from ripe, dried seeds of the horse chestnut tree *Aesculus hippocastanum* L. (Hippocastanaceae), originally native to Albania and Greece, by percolation with 60–80% ethanol in water. It consists of a complex mixture of triterpene ester saponins (more than 30 individual saponins; Wulff and Tschesche, 1969), containing the following components:

- – β-aescine: a mixture of C-21 and C-22 diesters
- – kryptoaescine: a mixture of C-21 and C-28 diesters
- – α-aescine: a 4:6 mixture of β-aescine and kryptoaescine
- – aescinols: artifacts which contain free C-21, C-22 and C-28 hydroxyl groups from the hydrolysis of aescines.

β-Aescine. This contains the aglycones protoaescigenin or barringtogenol C (see Fig. 7.2), esterified with tiglic acid (40%), angelic acid (33%), α-methylbutyric acid (15%) or isobutyric acid (10%). The main saponins are 3-O-[β-D-glucopyranosyl-(1→2)-β-D-glucopyranosyl-(1→4)-β-D-glucuronopyranosyl]-21β-tigloyl-22α-acetylprotoaescigenin (R_1 = tigloyl, R_2 = CH_2OH, R_3 = R_4 = Glc) (**1339**) and the 21β-angeloyl analogue (R_1 = angeloyl, R_2 = CH_2OH, R_3 = R_4 = Glc) (**1340**) (Fig. 7.2). β-Aescine is haemolytic, with a haemolytic index of 40 000. It is crystalline and sparingly soluble in water. On heating an aqueous solution, formation of α-aescine occurs, during which there is migration of the acetyl group from C-22 to C-28.

Kryptoaescine. This component corresponds to β-aescine, with the exception of a hydroxyl group at C-22 and an acetyl group at C-28. There are at least seven components of kryptoaescine (Tschesche and Axen, 1964). It is haemolytically inactive and relatively unstable when heated in aqueous solution. Kryptoaescine can be obtained by counter-current distribution of aescine and is soluble in water.

α-Aescine. α-Aescine is a mixture of β-aescine and kryptoaescine; it is amorphous, more soluble in water than β-aescine and has

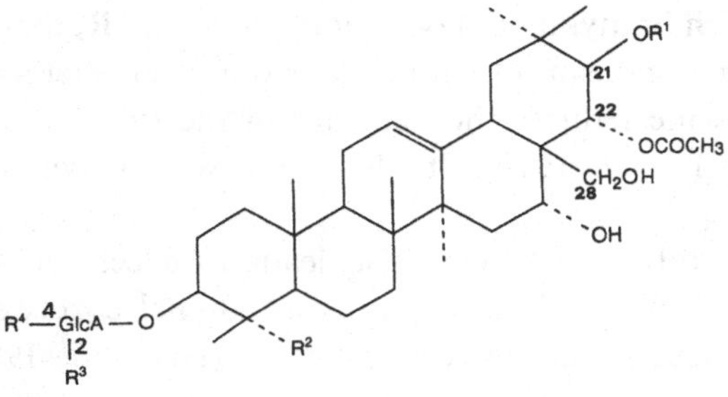

R^1 = Tigloyl, angeloyl, α-methylbutyryl, isobutyryl

R^2 = CH$_2$OH (protoaescigenin)
or CH$_3$ (barringtogenol C)

R^3, R^4 = Glc, Gal, Xyl

Fig. 7.2. Composition of β-aescine.

greater stability in solution than β-aescine. It has a haemolytic index of 20 000 and is obtained from the mother liquors after crystallization of β-aescine.

7.3.1 Pharmacokinetics and pharmacodynamics

As α- and β-aescine have different water solubilities, their absorption is different on oral administration: α-aescine is approximately 11% absorbed in rats and β-aescine 5% (Henschler *et al.* 1971). Maximum levels in the blood are reached after about 1 h and elimination is also very rapid – one third via the bile duct and one third via the kidneys. Aescine has been reported to have a long half-life on blood vessel endothelia, about 4–5 days (Felix, 1985). Levels of β-aescine in plasma (human) have been measured by RIA. It was found that the bioavailability of Venoplant-N^R Retard was 1.8 times higher than that of quick-release tablets (Kunz *et al.* 1991). The main metabolite of aescine in human serum and urine is aescinol, the triglycoside derivative which lacks the two ester groups (Lang, 1977).

In animal experiments, aescine has a number of anti-inflammatory effects (Hiai *et al.* 1981). It reduces capillary and membrane permeability, resulting in antiexudative (Rothkopf and Vogel, 1976) and oedema-inhibitory (or oedema-protective) properties (Lorenz and Marek, 1960; Vogel and Marek, 1962; Vogel *et al.* 1968a, 1970). Of the three phases of inflammation (namely (i) increased membrane permeability; (ii) emigration of leucocytes and tissue detection; and (iii) increased macrophage production and fibroblast proliferation), aescine is only effective in the first, the phase in which bradykinin plays a dominant rôle (Rothkopf and Vogel, 1976). The mechanism of action is poorly understood; one possibility is that aescine reduces the diameter of the tiny holes in the capillary walls (Vogel *et al.* 1970). It also increases the viscosity of blood.

Parenteral administration of aescine also leads to effects on sodium transport and the activation of the hypophysis–adrenal cortex system. The result is an increased release of corticosteroids (Hiai *et al.* 1981).

In the case of preparations of horse chestnut extracts, it cannot be excluded that other constituents (flavonoids, coumarins and tannins) may also contribute to the pharmacological activity.

7.3.2 Applications

Aescine (and horse chestnut extracts) finds widespread use in the prevention and treatment of peripheral vascular disorders (Rothkopf and Vogel, 1976; Annoni *et al.* 1979) and in the treatment of venous stasis, *Ulcus cruris*, dysmenorrhoea and haemorrhoids. The antioedematous action of aescine is most pronounced and it exerts a marked and prolonged antagonism against several types of local oedema induced by formalin, dextran and ovalbumin. An example of use is the application of aescine to the treatment of traumatic oedema of the brain (Diemath, 1981; Hemmer, 1985). There is usually a long latent period before aescine exerts its maximum effect (after approximately 16 h in the rat paw oedema test) (Vogel, 1963).

While it can be administered intravenously or orally, aescine is also applied locally in the form of gels and salves.

Other areas of application include the general strengthening of blood vessels (increased capillary resistance and decreased capillary permeability, i.e. capillaro-protective or venotonic properties); the treatment of varicose veins, bruises and haematomas (Crielaard and Franchimont, 1988); and the prevention of thrombosis. One study, for example, found a clear improvement of venous tone (reduced dilation capacity) after oral administration of 90 mg horse chestnut extract to human volunteers (Ehringer, 1968).

Commercially available products include Venostasin[R] and Vasotonin[R] (total extracts of horse chestnut seeds), Reparil[R] (a standardized aescine preparation used as an antiphlogistic) and Venoplant[R] (a combination of aescine and heparin).

The antiulcerogenic activity of aescine has been studied: the drug protects the gastric mucosa against lesions induced by absolute ethanol and also by pyloric ligature. At a dose of 50 mg/kg body weight, the antiulcer activity of aescine is comparable with that of carbenoxolone (80 mg/kg) in rats, after oral administration (Marhuenda *et al.* 1993).

β-Aescine is used in the cosmetic field for the prevention and treatment of panniculopatia oedemato-fibrosclerotica (cellulitis) (Proserpio *et al.* 1980).

7.3.3   Toxicity

Aescine has local irritant activity (notably on mucous membranes) and cannot be administered subcutaneously or intramuscularly. On intravenous administration in humans, aescine binds readily to serum albumin. However, care has to be taken to avoid damage to the kidneys and possible haemolysis.

Acute toxicity values for aescine are as follows (Boiteau *et al.* 1964): LD_{50} mice (i.v.) 4.0–22.0 mg/kg; LD_{50} rats (i.v.) 10.0–15.0 mg/kg; LD_{50} rats (i.p.) 25.0–50.0 mg/kg.

7.4   *Hederae folium* (ivy leaves)

Together with *Liquiritiae radix* and *Primulae radix*, *Hederae folium* is one of the most frequently used saponin-containing plant drugs for the treatment of coughs.

The leaves of *Hedera helix* L. (Araliaceae) contain the bidesmosidic saponin hederacoside C (**372**) and the corresponding monodesmoside α-hederin (**368**) (Fig. 7.3). Hederacoside B (**156**), a glycoside of oleanolic acid, is only present in small quantities (Tschesche *et al.* 1965).

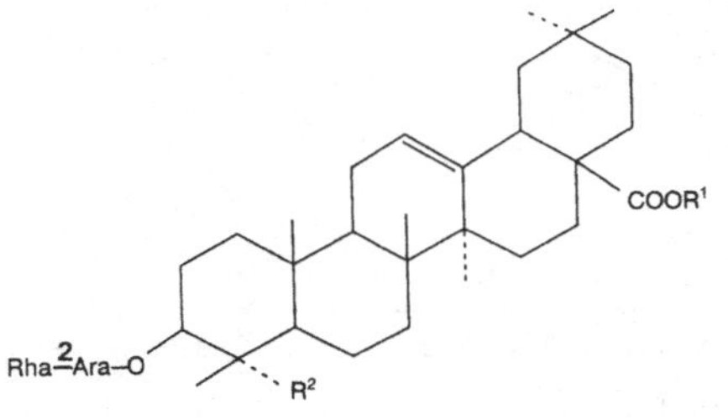

	R^1	R^2
368 α-Hederin	-H	$-CH_2OH$
156 Hederacoside B	$-Glc^6-Glc^4$-Rha	$-CH_3$
372 Hederacoside C	$-Glc^6-Glc^4$-Rha	$-CH_2OH$

Fig. 7.3. Constituents of ivy leaves.

α-Hederin, in contrast to hederacosides B and C, has strong haemolytic activity (haemolytic index = 150 000) (Wulff, 1968; see Section 5.1.3) and molluscicidal activity (see Section 5.3.4). *Hedera* extracts have expectorant properties (Vogel, 1963) and the aqueous alcohol preparations (e.g. Prospan[R]) have a weak spasmolytic activity (Wagner and Reger, 1986a). *Hedera* saponins are also known to be antiexudative (Vogel, 1963).

7.5   Tea

The haemolytic properties of the seed of tea, *Thea sinensis* L. (syn. *Camellia sinensis* O. KUNTZE; Theaceae) were first reported by Weil (1901). Only one saponin ('theasaponin', **804**, Table 2.6, p. 60) appears to have been partially characterized (Tschesche *et al.* 1969a). The reason for this is the complexity of the saponin mixture present. A number of highly hydroxylated aglycones have been discovered (Fig. 7.4) and these are not only glycosylated but also esterified by different organic acids: tiglic, angelic, isobutyric, α-methylbutyric, cinnamic and acetic. Theasaponin is principally a mixture of three aglycones, theasapogenols A, B and E, each with the same tetrasaccharide chain (branched at the glucuronic acid moiety) and esterified with tiglic (or angelic) acid at C-21 and acetic acid at C-22 (Tschesche *et al.* 1969a).

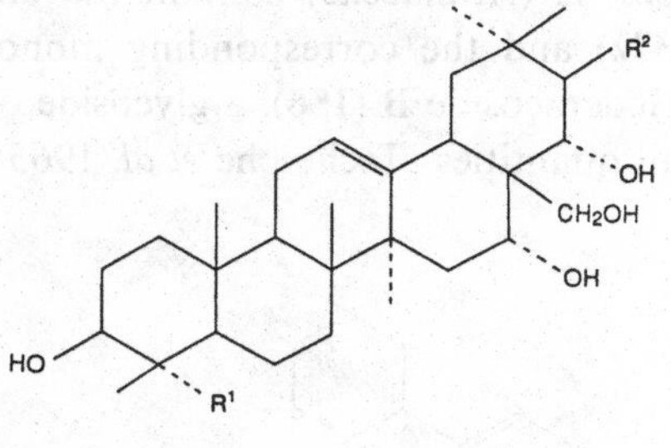

		R^1	R^2
69	Theasapogenol A	-CH$_2$OH	-OH
64	Theasapogenol B (barringtogenol C)	-CH$_3$	-OH
65	Theasapogenol C (camelliagenin C)	-CH$_2$OH	-H
	Theasapogenol D (camilliagenin A)	-CH$_3$	-H
	Theasapogenol E	-CHO	-OH

Fig. 7.4. Tea seed aglycones.

The antiexudative and anti-inflammatory properties of tea seed saponins have been reported (Chapter 5) (Vogel *et al.* 1968b). They also have expectorant, insecticidal, fungicidal and uterus-contracting activities (Tschesche *et al.* 1969a).

The major sapogenins from tea leaf are barringtogenol C, R_1-barringenol, camelliagenin A and A_1-barringenol (Hashizume, 1973); these are both glycosylated and esterified (with angelic, tiglic and cinnamic acids).

7.6   *Senegae radix* (**senega root**)

Senega root consists of the dried rootstock and root of *Polygala senega* L. from the USA and Canada or *P. senega* var. *latifolia* TORR. et GRAY from Japan (Polygalaceae). These are perennial herbs about 20–30 cm high, with white flowers. Originally senega was used by the North American Indians as a snake-bite remedy.

The powdered drug tastes at first sweet and subsequently unpleasant. It provokes sneezing.

The roots contain 6–10 saponins, with presenegin as aglycone. Those in largest amounts are senegins II (**830**), III (**831**) and IV (**834**, Table 2.7, p. 64), bidesmosidic ester glycosides.

Like primula root, senega root is an expectorant and is used in cough mixtures and instant teas. It is often prescribed with other expectorants such as ipecacuanha.

7.7   *Primulae radix* (**primula root**)

Primula root is obtained from the underground parts of the cowslip *Primula veris* L. (syn. *P. officinalis* (L.) HILL.) and the oxslip *Primula elatior* (L.) HILL. (Primulaceae). Both plants are small perennial herbs, *P. elatior* with sulphur-yellow flowers and *P. veris* with golden-yellow flowers (containing orange spots at the centre). The two species are widely distributed over Europe and Asia. The plant drug, primula root, contains 5–10% saponins.

Primula elatior roots. The saponin mixture is composed of approximately 90% 'primulic acid A' (**531**), the remaining 10% being glycosides of echinocystic acid and 28-dehydroprimulagenin A. It was originally thought that the aglycone of the major saponin in *Primula elatior* was primulagenin A (**35**, Table 2.1, p. 21) (Tschesche and Ziegler, 1964). However, it is now known that the true aglycone is proto-primulagenin A (**71**, Table 2.2, p. 24) and that primulagenin A is an artifact (Tschesche *et al.* 1983). Another glycoside of protoprimulagenin A has since been isolated from *Primulae radix*. This contains an additional xylose moiety but its substitution position has not yet been established (Wagner and Reger, 1986b).

Primula veris roots. The saponins consist of approximately 50% 'primulic acid A' (**531**) and a complex, poorly defined mixture which includes ester saponins or priverogenin A 16-acetate and priverogenin B

22-acetate. Primula root is an expectorant (a so-called reflex expectorant because it stimulates sensitive nerves in the mucous membranes), increases the solubility of mucus (Vogel, 1963) and is slightly laxative. It is used in the treatment of colds which are accompanied by coughs and in chronic bronchitis. For these purposes, the drug is available in the form of cough mixtures and instant teas. Primula saponins are anthelmintic (Jentzsch *et al.* 1961) and have antimicrobial properties (Tschesche and Wulff, 1965).

7.8   Asiaticoside

Asiaticoside from *Hydrocotyle asiatica* L. (syn. *Centella asiatica* [L.] URBAN) (Umbelliferae) (Madagascar) is used in crystalline form for different medical purposes in France (Chausseaud *et al.* 1971). The evergreen, creeping herb has large, kidney-shaped leaves and small purple flowers. *H. asiatica* is also found in India, where it has a reputation as a diuretic and tonic. An infusion of the plant is used in India and Madagascar for the treatment of leprosy and syphilis; the major glycoside, asiaticoside (**597**, Appendix 2), has antileprotic (Polonsky, 1968), wound-healing, antimycotic and antiphlogistic properties. It is claimed that scar formation is reduced when asiaticoside is applied during wound healing. Other properties include the healing of leg ulcers, the treatment of lesions of the cornea and the treatment of eczema and alopecia (Boiteau *et al.* 1964). However, reports of tuberculostatic activity have not been confirmed (Boiteau *et al.* 1964).

Ointments and liniments (such as Emdecassol[R] and Cothyline[R]) containing purified extracts of the herb are used externally for light burns, accelerated wound healing, eczemas and abscesses. An evaluation of the wound-healing properties of hydrocotyle tincture (Cothyline[R]) has shown that 64% of patients in a clinical study had completely healed lesions after treatment (Morisset *et al.* 1987).

The saponins asiaticoside-A (madecassoside, **607**) and asiaticoside-B (**1341**) (Appendix 2), trisaccharides of 6β-hydroxyasiatic acid (madecassic acid, **102**, Table 2.2, p. 24) and terminolic acid, respectively, have been isolated from the leaves but no details of their biological activities are yet known (Pinhas and Bondiou, 1967; Sahu *et al.* 1989).

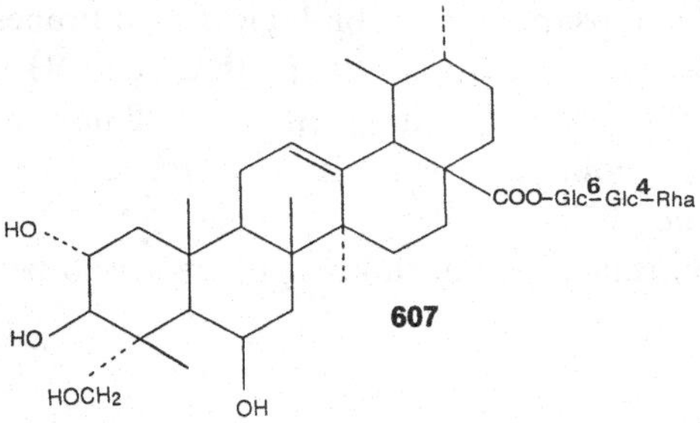

607

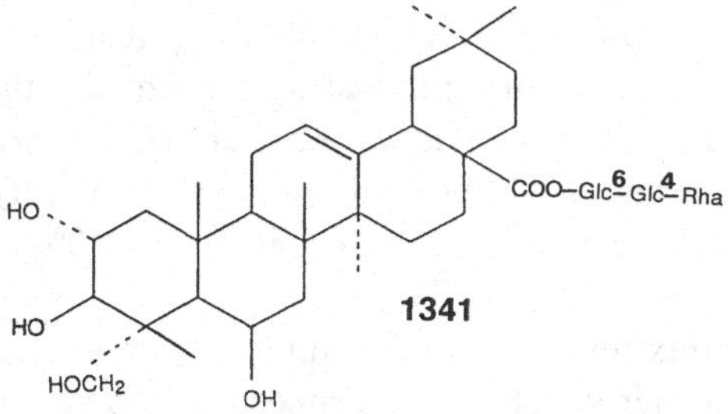

1341

7.9   *Bupleuri radix*

The roots of *Bupleurum falcatum* L. (Umbelliferae) and related species are a very important crude drug (Japanese name *saiko*) in oriental traditional medicine ('Kampo'). The chemical evaluation of these species has been reported (Ding *et al.* 1986) and saikosaponin-a (**547**) and saikosaponin-d (**548**) (Appendix 2) have been found to comprise the most pharmacologically important compounds. Saikosaponin-a and saikosaponin-d have anti-inflammatory action, corticosterone secretion-inducing activity, haemolytic activity, effects on membrane fluidity (Kimata *et al.* 1979), plasma cholesterol-lowering action and inhibitory action against hepatic injury by D-galactosamine and carbon tetrachloride

(Kimata *et al.* 1985). The crude drug is able to prevent the side effects of glucocorticoids (Hanawa *et al.* 1987).

Concerning the anti-inflammatory activity, oral and intra-muscular injection of saikosaponin-a and saikosaponin-d lead to antiexudative (granuloma pouch method) and antigranulomatous (cotton pellet method) effects in rats (Yamamoto *et al.* 1975). Antihepatotoxic actions of the saikosaponins are possibly caused by decreases of microsomal enzyme activities and inhibition of lipid peroxidation (Hikino and Kiso, 1988). Further results demonstrate that the biological actions of saikosaponins are quite similar to those of glycyrrhizin.

Saikosaponin-a possesses antiviral activity against influenza virus (Rao *et al.* 1974).

Studies on the effects of saikosaponins on biological membranes have shown that saikosaponin-a and saikosaponin-b$_1$ (with a 16β-OH group) give rise to a considerable decrease in membrane fluidity, while saikosaponin-d and saikosaponin-b$_2$ (with a 16α-OH group) have virtually no effect (Abe *et al.* 1978).

Further details of the pharmacology of saikosaponins are to be found in Chapter 6.

7.10   *Gypsophila* species

Commercial white saponin (Merck) is obtained from the roots of various *Gypsophila* species (Caryophyllaceae), such as the east European *G. paniculata* L. and the Middle Eastern *G. arrosti* GUSS. Members of the genus are herbs or small shrubs, with small, mainly white flowers. The content in saponins is very high, reaching over 10% in some species.

The triterpene glycoside mixture, gypsophila-saponin, from *G. paniculata* is a white, bitter-tasting powder which induces sneezing. It is very soluble in water and sparingly soluble in 96% ethanol. Even in dilutions of up to 1:1000 in water, there is formation of a stable foam and consequently it is added to fire extinguishers and detergents as a foaming agent. Gypsophila-saponin is used as reference substance in pharmacopoeia for the determination of haemolytic indices. It is also employed in certain expectorants.

Structure determination of all the components of the saponin mixture is not yet complete. However, the gypsogenin bidesmosides gypsoside A (**292**), saponaside D (saponaroside D, **295**), saponins **285** and **290**, and quillaic acid bidesmosides **430** and **431** (structures in Appendix 2) have been characterized from *G. paniculata*, while the gypsogenin 3-*O*-glycoside **1342** has been obtained from commercial white saponin (Higuchi *et al.* 1985).

Saponins from the genus *Gypsophila* have recently been shown to possess hypocholesterolaemic effects (Velieva *et al.* 1988).

7.11   Quillaia bark

Quillaia bark (soap bark, Panama wood) is from *Quillaja saponaria* MOL. (Rosaceae) and contains 9–10% saponins. *Q. saponaria* is an evergreen tree, found in Chile, Peru and Bolivia. The generic name is from the Chilean word *quillean,* to wash.

Quillaia-saponin is a white powder which provokes sneezing. The taste is at first sweet and then bitter. Stable foams are produced at high dilutions in water.

The saponin mixture possesses immunoadjuvant properties (see Chapter 5) and has pharmaceutical applications as a suspension stabilizer. It is used as a shampoo additive in the cosmetic industry to slow down grease formation in hair. Quillaia bark is a source of commercial saponin and is used as a foaming agent in beverages, confectionery, baked goods and dairy desserts. An additional property is the cholesterol-lowering effect in plasma (Chapter 5) (Topping *et al.* 1980).

The composition of quillaia-saponin is poorly defined but it is known that saponins of quillaic acid (**28**) and gypsogenic acid (**8**) (Table 2.1, p. 21) are present. It is only recently that one component, QS-III, an acylated quillaic acid 3,28-bidesmoside, has been characterized from this complex and poorly separable mixture (Higuchi *et al.* 1987, 1988). QS-III (**1343**) contains 10 sugar residues, including the monosaccharides fucose and apiose, and two acyl moieties, which are composed of the organic acid 3,5-dihydroxy-6-methyl-octanoic acid (Higuchi *et al.* 1988).

Two of the deacylated saponins (**1344** and **1345**) from quillaia, with structures similar to glycosides **430** and **431** (Appendix 2) from *Gypsophila paniculata*, have been assigned antiulcer properties (Muto *et al.* 1987).

As an *adjuvant*, the saponin mixture from *Q. saponaria* has found a wide application in veterinary vaccines, particularly against foot-and-mouth disease, and more recently in antiparasite vaccines for malaria, babesiosis

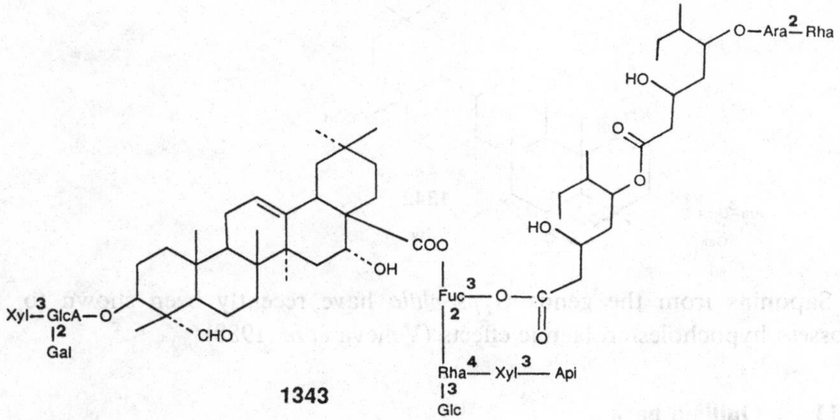

O—Ara$\overset{2}{-}$Rha

HO

Xyl$\overset{3}{-}$GlcA—O
|2
Gal

COO
OH
CHO

Fuc$\overset{3}{-}$O—C
|2   ‖
                     O

Rha$\overset{4}{-}$Xyl$\overset{3}{-}$Api
|3
Glc

1343

Xyl$\overset{3}{-}$GlcA—O
|2
Gal

COO
OH
CHO

Fuc
|2

Rha$\overset{4}{-}$Xyl$\overset{3}{-}$Api
|3
R

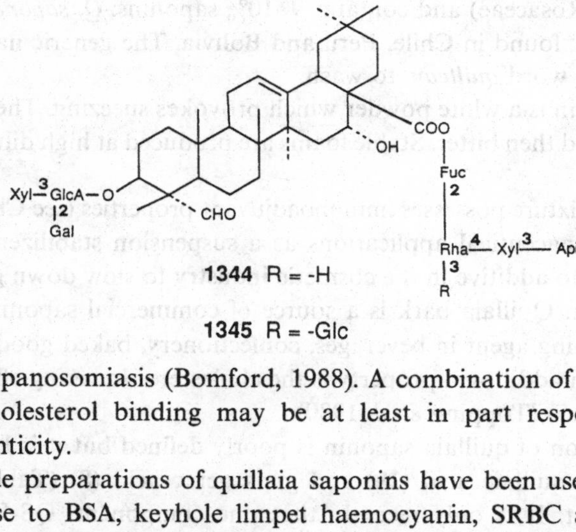

1344 R = -H

1345 R = -Glc

and trypanosomiasis (Bomford, 1988). A combination of surface activity and cholesterol binding may be at least in part responsible for the adjuvanticity.

Crude preparations of quillaia saponins have been used to boost the response to BSA, keyhole limpet haemocyanin, SRBC and aluminium hydroxide-based vaccines (Dalsgaard, 1974; Kensil *et al.* 1991). They also form antigen/lipid/saponin complexes known as ISCOMs (see Chapter 5).

In the search for adjuvants which are both safe and efficacious, the four major saponins from a water extract of *Q. saponaria* bark were purified (but not completely characterized) and tested for activity in mice. They boosted antibody levels by 100-fold or more with BSA and beef liver cytochrome b_5 antigens. As significant levels of IgG_{2a}, IgG_{2b} and G_1 antibodies were induced, the high level of protection on using saponins with vaccines in mice may in part be caused by the ability of saponins to induce an isotype profile similar to that observed in natural immunity arising from a viral or bacterial infection.

The lethality of these saponins and of a crude saponin preparation Quil-A, from *Q. saponaria*, to CD-1 mice was tested; while the major saponin caused death at doses as low as 25 µg, two of the other saponins showed no or very low toxicity (Kensil *et al.* 1991).

7.12   Ginseng

Ginseng belongs to the group of products which act as *adaptogens* (they stimulate the non-specific resistance of an organism and build up general vitality). The drug has been obtained for over 2000 years from roots of *Panax ginseng* C.A. MEYER (Araliaceae) (*panax* means 'cure all'), a perennial bush found in the mountain forests of Manchuria and North Korea (Hu, 1976, 1977). Because of the difficulty in finding sufficient wild-plant material, most of the ginseng preparations now available derive from plants cultivated in South Korea. Farming has also been initiated in Russia and China. The branching roots are 8–20 cm long and about 2 cm thick. The thin ends of main and secondary roots ('slender tails') are traditionally removed in Chinese medicine so that the human form of certain highly prized roots can be better brought into evidence (the name ginseng is derived from the Chinese *jen-shen*, meaning 'image of man'). The 'slender tails' are themselves used in certain ginseng extracts.

Two types of ginseng are available:

(1) White ginseng. Freshly harvested roots are washed and the secondary roots removed. After scraping and blanching with sulphur dioxide, the roots are dried in the sun or in ovens at 100–200 °C. By this means, the outer, darker layers of the root bark are lost.

(2) Red ginseng. Fresh roots are treated with steam at 120–130 °C for about 2–3 h and then dried. They assume a glassy, reddish appearance.

The first reports on ginseng saponins date from 1854 (Tanaka and Kasai, 1984). Ginseng root contains 0.5–3% of ginsenosides (or 'panaxosides'), a complex mixture of dammarane glycosides. Many of these are based on protopanaxadiol or protopanaxatriol skeletons (see Table 2.4, p. 52). Ginsenosides Rb_1 and Rg_1 are found in the largest amounts (Table 7.2). In addition, certain oleanolic acid glycosides are present.

'Slender tails' can contain over 12% of ginsenosides but the relative proportions of the individual saponins do not vary significantly from those of the main root preparations. It has been found that the saponin contents

Table 7.2. *Ginsenoside content of ginseng preparations from different sources*

	Total content (%)	Distribution (relative to total content)					
		$Rg_1 + Rf$	Re	Rd	Ro	Rb_2	Rb_1
White ginseng (Korea, Sam Geon Sam)	2.09	18.8	15.4	3.7	13.3	15.6	34.1
White ginseng (Korea, Kiboshi)	1.86	15.1	37.3	1.6	9.1	9.7	26.9
Red ginseng (Korea)	1.02	17.5	7.8	1.7	19.3	24.0	29.6
White ginseng ('slender tails')	10.6	4.9	13.0	3.9	22.7	22.8	32.5
American ginseng	6.01	3.4	18.8	6.3	6.8	0.0	64.7

From Bombardelli *et al.* 1976.

in the rhizome and lateral roots are higher than in the main root (Yamaguchi *et al.* 1988b). There are no great differences between white and red ginseng except for the content of ginsenoside Ra (white: 20%; red: 3%) and the presence of malonyl ginsenosides exclusively in white ginseng (Tanaka, 1987). A systematic investigation has also shown that elimination of the C-20 glycosyl residue and isomerization of the hydroxyl configuration at C-20 may occur during the preparation of red ginseng (Kitagawa *et al.* 1987b).

One group of *Panax* species possesses carrot-like roots that contain mainly dammarane saponins: *P. ginseng*, *P. quinquefolium* and *P. trifolius*. The second group, including *P. japonicus*, *P. zingiberensis* and *P. pseudoginseng*, contain long horizontal rhizomes possessing large amounts of oleanolic acid saponins as well as species-specific dammarane saponins.

P. quinquefolium L. (American ginseng, Kanton ginseng) differs from *P. ginseng* only in the absence of ginsenoside Rb_2 (Table 7.2).

P. notoginseng (BURK.) F.H. CHEN (Sanchi-ginseng) is a species of ginseng cultivated in China. Ginsenoside Rg_1 is the predominant dammarane glycoside and the oleanolic glycoside ginsenoside Ro is absent. The drug is used as a tonic, a haemostatic and in the prevention and cure of heart disease.

P. japonicus C.A. MEYER (syn. *P. pseudoginseng* WALL ssp. *japonicus* (MEYER) HARA) grows in Japan and south China ('Chikusetsu-ninjin'). The rhizome is used in traditional medicine as a stomachic, expectorant and antipyretic. Recent results have shown that *P. japonicus* rhizome extract and the corresponding saponin fractions have an antiulcer effect (Yamahara *et al.* 1987a). The content of saponins reaches 20% (chikustsu-

and chikusetsu-saponins) but the principal constituents are oleanolic acid glycosides, such as ginsenoside Ro (chikusetsusaponin V, **184**) (Appendix 2).

Various subspecies and varieties of *P. pseudoginseng* are also found in the Himalaya region. These include *P. pseudoginseng* ssp. *himalaicus* HARA and its two varieties *angustifolius* (BURK) LI and *bipinnatifidus* (SEEM) LI.

P. trifolius (dwarf ginseng) is found in northern USA and southern Canada. The saponin content of this species is very low (Lee and der Marderosian, 1981).

P. zingiberensis, which grows in China, contains oleanoic acid saponins and the dammarane saponins ginsenoside Rg_1 and Rh_1 (Tanaka and Kasai, 1984).

Noteworthy is the presence of dammarane saponins in aerial parts (leaves, flowers, stems) of *Panax* species – in certain cases at higher levels than the roots. The leaves of *P. pseudoginseng* and *P. quinquefolium* also contain glycosides of ocotillol-type triterpenes, e.g. (20*S*)-protopanaxatriol oxide I (**1346**) (Tanaka and Kasai, 1984).

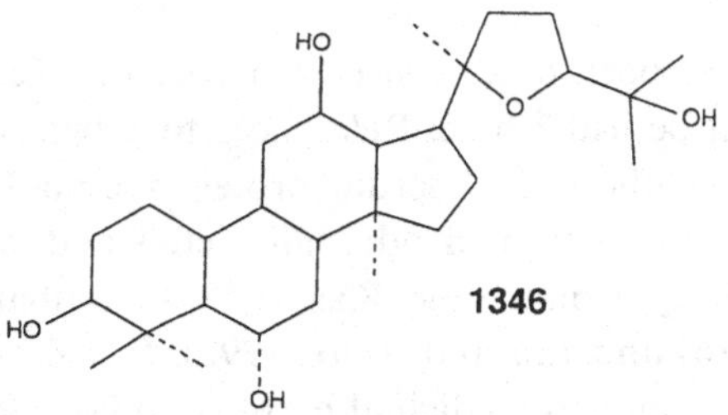

1346

In view of the high price of wild and cultivated ginseng, investigations of tissue culture as a viable alternative source of ginsenosides have been carried out (Furuya *et al.* 1983). Callus tissue of *P. japonicus* has also been shown to produce useful amounts of dammarane saponins (Fujioka *et al.* 1989).

Chinese traditional medicine ascribes to ginseng the properties of a stimulant, tonic, diuretic and stomachic agent. In the former USSR, ginseng is reputed to have a wide range of therapeutic action: it increases capacity for work and normalizes numerous pathological states (Baranov, 1982). The efficacy of ginseng is, however, still under discussion, even though it is an official drug in Korea, China, Japan, Russia, Austria, Switzerland and Germany (Sonnenborn, 1989). Monographs on the root appear in the German (*DAB 9*), Swiss (*Pharmacopoea Helvetica VII*) and Austrian pharmacopoeia. In other countries, ginseng is recognised only as a food supplement ('health food').

Table 7.3. *Pharmacological activities of ginseng extracts*

CNS-stimulating action
CNS-depressant or tranquillizing action
Cholinergic action
Histamine-like action
Non-antihistaminic action
Antihistaminic action
Hypotensive effects
Hypertensive effects
Papaverine-like action
Serotonin-like action
Ganglion-stimulant action
Analgesic, antipyretic and anti-inflammatory actions

From Shibata, 1986.

The ginsenosides are claimed to be responsible for most, if not all, of the pharmacological activities of ginseng, although the properties of other constituents (sesquiterpenes, alkyne alcohols, etc.) need to be investigated in more detail.

A multitude of effects have been described (summarized in Table 7.3) and only a few of these will be listed here. References to other activities (effects on nucleic acid biosynthesis and serum protein metabolism, on cholesterol metabolism, on the action of adrenalin, etc.) and pharmacology are given in reviews by Tanaka and Kasai (1984), Shibata *et al.* (1985), Ng and Yeung (1986) and Liu and Xiao (1992). In fact, ginseng is one of the most intensively studied medicinal herbs of today. However, many pharmacological studies must be regarded with caution because (i) the route of application (mostly i.p.) is often different from the oral route preferred in human practice; (ii) treatment with very high doses is common, with toxic levels being considered in some cases; (iii) pure ginsenosides may exhibit antagonistic effects on the model system tested.

It is only recently that the *immunomodulatory* and adaptogenic properties of ginseng have been satisfactorily investigated through the use of suitable immunological models (Sonnenborn, 1987). Increased macrophage activity, enhanced phagocytosis, increased interferon production and other effects are now well documented (Jie *et al.* 1984), influencing both cellular and humoral immunity. *P. ginseng* extracts are capable in mice of enhancing the antibody plaque-forming cell response and the circulating antibody titre against sheep erythrocytes, and of enhancing cell-mediated immunity against Semliki forest virus antigen and natural killer cell activity (Singh *et al.* 1984).

The possible use of ginseng as an adjuvant in cancer therapy and, in addition, direct cytotoxic effects on tumour cells have been investigated (Sonnenborn, 1987).

The antifatigue and antistress actions of dammarane saponins have been described by Kaku *et al.* (1975), Shibata (1986) and Chong and Oberholzer (1988). Anxiolytic activity of red and white ginseng has been confirmed in rats and mice (Bhattacharya and Mitra, 1991).

As far as the influence on energy turnover in skeletal muscle is concerned, some results from animal tests are as follows:

- Swimming test. It has been shown that the length of time a mouse can swim before the onset of exhaustion increases by about 50% when ginseng extract is administered in a daily oral dose of 3 mg/kg body weight (Shibata *et al.* 1985).
- Tolerance to physical stress is increased (Kim *et al.* 1970; Sandberg, 1973). Furthermore, administration of ginseng extract (0.4 µg i.v.) leads within several minutes to an increase in the reduced glutathione level in rat liver.

Clinical tests have shown:

- Increased physical performance. Treatment over 9 weeks with ginseng extract (200 mg/day) leads to a slower tiring of sportsmen (Forgo and Kirchdorfer, 1980).
- Increased mental performance. This can be achieved with a twice daily dose of 200 mg ginseng extract (Schöpfer, 1976).

Ginsenosides potentiate the activity of nerve growth factor (NGF) (Shibata, 1986). Since NGF is a hormonal protein which promotes regeneration and maintenance of sympathetic nerve cells, this might go some way to explaining the administration of ginseng for the alleviation of gerontological problems. In addition, ginsenosides Rb_1 and Rg_1 exert survival-promoting effects on chick and rat cerebral cortex neurons in cell cultures – effects which might have some influence on the process of learning and on memory enhancement (Himi *et al.* 1989).

Ginsenoside Rg_2 inhibits platelet aggregation induced by aggregating agents and ginsenoside Ro inhibits the thrombin-induced conversion of fibrinogen to fibrin. Consequently, ginseng might have some preventive effects on thrombosis or arteriosclerosis (Matsuda *et al.* 1986a,b).

The *antihepatotoxic* effects of ginsenosides have been investigated. Ginsenosides Rh_2 and Rg_3 and the prosapogenins of ginsenosides Ro and Rs were found to exert marked protection against carbon tetrachloride-induced cytotoxicity in primary cultured rat hepatocytes. Ginsenoside Rh_1

and the prosapogenin of ginsenoside Rs were effective in preventing galactosamine-induced liver cell damage. The oleanolic acid-based chikusetsu-saponins also showed antihepatotoxic activity (Hikino *et al.* 1985). Rats given 200 mg of ginsenoside Ro orally per kg of body weight 1 h before i.p. administration of 400 mg galactosamine/kg were protected against hepatotoxic effects when compared with control animals administered galactosamine alone (Jap. Pat. 92 05,235).

To summarize, it would appear that at least some of the claims for the effectiveness of ginseng are justified (Phillipson and Anderson, 1984). Triterpene glycosides from ginseng (Korean, Japanese and American varieties) are known to provide resistance to stress and to prevent and/or treat exhaustion, insomnia and depression. They lower cholesterol and blood sugar levels and are a tonic for strengthening and revitalizing the body, to help prevent illness and infection (perhaps via non-specific stimulation of the immune system). In addition, ginsenosides Rb_1 and Rg_1 have been suggested to alleviate the symptoms of Alzheimer-type senile dementia. They may do this by enhancing the availability of acetylcholine in the cortical and hippocampal regions of the brain (Pang *et al.* 1991).

Indian *P. pseudoginseng* has been investigated and compared with Korean ginseng. The adaptogenic activity, measured by the swimming performance test, was claimed to be even better than the usual *P. ginseng*. Gastric ulceration induced by swimming stress was inhibited. A semi-purified saponin extract of *P. pseudoginseng* was a more powerful immunostimulant than Korean ginseng saponins but, in both cases, only weak anti-inflammatory activity was observed (Dua *et al.* 1989).

It has been reported that the pharmacological activities of saponins of (20*S*)-protopanaxadiol (**113**) differ somewhat from those of (20*S*)-protopanaxatriol (**114**) (Table 2.2, p. 24). For example, saponins of (20*S*)-protopanaxadiol exhibit a sedative action while those of **114** stimulate the central nervous system (Nabata *et al.* 1973).

Red ginseng is sold in Europe as a 'spissum' extract and in the form of instant teas. Most preparations, however, originate from the cheaper white ginseng. The powdered drug is formed into pills; dried and spray-dried extracts, *spissum* and *fluidum* extracts are also employed.

A number of patents (comprising ginseng preparations or pure ginsenosides) have been brought out. One covers the neoplasm inhibition of ginsenosides Rg_1, Rg_2, Rf, Rb_1, Rb_2, Ra and Ro, and the sarcoma-growth inhibition of an extract (Jap. Pat. 81 46,817). Another concerns the use of pharmaceuticals containing ginsenoside Ro as antithrombotic agents (Jap. Pat. 82 163,315).

7.12.1   *Pharmacokinetics and pharmacodynamics*

This subject has not yet been thoroughly investigated for the ginsenosides. However, it is known that the relatively lipophilic ginsenoside Rg_1 is only 20% resorbed after peroral application; further studies on its absorption and distribution have been carried out in the rat (Odani *et al.* 1983). In rats, a dose of 100 mg/kg p.o. gives a maximum serum content of 0.9 μg/ml after 30 min. Neither the unchanged glycoside nor its metabolites cross the blood–brain barrier into the CNS (Strömbom *et al.* 1985). Thus the effects of ginseng on the CNS must be of an indirect nature.

The above pharmacokinetic data show that results on the pharmacological effects of ginseng should be treated with caution, especially as the doses involved are between 10 and 1000 mg/kg i.p. By way of contrast, a single 2 g dose of ginseng p.o. in humans gives a maximal resorbable amount of 0.2 mg/kg.

While ginsenosides Rb_1, Re and Rg_1 are thought to be transformed under the acidic conditions of the stomach into certain metabolites whose structures are not yet known (Han *et al.* 1982), some data is available on the metabolism of ginsenosides in the rat large intestine after oral administration. Ginsenoside Rg_1 (**758**) is transformed into ginsenosides Rh_1 (**757**) and F_1, while ginsenoside Rb_1 (**749**) is decomposed to ginsenoside Rd (**743**) (Table 2.4, p. 52). Ginsenoside Rb_2 (**745**) (which has beneficial effects in the treatment of arteriosclerosis) gives ginsenoside Rd (**743**), 3-*O*-β-D-glucopyranosyl-20-*O*-[α-L-arabinopyranosyl(1→6)-β-D-glucopyranosyl]-(20*S*)-protopanaxadiol (**1347**), ginsenoside F_2 (**1348**), 20-*O*-[α-L-arabinopyranosyl(1→6)-β-D-glucopyranosyl]-(20*S*)-protopanaxadiol (**1349**) and 20-*O*-β-D-glucopyranosyl-(20*S*)-protopanaxadiol (**1350**) (Fig. 7.5). Since decomposition began with cleavage of the glucosyl moieties at the 3-OH group, the presence of a β-glucosidase was suspected (Karikura *et al.* 1990). Ginsenoside Rb_2 (**745**) is little decomposed in rat stomachs after oral administration. It tends to undergo oxygenation rather than hydration of the side chain and a small quantity of the hydroxylated products **1351** and **1353** and the hydroperoxylated products **1352** and **1354** are formed (Karikura *et al.* 1991). The biotransformation of (20*S*)-ginsenoside Rg_2 has been studied in China by Chen *et al.* (1987c). A complex mixture of products is produced in the intestine.

7.12.2   *Toxicity*

Numerous studies with ginseng indicate its safety for human use. For example, in Japan, ginseng has been given to 500 patients in the course of two studies without any apparent side effects (Okuda and Yoshida, 1980; Yamamoto and Uemura, 1980). However, a few rare cases

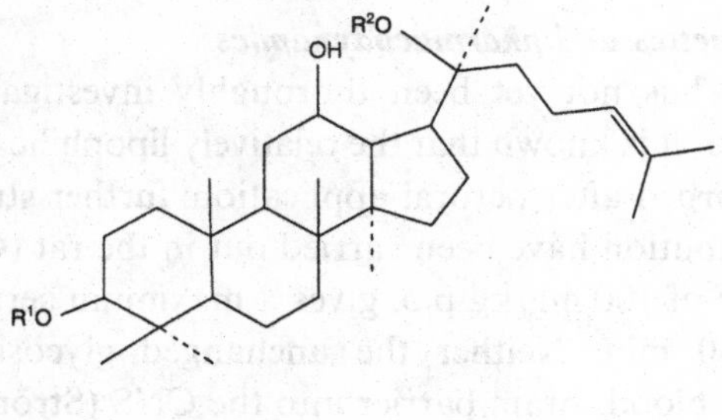

	R^1	R^2
745 Ginsenoside Rb$_2$	-Glc^2-Glc	-Glc^6-Ara$f$
743 Ginsenoside Rd	-Glc^2-Glc	-Glc
1347	-Glc	-Glc^6-Ara$f$
1348 Ginsenoside F$_2$	-Glc	-Glc
1349	-H	-Glc^6-Ara$f$
1350	-H	-Glc

Fig. 7.5. Structures of ginsenoside Rb$_2$ and its metabolites in rat large intestine.

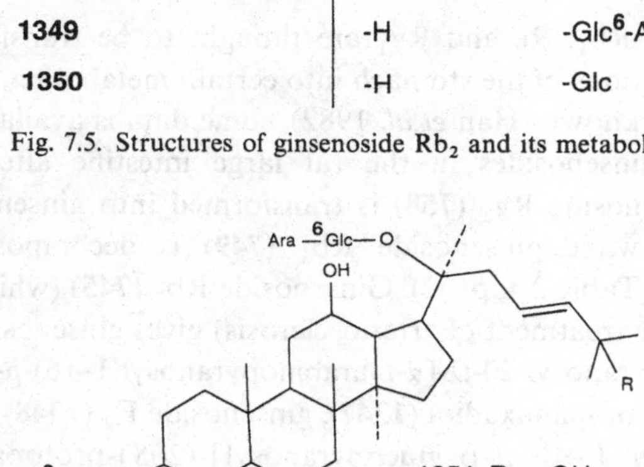

1351 R = -OH
1352 R = -OOH

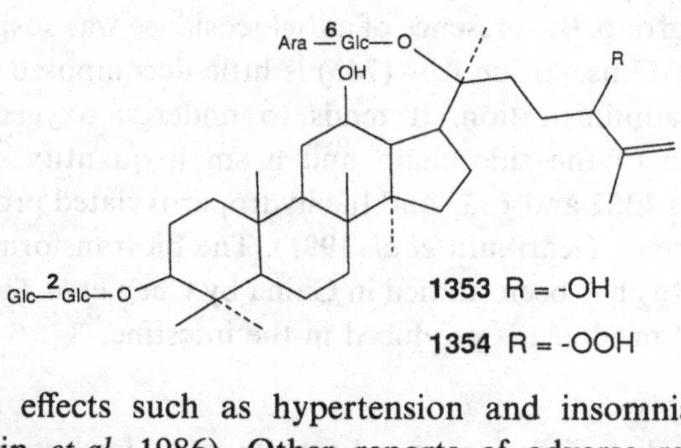

1353 R = -OH
1354 R = -OOH

of side effects such as hypertension and insomnia have been noted (Baldwin *et al.* 1986). Other reports of adverse reactions often show methodological flaws, as illustrated by Sonnenborn and Hänsel (1992), and there is no real evidence for any risk involved in the consumption of ginseng-based products.

Toxicity has been studied in a number of animal species but no reports of toxic effects have been documented (Baldwin *et al.* 1986). Acute toxicity experiments have used extremely high doses; for example, the LD_{50} of a crude neutral saponin fraction from *P. ginseng* was more than 5000 mg/kg p.o. for mice (Sonnenborn and Hänsel, 1992).

7.13   *Eleutherococcus senticosus*

The genus *Eleutherococcus* (syn. *Acanthopanax*) is closely botanically related to the genus *Panax*. Both are members of the family Araliaceae. *Eleutherococcus* species are used in Korean and Chinese folk medicine as tonics, antirheumatics, anti-inflammatories and prophylactics for chronic bronchitis, hypertension and ischaemic heart disease.

Eleutherococcus senticosus MAXIM. (syn. *Acanthopanax senticosus* (RUPR. et MAXIM.) HARMS) is a shrub from central and north China, North and South Korea, Japan and eastern Russia. It is, therefore, more widely distributed than ginseng.

In addition to other constituents such as phenylpropanoids and lignans (Slacanin *et al.* 1991), the plant drug contains the oleanolic glycosides eleutherosides I, K, L and M (Fig. 7.6) (Frolova and Ovodov, 1971; Frolova *et al.* 1971) but *no* dammarane saponins.

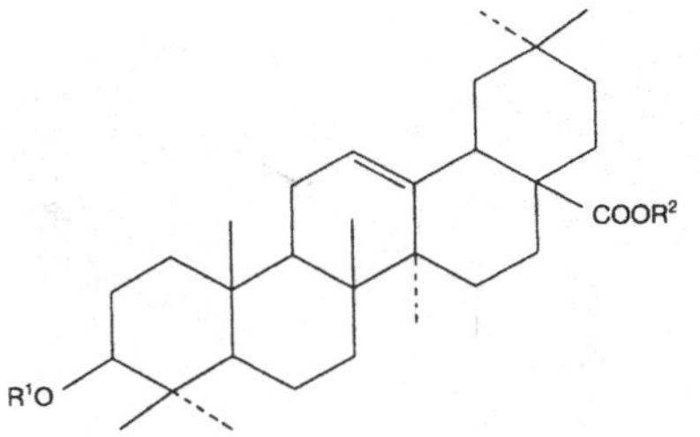

	R^1	R^2
157 Eleutheroside I (mubenin B)	-Ara^4-Rha	-H
154 Eleutheroside K	-Ara^2-Rha	-H
1355 Eleutheroside L	-Ara^4-Rha	-Glc^6-Glc^4-Rha
156 Eleutheroside M (hederasaponin B)	-Ara^2-Rha	-Glc^6-Glc^4-Rha

Fig. 7.6. Oleanolic acid glycosides from *Eleutherococcus senticosus* (Frolova and Ovodov, 1971; Frolova *et al.* 1971).

Recent work on leaves of the plant, collected in China, has led to the isolation of four new oleanolic acid derivatives (146, 1356–1358), three new 29-hydroxyhederagenin derivatives (1359–1361) and six new 30-norolean-12,20(29)-dien-28-oic acid saponins (1362–1367) (Shao *et al.* 1988, 1989a,b) (Fig. 7.7). Six of these saponins (1357, 1358, 1360, 1361, 1365, 1366) are acetylated at one of the sugar moieties.

Roots of a batch of *E. senticosus* from Russia have been reported to contain a tetrasaccharide (534) and a pentasaccharide (535) (Appendix 2) of protoprimulagenin A (Segiet-Kujawa and Kaloga, 1991).

Three different types of preparation exist:

- powdered root bark, from nothern China (also known as Siberian ginseng)
- powdered roots (known as Siberian ginseng).
- 'taiga root' or Russian ginseng consists of *E. senticosus* roots from eastern Russia.

The leaves are also used for medicinal purposes; some Soviet authors suggested that they have the same therapeutic properties as the roots (Baranov, 1982) but conclusive evidence is required.

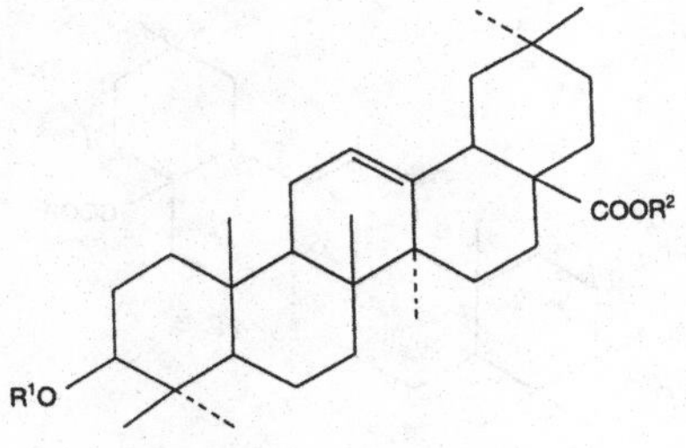

	R^1	R^2
146 Ciwujianoside A$_1$	-Ara^2-Glc	-Glc^6-Glc^4-Rha
1356 Ciwujianoside C$_3$	-Ara	-Glc^6-Glc^4-Rha
1357 Ciwujianoside C$_4$	-Ara^2-Rha	-Glc^6-Glc^4-Rha $\mid$6 OCOCH$_3$
1358 Ciwujianoside D$_1$	-Ara	-Glc^6-Glc^4-Rha $\mid$6 OCOCH$_3$

Fig. 7.7. Saponins from the leaves of *Eleutherococcus senticosus*.

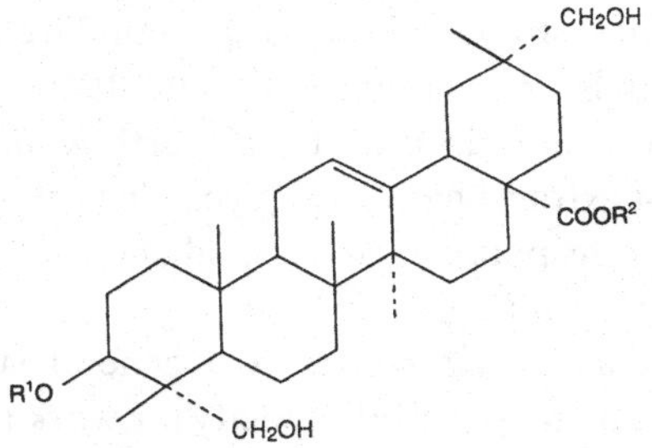

	R^1	R^2
1359 Ciwujianoside A$_3$	-Ara^2-Rha	-Glc^6-Glc^4-Rha
1360 Ciwujianoside D$_3$	-Ara	-Glc^6-Glc^4-Rha $\mid$6 OCOCH$_3$
1361 Ciwujianoside A$_4$	-Ara^2-Glc	-Glc^6-Glc^4-Rha $\mid$6 OCOCH$_3$

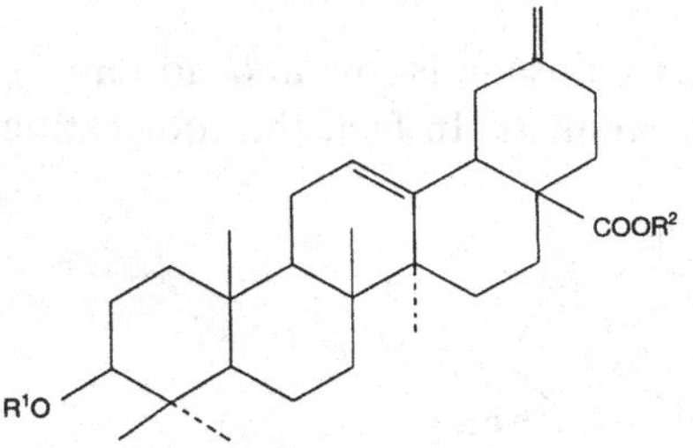

	R^1	R^2
1362 Ciwujianoside A$_2$	-Ara^2-Glc	-Glc^6-Glc^4-Rha
1363 Ciwujianoside B	-Ara^2-Rha	-Glc^6-Glc^4-Rha
1364 Ciwujianoside C$_1$	-Ara	-Glc^6-Glc^4-Rha
1365 Ciwujianoside C$_2$	-Ara^2-Rha	-Glc^6-Glc^4-Rha $\mid$6 OCOCH$_3$
1366 Ciwujianoside D$_2$	-Ara	-Glc^6-Glc^4-Rha $\mid$6 OCOCH$_3$
1367 Ciwujianoside E	-Ara^2-Rha	-H

Fig. 7.7. (*Cont.*)

Extensive experimental studies with a 33% ethanolic extract of the roots have been carried out in Russia since the late 1950s. A large portion of this work has been summarized by Farnsworth *et al.* (1985).

Eleutherococcus root extracts have hypoglycaemic action, an antioedema and anti-inflammatory response in mice, and increase the resistance of mice and rabbits to infection.

Like ginseng, *Eleutherococcus* possesses a general strengthening and *tonic* effect, raises mental and physical capacity for work (stimulant action) and alters carbohydrate and albumin metabolism (anabolic effect). The wide range of efficacy may partly be caused by an adaptogenic action (Brekhman and Dardymov, 1969a).

It has been claimed that *Eleutherococcus* has some advantages over ginseng (Brekhman and Dardymov, 1969a):

- the stimulant and tonic effects of *Eleutherococcus* are longer and stronger
- sleep disturbances are rarely produced
- excitation is never observed in patients.

Oral toxicity of Siberian ginseng is low and no teratogenic effects are reported in experimental animals. In fact, the root extract decreases the

690   Divaroside   R = -Glc^6-Glc
1368 Chiisanoside   R = -Glc^6-Glc^4-Rha

1369 Isochiisanoside

toxicity of certain antitumour agents and delays the induction of chemically induced or spontaneous tumours in rodents (Farnsworth *et al.* 1985). Very few side effects have been reported in studies involving humans (Farnsworth *et al.* 1985).

Studies on other *Acanthopanax* species have led to the isolation of different saponins: the leaves of *A. chiisanensis* and *A. divaricatus* contain novel 3,4-secolupane glycosides (**690**, **1368**, **1369**) (Matsumoto *et al.* 1987).

Appendix 1

Appendix 1. *Species of plants containing characterized triterpene saponins*

Family	Species	Reference
Lycopodiaceae (Pteridophyta)	*Lycopodium inundatum*	Tsuda *et al.* 1981
Acanthaceae	*Acanthus ilicifolius*	Minocha and Tiwari, 1981
	Justicia simplex	Ghosal *et al.* 1981
Aceraceae	*Acer negundo*	Kupchan *et al.* 1970
Amaranthaceae	*Achyranthes aspera*	Seshadri *et al.* 1981
	A. bidentata	Gedeon and Kincl, 1956
	Alternanthera hypochondriacus	Kohda *et al.* 1991
	A. philoxeroides	Parkhurst *et al.* 1980
	A. sessilis	Penders *et al.* 1992
	Amaranthus spinosus	Banerji, 1979
	Cornulaca monacantha	Amer *et al.* 1974
	Deeringia amaranthoides	Sati *et al.* 1990
	Pfaffia iresinoides	Nishimoto *et al.* 1987
	P. paniculata	Nakai *et al.* 1984
	P. pulverulenta	Shiobara *et al.* 1992
Anacardiaceae	*Mangifera indica*	Khan *et al.* 1993
Apocynaceae	*Ichnocarpus frutescens*	Minocha and Tandon, 1980
	Trachelospermum asiaticum	Abe and Yamauchi, 1987
Aquifoliaceae	*Ilex chinensis*	Inada *et al.* 1987b
	I. cornuta	Nakanishi *et al.* 1982
	I. crenata	Kakuno *et al.* 1991a
	I. hainanensis	Min and Qin, 1984
	I. integra	Yano *et al.* 1993
	I. mitis	Razafintsalama *et al.* 1987
	I. opaca	Potter and Kimmerer, 1989
	I. paraguariensis	Gosmann *et al.* 1989
	I. pubescens	Hidaka *et al.* 1987a
	I. rotunda	Nakatani *et al.* 1989
Araliaceae	*Acanthopanax chiisanensis*	Kasai *et al.* 1986b
	A. divaricatus	Matsumoto *et al.* 1987

342

Appendix 1. (*Cont.*)

Family	Species	Reference
Araliaceae	*A. sessiliflorus*	Kong *et al.* 1988
(*cont.*)	*A. sieboldianus*	Sawada *et al.* 1993
	A. spinosus	Miyakoshi *et al.* 1993
	Aralia bipinnatifida	Chu and Chou, 1941
	A. chinensis	Kuwada, 1931
	A. cordata	Kawai *et al.* 1989
	A. decaisneana	Fang *et al.* 1992
	A. elata	Saito *et al.* 1990b
	A. japonica	Winterstein and Stein, 1932
	A. mandshurica	Kochetkov *et al.* 1962
	A. montana	Winterstein and Stein, 1932
	A. quinquaefolia	Boiteau *et al.* 1964
	A. schmidtii	Elyakov *et al.* 1962
	A. spinifolia	Yu *et al.* 1992
	Cussonia barteri	Dubois *et al.* 1986
	C. spicata	Gunzinger *et al.* 1986
	Eleutherococcus senticosus (*Acanthopanax senticosus*)	Frolova *et al.* 1971
	Fatsia japonica	Kemoklidze *et al.* 1987
	Hedera caucasigena	Dekanosidze *et al.* 1970a
	H. colchica	Dekanosidze *et al.* 1985
	H. helix	Tschesche *et al.* 1965
	H. japonica	Kuwada and Matsukawa, 1934
	H. nepalensis	Kizu *et al.* 1985c
	H. pastuchovii	Iskenderov, 1971
	H. rhombea	Kizu *et al.* 1985a
	H. taurica	Grishkovets *et al.* 1991
	Kalopanax pictus	Sano *et al.* 1991
	K. ricinifolius	Kotake and Taguchi, 1932
	K. septemlobus	Shao *et al.* 1989b
	Macropanax disperum	Srivastava and Jain, 1989
	Nothopanax davidii	Yu and Xiao, 1992
	N. delavayi	Kasai *et al.* 1987a
	Panax ginseng	Kitagawa *et al.* 1989d
	P. japonicus	Lin *et al.* 1976
	P. notoginseng	Namba *et al.* 1986
	P. pseudoginseng	Kondo and Shoji, 1975
	P. quinquefolius	Xu *et al.* 1987
	P. repens	Kitasato and Sone, 1932
	P. stipuleanatus	Yang *et al.* 1985
	P. trifolius	Lee and der Marderosian, 1988
	P. zingiberensis	Namba *et al.* 1986
	Polyscias dichroostachya	Gopalsamy *et al.* 1990
	P. elegans	Simes *et al.* 1959

(*cont.*)

Appendix 1. (*Cont.*)

Family	Species	Reference
Araliaceae	*P. nodosa*	Hiller *et al.* 1966
(*cont.*)	*P. scutellaria*	Paphassarang *et al.* 1990
	Schefflera capitata	Jain and Khanna, 1982
	S. impressa	Srivastava and Jain, 1989
	S. octophylla	Sung *et al.* 1992
	S. venulosa	Purohit *et al.* 1991
	Tetrapanax papyriferum	Takabe *et al.* 1981
Asclepiadaceae	*Gymnema sylvestre*	Yoshikawa *et al.* 1989a
Asteraceae	*Aspilia kotschyi*	Kapundu *et al.* 1988
(Compositae)	*Aster baccharoides*	Hui *et al.* 1971
	A. scaber	Nagao and Okabe, 1992
	A. tataricus	Nagao *et al.* 1989b
	Bellis perennis	Hiller *et al.* 1988
	Calendula arvensis	Pizza *et al.* 1987
	C. officinalis	Vidal-Ollivier *et al.* 1989b
	Celmisia petriei	Rowan and Newman, 1984
	Centipeda minima	Gupta and Singh, 1989
	Chrysanthellum procumbens	Becchi *et al.* 1979
	Cynara cardunculus	Shimizu *et al.* 1988
	Elvira bifolia	Oliveira and Andrade, 1961
	Gundelia tournefortii	Wagner *et al.* 1984b
	Helianthus annuus	Bader *et al.* 1991
	Rhaponticum carthamoides	Vereskovskii and Chekalinskaya, 1980
	Silphium perfoliatum	Davidyants *et al.* 1986
	Solidago canadensis	Hiller *et al.* 1991
	S. gigantea	Hiller *et al.* 1991
	S. virgaurea	Hiller *et al.* 1991
	Spilanthes acmella	Mukharya and Ansari, 1987
	Tragopogon porrifolius	Warashina *et al.* 1991
	T. pratensis	Miyase *et al.* 1992
	Vicoa indica	Vasanth *et al.* 1991
	Wedelia scaberrima	Matos and Tomassini, 1983
	Zexmenia buphthalmiflora (*Wedelia buphthalmiflora*)	Schteingart and Pomilio, 1984
Basellaceae	*Boussingaultia baselloides*	Espada *et al.* 1990
	Ullucus tuberosus	Dini *et al.* 1991
Berberidaceae	*Caulophyllum robustum*	Chandel and Rastogi, 1980
	C. thalictroides	McShefferty and Stenlake, 1956
	Leontice eversmannii	Agarwal and Rastogi, 1974
	L. leontopetalum	McShefferty and Stenlake, 1956
Betulaceae	*Alnus pendula*	Aoki *et al.* 1990
	A. serrulatoides	Aoki *et al.* 1990
Bignoniaceae	*Macfadyena unguis-cati*	Ferrari *et al.* 1981
	Mussatia hyacintha	Jimenez *et al.* 1989
Bombacaceae	*Caryocar glabrum*	Boiteau *et al.* 1964

Appendix 1. (*Cont.*)

Family	Species	Reference
Boraginaceae	*Anchusa officinalis*	Romussi *et al.* 1980
	Caccinia glauca	Tewari *et al.* 1970
	Cordia obliqua	Chauhan and Srivastava, 1978
	Symphytum officinale	Ahmad *et al.* 1993a
Cactaceae	*Escontria chiotilla*	Djerassi *et al.* 1956a
	Lemaireocereus beneckei	Djerassi *et al.* 1956b
	L. chichipe	Khong and Lewis, 1975
	L. dumortieri	Djerassi *et al.* 1954a
	L. griseus	Djerassi *et al.* 1956a
	L. hystrix	Djerassi and Lippman, 1954
	L. longispinus	Djerassi *et al.* 1953
	L. pruinosus	Djerassi *et al.* 1955
	L. queretaroensis	Djerassi *et al.* 1956b
	L. quevedonis	Djerassi *et al.* 1956a
	L. stellatus	Djerassi *et al.* 1955
	L. thurberi	Kircher, 1977
	L. treleasei	Djerassi *et al.* 1956a
	Machaerocereus eruca	Djerassi *et al.* 1957
	M. gummosus	Djerassi *et al.* 1954b
	Myrtillocactus cochal	Djerassi *et al.* 1957
	M. eichlamii	Djerassi *et al.* 1957
	M. geometrizans	Djerassi *et al.* 1957
	M. grandiareolatus	Djerassi *et al.* 1957
	M. schenckii	Djerassi *et al.* 1957
	Pereskia grandiflora	Sahu *et al.* 1974
	Rhipsalis mesembryanthemoides	Tursch *et al.* 1965
Campanulaceae	*Platycodon grandiflorum*	Tada *et al.* 1975
Caprifoliaceae	*Lonicera japonica*	Kawai *et al.* 1988
	L. nigra	Domon and Hostettmann, 1983
	Viburnum nervosum	Jain *et al.* 1978
Caryophyllaceae	*Acanthophyllum adenophorum*	Amanmuradov and Tanyurcheva, 1969
	A. gypsophiloides	Kondratenko *et al.* 1981
	A. paniculata	Putieva *et al.* 1974
	A. squarrosum	Lacaille-Dubois *et al.* 1993
	A. subglabrum	Basu and Rastogi, 1967
	Agrostemma githago	Tschesche and Schulze, 1974
	Arenaria graminifolia	Bukharov and Shcherbak, 1973
	Dianthus barbatus	Cordell *et al.* 1977
	D. chinensis	Li *et al.* 1993
	D. deltoides	Kondratenko *et al.* 1981
	D. inacotinus	Chirva and Kintya, 1971
	D. superbus	Oshima *et al.* 1984a
	Gypsophila acutifolia	Kondratenko *et al.* 1981

(*cont.*)

Appendix 1. (*Cont.*)

Family	Species	Reference
Caryophyllaceae (*cont.*)	*G. arrostii*	Frechet *et al.* 1991
	G. bicolor	Tagiev and Ismailov, 1978
	G. capitata	Kondratenko *et al.* 1981
	G. elegans	Kondratenko *et al.* 1981
	G. oldhamiana	Kondratenko *et al.* 1981
	G. pacifica	Kondratenko *et al.* 1981
	G. paniculata	Frechet *et al.* 1991
	G. patrinii	Kondratenko *et al.* 1981
	G. struthium	Kondratenko *et al.* 1981
	G. trichotoma	Kondratenko *et al.* 1981
	Herniaria glabra	Cart *et al.* 1992
	H. hirsuta	Boiteau *et al.* 1964
	Melandrium firmum	Woo *et al.* 1992
	Polycarpone loeflingiae	Bhandari *et al.* 1990
	Saponaria officinalis	Chirva and Kintya, 1969
	Silene nutans	Kondratenko *et al.* 1981
	Spergularia arbuscula	Appel and Haleby, 1965
	S. marginata	Bouché, 1955
	Vaccaria segetalis	Kondratenko *et al.* 1981
	Viscaria viscosa	Kondratenko *et al.* 1981
	V. vulgaris	Bukharov and Bukharova, 1974
Cecropiaceae	*Musanga cecropioides*	Lontsi *et al.* 1990
Chenopodiaceae	*Anabasis articulata*	Segal *et al.* 1969
	A. setifera	Sandberg and Shalaby, 1960
	Atriplex canescens	Boiteau *et al.* 1964
	A. nummularia	Christensen and Omar, 1985
	Beta vulgaris	Wagner and Sternkopf, 1958
	Chenopodium ambrosioides	Bogacheva *et al.* 1972
	C. anthelmintica	Chirva *et al.* 1971
	C. quinoa	Mizui *et al.* 1988
	Climacoptera transoxana	Annaev and Abubakirov, 1984
	Salsola micranthera	Annaev *et al.* 1983
	Spinacea oleracea	Tschesche *et al.* 1969e
Combretaceae	*Combretum edwardsii*	Rogers, 1989
	C. elaeagnoides	Osborne and Pegel, 1984
	C. imberbe	Rogers, 1988
	C. molle	Rogers and Thevan, 1986
	Terminalia arjuna	Honda *et al.* 1976
	T. bellerica	Nandy *et al.* 1989
	T. chebula	Kundu and Mahato, 1993
Cornaceae	*Griselinia scandens*	Silva *et al.* 1968
Cucurbitaceae	*Actinostemma lobatum*	Iwamoto *et al.* 1987
	Bryonia dioica	Hylands and Kosugi, 1982
	Bulbostemma paniculatum	Kasai *et al.* 1986a
	Cucurbita foetidissima	Dubois *et al.* 1988b

Appendix 1. (*Cont.*)

Family	Species	Reference
Cucurbitaceae (*cont.*)	*Echinocystis fabacea*	Bergsteinsson and Noller, 1934
	Fevillea cordifolia	Achenbach *et al.* 1987
	Gynostemma compressum	Ding and Zhu, 1993
	G. pentaphyllum	Kuwahara *et al.* 1989
	Hemsleya carnosiflora	Kasai *et al.* 1987c
	H. chinensis	Nie *et al.* 1984
	H. graciliflora	Kasai *et al.* 1990
	H. macrosperma	Nie *et al.* 1984
	Lagenaria breviflora	Elujoba *et al.* 1990
	Luffa acutangula	Nagao *et al.* 1991
	L. aegyptica	Varshney and Beg, 1977
	L. cylindrica	Yoshikawa *et al.* 1991c
	L. echinata	Khorana and Raisinghani, 1961
	L. graveolens	Sehgal *et al.* 1961
	L. operculata	Okabe *et al.* 1989
	L. purgans	Boiteau *et al.* 1964
	Marah macrocarpus	Hylands and Salama, 1979
	Momordica charantia	Okabe *et al.* 1980
	M. cochinchinensis	Kawamura *et al.* 1988
	M. grosvenorii	Takemoto *et al.* 1983c
	Neoalsomitra integrifolia	Minghua *et al.* 1992
	Thladiantha dubia	Nagao, 1989a
	T. hookeri	Nie *et al.* 1989
Cyperaceae	*Cyperus rotundus*	Singh and Singh, 1980
Dilleniaceae	*Dillenia pentagyna*	Tiwari *et al.* 1980
Dipsacaceae	*Cephalaria gigantea*	Zviadadze *et al.* 1981
	C. kotchyi	Aliev and Movsumov, 1976
	Dipsacus asper	Kouno *et al.* 1990
	D. azureus	Mukhamedziev *et al.* 1971
	D. laciniatus	Alimbaeva *et al.* 1977
	Pterocephalus bretschneidri	Tian *et al.* 1993a
	P. hookeri	Tian *et al.* 1993b
	Scabiosa bipinnata	Kurilchenko *et al.* 1971
	S. micrantha	Alimbaeva *et al.* 1977
	S. ochroleuca	Alimbaeva *et al.* 1977
	S. oliveri	Alimbaeva *et al.* 1977
	S. soongorica	Akimaliev *et al.* 1976
	Triplostegia grandiflora	Ma *et al.* 1991
Ebenaceae	*Diospyros zombensis*	Gafner *et al.* 1987
Empetraceae	*Empetrum sibiricum*	Bukharov and Karneeva, 1970b
Ericaceae	*Lyonia ovalifolia*	Sakakibara *et al.* 1974
	Oxycoccus quadripetalus	Shatilo, 1971
	Rhododendron ungernii	Dekanosidze, 1973
	Vaccinium macrocarpon	Shatilo, 1971

(*cont.*)

Appendix 1. (*Cont.*)

Family	Species	Reference
Euphorbiaceae	*Euphorbia geniculata*	Tripathi and Tiwari, 1980
	E. sequieriana	Soboleva and Boguslavskaya, 1987
	Glochidion heyneanum	Srivastava and Kulshreshtha, 1988
	Putranjiva roxburghii	Hariharan, 1974
Eupteleaceae	*Euptelea polyandra*	Murata *et al.* 1970
Fagaceae	*Castanopsis cuspidata*	Ageta *et al.* 1988
	C. hystrix	Chen *et al.* 1993a
	Fagus sylvatica	Romussi *et al.* 1987
	Quercus cerris	Romussi *et al.* 1988b
	Q. ilex	Romussi *et al.* 1983
	Q. laurifolia	Romussi *et al.* 1993
	Q. suber	Romussi *et al.* 1991
Flacourtiaceae	*Aphloia madagascariensis*	Dijoux *et al.* 1993
	A. theiformis	Gopalsamy *et al.* 1988
Fouquieriaceae	*Fouquieria peninsularis*	Giral and Hahn, 1960
Gentianaceae	*Swertia cincta*	Zhang and Mao, 1984
Gramineae	*Avena sativa*	Begley *et al.* 1986
Hippocastanaceae	*Aesculus hippocastanum*	Wagner and Hoffmann, 1967
	A. indica	Singh *et al.* 1987
	A. turbinata	Yosioka *et al.* 1971
Hypoxidaceae	*Curculigo orchioides*	Xu *et al.* 1992a
Iridaceae	*Crocosmia crocosmiiflora*	Nagamoto *et al.* 1988
	C. masonorum	Asada *et al.* 1989b
Juglandaceae	*Cyclocarya paliurus* (*Pterocarya paliurus*)	Yang *et al.* 1992
Labiatae	*Clinopodium chinense*	Xue *et al.* 1992a
	C. micranthum	Yamamoto *et al.* 1993
	C. polycephalum	Xue *et al.* 1992b
	Collinsonia canadensis	Joshi *et al.* 1992
	Leucas nutans	Hasan *et al.* 1991
	Micromeria chamissonis	Boiteau *et al.* 1964
	Orthosiphon stamineus	Trofimova *et al.* 1985
	Prunella grandiflora	Sendra, 1963
	P. vulgaris	Sendra, 1963
Lardizabalaceae	*Akebia quinata*	Fujita *et al.* 1974
	Boquilla trifoliata	Silva and Stück, 1962
	Decaisnea fargesii	Kong *et al.* 1993
	Holboellia angustifolia	Mitra and Karrer, 1953
	H. latifolia	Mitra and Karrer, 1953
	Lardizabala biternata	Silva and Mancinelli, 1961
	Stauntonia chinensis	Wang *et al.* 1989c
	S. hexaphylla	Takemoto and Kometani, 1965
Lecythidaceae	*Barringtonia acutangula*	Pal *et al.* 1991
	Careya arborea	Mahato *et al.* 1973

Appendix 1. (*Cont.*)

Family	Species	Reference
	Petersianthus macrocarpus	Massiot *et al.* 1992a
Leguminosae	*Abrus cantoniensis*	Sakai *et al.* 1990
	A. precatorius	Choi *et al.* 1989
	Acacia auriculiformis	Mahato *et al.* 1989
	A. concinna	Sharma and Walia, 1983
	A. intsia	Farooq *et al.* 1961
	A. leucophloea	Mishra and Srivastava, 1985
	A. myrtifolia	Eade *et al.* 1973
	A. suaveolens	Boiteau *et al.* 1964
	Adenanthera pavonina	Yadav *et al.* 1976
	Albizia amara	Varshney and Shamsuddin, 1962
	A. anthelmintica	Tschesche and Kämmerer, 1969
	A. chinensis	Rawat *et al.* 1989
	A. julibrissin	Sergienko *et al.* 1977
	A. lebbek	Shrivastava and Saxena, 1988
	A. lucida	Orsini *et al.* 1991
	A. odoratissima	Varshney and Khan, 1961
	A. procera	Banerji *et al.* 1979b
	Arachis hypogaea	Price *et al.* 1988
	Astragalus alopecurus	Agzamova and Isaev, 1991
	A. amarus	Isaev and Imomnazarov, 1991
	A. coluteocarpus	Imomnazarov and Isaev, 1992
	A. complanatus	Cui *et al.* 1992a
	A. ernestii	Wang *et al.* 1989b
	A. falcatus	Alaniya, 1988
	A. galegiformis	Alaniya *et al.* 1984
	A. glycyphyllos	Elenga *et al.* 1987
	A. iliensis	Chen *et al.* 1990b
	A. membranaceus	Kitagawa *et al.* 1983a
	A. mongholicus	Zhu *et al.* 1992
	A. orbiculatus	Agzamova *et al.* 1987a
	A. pamirensis	Agzamova *et al.* 1986
	A. pterocephalus	Agzamova *et al.* 1986
	A. schachirudensis	Isaev *et al.* 1988
	A. sieversianus	Gan *et al.* 1986b
	A. sinicus	Cui *et al.* 1992b
	A. spinosus	El-Sebakhy *et al.* 1990
	A. taschkendicus	Isaev *et al.* 1983c
	A. tomentosus	El-Sebakhy *et al.* 1990
	A. tragacantha	Fadeev *et al.* 1988

(*cont.*)

Appendix 1. (*Cont.*)

Family	Species	Reference
Leguminosae (*cont.*)	*A. trigonus*	El-Sebakhy and Waterman, 1985
	A. villosissimus	Isaev and Imomnazarov, 1991
	Canavalia ensiformis	Mukharya, 1985
	Caragana sinica	Lee *et al.* 1992
	Castanospermum australe	Rao *et al.* 1969
		Ahmad *et al.* 1992
	Cicer arietium	Ireland and Dziedzic, 1987
	Crotalaria albida	Ding *et al.* 1991a
	Cyamopsis tetragonoloba	Curl *et al.* 1986
	Cylicodiscus gabunensis	Pambou Tchivounda *et al.* 1990
	Dalbergia hupeana	Yahara *et al.* 1989a
	Derris urucu	Parente and Mors, 1980
	Desmodium styracifolium	Kubo *et al.* 1989
	Dolichos falcata	Feng *et al.* 1985
	D. kilimandscharicus	Marston *et al.* 1988a
	Entada phaseoloides	Okada *et al.* 1987
	Enterolobium contorstisiliquum	Delgado *et al.* 1986
	E. cyclocarpum	Dominguez *et al.* 1979
	Galega officinalis	Fukunaga *et al.* 1987
	Gleditsia australis	Ciulei *et al.* 1966
	G. fera	Dang Tam, 1967
	G. horrida	Kubota and Matsushima, 1928
	G. japonica	Okada *et al.* 1980
	G. triacanthus	Badalbaeva *et al.* 1974
	Glycine max	Kudou *et al.* 1992
	Glycyrrhiza aspera	Kiryalov *et al.* 1973
	G. echinata	Semenchenko and Muravev, 1968
	G. glabra	Spinks and Fenwick, 1990
	G. inflata	Kitagawa *et al.* 1989b
	G. macedonica	Kiryalov and Bogatkina, 1969
	G. pallidiflora	Kiryalov, 1969
	G. uralensis	Kitagawa *et al.* 1988c
	G. yunnanensis	Ohtani *et al.* 1992b
	Gymnocladus chinensis	Konoshima *et al.* 1985
	G. dioica	Parkhurst *et al.* 1980
	Lens culinaris	Price *et al.* 1988
	Lotus corniculatus	Walter, 1961
	Lupinus verbasciformis	Nakano *et al.* 1976
	Medicago hispida	Mahato, 1991
	M. sativa	Massiot *et al.* 1988b
	Melilotus alba	Nicollier and Thompson, 1983

Appendix 1. (*Cont.*)

Family	Species	Reference
Leguminosae	*M. officinalis*	Kang *et al.* 1987
(*cont.*)	*Meristotropis triphylla*	Shnulin *et al.* 1972
	M. xanthoides	Chirva *et al.* 1970e
	Milletia laurentii	Kapundu *et al.* 1978
	Mimosa tenuiflora	Jiang *et al.* 1991
	Mora excelsa	Barton and Brooks, 1951
	M. gonggrijpii	Laidlaw, 1954
	Mucuna birdwoodiana	Ding *et al.* 1991b
	Ononis repens	Boiteau *et al.* 1964
	O. spinosa	Boiteau *et al.* 1964
	Ougeinia dalbergioides	Ghosh and Dutta, 1965
	Oxytropis bicolor	Sun *et al.* 1991
	O. glabra	Sun *et al.* 1991
	O. ochrocephala	Sun *et al.* 1991
	Parkia biglandulosa	Garg and Oswal, 1993
	Periandra dulcis	Hashimoto *et al.* 1983
	Phaseolus aureus	Price *et al.* 1988
	P. coccineus	Curl *et al.* 1988a
	P. lunatus	Price *et al.* 1988
	P. radiatus	Chirva *et al.* 1970f
	P. vulgaris	Jain *et al.* 1988
	Pisum sativum	Tsurumi *et al.* 1992
	Pithecellobium arboreum	Ripperger *et al.* 1981
	P. cubense	Ripperger *et al.* 1981
	P. saman	Varshney and Khanna, 1978
	Podalyria tinctoria	Boiteau *et al.* 1964
	Pueraria lobata	Kinjo *et al.* 1988
	Robinia pseudo-acacia	Cui *et al.* 1992c
	Rothia trifoliata	Kitagawa *et al.* 1983f
	Sesbania aculeata	Simes *et al.* 1959
	S. aegyptica	Farooq *et al.* 1959
	S. sesban	Dorsaz *et al.* 1988
	Soja hispida	Kitagawa *et al.* 1976b
	Sophora flavescens	Yoshikawa *et al.* 1985
	S. japonica	Kitagawa *et al.* 1988e
	S. subprostrata	Ding *et al.* 1992b
	Spartium junceum	Bilia *et al.* 1993
	Strongylodon macrobotrys	Inada *et al.* 1987a
	Stryphnodendron coriaceum	Tursch *et al.* 1963
	Swartzia madagascariensis	Borel and Hostettmann, 1987
	S. simplex	Borel *et al.* 1987
	Taverniera aegyptiaca	Hasanean *et al.* 1992b
	Tetrapleura tetraptera	Maillard *et al.* 1989
	Trifolium alpinum	Boiteau *et al.* 1964
	T. ragiferum	Walter, 1960
	T. repens	Sakamoto *et al.* 1992

(*cont.*)

Appendix 1. (*Cont.*)

Family	Species	Reference
Leguminosae	*Vicia faba*	Price *et al.* 1988
(*cont.*)	*Vigna angularis*	Kitagawa *et al.* 1983d
	Wisteria brachybotrys	Konoshima *et al.* 1991
	W. sinensis	Tiwari and Choudhary, 1979
Liliaceae	*Chionodoxa gigantea*	Mimaki *et al.* 1992
	C. luciliae	Adinolfi *et al.* 1993
	Eucomis bicolor	Mimaki *et al.* 1992
	Ophiopogon japonicus	Adinolfi *et al.* 1990
	Scilla peruviana	Mimaki *et al.* 1992
Loganiaceae	*Buddleja japonica*	Yamamoto *et al.* 1991
	B. officinalis	Ding *et al.* 1992a
	Desfontainia spinosa	Houghton and Lian, 1986b
Malvaceae	*Pavonia zeylanica*	Tiwari and Minocha, 1980
Meliaceae	*Amoora rohituka*	Jain and Srivastava, 1984
	Aphanamixis polystachya	Srivastava and Agnihotri, 1983
	Dysoxylum cumingianum	Kashiwada *et al.* 1992
	Lansium domesticum	Nishizawa *et al.* 1983
Melianthaceae	*Bersama yangambiensis*	Vanhaelen, 1972
Menispermaceae	*Diploclisia glaucescens*	Bandara *et al.* 1990
Menyanthaceae	*Menyanthes trifoliata*	Janeczko *et al.* 1990
Molluginaceae	*Mollugo hirta*	Barua *et al.* 1976
	M. nudicaulis	Sosa, 1962
	M. spergula	Barua *et al.* 1986
Moraceae	*Streblus asper*	Chaturvedi and Saxena, 1985
Myristicaceae	*Myristica fragrans*	Varshney and Sharma, 1968
Myrsinaceae	*Aegiceras corniculatum*	Hensens and Lewis, 1965
	A. majus	Hensens and Lewis, 1965
	Ardisia crenata	Wang *et al.* 1992
	A. crispa	Jansakul *et al.* 1987
	A. japonica	De Tommasi *et al.* 1993
	A. neriifolia	Malaviya *et al.* 1989
	Maesa chisia	Gomes *et al.* 1987
	Myrsine australis	Cambie and Couch, 1967
	Rapanea melanophloeos	Ohtani and Hostettmann, 1992
Myrtaceae	*Barringtonia acutangula*	Barua *et al.* 1972
	B. asiatica	Ito *et al.* 1967
	B. racemosa	Ito *et al.* 1967
	Callistemon lanceolatus	Younes, 1975
Olacaceae	*Olax andronensis*	Forgacs and Provost, 1981
	O. angustifolia	Delaude and Davreux, 1973
	O. glabriflora	Forgacs and Provost, 1981
	O. psittacorum	Forgacs and Provost, 1981
Opiliaceae	*Opilia celtidifolia*	Shihata *et al.* 1977
Passifloraceae	*Passiflora edulis*	Bombardelli *et al.* 1975

Appendix 1. (*Cont.*)

Family	Species	Reference
	P. quadrangularis	Orsini *et al.* 1986
Pedaliaceae	*Sesamum alatum*	Potterat *et al.* 1992
Phytolaccaceae	*Phytolacca acinosa*	Harkar *et al.* 1984
	P. americana	Kang and Woo, 1987
	P. bogotensis	Haraguchi *et al.* 1988
	P. dirceflora	Texeira *et al.* 1984
	P. dodecandra	Parkhurst *et al.* 1974
	P. esculenta	Yi, 1990
	P. rivinoides	Haraguchi *et al.* 1988
	P. thyrsiflora	Haraguchi *et al.* 1988
Pittosporaceae	*Pittosporum nilghrense*	Jain *et al.* 1980
	P. phyllyraeoides	Knight and White, 1961
	P. tobira	Yosioka *et al.* 1972
	P. undulatum	Higuchi *et al.* 1983
Polemoniaceae	*Polemonium acutifolium*	Stecka, 1968
	P. caeruleum	Hiller *et al.* 1979
	P. occidentale	Stecka, 1968
	P. pauciflorum	Stecka, 1968
	P. reptans	Jurenitsch *et al.* 1979
Polygalaceae	*Bredemeyera floribunda*	Tschesche and Striegler, 1965
	Muraltia ononidifolia	Delaude, 1990b
	Nylandtia spinosa	Delaude, 1990c
	Polygala chamaebuxus	Hamburger and Hostettmann, 1986
	P. chinensis	Brieskorn and Kilbinger, 1975
	P. exelliana	Delaude, 1973
	P. japonica	Fang and Ying, 1986
	P. myrtifolia	Delaude, 1990a
	P. paenea	Polonsky *et al.* 1960
	P. senega	Tsukitani and Shoji, 1973
	P. tenuifolia	Sakuma and Shoji, 1982
	Xanthophyllum octandrum	Simes *et al.* 1959
Polypodiaceae	*Polypodium occidentale*	Fischer and Goodrich, 1930
	P. vulgare	Fischer and Goodrich, 1930
Portulacaceae	*Talinum tenuissimum*	Gafner *et al.* 1985
	T. triangulare	Kohda *et al.* 1992
Primulaceae	*Anagallis arvensis*	Glombitza and Kurth, 1987b
	Androsace saxifragifolia	Waltho *et al.* 1986
	A. septentrionalis	Kintia and Pirozhkova, 1982
	Cyclamen africanum	Reznicek *et al.* 1989a
	C. coum	Reznicek *et al.* 1989a
	C. europaeum	Tschesche *et al.* 1969c
	C. graecum	Reznicek *et al.* 1989a

(*cont.*)

Appendix 1. (*Cont.*)

Family	Species	Reference
Primulaceae (*cont.*)	*C. neapolitanum* (*C. hederifolium*)	Reznicek *et al.* 1989a
	Lysimachia clethroides	Kitagawa *et al.* 1972b
	L. japonica	Kitagawa *et al.* 1972b
	L. mauritiana	Kitagawa *et al.* 1972b
	L. paridiformis	Han *et al.* 1987
	L. sikokiana	Kohda *et al.* 1989
	L. thyrsiflora	Karpova *et al.* 1975
	Primula auriculata	Calis, 1987, 1989
	P. denticulata	Ahmad *et al.* 1988b
	P. elatior	Calis *et al.* 1992
	P. japonica	Kitagawa *et al.* 1972b
	P. longipes	Calis, 1987, 1989
	P. macrophylla	Ahmad *et al.* 1993b
	P. megaseifolia	Calis *et al.* 1987, 1989
	P. sieboldi	Kitagawa *et al.* 1980b
	P. turkestanica	Zakharov *et al.* 1968
	P. macrocalyx	Zakharov *et al.* 1971
	P. veris	Calis *et al.* 1992
Ranunculaceae	*Actaea racemosa*	Corsano *et al.* 1969
	A. spicata	Fardella and Corsano, 1973
	Anemone chinensis (*Pulsatilla chinensis*)	Huang *et al.* 1962
	A. coronaria	Janeczko *et al.* 1992
	A. flaccida	Zhao *et al.* 1990
	A. hupehensis	Wang *et al.* 1993
	A. narcissiflora	Masood *et al.* 1981
	A. obtusiloba	Tiwari and Masood, 1980
	A. raddeana	Wu *et al.* 1989
	A. ranunculoides	Figurkin and Ogurtsova, 1975
	A. rivularis	Mizutani *et al.* 1984
	Beesia calthaefolia	Inoue *et al.* 1985
	Caltha palustris	Bhandari *et al.* 1987
	C. polypetala	Vugalter *et al.* 1987
	C. silvestris	Strigina *et al.* 1972
	Cimicifuga acerina	Takemoto *et al.* 1970
	C. dahurica	Sakurai *et al.* 1972
	C. japonica	Takemoto *et al.* 1970
	C. racemosa	Linde, 1964
	C. simplex	Takemoto *et al.* 1970
	Clematis apiifolia	Woo *et al.* 1976
	C. chinensis	Kizu and Tomimori, 1982
	C. grata	Uniyal and Sati, 1992
	C. mandshurica	Kizu and Tomimori, 1982
	C. montana	Bahuguna *et al.* 1989
	C. orientalis	Dekanosidze *et al.* 1979
	C. paniculata	Boiteau *et al.* 1964

Appendix 1. (*Cont.*)

Family	Species	Reference
Ranunculaceae	*C. songarica*	Krokhmalyuk *et al.* 1975
(*cont.*)	*C. vitalba*	Chirva *et al.* 1978
	Ficaria ranunculoides	Texier *et al.* 1984
	Nigella sativa	Ansari *et al.* 1988
	Pulsatilla campanella	Li *et al.* 1990c
	P. cernua	Shimizu *et al.* 1978
	P. chinensis (*Anemone chinensis*)	Huang *et al.* 1962
	P. dahurica	Zinova *et al.* 1992
	P. nigricans	Kurilenko, 1968
	Souliea vaginata	Inoue *et al.* 1985
	Thalictrum aquilegifolium	Ina *et al.* 1985
	T. foeniculaceum	Yi and Wu, 1991
	T. foetidum	Ganenko *et al.* 1986a
	T. minus	Gromova *et al.* 1985
	T. squarrosum	Khamidullina *et al.* 1989
	T. thunbergii	Yoshimitsu *et al.* 1992
Rhamnaceae	*Alphitonia excelsa*	Boiteau *et al.* 1964
	A. zizyphoides	Perera *et al.* 1993
	Ampelozizyphus amazonicus	Lins Brandao *et al.* 1992
	Colubrina asiatica	Wagner *et al.* 1983
	Emmenospermum alphitonioides	Eade *et al.* 1965
	Gouania lupuloides	Kennelly *et al.* 1993
	Hovenia dulcis	Inoue *et al.* 1978
	Paliurus ramosissimus	Lee *et al.* 1991
	Ziziphus joazeiro (*Zizyphus joazeiro*)	Higuchi *et al.* 1984
	Z. jujuba	Yoshikawa *et al.* 1991b
	Z. mauritiana	Sharma and Kumar, 1982
	Z. nummularia	Sharma and Kumar, 1983
	Z. rugosa	Kulshreshtha and Rastogi, 1972
	Z. spinosus	Zeng *et al.* 1987
	Z. vulgaris	Ikram *et al.* 1981
Rosaceae	*Agrimonia asiatica*	Ibragimov and Khazanovich, 1972
	Bencomia caudata	Reher and Budesinsky, 1992
	Crataegus pentagyna	Iskenderov and Isaev, 1974
	Dendriopoterium menendezii	Reher and Budesinsky, 1992
	D. pulidoi	Reher and Budesinsky, 1992
	Geum japonicum	Shigenaga *et al.* 1985
	Marcetalla moquiniana	Reher and Budesinsky, 1992
	M. maderensis	Reher and Budesinsky, 1992
	Potentilla anserina	Pailer and Berner, 1967
	P. tormentilla	Bilia *et al.* 1992
	Poterium lasiocarpum	Shukyurov *et al.* 1974
	P. polygamum	Shukyurov *et al.* 1974
	P. sanguisorba	Pourrat and Pourrat, 1961
	Quillaja saponaria	Higuchi *et al.* 1988

(*cont.*)

Appendix 1. *(Cont.)*

Family	Species	Reference
Rosaceae *(cont.)*	*Rosa canina*	Pourrat and Pourrat, 1961
	R. gallica	Bugorskii and Melnikov, 1977
	R. multiflora	Du *et al.* 1983
	R. rugosa	Young *et al.* 1987
	Rubus accuminatus	Zhou *et al.* 1992
	R. babae	Seto *et al.* 1984
	R. chorchorifolius	Seto *et al.* 1984
	R. coreanus	Ohtani *et al.* 1990
	R. crataegifolius	Ohtani *et al.* 1990
	R. ellipticus	Zhou *et al.* 1992
	R. illecebrosus	Seto *et al.* 1984
	R. koehneanus	Ohtani *et al.* 1990
	R. medicus	Seto *et al.* 1984
	R. microphyllus	Ohtani *et al.* 1990
	R. multibreatus	Zhou *et al.* 1992
	R. nishimuranus	Seto *et al.* 1984
	R. parvifolius	Ohtani *et al.* 1990
	R. pseudoacer	Seto *et al.* 1984
	R. pseudojaponicus	Seto *et al.* 1984
	R. sieboldii	Seto *et al.* 1984
	R. suavissimus	Ohtani *et al.* 1992a
	R. trifidus	Seto *et al.* 1984
	R. utchinensis	Seto *et al.* 1984
	Sanguisorba alpina	Jia *et al.* 1993
	S. minor	Reher and Budesinsky, 1992
	S. officinalis	Reher and Budesinsky, 1992
	Sarcopoterium spinosum	Reher *et al.* 1991
Rubiaceae	*Adina pilulifera*	Hui and Yee, 1968
	Anthocephalus cadamba	Banerji, 1977
	Antirrhoea chinensis	Hui and Szeto, 1967
	Brenania brieyi	Menu *et al.* 1976
	Canthium diococcum	Mukherjee and Bose, 1975
	Cinchona calisaya	Tschesche *et al.* 1963
	Crossopteryx febrifuga	Babady-Bila *et al.* 1991
	Gardenia latifolia	Reddy *et al.* 1975
	G. lutea	Ahmed *et al.* 1985
	Guettarda angelica	Matos *et al.* 1986
	G. platypoda	Aquino *et al.* 1988
	Isertia haenkeana	Javier Arriaga *et al.* 1990
	Mussaenda pubescens	Xu *et al.* 1992b
	Nauclea orientalis	Fujita *et al.* 1967
	Randia brandis	Gedeon, 1952
	R. canthioides	Hui and Ho, 1968
	R. dumetorum (Xeromphis spinosa)	Sotheeswaran *et al.* 1989
	R. sinensis	Hui and Ho, 1968
	R. tetrasperma	Quershi and Thakur, 1977

Appendix 1. (*Cont.*)

Family	Species	Reference
Rubiaceae (*cont.*)	*R. uliginosa*	Sati *et al.* 1989
	Uncaria guianensis	Yepez *et al.* 1991
	U. tomentosa	Aquino *et al.* 1989b
	Vangueria spinosa	Boiteau *et al.* 1964
	V. tomentosa	Brieskorn and Wunderer, 1967
	Xeromphis spinosa (*Randia dumetorum*)	Sati *et al.* 1987
Santalaceae	*Exocarpus cupressiformis*	Simes *et al.* 1959
Sapindaceae	*Blighia sapida*	Garg and Mitra, 1967
	B. welwitschii	Penders *et al.* 1989
	Chytranthus macrobotrys	Delaude *et al.* 1976
	Dodonaea viscosa	Wagner *et al.* 1987
	Jagera pseudorhus	Simes *et al.* 1959
	Koelreuteria paniculata	Chirva *et al.* 1970a
	Lecaniodiscus cupanioides	Encarnacion *et al.* 1981
	Majidea fosteri	Kapundu *et al.* 1984
	Sapindus delavayi	Nakayama *et al.* 1986a
	S. emarginatus	Row and Rukmini, 1966
	S. laurifolius	Maiti *et al.* 1968
	S. mukurossi	Kimata *et al.* 1983
	S. rarak	Hamburger *et al.* 1992
	S. saponaria	Boiteau *et al.* 1964
	S. surinamensis	Boiteau *et al.* 1964
	S. trifoliatus	Kasai *et al.* 1988b
	S. utilis	Boiteau *et al.* 1964
	Serjania lethalis	Texeira *et al.* 1984
	Thinouia coriacea	Schenkel *et al.* 1991
	Xanthoceras sorbifolia	Chen *et al.* 1985
	Zanha golungensis	Dimbi *et al.* 1987
Sapotaceae	*Argania spinosa*	Charrouf *et al.* 1992
	Bassia butyracea	Boiteau *et al.* 1964
	B. parkii	Heywood *et al.* 1939
	Chrysophyllum glycyphloeum	Boiteau *et al.* 1964
	Dumoria heckelii	Heywood and Kon, 1940
	Madhuca butyracea	Nigam *et al.* 1992
	M. longifolia (*Bassia longifolia*)	Kitagawa *et al.* 1978b
	Mimusops djave	Heywood and Kon, 1940
	M. elengi	Varshney and Badhwar, 1971
	M. heckelii	King *et al.* 1955
	M. hexandra	Heywood and Kon, 1940
	Palaquium gutta	Heywood and Kon, 1940
	Payena lucida	Heywood and Kon, 1940
	Planchonella pohlmanianum	Eade *et al.* 1969
	Sideroxylon cubense	Jiang *et al.* 1993
	S. tomentosum	Gedeon and Kincl, 1956

(*cont.*)

Appendix 1. (*Cont.*)

Family	Species	Reference
Sapotaceae (*cont.*)	*Tridesmostemon claessenssi*	Massiot *et al.* 1990
Sargentodoxaceae	*Sargentodoxa cuneata*	Rücker *et al.* 1991
Saxifragaceae	*Deutzia corymbosa*	Malaviya *et al.* 1991
Scrophulariaceae	*Bacopa monniera*	Kawai and Shibata, 1978
	Digitalis ciliata	Kemertelidze *et al.* 1992
	Gratiola officinalis	Tschesche and Heesch, 1952
	Scrophularia smithii	Breton and Gonzalez, 1963
	Verbascum lychnitis	Klimek *et al.* 1992
	V. nigrum	Klimek *et al.* 1992
	V. phlomoides	Klimek *et al.* 1992
	V. songaricum	Seifert *et al.* 1991
Sterculiaceae	*Asteropus sarasinosum*	Kitagawa *et al.* 1987a
Styracaceae	*Styrax japonica*	Kitagawa *et al.* 1980b
	S. officinalis	Anil, 1979
Symplocaceae	*Symplocos spicata*	Higuchi *et al.* 1982
Theaceae	*Camellia japonica*	Nagata *et al.* 1985
	C. oleifera	Sokolski *et al.* 1975
	C. sasanqua	Sokolski *et al.* 1975
	C. sinensis (*Thea sinensis*)	Tschesche *et al.* 1969a
	Schima kankoensis	Cole *et al.* 1955
	S. mertensiana	Kitagawa *et al.* 1980b
	Stewartia monadelpha	Ogino *et al.* 1968
	Ternstroemia japonica	Saeki *et al.* 1968
Theophrastaceae	*Jacquinia armillaris*	Maheas, 1961
	J. pungens	Hahn *et al.* 1965
Tiliaceae	*Corchorus acutangulus*	Mahato and Pal, 1987
	C. capsularis	Quader *et al.* 1990
Umbelliferae	*Astrantia major*	Hiller *et al.* 1990b
	Bupleurum chinensis	Kimata *et al.* 1979
	B. falcatum	Kimata *et al.* 1979
	B. fruticosum	Pistelli *et al.* 1993
	B. kunmingense	Luo *et al.* 1987
	B. longeradiatum	Kimata *et al.* 1979
	B. marginatum	Ding *et al.* 1986
	B. polyclonum	Seto *et al.* 1986
	B. rockii	Ding *et al.* 1986
	B. rotundifolium	Akai *et al.* 1985b
	B. smithii	Chen *et al.* 1993b
	B. wenchuanense	Luo *et al.* 1993
	Centella asiatica (*Hydrocotyle asiatica*)	Mahato *et al.* 1987
	Eryngium biebersteinianum	Ikramov *et al.* 1971
	E. bromeliifolium	Hiller *et al.* 1987b
	E. giganteum	Hiller *et al.* 1975
	E. incognitum	Serova, 1961
	E. macrocalyx	Ikramov *et al.* 1976
	E. maritimum	Hiller *et al.* 1976

Appendix 1. (*Cont.*)

Family	Species	Reference
Umbelliferae	*E. octophyllum*	Ikramov *et al.* 1971
(*cont.*)	*E. planum*	Voigt *et al.* 1985
	Hydrocotyle vulgaris	Hiller *et al.* 1981
	Ladyginia bucharica	Adler and Hiller, 1985
	Peucedanum decursivum	Okuyama *et al.* 1989
	P. praeruptorum	Takata *et al.* 1990
	Sanicula europaea	Kühner *et al.* 1985
	Steganotaenia araliacea	Lavaud *et al.* 1992
Valerianaceae	*Patrinia intermedia*	Bukharov *et al.* 1969
	P. scabiosaefolia	Choi and Woo, 1987
	P. sibirica	Bukharov and Karlin, 1970d
	P. villosa	Fujita and Furuya, 1954
Verbenaceae	*Clerodendrum serratum*	Rangaswami and Sarangan, 1969
	C. wildii	Toyota *et al.* 1990
	Lippia rehmanni	Boiteau *et al.* 1964
Zygophyllaceae	*Fagonia indica*	Ansari *et al.* 1987
	F. cretica	Rimpler and Rizk, 1969
	Guaiacum officinale	Ahmad *et al.* 1986b
	Larrea divaricata	Habermehl and Möller, 1974
	Putranjiva roxburghii	Seshadri and Rangaswami, 1975
	Zygophyllum album	Hasanean *et al.* 1992a
	Z. fabago	Kerimova *et al.* 1971
	Z. obliquum	Apsamatova *et al.* 1979
	Z. propinquum	Ahmad *et al.* 1990

Appendix 2

Appendix 2. *Structures of triterpene saponins (cf. p. 21, Table 2.1)*

No.	Name	Structure	Plant	Reference
Oleanolic acid aglycone (2)				
127	Fatsiaside A$_1$	3-Ara	*Fatsia japonica* (Araliaceae)	Kemoklidze *et al.* 1982
			Thinouia coriacea (Sapindaceae)	Schenkel *et al.* 1991
128		3-Ara	*Lonicera japonica* (Caprifoliaceae)	Kawai *et al.* 1988
		28-Glc^6-Glc		
129		3-Gal	*Xeromphis spinosa* (Rubiaceae)	Saharia and Seshadri, 1980
130		3-Gal	*Lagenaria breviflora* (Cucurbitaceae)	Elujoba *et al.* 1990
		28-Ara^3-Xyl^3-Rha^4-Xyl		
131		3-Gal	*L. breviflora* (Cucurbitaceae)	Elujoba *et al.* 1990
		28-Ara^3-Xyl^3-Rha^4-Gal		
132		3-Gal	*L. breviflora* (Cucurbitaceae)	Elujoba *et al.* 1990
		28-Ara^3-Xyl^3-Rha^4-Gal^6-Ara		
133		3-GalA	*Wedelia scaberrima* (Asteraceae)	Matos and Tomassini, 1983
		28-Glc	*Beta vulgaris* (Chenopodiaceae)	Kretsu and Lazurevskii, 1970
134	Androseptoside	3-Glc	*Chenopodium quinoa* (Chenopodiaceae)	Ma *et al.* 1989
			Androsace septentrionalis (Primulaceae)	Kintia and Pirozhkova, 1982
			Calendula officinalis (Asteraceae)	Lutomski, 1983

No.	Name	Substituents	Source (Family)	Reference
135	Lucyoside H	3-Glc, 28-Glc	*Luffa cylindrica* (Cucurbitaceae)	Takemoto et al. 1985
			L. cylindrica (Cucurbitaceae)	Takemoto et al. 1984
			Silphium perfoliatum (Asteraceae)	Davidyants et al. 1984a
			Anchusa officinalis (Boraginaceae)	Romussi et al. 1980
				Romussi and de Tommasi, 1993
136	Hypoleucoside A	3-Glc, 11-OMe, 28-Glc	*Acanthopanax hypoleucus* (Araliaceae)	Kohda et al. 1990
138	Clemontanoside A	3-Glc, 28-Glc^6-Rha $\overset{|^2}{\text{Glc}}$	*Clematis montana* (Ranunculaceae)	Bahuguna et al. 1989
139		28-Glc	*Hemsleya macrosperma* (Cucurbitaceae)	Nie et al. 1984
			H. chinensis (Cucurbitaceae)	Nie et al. 1984
			Macropanax disperum (Araliaceae)	Srivastava and Jain, 1989
140	Calenduloside E (saponoside D)	3-GlcA	*Beta vulgaris* (Chenopodiaceae)	Marsh and Levvy, 1956
				Wagner and Sternkopf, 1958
			Lonicera nigra (Caprifoliaceae)	Domon and Hostettmann, 1983
				Lutomski, 1983
			Calendula officinalis (Asteraceae)	Vidal-Ollivier et al. 1989b

(cont.)

Appendix 2. (*Cont.*)

No.	Name	Structure	Plant	Reference
Oleanolic acid aglycone (2)				
141	Chikusetsu-saponin IVa	3-GlcA 28-Glc	*Hedera nepalensis* (Araliaceae)	Kizu *et al.* 1985c
			Polyscias scutellaria (Araliaceae)	Paphassarang *et al.* 1988, 1989b
			Boussingaultia baselloides (Basellaceae)	Espada *et al.* 1991
			Hemsleya chinensis (Cucurbitaceae)	Nie *et al.* 1984
			Panax spp. (Araliaceae)	Tanaka and Kasai, 1984
			Swartzia simplex (Leguminosae)	Borel *et al.* 1987
			Calendula officinalis (Asteraceae)	Vidal-Ollivier *et al.* 1989a,b
			Chenopodium quinoa (Chenopodiaceae)	Mizui *et al.* 1990
			Boussingaultia baselloides (Basellaceae)	Espada *et al.* 1991
			Cynara cardunculus (Asteraceae)	Shimizu *et al.* 1988
142		3-Xyl	*Xeromphis spinosa* (Rubiaceae)	Saluja and Santani, 1986
143	Hederasaponin F	3-OSO_3^- $28\text{-Glc}^6\text{-Glc}^4\text{-Rha}$	*Hedera helix* (Araliaceae)	Elias *et al.* 1991
144		$3\text{-Ara}^2\text{-Glc}$	*Lonicera nigra* (Caprifoliaceae)	Domon and Hostettmann, 1983

No.	Name	Structure	Source	Reference
145		3-Ara^2-Glc 28-Glc^6-Glc	*L. japonica* (Caprifoliaceae) *L. nigra* (Caprifoliaceae)	Kawai et al. 1988 Domon and Hostettmann, 1983
146	Ciwujianoside A	3-Ara^2-Glc 28-Glc^6-Glc^4-Rha	*L. japonica* (Caprifoliaceae) *Acanthopanax senticosus* (Araliaceae)	Kawai et al. 1988 Shao et al., 1989a
147	Guaiacin B	3-Ara^3-Glc 28-Glc	*Guaiacum officinale* (Zygophyllaceae)	Ahmad et al. 1989a
148	Scabioside B	3-Ara^4-Glc	*Patrinia scabiosaefolia* (Valerianaceae) *Thinouia coriacea* (Sapindaceae)	Bukharov et al. 1970 Schenkel et al. 1991
149	Scabioside D	3-Ara^4-Glc 28-Xyl	*Patrinia scabiosaefolia* (Valerianaceae)	Bukharov and Karlin, 1970a
150	Scabioside E	3-Ara^4-Glc 28-Xyl^4-Rha	*P. scabiosaefolia* (Valerianaceae)	Bukharov and Karlin, 1970a
151	Scabioside F	3-Ara^4-Glc 28-Xyl^4-Rha^3-Xyl	*P. scabiosaefolia* (Valerianaceae)	Bukharov and Karlin, 1970b
152	Scabioside G	3-Ara^4-Glc 28-Xyl^4-Rha^3-Xyl^4-Glc	*P. scabiosaefolia* (Valerianaceae)	Bukharov and Karlin, 1970c
153	Tauroside C (eleutheroside K)	3-Ara^2-Man	*Hedera taurica* (Araliaceae)	Loloiko et al. 1988
154	β-Hederin	3-Ara^2-Rha	*H. helix* (Araliaceae) *H. nepalensis* (Araliaceae) *Eleutherococcus senticosus* (Araliaceae) *Caltha palustris* (Ranunculaceae)	Tschesche et al. 1965 Kizu et al. 1985c Frolova and Ovodov, 1971 Bhandari et al. 1987

(cont.)

Appendix 2. (*Cont.*)

No.	Name	Structure	Plant	Reference
Oleanolic acid aglycone (2)				
155		3-Ara²-Rha 28-Glc⁶-Glc	*Thinouia coriacea* (Sapindaceae) *Lonicera japonica* (Caprifoliaceae)	Schenkel *et al.* 1991 Kawai *et al.* 1988
156	Hederasaponin B (hederacoside B) (eleutheroside M)	3-Ara²-Rha 28-Glc⁶-Glc⁴-Rha	*Hedera helix* (Araliaceae) *Eleutherococcus senticosus* (Araliaceae)	Tschesche *et al.* 1965 Frolova and Ovodov, 1971
157	Mubenin B	3-Ara⁴-Rha	*Stauntonia hexaphylla* (Lardizabalaceae) *Eleutherococcus senticosus* (Araliaceae)	Takemoto and Kometani, 1965 Frolova and Ovodov, 1971
158		3-Gal²-Fuc	*Tetrapanax papyriferum* (Araliaceae)	Takabe *et al.* 1985
159		3-Gal³-Glc	*Xeromphis spinosa* (Rubiaceae)	Sati *et al.* 1987
160		3-Gal⁴-Rha	*Spilanthes acmella* (Asteraceae)	Mukharya and Ansari, 1987
161	Arvensoside B	3-Glc³-Gal	*Calendula arvensis* (Asteraceae)	Chemli *et al.* 1987; Pizza *et al.* 1987
162	Arvensoside A	3-Glc³-Gal 28-Glc	*C. arvensis* (Asteraceae)	Chemli *et al.* 1987; Pizza *et al.* 1987
163	Calenduloside A	3-Glc⁴-Gal	*C. officinalis* (Asteraceae)	Wojciechowski *et al.* 1970
164	Calenduloside B	3-Glc⁴-Gal 28-Glc	*C. officinalis* (Asteraceae)	Vecherko *et al.* 1971

No.	Name	Structure	Source (Family)	Reference
165		3-Glc^2-Glc	*Passiflora quadrangularis* (Passifloraceae)	Orsini *et al.* 1987
			Silphium perfoliatum (Asteraceae)	Davidyants *et al.* 1984b
			Luffa acutangula (Cucurbitaceae)	Nagao *et al.* 1991
166	Acutoside B	3-Glc^2-Glc 28-Ara^2-Rha^4-Xyl	*L. acutangula* (Cucurbitaceae)	Nagao *et al.* 1991
167	Acutoside D	3-Glc^2-Glc 28-Ara^2-Rha^4-Xyl^3-Xyl	*L. acutangula* (Cucurbitaceae)	Nagao *et al.* 1991
168	Acutoside E	3-Glc^2-Glc 28-Ara^2-Rha^4-Xyl^3-Ara	*L. acutangula* (Cucurbitaceae)	Nagao *et al.* 1991
169	Acutoside F	3-Glc^2-Glc 28-Ara^2-Rha^4-Xyl $\quad\quad\quad\ \overset{\mid^3}{\text{Xyl}}$	*L. acutangula* (Cucurbitaceae)	Nagao *et al.* 1991
170	Acutoside G	3-Glc^2-Glc 28-Ara^2-Rha^4-Xyl^3-Ara $\quad\quad\quad\ \overset{\mid^3}{\text{Xyl}}$	*L. acutangula* (Cucurbitaceae)	Nagao *et al.* 1991
171		3-Glc^3-Glc	*Randia dumetorum* (Rubiaceae)	Sotheeswaran *et al.* 1989
172	Anchusoside-7	3-Glc^3-Glc 28-Glc	*Anchusa officinalis* (Boraginaceae)	Romussi *et al.* 1980; Romussi and De Tommasi, 1993
173	Cynarasaponin H	3-GlcA^2-Ara 28-Glc	*Cynara cardunculus* (Asteraceae)	Shimizu *et al.* 1988
174	Momordin I	3-GlcA^3-Ara	*Momordia cochinchinensis* (Cucurbitaceae)	Kawamura *et al.* 1988

(cont.)

Appendix 2. (*Cont.*)

No.	Name	Structure	Plant	Reference
Oleanolic acid aglycone (2)				
175	Hemsloside Ma1 (Momordin II)	3-GlcA³-Ara 28-Glc	*Hemsleya macrosperma* (Cucurbitaceae)	Nie *et al.* 1984
			H. chinensis (Cucurbitaceae)	Nie *et al.* 1984
			Momordica cochinchinensis (Cucurbitaceae)	Kawamura *et al.* 1988
176		3-GlcA⁴-Ara*f*	*Anemone narcissiflora* (Ranunculaceae)	Masood *et al.* 1981
			Cussonia spicata (Araliaceae)	Gunzinger *et al.* 1986
177	Chikusetsusaponin IV (araloside A)	3-GlcA⁴-Ara*f* 28-Glc	*Aralia mandschurica* (Araliaceae)	Kochetkov *et al.* 1962
			Panax japonicum (Araliaceae)	Kondo *et al.* 1969
			Other *Panax* species (Araliaceae)	Tanaka and Kasai, 1984
178		3-GlcA⁴-Ara*f* 28-Glc⁶-Glc⁴-Rha	*Tetrapanax papyriferum* (Araliaceae)	Takabe *et al.* 1985
179		3-GlcA²-Gal 28-Glc	*T. papyriferum* (Araliaceae)	Takabe *et al.* 1985
180	Saponoside D	3-GlcA³-Gal	*Calendula arvensis* (Asteraceae)	Pizza.*et al.* 1987
			C. officinalis	Vidal-Ollivier *et al.* 1989b
181	Saponoside C	3-GlcA³-Gal 28-Glc	*C. officinalis* (Asteraceae)	Kasprzyk and Wojciechowski, 1967
			C. arvensis	Pizza *et al.* 1987

182	Hederasaponin H	3-GlcA[4]-Gal 28-Glc[6]-Glc[4]-Rha	*Hedera helix* (Araliaceae)	Elias *et al.* 1991
183	Zingibroside R$_1$	3-GlcA[2]-Glc	*Panax zingiberensis* (Araliaceae)	Yang *et al.* 1984
			Polyscias scutellaria (Araliaceae)	Paphassarang *et al.* 1989a, 1990
184	Chikusetsusaponin V (ginsenoside Ro)	3-GlcA[2]-Glc 28-Glc	*Panax japonicus* (Araliaceae) And other *Panax* species	Kondo *et al.* 1971 Tanaka and Kasai, 1984
			Fagus silvatica (Fagaceae)	Romussi *et al.* 1987
			Polyscias scutellaria (Araliaceae)	Paphassarang *et al.* 1989c
185	Spinasaponin A	3-GlcA[3]-Glc	*Spinacea oleracea* (Chenopodiaceae)	Tschesche *et al.* 1969e
			Polyscias scutellaria (Araliaceae)	Paphassarang *et al.* 1989b
			Kalopanax septemlobus (Araliaceae)	Shao *et al.* 1989b
186		3-GlcA[3]-Glc 28-Glc	*Fagus silvatica* (Fagaceae)	Romussi *et al.* 1987
			Polyscias scutellaria (Araliaceae)	Paphassarang *et al.* 1989b
187		3-GlcA[4]-Glc	*Calendula officinalis* (Asteraceae)	Lutomski, 1983
			Ladyginia bucharica (Umbelliferae)	Patkhulaeva *et al.* 1972
188		3-GlcA[4]-Glc 28-Glc	*Swartzia simplex* (Leguminosae)	Borel *et al.* 1987
189		3-GlcA[3]-Rha	*Swartzia madagascariensis* (Leguminosae)	Borel and Hostettmann, 1987

(cont.)

Appendix 2. (*Cont.*)

No.	Name	Structure	Plant	Reference
Oleanolic acid aglycone (2)				
			Swartzia simplex (Leguminosae)	Borel *et al.* 1987
			Diospyros zombensis (Ebenaceae)	Gafner *et al.* 1987
			Putranjiva roxburghii (Euphorbiaceae)	Hariharan, 1974
			Zexmenia buphthalmiflora (Asteraceae)	Schteingart and Pomilio, 1984
			Sesbania sesban (Leguminosae)	Dorsaz *et al.* 1988
			Deeringia amaranthoides (Amaranthaceae)	Sati *et al.* 1990
190		3-GlcA³-Rha 28-Glc	*Putranjiva roxburghii* (Euphorbiaceae)	Hariharan, 1974
			Zexmania buphthalmiflora (Asteraceae)	Schteingart and Pomilio, 1984
			Diospyros zombensis (Ebenaceae)	Gafner *et al.* 1987
			Swartzia simplex (Leguminosae)	Borel *et al.* 1987
			Sesbania sesban (Leguminosae)	Dorsaz *et al.* 1988
			Deeringia amaranthoides (Amaranthaceae)	Sati *et al.* 1990

No.	Compound	Structure	Source (Family)	Reference
191		3-GlcA^3-Rha 28-Glc^2-Xyl	*Deeringia amaranthoides* (Amaranthaceae)	Sati *et al.* 1990
192	Olaxoside	3-GlcA^4-Rha 28-Glc	*Olax andronensis* (Olacaceae)	Forgacs and Provost, 1981
			O. glabriflora (Olacaceae)	Forgacs and Provost, 1981
			O. psittacorum (Olacaceae)	Forgacs and Provost, 1981
193	Pseudo-ginsenoside RT$_1$	3-GlcA^2-Xyl 28-Glc	*Panax pseudoginseng* (Araliaceae)	Morita *et al.* 1986a
194		3-GlcA^3-Xyl	*Talinum tenuissimum* (Portulacaceae)	Gafner *et al.* 1985
			Chenopodium quinoa (Chenopodiaceae)	Ma *et al.* 1989
195		3-GlcA^3-Xyl 28-Glc	*Talinum tenuissium* (Portulacaceae)	Gafner *et al.* 1985
			Chenopodium quinoa (Chenopodiaceae)	Ma *et al.* 1989; Mizui *et al.* 1990
196	Salsoloside D	3-GlcA^4-Xyl 28-Glc	*Salsola micranthera* (Chenopodiaceae)	Annaev *et al.* 1983
197	Hishoushi saponin A	3-Ara^2-Rha^3-Ara	*Sapindus delavayi* (Sapindaceae)	Nakayama *et al.* 1986a
			Patrinia scabiosaefolia (Valerianaceae)	Choi and Woo, 1987
198		3-Ara^3-Glc^2-Glc 28-Glc	*Chenopodium quinoa* (Chenopodiaceae)	Mizui *et al.* 1990
199	Guaianin C	3-Ara^3-Glc^3-Rha 28-Glc	*Guaiacum officinale* (Zygophyllaceae)	Ahmad *et al.* 1989b
200	Thalicoside B	3-Ara^3-Glc^2-Glc 28-Glc	*Thalictrum minus* (Ranunculaceae)	Gromova *et al.* 1985

(*cont.*)

Appendix 2. (*Cont.*)

No.	Name	Structure	Plant	Reference
Oleanolic acid aglycone (2)				
201		3-Ara^2-Rha^3-Glc	*Thinouia coriacea* (Sapindaceae)	Schenkel *et al.* 1991
202		3-Ara^4-Glc $\vert^2$ Rha	*T. coriacea* (Sapindaceae)	Schenkel *et al.* 1991
203		3-Ara^2-Rha^3-Glc	*Patrinia scabiosaefolia* (Valerianaceae)	Choi and Woo, 1987
204	Kalopanax saponin-D	3-Ara^2-Rha $\vert^3$ Glc 28-Glc^6-Glc	*Kalopanax septemlobus* (Araliaceae)	Shao *et al.* 1989b
205	Hederacholchiside E	3-Ara^2-Rha $\vert^4$ Glc 28-Glc^6-Glc^4-Rha	*Hedera colchica* (Araliaceae)	Dekanosidze *et al.* 1970a
206		3-Ara^2-Rha^3-Rib	*Anemone rivularis* (Ranunculaceae)	Mizutani *et al.* 1984
207	Huzhangoside B	3-Ara^2-Rha^3-Rib 28-Glc^6-Glc^4-Rha	*A. rivularis* (Ranunculaceae)	Mizutani *et al.* 1984
208		3-Ara^2-Rha^3-Xyl	*Sapindus delavayi* (Sapindaceae)	Nakayama *et al.* 1986a
209	Foetoside C	3-Ara^2-Rha^3-Xyl 28-Glc-Glc	*Thalictrum foetidum* (Ranunculaceae)	Ganenko *et al.* 1988
210		3-Gal^3-Glc^2-Glc	*Xeromphis spinosa* (Rubiaceae)	Sati *et al.* 1987

211	Hypoleucoside B	$3\text{-Glc}^4\text{-Ara}^2\text{-Glc}$ $28\text{-Glc}^6\text{-Glc}$	*Acanthopanax hypoleucus* (Araliaceae)	Kohda *et al.* 1990
212		$3\text{-Glc}^4\text{-Gal}$ $\mid^3$ Glc	*Calendula officinalis* (Asteraceae)	Lutomski, 1983
213		$3\text{-Glc}^4\text{-Glc}$ $\mid^3$ Gal 28-Glc	*C. arvensis* (Asteraceae)	De Tommasi *et al.* 1991
214	Calenduloside C	$3\text{-Glc}^3\text{-Gal}$ $\mid^2$ Glc	*C. officinalis* (Asteraceae) *C. arvensis*	Vecherko *et al.* 1975 Vidal-Ollivier *et al.* 1989c
215	Calenduloside D	$3\text{-Glc}^3\text{-Gal}$ $\mid^2$ Glc 28-Glc	*C. officinalis* (Asteraceae) *C. arvensis*	Vecherko *et al.* 1975 Vidal-Ollivier *et al.* 1989c
216	Anchusoside-2	$3\text{-Glc}^3\text{-Glc}$ $\mid^2$ Glc	*Anchusa officinalis* (Boraginaceae)	Romussi *et al.* 1980
217	Oleanoglycotoxin-A	$3\text{-Glc}^4\text{-Glc}$ $\mid^2$ Glc	*Phytolacca dodecandra* (Phytolaccaceae)	Parkhurst *et al.* 1973a Domon and Hostettmann, 1984
218		$3\text{-Glc}^4\text{-Glc}$ $\mid^2$ Glc 28-Glc	*P. dodecandra* (Phytolaccaceae)	Domon and Hostettmann, 1984
219	Lemmatoxin	$3\text{-Glc}^4\text{-Glc}$ $\mid^3$ Gal	*P. dodecandra* (Phytolaccaceae)	Parkhurst *et al.* 1974

(*cont.*)

Appendix 2. (*Cont.*)

No.	Name	Structure	Plant	Reference
Oleanolic acid aglycone (2)				
220	Lemmatoxin C	3-Glc^2-Glc^2-Rha	*P. dodecandra* (Phytolaccaceae)	Parkhurst *et al.* 1973b
221	Hederacaucaside B	28-Glc^6-Glc^4-Rha	*Hedera caucasigena* (Araliaceae)	Dekanosidze *et al.* 1970b
			Cussonia barteri (Araliaceae)	Dubois *et al.* 1986
222		3-Glc^6-Rha^2-Glc	*Xeromphis spinosa* (Rubiaceae)	Sati *et al.* 1986
223	Helianthoside 1	3-Glc^4-Xyl $\vert^2$ Rha 28-Ara^2-Rha^4-Glc	*Helianthus annuus* (Asteraceae)	Bader *et al.* 1991
224	Araloside B	3-GlcA^4-Ara$f$ $\vert^3$ Ara$f$ 28-Glc	*Aralia mandschurica* (Araliaceae)	Kochetkov *et al.* 1962
225		3-GlcA^4-Ara$f$ $\vert^2$ Gal	*Tetrapanax papyriferum* (Araliaceae)	Takabe *et al.* 1981
			Cussonia spicata (Araliaceae)	Gunzinger *et al.* 1986
226		3-GlcA^4-Ara$f$ $\vert^2$ Gal 28-Glc	*Tetrapanax payriferum* (Araliaceae)	Takabe *et al.* 1985

No.	Name	Structure	Source	Reference
227		3-GlcA^4-Gal $\vert^3$ Ara	*Randia uliginosa* (Rubiaceae)	Sati *et al.* 1989
228	Hemsloside Ma3	3-GlcA^3-Ara $\vert^2$ Glc 28-Glc	*Hemsleya macrosperma* (Cucurbitaceae) *H. chinensis* (Cucurbitaceae)	Nie *et al.* 1984 Nie *et al.* 1984
229	Kalopanax saponin-F	3-GlcA^2-Ara $\vert^3$ Glc 28-Glc	*Kalopanax septemlobus* (Araliaceae)	Shao *et al.*, 1989b
230	Hemsloside H$_1$	3-GlcA^3-Ara $\vert^2$ Glc 28-Glc^6-Glc	*Hemsleya chinensis* (Cucurbitaceae)	Morita *et al.* 1986b
231	Stipuleanoside R$_1$	3-GlcA^4-Ara $\vert^3$ Glc	*Panax stipuleanatus* (Araliaceae)	Yang *et al.* 1985
232	Stipuleanoside R$_2$	3-GlcA^4-Ara $\vert^3$ Glc 28-Glc	*P. stipuleanatus* (Araliaceae)	Yang *et al.* 1985
233	Hemsloside Ma2	3-GlcA^3-Ara $\vert^2$ Xyl 28-Glc	*Hemsleya macrosperma* (Cucurbitaceae)	Nie *et al.* 1984
234	Saponoside B	3-GlcA^4-Glc $\vert^3$ Gal	*Calendula officinalis* (Asteraceae)	Lutomski, 1983 Vidal-Ollivier *et al.* 1989a,b

(cont.)

Appendix 2. (*Cont.*)

No.	Name	Plant	Structure	Reference
Oleanolic acid aglycone (2)				
235	Saponoside A	*C. officinalis* (Asteraceae)	3-GlcA⁴-Glc |³ Gal 28-Glc	Kasprzyk and Wojciechowski, 1967; Vidal-Ollivier et al. 1989b
236	Araloside C	*Aralia mandschurica* (Araliaceae)	3-GlcA⁴-Xyl |³ Gal 28-Glc	Kochetkov and Khorlin, 1966
237	Salsoloside E	*Salsola micranthera* (Amaranthaceae)	3-GlcA⁴-Xyl |² Glc 28-Glc	Annaev et al. 1984
238	Polysciasaponin P₂	*Polyscias scutellaria* (Araliaceae)	3-GlcA²-Glc⁴-Glc 28-Glc	Paphassarang et al. 1989a–c
239	Polysciasaponin P₁	*P. scutellaria* (Araliaceae)	3-GlcA²-Glc⁴-Glc 28-Glc	Paphassarang et al., 1990
240	Putranjiasaponin III	*Putranjiva roxburghii* (Zygophyllaceae)	3-GlcA³-Glc²-Rha 28-Glc	Rangaswami and Seshadri, 1971
241		*Swartzia madagascariensis* (Leguminosae)	3-GlcA³-Rha |² Glc	Borel and Hostettmann, 1987
242		*S. madagascariensis* (Leguminosae) *Sesbania sesban* (Leguminosae)	3-GlcA³-Rha |² Glc 28-Glc	Borel and Hostettmann, 1987 Dorsaz et al. 1988

No.	Name	Structure	Species (Family)	Reference

243		3-GlcA^3-Rha
		   |^2
		   Xyl
		28-Glc
			Diospyros zombensis
			(Ebenaceae)
			Swartzia simplex
			(Leguminosae)
				Gafner *et al.* 1987
				Borel *et al.* 1987

244	Momordin Id	3-GlcA^3-Xyl
		   |^2
		   Xyl
		28-Glc
			Momordica cochinchinensis
			(Cucurbitaceae)
				Kawamura *et al.* 1988

245	Momordin IId	3-GlcA^3-Xyl
		   |^2
		   Xyl
		28-Glc
			M. cochinchinensis
			(Cucurbitaceae)
				Kawamura *et al.* 1988

246	Patrinoside Cl	3-Rha^2-Glcf^2-Xyl	*Patrinia intermedia* (Valerianaceae)	Bukharov *et al.* 1969
247	Patrinoside C	3-Rha^3-Glcf^2-Xyl 28-Glc^4-Glcf^2-Glc	*P. intermedia* (Valerianaceae)	Bukharov *et al.* 1969
248	Huzhangoside A	3-Xyl^2-Rha^3-Rib	*Anemone rivularis* (Ranunculaceae)	Mizutani *et al.* 1984
249	Huzhangoside C	3-Xyl^2-Rha^3-Rib 28-Glc^6-Glc^4-Rha	*A. rivularis* (Ranunculaceae)	Mizutani *et al.* 1984
250	Mubenin A	3-Ara^4-Rha^2-Glc^6-Glc	*Stauntonia hexaphylla* (Lardizabalaceae)	Takemoto and Kometani, 1965

251		3-Ara^4-Glc
		   |^2
		   Rha^3-Xyl
			Thinouia coriacea
			(Sapindaceae)
				Schenkel *et al.* 1991

252		3-Ara^4-Glc
		   |^2
		   Rha^3-Glc
			T. coriacea (Sapindaceae)	Schenkel *et al.* 1991

(cont.)

Appendix 2. (*Cont.*)

No.	Name	Structure	Plant	Reference
Oleanolic acid aglycone (2)				
253	Guaianin G	3-Ara^3-Glc $\vert^2$ Rha^3-Rha 28-Glc^6-Glc	*Guaiacum officinale* (Zygophyllaceae)	Ahmad *et al.* 1989d
254		3-Glc^4-Glc $\vert^2$ Glc^2-Rha	*Phytolacca dodecandra* (Phytolaccaceae)	Dorsaz and Hostettmann, 1986
255		3-Glc^4-Glc $\vert^2$ Glc^2-Rha 28-Glc	*P. dodecandra* (Phytolaccaceae)	Dorsaz and Hostettmann, 1986
256		3-Glc^6-Rha^2-Glc $\vert^2$ Glc	*Xeromphis spinosa* (Rubiaceae)	Sati *et al.* 1986
257	Putranjia-saponin IV	3-GlcA^3-Glc^2-Rha $\vert^4$ Xyl 28-Glc	*Putranjiva roxburghii* (Zygophyllaceae)	Rangaswami and Seshadri, 1971
258	Patrinoside D$_1$	3-Xyl^4-Rha^3-Glc $\vert^2$ Xyl	*Patrinia intermedia* (Valerianaceae)	Bukharov *et al.* 1969
259	Patrinoside D	3-Xyl^4-Rha^3-Glc $\vert^2$ Xyl 28-Glc^4-Glcf^2-Glc	*P. intermedia* (Valerianaceae)	Bukharov *et al.* 1969

	Name	Structure	Species (Family)	Reference
260	Clematoside A′	3-Rha^4-Ara^2-Ara^2-Xyl^4-Glc	*Clematis mandschurica* (Ranunculaceae)	Chirva and Konyukhov, 1969
261	Clematoside A	3-Rha^4-Ara^2-Ara^2-Xyl^4-Glc 28-Glc^6-Glc^4-Glc^4-Rha	*C. mandschurica* (Ranunculaceae)	Chirva and Konyukhov, 1968a
262		3-Xyl^2-Glc^6-Glc^6-Glc $\quad\quad\quad\quad\mid^3$ $\quad\quad\quad\quad$Ara	*Gundelia tournefortii* (Asteraceae)	Wagner *et al.* 1984b
263	Clematoside B	3-Rha^4-Ara^2-Ara^2-Xyl^4-Glc 28-Glc^6-Glc^4-Glc^4-Rha	*Clematis mandschurica* (Ranunculaceae)	Chirva and Konyukhov, 1968b
264		3-Xyl^2-Glc^6-Glc^6-Glc $\quad\quad\quad\quad\mid^3$ $\quad\quad\quad\quad$Ara	*Gundelia tournefortii* (Asteraceae)	Wagner *et al.* 1984b
265	Clematoside C	3-Rha^4-Ara^2-Ara^2-Xyl^4-Glc^4-Glc^6-Rha 28-Glc^6-Glc^4-Glc^4-Rha	*Clematis mandschurica* (Ranunculaceae)	Khorlin *et al.* 1967
266	Mimonoside A	3-Glc^2-Xyl^3-Glc^2-Rha $\quad\quad\mid^4\quad\mid^4$ $\quad\quad$Xyl Ara $\quad\quad$28-Rha	*Mimosa tenuiflora* (Leguminosae)	Jiang *et al.* 1991
267	Mimonoside B	3-Glc^2-Xyl^3-Glc^2-Rha $\quad\quad\mid^4\quad\mid^4$ $\quad\quad$Xyl Ara	*M. tenuiflora* (Leguminosae)	Jiang *et al.* 1991
Epikatonic acid aglycone (3)				
268		3-Glc^2-Glc^4-Rha $\quad\quad\quad\mid^2$ $\quad\quad\quad$Rha 29-Glc^2-Glc	*Cyamopsis tetragonobola* (Leguminosae)	Curl *et al.* 1986
Glycyrrhetinic acid aglycone (6)				
269	Glycyrrhizin	3-GlcA^2-GlcA	*Glycyrrhiza glabra* (Leguminosae)	Lythgoe and Trippett, 1950

(cont.)

Appendix 2. (*Cont.*)

No.	Name	Structure	Plant	Reference
Glycyrrhetinic acid aglycone (6)				
270	Licorice-saponin A3	3-GlcA^2-GlcA 30-Glc	*G. inflata* (Leguminosae) *G. uralensis* (Leguminosae)	Kitagawa *et al.* 1989b Kitagawa *et al.* 1988c
271	Uralsaponin A	3-GlcA^2-GlcA	*G. uralensis* (Leguminosae)	Zhang *et al.* 1986
272	Uralsaponin B	3-GlcA^3-GlcA	*G. uralensis* (Leguminosae)	Zhang *et al.* 1986
273	Apioglycyrrhizin	3-GlcA^2-Api.f	*G. inflata* (Leguminosae)	Kitagawa *et al.* 1989b
274	Araboglycyrrhizin	3-GlcA^2-Ara	*G. inflata* (Leguminosae)	Kitagawa *et al.* 1989b
Gypsogenin aglycone (7)				
275		3-Glc	*Dianthus superbus* (Caryophyllaceae) *Luffa cylindrica* (Cucurbitaceae)	Shimizu and Takemoto, 1967 Takemoto *et al.* 1985
276	Lucyoside F	3-Glc 28-Glc	*L. cylindrica* (Cucurbitaceae)	Takemoto *et al.* 1984
277		3-GlcA		
		3-GlcA 28-Glc	*Saponaria officinalis* (Caryophyllaceae) *Swartzia simplex* (Leguminosae)	Chirva and Kintya, 1969 Borel *et al.* 1987
278				
279	Saponoside A	3-GlcA 28-Glc^6-Glc $\mid^2$ Glc	*Saponaria officinalis* (Caryophyllaceae)	Chirva and Kintya, 1969
280	Lucyoside L	3-Glc^2-Glc 28-Glc	*Luffa cylindrica* (Cucurbitaceae)	Takemoto *et al.* 1985

281	Lucyoside P	3-GlcA^2-Gal 28-Ara^2-Rha^4-Xyl $\mid^3$ Glc	*L. cylindrica* (Cucurbitaceae)	Yoshikawa *et al.* 1991c
282	Thladioside H1	3-GlcA^2-Gal 28-Xyl^2-Rha^4-Xyl^3-Xyl	*Thladiantha hookeri* (Cucurbitaceae)	Nie *et al.* 1989
283	Trichoside A	3-GlcA^3-Glc 28-Gal	*Gypsophila trichotoma* (Caryophyllaceae)	Luchanskaya *et al.* 1971a
284	Trichoside B	3-GlcA^3-Glc 28-Gal^4-Glc	*G. trichotoma* (Caryophyllaceae)	Luchanskaya *et al.* 1971b
285		3-GlcA^2-Glc 28-Fuc^2-Rha^4-Xyl $\mid^3$ Glc	*G. paniculata* (Caryophyllaceae) *G. arrostii* (Caryophyllaceae)	Frechet *et al.* 1991 Frechet *et al.* 1991
286	Trichoside C	3-GlcA^3-Glc 28-Fuc^4-Rha^4-Gal	*G. trichotoma* (Caryophyllaceae)	Luchanskaya *et al.* 1971c
287		3-GlcA^3-Rha	*Swartzia madagascariensis* (Leguminosae)	Borel and Hostettmann, 1987
288		3-GlcA^3-Rha 28-Glc	*S. madagascariensis* (Leguminosae)	Borel and Hostettmann, 1987
289	Luperoside I	3-GlcA^3-Ara $\mid^2$ Gal 28-Qui^2-Rha^4-Xyl $\mid^3$ Rha	*Luffa operculata* (Cucurbitaceae)	Okabe *et al.* 1989

(cont.)

Appendix 2. (*Cont.*)

No.	Name	Structure	Plant	Reference
	Gypsogenin aglycone (7)			
290		3-GlcA^3-Xyl $\vert^2$ Gal 28-Fuc^2-Rha^4-Xyl $\vert^3$ Glc	*Gypsophila paniculata* (Caryophyllaceae) *G. arrostii* (Caryophyllaceae)	Frechet *et al.* 1991 Frechet *et al.* 1991
291	Dianthoside C	3-GlcA^3-Xyl^3-α-Gal Gal $\vert^3$ 28-α-Fuc^4-α-Xyl^2-Rha Rha^2-Ara	*Dianthus deltoides* (Caryophyllaceae)	Bukharov and Shcherbak, 1971
292	Gypsoside A	3-GlcA^4-Glc^4-Gal $\vert^3$ Ara 28-Rha^4-Fuc^3-Xyl $\vert^2$ Xyl^3-Xyl	*Gypsophila paniculata*, *G. pacifica* and other *Gypsophila* species (Caryophyllaceae); *Acanthophyllum* species (Caryophyllaceae)	Kochetkov and Khorlin, 1966; Yukananov *et al.* 1971
293	Githagoside	3-Fuc^2-Rha^4-Xyl $\vert^3$ Glc	*Agrostemma githago* (Caryophyllaceae)	Tschesche and Schultz, 1974

294	Nutanoside	3-GlcA^3-Gal^6-Ara $\quad\vert^4$ $\quad$Xyl 28-Rha^3-Gal^2-Fuc^4-Glc $\quad\quad\quad\quad\vert^4$ $\quad\quad\quad\quad$Glc	*Silene nutans* (Caryophyllaceae)	Bukharov and Karneeva, 1971
295	Saponaside D	3-GlcA^3-Xyl^2-α-Gal $\quad\quad\vert^4\quad\vert^4$ $\quad\quad$Rha$\quad$Ara 28-Fuc^3-Rha^2-Glc $\quad\vert^4\quad\quad\vert^4$ $\quad$Xyl$\quad\quad\alpha$-Gal	*Saponaria officinalis* (Caryophyllaceae) *Gypsophila paniculata* (Caryophyllaceae)	Chirva and Kintya, 1970 Yukananov *et al.* 1971
Gypsogenic acid aglycone (8)				
296	Dianthus-saponin A	3-Glc	*Dianthus deltoides* (Caryophyllaceae)	Bukharov *et al.* 1971
297	Dianoside H	28-Glc	*D. superbus* (Caryophyllaceae)	Oshima *et al.* 1984c
298	Dianoside A	3-Glc 28-Glc	*D. superbus* (Caryophyllaceae)	Oshima *et al.* 1984a
299	Azukisaponin IV	3-Glc 28-Glc^6-Glc	*Vigna angularis* (Leguminosae) *Dianthus superbus* (Caryophyllaceae)	Kitagawa *et al.* 1983c Oshima *et al.* 1984c
300	Dianoside B	3-Glc 28-Glc^6-Glc $\quad\quad\quad\vert^3$ $\quad\quad\quad$Glc	*D. superbus* (Caryophyllaceae)	Oshima *et al.* 1984a
301		3-GlcA	*Climacoptera transoxana* (Chenopodiaceae)	Annanaec and Abubakirov, 1984

(cont.)

Appendix 2. (*Cont.*)

No.	Name	Structure	Plant	Reference
Gypsogenic acid aglycone (8)				
302		3-GlcA 28-Glc	*Swartzia simplex* (Leguminosae)	Borel et al. 1987
303	Saponaroside	3-Xyl	*Saponaria officinalis* (Caryophyllaceae)	Bukharov and Shcherbak, 1969
304	Dianthus-saponin B	3-Glc^6-Glc	*Dianthus deltoides* (Caryophyllaceae)	Bukharov et al. 1971
305	Trichoside A	3-GlcA^3-Glc 28-Gal	*Gypsophila trichotoma* (Caryophyllaceae)	Luchanskaya et al. 1971a
306		3-GlcA^4-Glc^4-Gal $\mid^3$ Ara 28-Rha^4-Fuc^3-Xyl $\mid^2$ Xyl^3-Xyl	*G. pacifica* (Caryophyllaceae)	Kochetkov et al. 1963
Cincholic acid aglycone (9)				
307		3-Qui 28-Glc	*Isertia haenkeana* (Rubiaceae)	Javier Arriaga et al. 1990
Serjanic acid aglycone (10) (30-*O*-methyl spergulagenate)				
308		3-Glc	*Phytolacca thyrsiflora* (Phytolaccaceae)	Haraguchi et al. 1988
309		3-Glc^2-Glc 28-Glc	*P. thyrsiflora* (Phytolaccaceae)	Haraguchi et al. 1988

No.	Name	Structure	Species (Family)	Reference
310		$3\text{-}Glc^3\text{-}Gal^2\text{-}Gal$	*P. thyrsiflora* (Phytolaccaceae)	Haraguchi et al. 1988
311		$3\text{-}Glc^4\text{-}Glc^4\text{-}All$ $28\text{-}Glc$	*P. thyrsiflora* (Phytolaccaceae)	Haraguchi *et al.* 1988
312		$3\text{-}Ara^3\text{-}Glc^2\text{-}Glc$ $28\text{-}Glc$	*Chenopodium quinoa* (Chenopodiaceae)	Mizui *et al.* 1990
Maniladiol aglycone (11)				
313	Sigmoiside A	$3\text{-}Gal$	*Erythrina sigmoidea* (Leguminosae)	Kouam *et al.* 1991
314	Sigmoiside B	$3\text{-}Glc$	*E. sigmoidea* (Leguminosae)	Kouam *et al.* 1991
Sophoradiol aglycone (12)				
315	Azukisaponin I	$3\text{-}GlcA^2\text{-}Glc$	*Vigna angularis* (Leguminosae)	Kitagawa *et al.* 1983c
315a		$3\text{-}GlcA^2\text{-}Ara^2\text{-}Rha$	*Pueraria lobata* (Leguminosae)	Kinjo et al. 1988
316	Kaikasaponin III	$3\text{-}GlcA^2\text{-}Gal^2\text{-}Rha$	*P. lobata* (Leguminosae) *Dalbergia hupeana* (Leguminosae)	Kinjo et al. 1988 Yahara *et al.* 1989a
317		$3\text{-}GlcA^2\text{-}Gal^2\text{-}Xyl$	*Crotalaria albida* (Leguminosae)	Ding *et al.* 1991a
3β,22β-Dihydroxyolean-12-en-29-oic acid aglycone (13)				
318		$3\text{-}GlcA^2\text{-}Gal^2\text{-}Rha$	*Dalbergia hupeana* (Leguminosae)	Yahara *et al.* 1989a
2β-Hydroxyoleanolic acid aglycone (14)				
319	Swericinctoside	$28\text{-}Glc^6\text{-}Glc^2\text{-}Glc$	*Swertia cincta* (Gentianaceae)	Zhang and Mao, 1984

(cont.)

Appendix 2. *(Cont.)*

No.	Name	Structure	Plant	Reference
Maslinic acid aglycone (15)				
320	Lucyoside G	3-Glc 28-Glc	*Luffa cylindrica* (Cucurbitaceae)	Takemoto *et al.* 1984
321		3-GlcA²-Ara*p*	*Mucuna birdwoodiama* (Leguminosae)	Ding *et al.* 1991b
322		3-Glc³-Ara²-Ara	*Cylicodiscus gabunensis* (Leguminosae)	Pambou Tchivounda *et al.* 1991
323		3-Glc³-Ara²-Ara 28-Rha²-Glc	*C. gabunensis* (Leguminosae)	Pambou Tchivounda *et al.* 1991
324		3-Glc³-Ara²-Ara 28-Rha²-Glc⁶-Glc	*C. gabunensis* (Leguminosae)	Pambou Tchivounda *et al.* 1991
Echinocystic acid aglycone (16)				
325		3-Glc	*Albizzia chinensis* (Leguminosae)	Rawat *et al.* 1989
326	Chrysantellin A	3-Glc 28-Xyl²-Rha⁴-Xyl³-Rha	*Chrysanthellum procumbens* (Asteraceae)	Becchi *et al.* 1979
327	Aster saponin Hd	3-Glc 28-Ara²-Rha⁴-Xyl³-Xyl |³ Api*f*	*Aster tataricus* (Asteraceae)	Tanaka *et al.* 1990
328	Foetidissimoside A	3-GlcA 28-Ara²-Rha⁴-Xyl	*Cucurbita foetidissima* (Cucurbitaceae)	Dubois *et al.* 1988b
329		3-Glc²-Rha	*Albizzia chinensis* (Leguminosae)	Rawat *et al.* 1989

No.	Name	Structure	Species	Reference
330	Aster saponin B	3-Glc⁶-Ara 28-Xyl²-Rha³-Apif $\mid$⁴ Ara³-Xyl	*Aster tataricus* (Asteraceae)	Nagao *et al.* 1989b
331	Aster saponin D	3-Glc⁶-Ara 28-Xyl²-Rha³-Apif $\mid$³ Rha Ara³-Xyl	*A. tataricus* (Asteraceae)	Nagao *et al.* 1989b
332	Aster saponin F	3-Glc⁶-Ara 28-Xyl²-Rha⁴-Ara³-Xyl	*A. tataricus* (Asteraceae)	Nagao *et al.* 1989b
333	Mar-saponin A	3-Ara²-Glc⁶-Xyl	*Chenopodium anthelmintica* (Chenopodiaceae)	Chirva *et al.* 1971
334	Mar-saponin B	3-Ara²-Glc⁶-Xyl 28-Rha⁴-Glc	*C. anthelmintica* (Chenopodiaceae)	Chirva *et al.* 1971
335	Helianthoside B	3-Rha⁴-Xyl 28-Ara²-Rha⁴-Glc	*Helianthus annuus* (Asteraceae)	Cheban and Chirva, 1969; Bader *et al.* 1991
336	Helianthoside 2	3-Glc⁴-Xyl $\mid$² Rha 28-Ara²-Rha⁴-Glc	*H. annuus* (Asteraceae)	Bader *et al.* 1991
337	Helianthoside 3	3-Glc⁴-Xyl $\mid$² Rha 28-Glc²-Rha⁴-Glc	*H. annuus* (Asteraceae)	Bader *et al.* 1991
338	Helianthoside A	3-Glc⁶-Xyl $\mid$³ Rha⁴-Rha	*H. annuus* (Asteraceae)	Cheban *et al.* 1969a; Bader *et al.* 1991

(*cont.*)

Appendix 2. (*Cont.*)

No.	Name	Structure	Plant	Reference
Echinocystic acid aglycone (16)				
339	Helianthoside C	3-Glc⁶-Xyl $\vert$³ Rha⁴-Rha 28-Ara²-Rha⁴-Glc	*H. annus* (Asteraceae)	Cheban *et al.* 1969b; Bader *et al.* 1991
340		3-Ara⁴-Ara	*Deutzia corymbosa* (Saxifragaceae)	Malaviya *et al.* 1991
341		3-Ara⁴-Rha⁴-Gal	*D. corymbosa* (Saxifragaceae)	Malaviya *et al.* 1991
342	Musennin A	3-Ara²-Ara²-Ara	*Albizzia anthelmintica* (Leguminosae)	Tschesche and Kämmerer, 1969
343		3-Glc⁶-Ara $\vert$² Glc	*A. lucida* (Leguminosae)	Orsini *et al.* 1991
344		3-Glc⁶-Ara²-Xyl $\vert$² Glc	*A. lucida* (Leguminosae)	Orsini *et al.* 1991
345	Musennin B	3-Ara²-Ara²-Ara³-Glc	*A. anthelmintica* (Leguminosae)	Tschesche and Kämmerer, 1969
Hederagenin aglycone (17)				
346	Koelreuteria-saponin A (fatsiaside B₁)	3-Ara	*Koelreuteria paniculata* (Sapindaceae) *Leontice eversmannii* (Berberidaceae) *Patrinia scabiosaefolia* (Valerianaceae)	Chirva *et al.* 1970a Mzhelskaya and Abubakirov, 1967a Bukharov *et al.* 1970

		Species (Family)	Reference
		Caulophyllum robustum (Berberidaceae)	Murakami *et al.* 1986
		Akebia quinata (Lardizabalaceae)	Higuchi *et al.* 1972
		Fatsia japonica (Araliaceae)	Kemoklidze *et al.* 1982
		Hedera helix (Araliaceae)	Hostettmann, 1980
		H. nepalensis (Araliaceae)	Kizu *et al.* 1985c
		Caltha palustris (Ranunculaceae)	Bhandari *et al.* 1987
		C. polypetala (Ranunculaceae)	Vugalter *et al.* 1986
		Lonicera japonica (Caprifoliaceae)	Kawai *et al.* 1988
		Polyscias dichroostachya (Araliaceae)	Gopalsamy *et al.* 1990
347	3-Ara 28-Glc	*Hedera nepalensis* (Araliaceae)	Kizu *et al.* 1985c
		Chenopodium quinoa (Chenopodiaceae)	Mizui *et al.* 1988
348	Akebia-saponin D 3-Ara 28-Glc^6-Glc	*Akebia quinata* (Lardizabalaceae)	Higuchi and Kawasaki, 1972
		Lonicera nigra (Caprifoliaceae)	Domon and Hostettmann, 1983
		L. japonica (Caprifoliaceae)	Kawai *et al.* 1988
		Patrinia scabiosaefolia (Valerianaceae)	Choi and Woo, 1987
		Acanthopanax hypoleucus (Araliaceae)	Kohda *et al.* 1990

(cont.)

Appendix 2. (Cont.)

No.	Name	Structure	Plant	Reference
Hederagenin aglycone (17)				
349		3-Ara 28-Glc^6-Glc^4-Rha	*Peucedanum praeruptorum* (Umbelliferae)	Takata *et al.* 1990
			Hedera nepalensis (Araliaceae)	Kizu *et al.* 1985c
			Caltha polypetala (Ranunculaceae)	Shashkov and Kemertelidze, 1988
350		3-Glc	*Hedera helix* (Araliaceae)	Elias *et al.* 1991
			H. helix (Araliaceae)	Hostettmann, 1980
			H. nepalensis (Araliaceae)	Kizu *et al.* 1985c
			Luffa cylindrica (Cucurbitaceae)	Takemoto *et al.* 1985
			Dolichos kilimandscharicus (Leguminosae)	Marston *et al.* 1988a
351	Sibiroside A	3-Glc 28-Ara	*Patrinia sibirica* (Valerianaceae)	Bukharov and Karlin, 1970d
352	Lucyoside E	3-Glc 28-Glc	*Luffa cylindrica* (Cucurbitaceae)	Takemoto *et al.* 1984
353		3-Glc 28-Rha	*Caltha polypetala* (Ranunculaceae)	Vugalter *et al.* 1986
354	Quinoside A	3-Glc 23-Glc 28-Ara^3-Glc	*Chenopodium quinoa* (Chenopodiaceae)	Meyer *et al.* 1990
355		3-GlcA	*Lonicera nigra* (Caprifoliaceae)	Domon and Hostettmann, 1983

No.	Compound	Structure	Source (Family)	Reference
356		3-GlcA 28-Glc	*Hedera nepalensis* (Araliaceae)	Kizu et al. 1985c
			Boussingaultia baselloides (Basellaceae)	Espada et al. 1991
357	Hederasaponin I	3-GlcA 28-Glc^6-Glc^4-Rha	*Chenopodium quinoa* (Chenopodiaceae)	Mizui et al. 1990
		3-GlcA^6-OMe	*Hedera helix* (Araliaceae)	Elias et al. 1991
358			*Schefflera impressa* (Araliaceae)	Srivastava and Jain, 1989
			Hedera pastuchovii (Araliaceae)	Iskenderov, 1970
359		3-Rha	*Caltha polypetala* (Ranunculaceae)	Vugalter et al. 1986
360	Caulophyllum-saponin B	3-Ara^2-Glc	*Caulophyllum robustum* (Berberidaceae)	Murakami et al. 1986
			Caltha silvestris (Ranunculaceae)	Strigina et al. 1972
			Akebia quinata (Lardizabalaceae)	Higuchi et al. 1972
			Lonicera nigra (Caprifoliaceae)	Domon and Hostettmann, 1983
			Polyscias dichroostachya (Araliaceae)	Gopalsamy et al. 1990
361		3-Ara^2-Glc 28-Glc	*P. dichroostachya* (Araliaceae)	Gopalsamy et al. 1990
362	Akebia-saponin F	3-Ara^2-Glc 28-Glc^6-Glc	*Akebia quinata* (Lardizabalaceae)	Higuchi and Kawasaki, 1972

(cont.)

Appendix 2. (*Cont.*)

No.	Name	Structure	Plant	Reference
Hederagenin aglycone (17)				
363	Cauloside G	3-Ara^2-Glc 28-Glc^6-Glc^4-Rha	*Lonicera nigra* (Caprifoliaceae) *Acanthopanax hypoleucus* (Araliaceae) *Caltha polypetala* (Ranunculaceae) *Hedera helix* (Araliaceae)	Domon and Hostettmann, 1983 Kohda *et al.* 1990 Shashkov and Kemertelidze, 1988 Elias *et al.* 1991
364	Scabioside C	3-Ara^3-Glc 28-Glc	*Chenopodium quinoa* (Chenopodiaceae)	Mizui *et al.* 1988
365	Scabioside C	3-Ara^4-Glc	*Patrinia scabiosaefolia* (Valerianaceae) *Leontice eversmannii* (Berberidaceae)	Bukharov *et al.* 1970 Mzhelskaya and Abubakirov, 1967a
366	Leontoside D	3-Ara^4-Glc 28-Glc^6-Glc^4-Rha	*L. eversmannii* (Berberidaceae) *Hedera helix* (Araliaceae)	Mzhelskaya and Abubakirov, 1968a Scheidegger and Cherbuliez, 1955
367	Hederasaponin A (hederacoside A)	3-Araf^5-Glc	*H. helix* (Araliaceae)	Tschesche *et al.* 1965; Hostettmann, 1980
368	α-Hederin (sapindoside A) (kalopanax saponin A)	3-Ara^2-Rha	*H. nepalensis* (Araliaceae) *H. taurica* (Araliaceae) *Kalopanax septemlobus* (Araliaceae) *Sapindus mukurossi* (Sapindaceae)	Kizu *et al.* 1985c Shashkov *et al.* 1987 Khorlin *et al.* 1964 Chirva *et al.* 1970c

No.	Saponin	Structure	Source (Family)	Reference
369		3-Ara^2-Rha 28-Glc	S. delavayi (Sapindaceae)	Nakayama et al. 1986a
			Caltha palustris (Ranunculaceae)	Bhandari et al. 1987
			Polyscias dichroostachya (Araliaceae)	Gopalsamy et al. 1990
370	Mukurozi saponin X	3-Ara^2-Rha 28-Glc^2-Glc	Lonicera japonica (Caprifoliaceae)	Kawai et al. 1988
			Sapindus mukurossi (Sapindaceae)	Kimata et al. 1983
371	Dipsacoside B	3-Ara^2-Rha 28-Glc^6-Glc	Dipsacus azureus (Dipsacaceae)	Mukhamedziev et al. 1971
			Lonicera japonica (Caprifoliaceae)	Kawai et al. 1988
			Hedera helix (Araliaceae)	Wagner and Reger, 1986a
			Astrantia major (Umbelliferae)	Hiller et al. 1990b
372	Hederasaponin C (hederacoside C kalopanaxsaponin B cauloside D)	3-Ara^2-Rha 28-Glc^6-Glc^4-Rha	Kalopanax septemlobus (Araliaceae)	Shao et al. 1989
			Hedera helix (Araliaceae)	Tschesche et al. 1965
			H. nepalensis (Araliaceae)	Kizu et al. 1985c
			Astrantia major (Umbelliferae)	Hiller et al. 1990b
373		3-Ara^2-Rha 28-Glc^6-Glc $\quad\mid^2$ $\quad$Rha	A. major (Umbelliferae)	Hiller et al. 1990b
374	Akebia-saponin B	3-Ara^2-Xyl	Akebia quinata (Araliaceae)	Higuchi et al. 1972
375	Akebia-saponin E	3-Ara^2-Xyl 28-Glc^6-Glc	A. quinata (Araliaceae)	Higuchi and Kawasaki, 1972

(cont.)

Appendix 2. (*Cont.*)

No.	Name	Structure	Plant	Reference
Hederagenin aglycone (17)				
376		3-Gal³-Glc 28-Glc	*Chenopodium quinoa* (Chenopodiaceae)	Mizui *et al.* 1988
377		3-Glc²-Glc	*Hedera helix* (Araliaceae)	Hostettmann, 1980
378		28-Glc⁶-Rha	*Caltha polypetala* (Ranunculaceae)	Vugalter *et al.* 1986
379	Spinasaponin B	3-GlcA³-Glc	*Spinacea oleracea* (Chenopodiaceae)	Tschesche *et al.* 1969e
380	Salsoloside C	3-GlcA⁴-Xyl 28-Glc	*Salsola micranthera* (Chenopodiaceae)	Annaev *et al.* 1983
381		3-GlcA³-Xyl 28-Glc	*Chenopodium quinoa* (Chenopodiaceae)	Mizui *et al.* 1990
382	Medicoside C	3-Ara²-Glc²-Ara	*Medicago sativa* (Leguminosae)	Timbekova and Abubakirov, 1985
383	Medicoside I	3-Ara²-Glc²-Ara 28-Glc	*M. sativa* (Leguminosae)	Timbekova and Abubakirov, 1986
384	Leontoside C	3-Ara⁴-Glc |³ Glc	*Leontice eversmannii* (Berberidaceae)	Mzhelskaya and Abubakirov, 1967b
385	Leontoside E	3-Ara⁴-Glc |³ Glc 28-Glc⁶-Glc⁴-Rha	*L. eversmannii* (Berberidaceae)	Mzhelskaya and Abubakirov, 1968b
386		3-Ara²-Rha³-Ara	*Sapindus mukurossi* (Sapindaceae) *S. delavayi* (Sapindaceae)	Kimata *et al.* 1983 Nakayama *et al.* 1986a

387	Mukurozi saponin Y$_2$	3-Ara^2-Rha^3-Ara 28-Glc^2-Glc	*S. mukurossi* (Sapindaceae)	Kimata *et al.* 1983
388		3-Ara^2-Rha^3-Ara$f$	*S. mukurossi* (Sapindaceae) *S. delavayi* (Sapindaceae)	Kimata *et al.* 1983 Nakayama *et al.* 1986a

389	Akebia-saponin G	3-Ara^4-Rha
		   $\mid^2$
		   Glc
			Akebia quinata
			(Lardizabalaceae)
				Higuchi and Kawasaki,
				1972

390	Kalopanax saponin C	28-Glc^6-Glc^4-Rha
		3-Ara^2-Rha
		   $\mid^3$
		   Glc
			Kalopanax septemlobus
			(Araliaceae)
				Shao *et al.* 1989b

391		28-Glc^6-Glc^4-Rha 3-Ara^2-Rha^3-Rib	*Anemone rivularis* (Ranunculaceae)	Mizutani *et al.* 1984
392	Huzhangoside D	3-Ara^2-Rha^3-Rib 28-Glc^6-Glc^4-Rha	*A. rivularis* (Ranunculaceae)	Mizutani *et al.* 1984
393	Sapindoside B	3-Ara^2-Rha^3-Xyl	*Sapindus mukurossi* (Sapindaceae)	Chirva *et al.* 1970c
394	Mukurozi saponin Y$_1$	3-Ara^2-Rha^3-Xyl 28-Glc^2-Glc	*S. delavayi* (Sapindaceae) *S. mukurossi* (Sapindaceae)	Nakayama *et al.* 1986a Kimata *et al.* 1983
395		3-Ara^2-Rha^3-Xyl 28-Glc^6-Glc^4-Rha	*Nigella sativa* (Ranunculaceae)	Ansari *et al.* 1988

396	Sapindoside E	3-Ara^{2}Rha^3-Xyl
		28-Ara^2-Rha^3-Xyl^4-Glc^6-Rha
		       $\mid^2$
		       Glc
			Sapindus mukurossi
			(Sapindaceae)
				Chirva *et al.* 1970b

397	Sibiroside C	3-Glc^4-Ara^4-Glc 28-Ara^4-Glc^4-Glc	*Patrinia sibirica* (Valerianaceae)	Bukharov and Karlin, 1970e

(*cont.*)

Appendix 2. *(Cont.)*

No.	Name	Structure	Plant	Reference
Hederagenin aglycone (17)				
398		3-Glc^4-Glc $\mid^2$ Glc	*Phytolacca dodecandra* (Phytolaccaceae)	Domon and Hostettmann, 1984
399		3-Glc^4-Glc $\mid^2$ Glc 28-Glc	*P. dodecandra* (Phytolaccaceae)	Domon and Hostettmann, 1984
400		28-Glc^6-Glc^4-Rha	*Hedera nepalensis* (Araliaceae)	Kizu *et al.* 1985c
			Cussonia barteri (Araliaceae)	Dubois *et al.* 1986
401	Pastuchoside B	3-Rha^3-α-Glc^4-α-Glc	*Hedera pastuchovii* (Araliaceae)	Iskenderov, 1970
402	Mubenin C	3-Ara^4-Rha^2-Glc^6-Gal	*Stauntonia hexaphylla* (Lardizabalaceae)	Takemoto and Kometani, 1965
403	Sapindoside C	3-Ara^2-Rha^3-Xyl^4-Glc	*Sapindus mukurossi* (Sapindaceae)	Chirva *et al.* 1970e
404		3-Ara^2-Rha^3-Glc^4-Ara^3-Xyl	*Blighia welwitschii* (Sapindaceae)	Penders *et al.* 1989
405	Sapindoside D	3-Ara^2-Rha^3-Xyl^4-Glc^6-Rha $\mid^2$ α-Glc	*Sapindus mukurossi* (Sapindaceae)	Chirva *et al.* 1970d
Phytolaccagenic acid aglycone (18)				
406		3-Ara 28-Glc	*Chenopodium quinoa* (Chenopodiaceae)	Mizui *et al.* 1988

407		3-Glc	*Diploclisia glaucescens* (Menispermaceae)	Bandara *et al.* 1990
408		3-Ara[3]-Glc 28-Glc	*Chenopodium quinoa* (Chenopodiaceae)	Mizui *et al.* 1988
409		3-Gal[3]-Glc 28-Glc	*C. quinoa* (Chenopodiaceae)	Mizui *et al.* 1988
410		3-Ara[3]-Glc[2]-Glc 28-Glc	*C. quinoa* (Chenopodiaceae)	Mizui *et al.* 1990
411	Esculentoside L	3-Xyl[4]-Glc 28-Glc	*Phytolacca esculenta* (Phytolaccaceae)	Yi, 1990
412	Esculentoside K	3-Glc[4]-Xyl[4]-Glc 28-Glc	*P. esculenta* (Phytolaccaceae)	Yi, 1990

Siaresinolic acid aglycone (19)

413	Ilexoside A	3-Xyl	*Ilex chinensis* (Aquifoliaceae)	Inada *et al.* 1987b

21β-Hydroxyoleanolic acid aglycone (20)

414	Lucyoside C	3-Glc 28-Glc	*Luffa cylindrica* (Cucurbitaceae)	Takemoto *et al.* 1984
415	Acutoside C	3-Glc[2]-Glc 28-Ara[2]Rha[4]-Xyl	*L. acutangula* (Cucurbitaceae)	Nagao *et al.* 1991
416	Cynarasaponin I	3-GlcA[2]-Ara	*Cynara cardunculus* (Asteraceae)	Shimizu *et al.* 1988
417	Cynarasaponin J	3-GlcA[2]-Ara	*C. cardunculus* (Asteraceae)	Shimizu *et al.* 1988

29-Hydroxyoleanolic acid aglycone (21)

418		3-Ara 28-Glc	*Ilex cornuta* (Aquifoliaceae)	Qin *et al.* 1986

(cont.)

Appendix 2. (*Cont.*)

No.	Name	Structure	Plant	Reference
29-Hydroxyoleanolic acid aglycone (21)				
419	Ciwujianoside A₃	3-Ara²-Rha 28-Glc⁶-Glc⁴-Rha	*Acanthopanax senticosus* (Araliaceae)	Shao *et al.* 1989a
Azukisapogenol aglycone (22)				
420	Azukisaponin III	3-GlcA²-Glc	*Vigna angularis* (Leguminosae) *Oxytropis glabra* (Leguminosae)	Kitagawa *et al.* 1983c Sun *et al.* 1991
421	Azukisaponin VI	3-GlcA²-Glc 29-Glc⁶-Glc	*Vigna angularis* (Leguminosae)	Kitagawa *et al.* 1983d
Soyasapogenol E aglycone (23)				
422	Dehydrosoyasaponin I	3-GlcA²-Gal²-Rha	*Medicago sativa* (Leguminosae) *Oxytropis glabra* (Leguminosae)	Kitagawa *et al.* 1988d Sun and Jia, 1990
423		3-GlcA⁴-Glc²-Rha		
3β,24-Dihydroxyolean-12,15-dien-28-oic acid aglycone (25)				
424	Phaseoloside D	3-GlcA³-Ara²-α-Gal⁴-Glc⁶-Rha |² α-Gal	*Phaseolus vulgaris* (Leguminosae)	Lazurevskii *et al.* 1971
425	Phaseoloside E	3-GlcA³-Ara²-α-Gal⁴-Glc⁶-Rha |² α-Gal⁴-Glc⁴-Glc	*P. vulgaris* (Leguminosae)	Lazurevskii *et al.* 1971

Quillaic acid aglycone (28)

No.	Name	Structure	Source	Reference
426	Dubioside A	3-GlcA²-Gal 28-Ara²-Rha	*Thladiantha dubia* (Cucurbitaceae)	Nagao *et al.* 1989a
427	Dubioside B	3-GlcA²-Gal 28-Ara²-Rha⁴-Xyl	*T. dubia* (Cucurbitaceae)	Nagao *et al.* 1989a
428	Lucyoside N	3-GlcA²-Gal 28-Ara²-Rha⁴-Xyl ⌐³ Glc	*Luffa cylindrica* (Cucurbitaceae)	Yoshikawa *et al.* 1991c
429	Luperoside K	3-GlcA³-Ara ⌐² Gal 28-Qui²-Rha⁴-Xyl ⌐³ Rha	*L. operculata* (Cucurbitaceae)	Okabe *et al.* 1989
430		3-GlcA³-Xyl ⌐² Gal 28-Fuc²-Rha⁴-Xyl ⌐³ Glc	*Gypsophila paniculata* (Caryophyllaceae) *G. arrostii* (Caryophyllaceae)	Frechet *et al.* 1991 Frechet *et al.* 1991
431		3-GlcA³-Xyl ⌐² Gal 28-Fuc²-Rha⁴-Xyl³-Ara⁴-Ara	*G. paniculata* (Caryophyllaceae) *G. arrostii* (Caryophyllaceae)	Frechet *et al.* 1991 Frechet *et al.* 1991
432	Luperoside L	3-GlcA³-Ara ⌐² Gal 28-Qui³-Rha⁴-Xyl³-Xyl ⌐³ Gal	*Luffa operculata* (Cucurbitaceae)	Okabe *et al.* 1989

(cont.)

Appendix 2. (Cont.)

No.	Name	Structure	Plant	Reference
21β-Hydroxygypsogenin aglycone (29)				
433	Lucyoside J	3-Glc 28-Glc	*L. cylindrica* (Cucurbitaceae)	Takemoto *et al.* 1985
Medicagenic acid aglycone (31)				
434	Medicoside A	3-Glc	*Medicago sativa* (Leguminosae) *Dolichos kilimandscharicus* (Leguminosae) *D. falcatus* (Leguminosae)	Timbekova and Abubakirov, 1984 Levy *et al.* 1986 Marston *et al.* 1988a Feng *et al.* 1985
435	Medicoside G	3-Glc 28-Glc	*Medicago sativa* (Leguminosae)	Timberkova and Abubakirov, 1984
436		3-Glc^4-Glc	*M. sativa* (Leguminosae)	Levy *et al.* 1989
437	Herniaria-saponin I	28-Glc^6-Glc	*Herniaria glabra* (Caryophyllaceae)	Bukharov and Shcherbak, 1970
438	Lucerne saponin	3-Glc^3-Glc^6-Glc	*Medicago sativa* (Leguminosae)	Gestetner, 1971
439	Herniaria-saponin II	28-Rha^2-Fuc $\mid^4$ Glc	*Herniaria glabra* (Caryophyllaceae)	Bukharov and Shcherbak, 1970
440		28-Ara^2-Rha^4-Xyl	*Medicago sativa* (Leguminosae)	Massiot *et al.* 1991b
441		3-Glc 28-Ara^2-Rha^4-Xyl	*M. sativa* (Leguminosae)	Massiot *et al.* 1991b

442		3-Glc^2-Glc 28-Ara^2-Rha^4-Xyl	*M. sativa* (Leguminosae)	Massiot *et al.* 1991b
443		3-GlcA 28-Ara^2-Rha^4-Xyl	*M. sativa* (Leguminosae)	Massiot *et al.* 1991b
Dianic acid aglycone (32)				
444	Dianoside C	3-Glc 28-Glc	*Dianthus superbus* (Caryophyllaceae)	Oshima *et al.* 1984b
445	Dianoside F	28-Glc	*D. superbus* (Caryophyllaceae)	Oshima *et al.,* 1984b
Soyasapogenol B aglycone (33)				
446		24-Glc	*Phaseolus vulgaris* (Leguminosae)	Jain *et al.* 1988
447	Soyasaponin IV	3-GlcA^2-Ara	*Glycine max* (Leguminosae)	Burrows *et al.* 1987
448	Soyasaponin III	3-GlcA^2-Gal	*G. max* (Leguminosae)	Kitagawa *et al.* 1982
449	Azukisaponin II	3-GlcA^2-Glc	*Vigna angularis* (Leguminosae)	Kitagawa *et al.* 1983c
			Medicago sativa (Leguminosae)	Kitagawa *et al.* 1988d
450	Phaseoluside A	3-Glc^4-Glc |^3 Glc	*Phaseolus vulgaris* (Leguminosae)	Jain *et al.* 1991
451	Soyasaponin II	3-GlcA^2-Ara^2-Rha	*Glycine max* (Leguminosae)	Kitagawa *et al.* 1982
452		3-GlcA^4-Ara^2-Rha	*Oxytropis ochrocephala* (Leguminosae)	Sun *et al.* 1987
			O. glabra (Leguminosae)	Sun *et al.* 1991

(cont.)

Appendix 2. (Cont.)

No.	Name	Structure	Plant	Reference
Soyasapogenol B aglycone (33)				
453	Soyasaponin V	3-GlcA^2-Gal^2-Glc	*Glycine max* (Leguminosae)	Taniyama *et al.* 1988b
			Phaseolus vulgaris (Leguminosae)	Curl *et al.*, 1988a
454	Soyasaponin I	3-GlcA^2-Gal^2-Rha	*Glycine max* (Leguminosae)	Kitagawa *et al.* 1976b
			Medicago sativa (Leguminosae)	Kitagawa *et al.* 1988d
			Pisum sativum (Leguminosae)	Price and Fenwick, 1984
455	Sophoraflavoside I	3-GlcA^2-Gal^2-Rha 28-Ara^2-Glc	*Sophora flavescens* (Leguminosae)	Yoshikawa *et al.* 1985
456	Azukisaponin V	3-GlcA^2-Glc^2-Rha	*Vigna angularis* (Leguminosae)	Kitagawa *et al.* 1983d
			Medicago sativa (Leguminosae)	Kitagawa *et al.* 1988d
457		3-GlcA^4-Glc^2-Rha	*Oxytropis ochrocephala* (Leguminosae)	Sun *et al.* 1987
458	Astragaloside VIII	3-GlcA^2-Xyl^2-Rha	*Astragalus membranaceus* (Leguminosae)	Kitagawa *et al.* 1983e
459		3-GlcA^4-Glc^2-Rha $\vert^2$ Glc	*Oxytropis glabra* (Leguminosae)	Sun and Jia, 1990
460		3-GlcA^2-Gal^2-Rha $\vert^6$ Glc	*Crotalaria albida* (Leguminosae)	Ding *et al.* 1991a

3β,22β,24-Trihydroxyolean-12-en-29-oic acid aglycone (34)
461 3-GlcA⁴-Glc²-Rha *Oxytropis glabra* (Leguminosae) Sun and Jia, 1990

461	3-GlcA^4-Glc^2-Rha	*Oxytropis glabra* (Leguminosae)	Sun and Jia, 1990

Primulagenin A aglycone (35)

462	Ardisioside A	3-Rha^4-Gal	*Ardisia neriifolia* (Myrsinaceae)	Malaviya *et al.* 1989
463	Rotundioside D	3-Glc^2-Glc^2-Rha	*Bupleurum rotundifolium* (Umbelliferae)	Akai *et al.* 1985b
464	Ardisioside B	3-Rha^4-Ara^2-Gal	*Ardisia neriifolia* (Myrsinaceae)	Malaviya *et al.* 1989

2β,3β,28-Trihydroxyolean-12-en aglycone (36)

465	Melilotin	2-Glc^4-Rha 3-Glc^4-Rha	*Melilotus alba* (Leguminosae)	Nicollier and Thompson, 1983

Caulophyllogenin aglycone (38)

466	Racemoside	3-Ara	*Cimicifuga racemosa* (Ranunculaceae)	Jap. Pat. 59 20,298
467	Chrysantellin B	3-Glc 28-Xyl^2-Rha^4-Xyl^3-Rha	*Chrysanthellum procumbens* (Asteraceae)	Becchi *et al.* 1980

21β-Hydroxyhederagenin aglycone (39)

468	Anchusoside-II	3-Xyl 21-Glc^2-Glc	*Anchusa officinalis* (Boraginaceae)	Romussi *et al.* 1988a
469	Lucyoside A	3-Glc 28-Glc	*Luffa cylindrica* (Cucurbitaceae)	Takemoto *et al.* 1984

3β,21β,22β-Trihydroxyolean-12-en-29-oic acid aglycone (40)

470	3-GlcA^2-Gal^2-Rha	*Dalbergia hupeana* (Leguminosae)	Yahara *et al.* 1989a

(cont.)

Appendix 2. (*Cont.*)

No.	Name	Structure	Plant	Reference
23-Hydroxyimberbic acid aglycone (41)				
471		23-Rha	*Combretum imberbe* (Combretaceae)	Rogers, 1988
Arjunolic acid aglycone (43)				
472	Lucyoside I	3-Glc	*Luffa cylindrica* (Cucurbitaceae)	Takemoto *et al.* 1985
473	Lucyoside B	3-Glc 28-Glc	*L. cylindrica* (Cucurbitaceae)	Takemoto *et al.* 1984
474	Saponin Pj$_1$	28-Glc^6-Glc^4-Rha^3-Xyl	*Akebia quinata* (Lardizabalaceae)	Higuchi and Kawasaki, 1976
Asterogenic acid aglycone (44)				
475	Aster saponin A	3-Glc^6-Ara 28-Xyl^2-Rha^3-Api$f$ $\vert^4$ Ara^3-Xyl	*Aster tataricus* (Asteraceae)	Nagao *et al.* 1989b
476	Aster saponin C	3-Glc^6-Ara 28-Xyl^2-Rha^3-Api$f$ $\vert^3$ $\vert^4$ Rha Ara^3-Xyl	*A. tataricus* (Asteraceae)	Nagao *et al.* 1989b
477	Aster saponin E	3-Glc^6-Ara 28-Xyl^2-Rha^4-Ara^3-Xyl	*A. tataricus* (Asteraceae)	Nagao *et al.* 1990
Bayogenin aglycone (45)				
478	Hederasaponin E	3-Ara 28-Glc^6-Glc^4-Rha	*Hedera helix* (Araliaceae)	Elias *et al.* 1991

No.	Compound	Structure	Source	Reference
479		3-Glc	*Dolichos kilimandscharicus* (Leguminosae)	Marston *et al.* 1988a
480		3-Glc[4]-Glc	*Phytolacca dodecandra* (Phytolaccaceae)	Domon and Hostettmann, 1984
481		3-Glc[4]-Glc 28-Glc	*P. dodecandra* (Phytolaccaceae)	Domon and Hostettmann, 1984
482		3-Glc[4]-Glc |[3] Gal 28-Glc	*P. dodecandra* (Phytolaccaceae)	Dorsaz and Hostettmann, 1986
483		28-Glc[2]-Glc[2]-Glc	*Polygala japonica* (Polygalaceae)	Fang and Ying, 1986
484	Canadensissaponin A	3-Glc[3]-Glc 28-Qui[2]-Rha[4]-Xyl[3]-Rha |[3] Api*f* Xyl	*Solidago canadensis* (Asteraceae)	Reznicek *et al.* 1990b, 1991
485	Canadensissaponin B	3-Glc[3]-Glc 28-Qui[2]-Rha[4]-Xyl[3]-Rha |[3] Rha Xyl	*S. canadensis* (Asteraceae)	Reznicek *et al.* 1990b, 1991
486	Canadensissaponin 2	3-Glc[3]-Glc 28-Ara[2]-Rha[4]-Xyl[3]-Rha |[3] Api*f* Xyl	*S. canadensis* (Asteraceae)	Reznicek *et al.* 1991
487	Canadensissaponin 4	3-Glc[3]-Glc 28-Ara[2]-Rha[4]-Xyl[3]-Rha |[3] Rha Xyl	*S. canadensis* (Asteraceae)	Reznicek *et al.* 1991

(*cont.*)

Appendix 2. (*Cont.*)

No.	Name	Structure	Plant	Reference
Bayogenin aglycone (45)				

488		3-Glc^3-Glc^3-Ara
		28-Qui^2-Rha^4-Xyl^3-Rha^2-Gal
		       |^3
		     Rha Xyl
			S. gigantea (Asteraceae)	Reznicek *et al.* 1989b

489	Giganteasaponin 1	3-Glc^3-Glc^2-Api*f*
		28-Qui^2-Rha^4-Xyl^3-Rha
		       |^3
		   Api*f* Xyl
			S. gigantea (Asteraceae)	Reznicek *et al.* 1990a

490	Giganteasaponin 2	3-Glc^3-Glc^3-Api*f*
		28-Qui^2-Rha^4-Xyl^3-Rha
		       |^3
		   Api*f* Rha
			S. gigantea (Asteraceae)	Reznicek *et al.* 1990a

491	Giganteasaponin 3	3-Glc^3-Glc^3-Api*f*
		28-Qui^2-Rha^4-Xyl^3-Rha^2-Gal
		       |^3
		   Api*f* Xyl
			S. gigantea (Asteraceae)	Reznicek *et al.* 1990a

492	Giganteasaponin 4	3-Glc^3-Glc^3-Api*f*
		28-Qui^2-Rha^4-Xyl^3-Rha^2-Gal
		       |^3
		   Rha Xyl
			S. gigantea (Asteraceae)	Reznicek *et al.* 1990a

Presenegenin aglycone (47)				
493	Tenuifolin	3-Glc	*Polygala senega* (Polygalaceae) *P. tenuifolia* (Polygalaceae)	Pelletier and Nakamura, 1967 Pelletier and Nakamura, 1967

Jaligonic acid aglycone (48)			
494	Phytolaccoside G	3-Xyl	*Phytolacca americana* (Phytolaccaceae)	Woo and Kang, 1977
Phytolaccagenin aglycone (49)			
495		3-Glc	*P. thyrsiflora* (Phytolaccaceae)	Haraguchi *et al.* 1988
			P. esculenta (Phytolaccaceae)	Yi and Wang, 1984
496	Phytolaccoside B	3-Xyl	*P. thyrsiflora* (Phytolaccaceae)	Haraguchi *et al.* 1988
			P. americana (Phytolaccaceae)	Woo *et al.* 1978
497	Phytolaccoside D_2	3-Xyl^2-Glc	*P. americana* (Phytolaccaceae)	Kang and Woo, 1987
498	Phytolaccoside E	3-Xyl^4-Glc	*P. thyrsiflora* (Phytolaccaceae)	Haraguchi *et al.* 1988
			P. americana (Phytolaccaceae)	Woo *et al.* 1978
499	Esculentoside H	3-Xyl^4-Glc 28-Glc	*P. esculenta* (Phytolaccaceae)	Yi and Wang, 1989
500	Phytolaccoside F	3-Xyl^2-Glc^2-Rha	*P. americana* (Phytolaccaceae)	Kang and Woo, 1987
Belleric acid aglycone (50)			
501	Bellericoside	28-Glc	*Terminalia bellerica* (Combretaceae)	Nandy *et al.* 1989
Protobassic acid aglycone (52)			
502	Mi-saponin A	3-Glc 28-Ara^2-Rha^4-Xyl^3-Rha	*Madhuca longifolia* (Sapotaceae)	Kitagawa *et al.* 1975
			Clerodendrum wildii (Verbenaceae)	Toyota *et al.* 1990

(cont.)

Appendix 2. (*Cont.*)

No.	Name	Structure	Plant	Reference
Protobassic acid aglycone (52)				
503	Mi-saponin B	3-Glc 28-Ara²-Rha⁴-Xyl³-Rha $\mid$³ Api*f*	*Madhuca longifolia* (Sapotaceae)	Kitagawa *et al.* 1975
504	Mi-saponin C	3-Glc 28-Ara²-Rha⁴-Xyl³-Rha $\mid$⁴ Glc	*M. longifolia* (Sapotaceae)	Kitagawa *et al.* 1978b
Polygalacic acid aglycone (54)				
505	Solidagosaponin I	16-Glc²-Ara	*Solidago virgaurea* (Asteraceae)	Inose *et al.* 1991
506	Solidagosaponin III	16-Glc²-Ara 28-Rha	*S. virgaurea* (Asteraceae)	Inose *et al.* 1991
507	Solidagosaponin V	16-Glc²-Ara 28-Xyl	*S. virgaurea* (Asteraceae)	Inose *et al.* 1991
508	Solidagosaponin VII	16-Glc²-Ara 28-Ara	*S. virgaurea* (Asteraceae)	Inose *et al.* 1991
509	Solidagosaponin IX	3-Glc⁴-Glc	*S. virgaurea* (Asteraceae)	Inose *et al.* 1991
Arjungenin aglycone (56)				
510	Arjunglucoside 1	28-Glc	*Terminalia bellerica* (Combretaceae)	Nandy *et al.* 1989
Esculentagenic acid aglycone (57)				
511	Esculentoside J	3-Xyl⁴-Glc 28-Glc	*Phytolacca esculenta* (Phytolaccaceae)	Yi and Dai, 1991

23-Hydroxylongispinogenin aglycone (58)

512	Gymnemasaponin I	28-Glc	*Gymnema sylvestre* (Asclepiadaceae)	Yoshikawa *et al.* 1991a
513	Gymnemasaponin II	23-Glc 28-Glc	*G. sylvestre* (Asclepiadaceae)	Yoshikawa *et al.* 1991a
514	Gymnemasaponin III	23-Glc 28-Glc⁶-Glc	*G. sylvestre* (Asclepiadaceae)	Yoshikawa *et al.* 1991a
515	Gymnemasaponin IV	23-Glc⁶-Glc 28-Glc	*G. sylvestre* (Asclepiadaceae)	Yoshikawa *et al.* 1991a
516	Gymnemasaponin V	23-Glc⁶-Glc 28-Glc⁶-Glc	*G. sylvestre* (Asclepiadaceae)	Yoshikawa *et al.* 1991a

Soyasapogenol A aglycone (60)

517	Soyasaponin A₂	3-GlcA²-Gal 22-Ara³-Glc	*Glycine max* (Leguminosae)	Kitagawa *et al.* 1985c
518	Soyasaponin A₁	3-GlcA²-Gal²-Glc 22-Ara³-Glc	*G. max* (Leguminosae)	Kitagawa *et al.* 1985a
519	Soyasaponin A₃	3-GlcA²-Gal²-Rha	*G. max* (Leguminosae)	Curl *et al.* 1988b

Oxytrogenol aglycone (61)

520		3-GlcA⁴-Glc²-Rha	*Oxytropis glabra* (Leguminosae)	Sun and Jia, 1990

3α,21β,22β,28-Tetrahydroxyolean-12-en aglycone (62)

521		3-Xyl	*Centipeda minima* (Asteraceae)	Gupta and Singh, 1990

3β,23,27,29-Tetrahydroxyoleanolic acid aglycone (63)

522		28-Ara	*Polygala chamaebuxus* (Polygalaceae)	Hamburger and Hostettmann, 1986
523		29-Glc	*P. chamaebuxus* (Polygalaceae)	Hamburger and Hostettmann, 1986

(cont.)

Appendix 2. (*Cont.*)

No.	Name	Structure	Plant	Reference
Barringtogenol C aglycone (64)				
524		28-Xyl	*Centipeda minima* (Asteraceae)	Gupta and Singh, 1990
16α-Hydroxyprotobassic acid aglycone (66)				
525	Tridesmosaponin A	3-Glc⁶-Glc 28-Xyl²-Rha⁴-Xyl²-Rha |³ Rha	*Tridesmostemon claessenssi* (Sapotaceae)	Massiot *et al.* 1990
526	Tridesmosaponin B	3-Rha 28-Xyl²-Rha-³Xyl³-Rha |² Rha	*T. claessenssi* (Sapotaceae)	Massiot *et al.* 1990
527		3-Glc 28-Ara²-Rha⁴-Xyl³-Rha	*Crossopteryx febrifuga* (Rubiaceae)	Gariboldi *et al.* 1990
528		3-Glc³-Apif 28-Ara²-Rha⁴-Xyl-Rha	*C. febrifuga* (Rubiaceae)	Gariboldi *et al.* 1990
Protoaescigenin aglycone (68)				
529	Aesculuside B	3-GlcA⁴-Glc |² Glc	*Aesculus indica* (Hippocastanaceae)	Singh *et al.* 1987
R₁-Barrigenol aglycone (70)				
530	Pittoside A	3-Rha⁴-Ara	*Pittosporum nilghrense* (Pittosporaceae)	Jain *et al.* 1980

Protoprimulagenin A aglycone (71)

531 PS4

$$3\text{-GlcA}^3\text{-Gal}^2\text{-Rha}$$
$$|^2$$
$$\text{Glc}$$

Primula elatior (Primulaceae)

Tschesche *et al.* 1983

532 PS3

$$3\text{-GlcA}^3\text{-Glc}$$
$$|^2$$
$$\text{Gal}^2\text{-Rha}$$

P. elatior (Primulaceae)

Tschesche *et al.* 1983

533 Lysikoianoside 1

$$3\text{-Ara}p^4\text{-Glc}^2\text{-Xyl}$$
$$|^2$$
$$\text{Glc}$$

Lysimachia sikokiana (Primulaceae)

Kohda *et al.* 1989

534

$$3\text{-GlcA}^4\text{-Gal}^3\text{-Glc}$$
$$|^2$$
$$\text{Rha}$$

Eleutherococcus senticosus (Araliaceae)

Segiet-Kujawa and Kaloga, 1991

535

$$3\text{-GlcA}^4\text{-Glc}^4\text{-Rha}^4\text{-Rha}$$
$$|^2$$
$$\text{Rha}$$

E. senticosus (Araliaceae)

Segiet-Kujawa and Kaloga, 1991

Cyclamiretin A aglycone (72)

536 Primulanin

$$3\text{-Ara}^4\text{-Glc}$$
$$|^2$$
$$\text{Xyl}$$

Primula denticulata (Primulaceae)

Ahmad *et al.* 1988b

537 Desglucocyclamin I (saxifragifolin B, ardisiacrispin A)

$$3\text{-Ara}^4\text{-Glc}$$
$$|^2 \quad |^2$$
$$\text{Glc} \quad \text{Xyl}$$

Cyclamen europaeum (Primulaceae)
C. graecum (Primulaceae)
Androsace saxifragifolia (Primulaceae)
Primula denticulata (Primulaceae)
Ardisia crispa (Myrsinaceae)

Tschesche *et al.* 1969c
Reznicek *et al.* 1989a
Waltho *et al.* 1986
Ahmad *et al.* 1988b
Jansakul *et al.* 1987

(cont.)

Appendix 2. *(Cont.)*

No.	Name	Structure	Plant	Reference
Cyclamiretin A aglycone (72)				
538	Cyclamin	$3\text{-Ara}^4\text{-Glc}^3\text{-Glc}$ $\quad\quad\quad\;\;\mid^2$ $\quad\quad\;\;\text{Glc}\;\;\text{Xyl}$	*Cyclamen europaeum* (Primulaceae)	Tschesche *et al.* 1969c
539	Isocyclamin	$3\text{-Ara}^4\text{-Glc}^6\text{-Glc}$ $\quad\quad\quad\;\;\mid^2$ $\quad\quad\;\;\text{Glc}\;\;\text{Xyl}$	*C. graecum* (Primulaceae) *C. graecum* (Primulaceae)	Reznicek *et al.* 1989a Reznicek *et al.* 1989a
540	Saxifragifolin D	$3\text{-Ara}^4\text{-Glc}^2\text{-Xyl}$ $\mid^2$ $\text{Glc}^4\text{-Glc}$	*Androsace saxifragifolia* (Primulaceae)	Pal and Mahato, 1987
542	Ardisiacrispin B	$3\text{-Ara}^4\text{-Glc}^2\text{-Rha}$ $\mid^2$ Glc	*Ardisia crispa* (Myrsinaceae)	Jansakul *et al.* 1987
Rotundiogenin A aglycone (73)				
543	Rotundioside G	$3\text{-Fuc}^2\text{-Glc}^2\text{-Xyl}$	*Bupleurum rotundifolium* (Umbelliferae)	Akai *et al.* 1985b
Saikogenin E aglycone (74)				
544	Saikosaponin c	$3\text{-Glc}^6\text{-Glc}$ $\quad\quad\;\mid^4$ $\quad\quad\;\text{Glc}$	*B. falcatum* (Umbelliferae)	Kubota and Hinoh, 1968
Anagalligenone aglycone (75)				
545	Anagallisin B	$3\text{-Ara}^4\text{-Glc}^2\text{-Xyl}$ $\mid^2$ $\text{Glc}^4\text{-Glc}$	*Anagallis arvensis* (Primulaceae)	Mahato *et al.* 1991

546 Anagallisin D

$$3\text{-Ara}^4\text{-Glc}^2\text{-Xyl}$$
$$|^2$$
$$\text{Glc}$$

A. arvensis (Primulaceae)

Mahato *et al.* 1991

Saikogenin F aglycone (76)
547 Saikosaponin a

$$3\text{-Fuc}^3\text{-Glc}$$

Bupleurum falcatum (Umbelliferae)

Kubota and Hinoh, 1968

Saikogenin G aglycone (77)
548 Saikosaponin d

$$3\text{-Fuc}^3\text{-Glc}$$

B. falcatum (Umbelliferae)

Kubota and Hinoh, 1968

549 Anagalloside B

$$3\text{-Ara}^4\text{-Glc}^4\text{-Glc}$$
$$|^2 \quad |^2$$
$$\text{Glc} \quad \text{Xyl}$$

Anagallis arvensis (Primulaceae)

Glombitza and Kurth, 1987a,b; Amoros and Girre, 1987

550 Desglucoanagalloside B

$$3\text{-Ara}^4\text{-Glc}$$
$$|^2$$
$$\text{Glc} \quad \text{Xyl}$$

A. arvensis (Primulaceae)

Glombitza and Kurth, 1987a,b; Amoros and Girre, 1987

551 Anagallisin A

$$3\text{-Ara}^2\text{-Glc}^4\text{-Glc}$$
$$|^4$$
$$\text{Glc}^2\text{-Xyl}$$

A. arvensis (Primulaceae)

Mahato *et al.* 1991

552 Anagallisin E

$$3\text{-Ara}^2\text{-Glc}$$
$$|^4$$
$$\text{Glc}$$

A. arvensis (Primulaceae)

Mahato *et al.* 1991

α-Amyrin aglycone (80)
553

$$3\text{-Glc}^4\text{-Rha}$$

Ichnocarpus frutescens (Apocynaceae)

Minocha and Tandon, 1980

Ursolic acid aglycone (81)
554 Matesaponin 1

$$3\text{-Ara}^3\text{-Glc}$$

Ilex paraguariensis (Aquifoliaceae)

Gosmann *et al.* 1989

(cont.)

Appendix 2. (*Cont.*)

No.	Name	Structure	Plant	Reference
Ursolic acid aglycone (81)				
555	Cynarasaponin C	3-GlcA 28-Glc	*Cynara cardunculus* (Asteraceae)	Shimizu *et al.* 1988
556	Cynarasaponin B	3-GlcA²-Ara	*C. cardunculus* (Asteraceae)	Shimizu *et al.* 1988
557	Cynarasaponin A	3-GlcA²-Ara 28-Glc	*C. cardunculus* (Asteraceae)	Shimizu *et al.* 1988
Quinovic acid aglycone (82)				
558		3-Fuc	*Uncaria guianensis* (Rubiaceae)	Yepez *et al.* 1991
559		28-Fuc	*Guettarda platypoda* (Rubiaceae)	Aquino *et al.* 1989a
560		3-Glc	*G. angelica* (Rubiaceae) *G. platypoda* (Rubiaceae)	Matos *et al.* 1986 Aquino *et al.* 1988
561		3-Qui	*Uncaria guianensis* (Rubiaceae)	Yepez *et al.* 1991
562		3-Rha	*Guettarda angelica* (Rubiaceae)	Matos *et al.* 1986
563		3-Fuc 27-Glc	*Uncaria guianensis* (Rubiaceae)	Yepez *et al.* 1991
564		3-Fuc 28-Glc	*U. tomentosa* (Rubiaceae)	Aquino *et al.* 1989b
565		3-Glc 27-Glc	*Guettarda platypoda* (Rubiaceae)	Aquino *et al.* 1988
566		3-Glc 28-Glc	*G. angelica* (Rubiaceae) *G. platypoda* (Rubiaceae)	Matos *et al.* 1986 Aquino *et al.* 1988

No.	Compound	Sugars	Species (Family)	Reference
567		3-Qui 27-Glc	Uncaria tomentosa (Rubiaceae)	Aquino et al. 1991
568		3-Qui 28-Glc	U. tomentosa (Rubiaceae) Isertia haenkeana (Rubiaceae)	Aquino et al. 1989b Javier Arriaga et al. 1990
569		28-Glc	Guettarda platypoda (Rubiaceae)	Aquino et al. 1989a
570		28-Glc-Glc	Uncaria tomentosa (Rubiaceae)	Aquino et al. 1989b
571		3-Fuc^3-Glc	U. tomentosa (Rubiaceae)	Cerri et al. 1988
572		3-Fuc^3-Glc 27-Glc	U. tomentosa (Rubiaceae) U. guianensis (Rubiaceal)	Cerri et al. 1988 Yepez et al. 1991
573		3-Fuc^3-Glc 28-Glc	U. tomentosa (Rubiaceae)	Cerri et al. 1988
574		3-Rha^3-Glc	Guettarda angelica (Rubiaceae) G. platypoda (Rubiaceae)	Matos et al. 1986 Aquino et al. 1989a

3β-Hydroxyurs-12,20(30)-dien-27,28-dioic acid aglycone (83)

No.	Compound	Sugars	Species (Family)	Reference
575		3-Rha 28-Glc	Crossopteryx febrifuga (Rubiaceae)	Babady-Bila et al. 1991

Pomolic acid aglycone (84)

No.	Compound	Sugars	Species (Family)	Reference
576	Ziyu-glycoside II	3-Ara	Ilex cornuta (Aquifoliaceae)	Qin et al. 1986
577	Ziyu-glycoside I	3-Ara 28-Glc	I. cornuta (Aquifoliaceae)	Qin et al. 1986
578	Ilexoside B	3-Xyl	I. chinensis (Aquifoliaceae)	Inada et al. 1987b
579	Ilexoside I	3-Ara^2-Glc	I. cornuta (Aquifoliaceae)	Nakanishi et al. 1982
580	Ilexoside II	3-Ara^2-Glc 28-Glc	I. cornuta (Aquifoliaceae)	Nakanishi et al. 1982

(cont.)

Appendix 2. (*Cont.*)

No.	Name	Structure	Plant	Reference
Ilexgenin B aglycone (85)				
581	Ilexsaponin B_1	3-Xyl^2-Glc	*I. pubescens* (Aquifoliaceae)	Hidaka *et al.* 1987a
582	Ilexsaponin B_3	3-Xyl^2-Glc	*I. pubescens* (Aquifoliaceae)	Hidaka *et al.* 1987a
		28-Glc		
583	Ilexsaponin B_2	3-Xyl^2-Glc^2-Rha	*I. pubescens* (Aquifoliaceae)	Hidaka *et al.* 1987a
Ilexgenin A aglycone (86)				
584	Ilexsaponin A_1	28-Glc	*I. pubescens* (Aquifoliaceae)	Hidaka *et al.* 1987b Qin *et al.* 1987
21β-Hydroxyursolic acid aglycone (87)				
585	Cynarasaponin F	3-$GlcA^2$-Ara	*Cynara cardunculus* (Asteraceae)	Shimizu *et al.* 1988
586	Cynarasaponin G	3-$GlcA^2$-Ara	*C. cardunculus* (Asteraceae)	Shimizu *et al.* 1988
		28-Glc		
23-Hydroxyursolic acid aglycone (88)				
587		3-$GlcA^6$-OMe	*Schefflera impressa* (Araliaceae)	Srivastava and Jain, 1989
588	Cynarasaponin E	3-GlcA	*Cynara cardunculus* (Asteraceae)	Shimizu *et al.* 1988
		28-Glc		
589	Cynarasaponin D	3-$GlcA^2$-Ara	*C. cardunculus* (Asteraceae)	Shimizu *et al.* 1988
		28-Glc		
3β,23-Dihydroxytaraxer-20-en-28-oic acid aglycone (89)				
590		3-Glc	*Fagonia indica* (Zygophyllaceae)	Ansari *et al.* 1987
		28-Glc		

No.	Compound	Glycosidation	Source (Family)	Reference
591		23-Glc 28-Glc	*F. indica* (Zygophyllaceae)	Ansari *et al.* 1987
Rotundic acid aglycone (90)				
592	Peduncloside	28-Glc	*Ilex rotunda* (Aquifoliaceae)	Nakatani *et al.* 1989
Rotungenic acid aglycone (91)				
593	Rotungenoside	28-Glc	*I. rotunda* (Aquifoliaceae)	Nakatani *et al.* 1989
Asiatic acid aglycone (93)				
594		3-GlcA	*Mucuna birdwoodiana* (Leguminosae)	Ding *et al.* 1991b
595		3-GlcA 28-Glc	*M. birdwoodiana* (Leguminosae)	Ding *et al.* 1991b
596		3-GlcA^2-Ara	*M. birdwoodiana* (Leguminosae)	Ding *et al.* 1991b
597	Asiaticoside	28-Glc^6-Glc^4-Rha	*Centella asiatica* (Umbelliferae)	Polonsky and Zylber (1961)
			Schefflera octophylla (Araliaceae)	Sung *et al.* 1992
Euscaphic acid aglycone (94)				
598	Kajiichigoside F1	28-Glc	*Rubus trifidus* (Rosaceae) *Sargentodoxa cuneata* (Sargentodoxaceae)	Seto *et al.* 1984 Rücker *et al.* 1991
Tormentic acid aglycone (95)				
599	Rosamultin	28-Glc	*Rosa multiflora* (Rosaceae) *Aphloia theiformis* (Flacourtiaceae)	Du *et al.* 1983 Gopalsamy *et al.* 1988

(cont.)

Appendix 2. (*Cont.*)

No.	Name	Structure	Plant	Reference
Tormentic acid aglycone (95)				
			Musanga cecropioides (Cecropiaceae)	Lontsi *et al.* 1990
			Sarcopoterium spinosum (Rosaceae)	Reher *et al.* 1991
			Sargentodoxa cuneata (Sargentodoxaceae)	Rücker *et al.* 1991
2α,3β,19α-Trihydroxyurs-12-en-23,28-dioic acid aglycone (96)				
600	Suavissimoside R$_1$	28-Glc	*Rubus coreanus* (Rosaceae)	Ohtani *et al.* 1990
6β-Hydroxytormentic acid aglycone (97)				
601		28-Glc	*Aphloia theiformis* (Flacourtiaceae)	Gopalsamy *et al.* 1988
7α-Hydroxytormentic acid aglycone (98)				
602		28-Glc	*Desfontainia spinosa* (Loganiaceae)	Houghton and Lian, 1986b
23-Hydroxytormentic acid aglycone (99)				
603		28-Glc	*D. spinosa* (Loganiaceae)	Houghton and Lian, 1986a,b
			Geum japonicum (Rosaceae)	Shigenaga *et al.* 1985
			Aphloia theiformis (Flacourtiaceae)	Gopalsamy *et al.* 1988
			Rubus coreanus (Rosaceae)	Ohtani *et al.* 1990; Kim and Kim, 1987
			Quercus suber (Fagaceae)	Romussi *et al.* 1991
			Q. ilex (Fagaceae)	Romussi *et al.* 1983

24-Hydroxytormentic acid aglycone (100)

604	28-Xyl	*Sarcopoterium spinosum* (Rosaceae)	Reher et al. 1991
		Sanguisorba minor (Rosaceae)	Reher et al. 1991
		Centipeda minima (Asteraceae)	Gupta and Singh, 1990
605	28-Glc	*Desfontainia spinosa* (Loganiaceae)	Houghton and Lian, 1986b

1α,3α,19α,23-Tetrahydroxyurs-12-en-28-oic acid aglycone (101)

606	28-Xyl	*Centipeda minima* (Asteraceae)	Gupta and Singh, 1989

Madecassic acid aglycone (102)

607 Madecassoside	28-Glc^6-Glc^4-Rha	*Hydrocotyle asiatica* (Umbelliferae)	Pinhas and Bondiou, 1967

6β,23-Dihydroxytormentic acid aglycone (103)

608	28-Glc	*Aphloia madagascariensis* (Flacourtiaceae)	Dijoux et al. 1993

Lupeol aglycone (104)

609	3-Rha	*Cordia obliqua* (Boraginaceae)	Srivastava et al. 1983
610	3-Glc^4-Xyl$f$	*Acanthus ilicifolius* (Acanthaceae)	Minocha and Tiwari, 1981
611	3-GlcA^4-Ara$f$	*A. ilicifolius* (Acanthaceae)	Minocha and Tiwari, 1981

(cont.)

Appendix 2. (Cont.)

No.	Name	Structure	Plant	Reference
Betulinic acid aglycone (106)				
612		3-Rha	*Dillenia pentagyna* (Dilleniaceae)	Tiwari *et al.* 1980
613		3-Glc²-Glc	*Schefflera venulosa* (Araliaceae)	Purohit *et al.* 1991
614		3-Glc⁴-Glc	*Acacia leucophloea* (Leguminosae)	Mishra and Srivastava, 1985
615		3-Glc⁶-Glc	*Eryngium bromeliifolium* (Umbelliferae)	Hiller et al. 1978a
616	Pavophyllin	3-Glc⁴-Rha²-Ara*f*	*Pavonia zeylanica* (Malvaceae)	Tiwari and Minocha, 1980
617	Menyanthoside	3-GlcA⁴-Gal 28-Glc⁶-Api*f*	*Menyanthes trifoliata* (Menyanthaceae)	Janeczko *et al.* 1990
3-*epi*-Betulinic acid aglycone (107)				
618		3-Glc	*Schefflera octophylla* (Araliaceae)	Kitajima and Tanaka, 1989
619		28-Rha⁴-Glc⁶-Glc	*S. octophylla* (Araliaceae)	Sung *et al.* 1991
3β,23-Dihydroxylup-20(29)-en-28-oic acid aglycone (108)				
620	Anemoside A₃	3-Ara²-Rha	*Pulsatilla chinensis* (Ranunculaceae)	Chen *et al.* 1990a
621	Anemoside B₄	3-Ara²-Rha 28-Glc⁶-Glc⁴-Rha	*P. chinensis* (Ranunculaceae)	Chen *et al.* 1990a

3α-Hydroxylup-20(29)-en-23,28-dioic acid aglycone (109)				
622		28-Rha^4-Glc^6-Glc	*Schefflera octophylla* (Araliaceae)	Kitajima and Tanaka, 1989; Sung *et al.* 1991
3α,11α-Dihydroxylup-20(29)-en-23,28-dioic acid aglycone (110)				
623		28-Rha^4-Glc^6-Glc	*S. octophylla*	Sung *et al.* 1991
Cylicodiscic acid aglycone (111)				
624		3-Ara^2-Ara^3-Glc	*Cylicodiscus gabunensis* (Leguminosae)	Pambou Tchivounda *et al.* 1990
625		3-Glc^3-Ara	*C. gabunensis* (Leguminosae)	Pambou Tchivounda *et al.*, 1991
Mollic acid aglycone (117)				
626		3-Glc	*Combretum molle* (Combretaceae)	Pegel and Rogers, 1976
627		3-Ara	*C. molle* (Combretaceae)	Rogers and Thevan, 1986
			C. edwardsii (Combretaceae)	Rogers, 1989
628		3-Xyl	*C. molle* (Combretaceae)	Pegel and Rogers, 1985; Rogers and Thevan, 1986
			C. edwardsii (Combretaceae)	Rogers, 1989
3β,21,26-Trihydroxy-9,19-cyclolanost-24-en aglycone (118)				
629	Quadranguloside	3-Glc^6-Glc 26-Glc^6-Glc	*Passiflora quadrangularis* (Passifloraceae)	Orsini *et al.* 1986
Thalicogenin aglycone (119)				
630		3-Gal 28-Glc	*Thalictrum minus* (Ranunculaceae)	Gromova *et al.* 1984

(cont.)

Appendix 2. (*Cont.*)

No.	Name	Structure	Plant	Reference
3β,16β,24,25-Tetrahydroxy-9,19-cyclolanostane aglycone (120)				
631	Cyclofoetoside A	3-Ara 16-Glc^6-Rha	*T. foetidum* (Ranunculaceae)	Ganenko *et al.* 1986
3β,6α,16β,24,25-Pentahydroxy-9,19-cyclolanostane aglycone (121)				
632	Askendoside C	3-Xyl^2-Ara	*Astragalus taschkendicus* (Leguminosae)	Isaev *et al.* 1983a
Cycloastragenol aglycone (122)				
633	Cycloaraloside A	3-Glc	*A. amarus* (Leguminosae)	Isaev *et al.* 1989
634	Cyclogaleginoside B	3-Xyl	*A. galegiformis* (Leguminosae)	Alaniya *et al.* 1984
			A. membranaceus (Leguminosae)	Cao *et al.* 1985
635	Astragaloside IV	3-Xyl 6-Glc	*A. membranaceus* (Leguminosae) *A. sieversianus* (Leguminosae)	Kitagawa *et al.* 1983a; Cao *et al.* 1985 Gan *et al.* 1986b
635a	Isoastragaloside IV	3-Xyl 25-Glc	*A. membranaceus* (Leguminosae)	He and Findlay, 1991
636	Astragaloside VII	3-Xyl 6-Glc 25-Glc	*A. membranaceus* (Leguminosae) *A. kuhitangi* (Leguminosae)	Kitagawa *et al.* 1983e Agzamova *et al.* 1988
637	Astrasieversianin X	3-Xyl 6-Xyl	*A. sieversianus* (Leguminosae)	Gan *et al.* 1986b
638	Astrailienin A	3-Glc^2-Apif	*A. iliensis* (Leguminosae)	Chen *et al.* 1990b

639	Askendoside D	3-Xyl^2-Ara 6-Xyl	*A. taschkendicus* (Leguminosae)	Isaev et al. 1983b
640	Cyclosieversioside G (astrasieversianin XV)	3-Xyl^2-Rha 6-Xyl	*A. sieversianus* (Leguminosae)	Svechnikova et al. 1983a; Gan et al. 1986a
641	Cyclosieversioside H	3-Xyl^2-Rha 6-Glc	*A. sieversianus* (Leguminosae)	Svechnikova et al. 1983b
642	Asernestioside A	3-Xyl^2-Rha 25-Glc	*A. ernestii* (Leguminosae)	Wang et al. 1989a

3β-Hydroxy-9,19-cyclolanost-24(28)-en aglycone (123)

643	Acanthoside K$_2$	3-Glc^4-Glc	*Acanthopanax sessiliflorus* (Araliaceae)	Kong et al. 1988

Jessic acid aglycone (124)

644		3-Ara	*Combretum elaeagnoides* (Combretaceae)	Osborne and Pegel, 1984

Appendix 3

Appendix 3. *Structures of spirostanol saponins (cf. p. 78, Table 2.8)*

No.	Name	Structure	Plant	Reference
915	Asparagoside A	Glc-^{3}[Sarsasapogenin (**870**)]	*Asparagus officinalis* (Liliaceae)	Goryanu et al. 1976
916		Gal-^{3}[Sarsasapogenin (**870**)]	*Cornus florida* (Cornaceae)	Hostettmann et al. 1978
917		Xyl-^{2}Gal-^{3}[Sarsasapogenin (**870**)]	*C. florida* (Cornaceae)	Hostettmann et al. 1978
918		Glc-^{2}Gal-^{3}[Sarsasapogenin (**870**)]	*C. florida* (Cornaceae) *Anemarrhena asphoderoides* (Liliaceae)	Hostettmann et al. 1978 Takeda et al. 1989
919	Asparanin A	Glc-^{2}Glc-^{3}[Sarsasapogenin (**870**)]	*Asparagus adscendens* (Liliaceae)	Sharma et al. 1982
920		Glc-^{3}Glc-^{3}[Sarsasapogenin (**870**)]	*A. officinalis* (Liliaceae)	Goryanu et al. 1976

921	Asparanin B	Rha-^{4}Glc-^{3}[Sarsasapogenin (**870**)]
		    |^2
		   Glc
		   Glc
			A. officinalis (Liliaceae)
			A. adscendens (Liliaceae)
				Pant et al. 1988d
				Sharma et al. 1982

922	As-1	Xyl-^{4}Glc-^{3}[Sarsasapogenin (**870**)]
		    |^2
		   Glc
		    |^6
		   Glc
			A. officinalis (Liliaceae)	Shimoyamada et al. 1990

923	Parillin	Rha-^{4}Glc-^{3}[Sarsasapogenin (**870**)]
		    |^2
		   Glc
			Smilax aristolochiaefolia (Liliaceae)	Tschesche et al. 1966b

No.	Structure	Source	Reference
924	Ara ↓6 Rha-⁴Glc-³[Sarsasapogenin (**870**)] ↓2 Glc	*Asparagus curillus* (Liliaceae)	Sati and Sharma, 1985
925 YS-I	Glc-²Glc-³[Smilagenin (**871**)]	*Yucca gloriosa* (Agavaceae)	Nakano *et al.* 1989b
926 YS-II	Glc-²Gal-³[Smilagenin (**871**)]	*Y. gloriosa* (Agavaceae)	Nakano *et al.* 1989b
927 Melongoside L	Glc-⁴Rha-³Glc-³[Smilagenin (**871**)] ↓2 Gal-⁴Glc	*Solanum melongena* (Solanaceae)	Kintia and Shvets, 1985a
928 YS-XI	Glc-²Gal-³[12β-Hydroxysmilagenin (**872**)]	*Yucca gloriosa* (Agavaceae)	Nakano *et al.* 1991b
929 YS-XII	Glc-³Glc-³[12β-Hydroxysmilagenin (**872**)] ↓2 Glc	*Y. gloriosa* (Agavaceae)	Nakano *et al.* 1991b
930	Glc-³[Rhodeasapogenin (**873**)]	*Rhodea japonica* (Liliaceae)	Miyahara *et al.* 1983
931	Rha-²Xyl-¹[Rhodeasapogenin (**873**)]	*R. japonica* (Liliaceae)	Miyahara *et al.* 1983
932 Convallasaponin D	Glc ⟍1 ⟋3 [Rhodeasapogenin (**874**)]	*Convallaria keisukei* (Liliaceae)	Kimura *et al.* 1968a
933 Convallasaponin C	Rha-²Xyl-³Rha Rha-³Rha-²Ara-³[Isorhodeasapogenin (**873**)]	*C. keisukei* (Liliaceae)	Kimura *et al.* 1966
934 YS-V	Glc-²Gal-³[Samogenin (**875**)]	*Yucca gloriosa* (Agavaceae)	Nakano *et al.* 1989a
935 YS-XIII	Glc-²Gal-³[12β-Hydroxysamogenin (**876**)]	*Y. gloriosa* (Agavaceae)	Nakano *et al.* 1991b
936	Glc-²Gal-³[Markogenin (**877**)]	*Anemarrhena asphoderoides* (Liliaceae)	Takeda *et al.* 1989
937 Yononin	Ara-²[Yonogenin (**878**)]	*Diosorea tokoro* (Dioscoreaceae)	Kawasaki and Miyahara, 1965
938 Convallasaponin A	Ara-³[Convallagenin A (**879**)]	*Convallaria keisukei* (Liliaceae)	Kimura *et al.* 1968a

(cont.)

Appendix 3. (Cont.)

No.	Name	Structure	Plant	Reference

939
940
	Convallasaponin B
	Glucoconvalla-
	saponin B
		Ara-⁵[Convallagenin B (**880**)]
		Ara
		    |⁵
		    [Convallagenin B (**880**)]
		    |³
		Glc
			C. keisukei (Liliaceae)
			C. majalis (Liliaceae)
				Kimura et al. 1968a
				Kimura et al. 1968b

941	Tokoronin	Ara-¹[Tokorogenin (**881**)]	Dioscorea tokoro (Dioscoreaceae)	Miyahara and Kawasaki, 1969
942	Tokorogenin- glucoside	Glc-¹[Tokorogenin (**881**)]	D. tokoro (Dioscoreaceae)	Miyahara et al. 1969

943
944
945
	Agaveside B	Glc-²Glc-³[Tigogenin (**882**)]
		Xyl-³Glc-²Xyl-²Glc-³Glc-³[Tigogenin (**882**)]
		Xyl-³Glc-³Glc-³[Tigogenin (**882**)]
		  |² |²
		  Xyl Gal
			Yucca aloifolia (Agavaceae)
			Y. aloifolia (Agavaceae)
			Agave cantala (Agavaceae)
				Bahuguna and Sati, 1990
				Bahuguna and Sati, 1990
				Uniyal et al. 1990

946	Agaveside A	Xyl-³Glc-³Glc-³[Tigogenin (**882**)]
		  |² |²
		  Xyl Gal³-Xyl
			A. cantala (Agavaceae)	Uniyal et al. 1990

947		Xyl-⁴Gal-³Glc-³[Tigogenin (**882**)]
		    |³
		    Xyl
			A. cantala (Agavaceae)	Kishor, 1990

948	Lanatigonin I	Xyl-³Glc-³[Tigogenin (**882**)]
		  |²
		  Glc-³Gal
			Digitalis lanata
			(Scrophulariaceae)
				Tschesche and Balle, 1963

949	Tigonin	Xyl-³Glc-⁴Gal-³[Tigogenin (**882**)]
		    |²
		    Glc-³Gal
			D. lanata (Scrophulariaceae)
			D. purpurea
			(Scrophulariaceae)
				Mahato et al. 1982
				Mahato et al. 1982

No.	Name	Structure	Source	Reference
950	Sativoside-R2	Xyl-^{3}Glc-^{4}Gal-^{3}[Tigogenin (**882**)] |^2 Glc-^{3}Glc	*Allium sativum* (Liliaceae)	Matsuura *et al.* 1989a
951	Desglucolanatigonin II	Xyl-^{3}Glc-^{4}Gal-^{3}[Tigogenin (**882**)] |^2 Gal	*Digitalis lanata* (Scrophulariaceae)	Tschesche *et al.* 1972
952	Desgalactotigonin (uttronin A)	Xyl-^{3}Glc-^{4}Gal-^{3}[Tigogenin (**882**)] |^2 Glc	*D. purpurea* (Scrophulariaceae) *Anemarrhena asphodeloides* (Liliaceae)	Mahato *et al.* 1982 Nagumo *et al.* 1991
953	Dongnoside E	Glc-^{3}Glc-^{4}Gal-^{3}[Tigogenin (**882**)] |^2 Xyl	*Agave sisalana* (Agavaceae)	Ding *et al.* 1989
954	Dongnoside D	Glc-^{3}Glc-^{4}Gal-^{3}[Tigogenin (**882**)] |^2 Xyl-^{3}Xyl	*A. sisalana* (Agavaceae)	Ding *et al.* 1989

955	Dongnoside C	Glc-^{3}Glc-^{4}Gal-^{3}[Tigogenin (**882**)]
		  |^2
		Rha-^{4}Xyl
		  |^4
		Glc
			A. sisalana (Agavaceae)	Ding *et al.* 1989

956	Yuccaloside B	Rha-^{4}Glc-^{3}Gal-^{3}[Tigogenin (**882**)]
		  |^2
		Glc
		Glc
		  |^4
		Glc
			Yucca aloifolia (Agavaceae)	Benidze *et al.* 1987

957	Yuccaloside C	Rha-^{4}Glc-^{3}Gal-^{3}[Tigogenin (**882**)] |^2 Glc-^{3}Glc	*Y. aloifolia* (Agavaceae)	Benidze *et al.* 1987

(*cont.*)

Appendix 3. (*Cont.*)

No.	Name	Structure	Plant	Reference
958	YG-1	Xyl-^{4}Xyl-^{3}Glc-^{4}Gal-^{3}[Tigogenin (**882**)]	*Y. gloriosa* (Agavaceae)	Nakano *et al.* 1988

959	Diuranthoside A	Xyl-^{3}Glc-^{4}Gal-^{3}[Neotigogenin (**883**)]
		    |^2
		   Glc
			Diuranthera major (Anthericaceae)	Li *et al.* 1990b

960	Chloromaloside C	Xyl-^{3}Glc-^{4}Gal-^{3}[Neotigogenin (**883**)]
		    |^2
		  Ara Rha
			Chlorophytum malayense (Anthericaceae)	Li *et al.* 1990a

961 **962**		Glc-^{3}Glc-^{3}[Gitogenin (**884**)] Rha-^{3}Ara-^{3}Xyl-^{3}Glc-^{3}[Gitogenin (**884**)]	*Agave cantala* (Agavaceae) *Yucca aloifolia* (Agavaceae)	Jain, 1987a Kishor and Sati, 1990

963	Gitonin	Xyl-^{3}Glc-^{4}Gal-^{3}[Gitogenin (**884**)]
		    |^2
		   Gal
			Digitalis purpurea (Scrophulariaceae)	Tschesche *et al.* 1972

964	F-gitonin	Xyl-^{3}Glc-^{4}Gal-^{3}[Gitogenin (**884**)]
		    |^2
		   Glc
			Digitalis purpurea (Scrophulariaceae)
			Allium sativum (Liliaceae)
			Anemarrhena asphodeloides (Liliaceae)
				Kawasaki *et al.* 1965
				Matsuura *et al.* 1989a
				Nagumo *et al.* 1991

965	YG-2	Xyl-^{4}Xyl-^{3}Glc-^{4}Gal-^{4}[Gitogenin (**884**)]
		    |^2
		   Glc
			Yucca gloriosa (Agavaceae)	Nakano *et al.* 1988

966 **967**		Xyl-^{3}Glc-^{3}Xyl-^{3}Glc-^{3}Glc-^{3}[Gitogenin (**884**)] Xyl-^{3}Gal-^{3}Glc-^{3}Glc-^{3}[Gitogenin (**884**)] |^2 Xyl-^{3}Rha-^{3}Xyl	*Y. aloifolia* (Agavaceae) *Y. aloifolia* (Agavaceae)	Bahuguna *et al.* 1991 Kishor *et al.* 1991

968	Agaveside C	Glc-^{3}Glc-^{3}Glc-^{3}[Gitogenin (884)] $\quad\vert^2 \quad \vert^2$ Rha Rha^4-Xyl	Agave cantala (Agavaceae)	Uniyal et al. 1991
969	Ampeloside Bs$_1$	Glc-^{4}Gal-^{3}[Agigenin (885)]	Allium ampeloprasum (Liliaceae)	Morita et al. 1988
970		Glc-^{3}Glc-^{4}Gal-^{3}[Agigenin (885)]	A. ampeloprasum (Liliaceae)	Morita et al. 1988
971	Aginoside	Xyl-^{3}Glc-^{4}Gal-^{3}[Agigenin (885)] $\qquad\quad\vert^2$ $\qquad\quad$Glc	A. porrum (Liliaceae) A. giganteum (Liliaceae)	Harmatha et al. 1987 Kawashima et al. 1991
972	Digitonin	Xyl-^{3}Glc-^{4}Gal-^{3}[Digitogenin (886)] $\qquad\quad\vert^2$ $\qquad\quad$Glc-^{3}Gal	Digitalis purpurea (Scrophulariaceae)	Tschesche and Wulff, 1963
973		Glc-^{6}[Chlorogenin (887)]	Camassia cusickii (Liliaceae)	Mimaki et al. 1991
974		Glc-^{2}Glc-^{6}[Chlorogenin (887)]	C. cusickii (Liliaceae)	Mimaki et al. 1991
975		Glc-^{3}Glc-^{6}[Chlorogenin (887)]	C. cusickii (Liliaceae)	Mimaki et al. 1991
976		Glc-^{3}Glc-^{6}[Chlorogenin (887)] $\qquad\quad\vert^2$ $\qquad\quad$Glc	C. cusickii (Liliaceae)	Mimaki et al. 1991
977	Paniculonin A	Xyl-^{3}Qui-^{6}[Paniculogenin (888)]	Solanum paniculatum (Solanaceae)	Ripperger and Schreiber, 1968
978	Paniculonin B	Rha-^{3}Qui-^{6}[Paniculogenin (888)]	S. paniculatum (Solanaceae)	Ripperger and Schreiber, 1968
979	Pardarinoside E	Ara-^{3}Glc-^{3}[Spirostan-3β,17α,21-triol (889)] $\qquad\quad\vert^2$ $\qquad\quad$Rha	Lilium pardarinum (Liliaceae)	Shimomura et al. 1989
980		Glc-^{2}[Alliogenin (890)]	Allium giganteum (Liliaceae)	Sashida et al. 1991a
981		Glc$\diagdown^2$[Alliogenin (890)] Ac$\diagup^3$	A. giganteum (Liliaceae)	Sashida et al. 1991a

(cont.)

Appendix 3. (*Cont.*)

No.	Name	Structure	Plant	Reference
982		Glc$\diagdown^2$[Alliogenin (**890**)] Bz$\diagup^3$	*A. giganteum* (Liliaceae)	Sashida *et al.* 1991a
983		Glc-^{2}[(25R)-5α-Spirostan- 2α,3β,5α,6α-tetraol (**891**)]	*A. aflatunense* (Liliaceae)	Kawashima *et al.* 1991
984		Glc-^{24}[(24S,25R)-5α-Spirostan- 2α,3β,5α,6β,24-pentaol (**892**)]	*A. giganteum* (Liliaceae)	Kawashima *et al.* 1991
985	Collettiside I	Glc-^{3}[Yamogenin (**893**)]	*Diosorea colletti* (Dioscoreaceae)	Xu and Lin, 1985
986	Collettiside III	Rha-^{4}Glc-^{3}[Yamogenin (**893**)] $\vert^2$ Rha	*D. colletti* (Dioscoreaceae)	Xu and Lin, 1985
987	Collettiside IV	Glc-^{3}Glc-^{3}[Yamogenin (**893**)] $\vert^2$ Rha	*D. colletti* (Dioscoreaceae)	Xu and Lin, 1985
988	Balanitin 2	Xyl-^{6}Glc-^{3}Glc-^{3}[Yamogenin (**893**)] $\vert^2$ Rha	*Balanites aegyptiaca* (Balanitaceae)	Liu and Nakanishi, 1982
989	Balanitin 3	Rha-^{2}Glc-^{4}Glc-^{3}[Yamogenin (**893**)] $\vert^2$ Rha	*B. aegyptiaca* (Balanitaceae)	Liu and Nakanishi, 1982
990	Balanitin 1	Rha-^{2}Glc-^{4}Glc-^{3}[Yamogenin (**893**)] $\vert^2$ Rha	*B. aegyptiaca* (Balanitaceae)	Liu and Nakanishi, 1982
991	Balanitin 4	Glc-^{3}Glc-^{4}Glc-^{3}[Yamogenin (**893**)] $\vert^2$ Rha	*B. aegyptiaca* (Balanitaceae)	Pettit *et al.* 1991

992	Balanitin 5	Rha-³Glc-⁴Glc-³[Yamogenin (**893**)] │² Rha	*B. aegyptiaca* (Balanitaceae)	Pettit *et al.* 1991
993	Balanitin 6	Glc-⁴Glc-³[Yamogenin (**893**)] │² Rha	*B. aegyptiaca* (Balanitaceae)	Pettit *et al.* 1991
994	Trillin	Glc-³[Diosgenin (**894**)]	*Paris polyphylla* (Liliaceae) *Yucca filamentosa* (Agavaceae)	Zhou, 1989 Kintia *et al.* 1972
995	Gracillin	Glc-³Glc-³[Diosgenin (**894**)] │² Rha	*Dioscorea gracillima* (Dioscoreaceae) and other *Dioscorea* spp. *Paris polyphylla* (Liliaceae)	Kawasaki and Yamauchi, 1962 Zhou, 1989
996		Rha-²Glc-³[Diosgenin (**894**)]	*P. vietnamensis* (Liliaceae) *P. polyphylla* (Liliaceae) *Trillium kamtschaticum* (Liliaceae) *Ophiopogon planiscapus* (Liliaceae) *Allium vineale* (Liliaceae)	Namba *et al.* 1989 Xu and Zhong, 1988b; Zhou, 1989 Nohara *et al.* 1975 Watanabe *et al.* 1983 Chen and Snyder, 1989
997 998 999	Polyphyllin C Convallasaponin E	Rha-²Glc-³[Diosgenin (**894**)]²⁶-OMe Rha-³Glc-³[Diosgenin (**894**)] Ara-²Ara-²Ara-³[Diosgenin (**894**)]	*Lilium speciosum* (Liliaceae) *Paris polyphylla* (Liliaceae) *Convallaria keisukei* (Liliaceae)	Mimaki and Sashida, 1991 Singh *et al.* 1982b Kimura *et al.* 1968a
1000		Glc-⁴Glc-³[Diosgenin (**894**)] │² Rha	*Ophiopogon planiscapus* (Liliaceae) *Allium vineale* (Liliaceae)	Watanabe *et al.* 1983 Chen and Snyder, 1989

(cont.)

Appendix 3. (*Cont.*)

No.	Name	Structure	Plant	Reference
1001	Floribundasaponin C	Rha-^{3}Rha-^{4}Glc-^{3}[Diosgenin (**894**)]	*Dioscorea floribunda* (Dioscoreaceae)	Mahato *et al.* 1978
1002	Dioscin	Rha-^{4}Glc-^{3}[Diosgenin (**894**)] $\mid^2$ Rha	*Dioscorea* spp. (Dioscoreaceae)	Kawasaki and Yamauchi, 1962
			Costus speciosus (Zingiberaceae)	Mahato *et al.* 1982
			Paris polyphylla (Liliaceae)	Mahato *et al.* 1982
			Solanum introsum (Solanaceae)	Mahato *et al.* 1982
			Trigonella foenum-graecum (Leguminosae)	Mahato *et al.* 1982
			Trachycarpus fortunei (Palmae)	Hirai *et al.* 1984
			Heloniopsis orientalis (Liliaceae)	Nakano *et al.* 1989a
			Trillium grandiflorum (Liliaceae)	Hufford *et al.* 1988
1003		Ara-^{3}Glc-^{3}[Diosgenin (**894**)] $\mid^2$ Rha	*Paris polyphylla* var. *chinensis* (Liliaceae)	Xu and Zhong, 1988a
1004	Polyphyllin D	Ara-^{4}Glc-^{3}[Diosgenin (**894**)] $\mid^2$ Rha	*P. polyphylla* (Liliaceae)	Xu and Zhong, 1988b; Zhou, 1989; Singh *et al.* 1982b; Ma and Lau, 1985
			P. vietnamesis	Namba *et al.* 1989

1005		Glc-^{3}Rha-^{4}Glc-^{3}[Diosgenin (**894**)] 　　　　　　　 $|^2$ 　　　　　　　Rha	*Allium vineale* (Liliaceae)	Chen and Snyder, 1987, 1989
1006	Melongoside M	Glc-^{4}Rha-^{3}Glc-^{3}[Diosgenin (**894**)] 　　　　　　　 $|^2$ 　　　　　　Gal-^{4}Glc	*Solanum melongena* (Solanaceae)	Kintia and Shvets, 1985a
1007	Pb	Rha-^{4}Rha-^{4}Glc-^{3}[Diosgenin (**894**)] 　　　　　　　 $|^2$ 　　　　　　　Rha	*Trachycarpus fortunei* (Palmae) *Paris polyphylla* (Liliaceae)	Hirai *et al.* 1984 Zhou, 1989; Xu and Zhong, 1988b
1008	Polyphyllin E	Rha-^{2}Rha-^{4}Glc-^{3}[Diosgenin (**894**)] 　　　　　　　 $|^3$ 　　　　　　　Rha	*P. polyphylla* (Liliaceae)	Singh *et al.* 1982b
1009		Glc-^{3}Glc-^{4}Glc-^{3}[Diosgenin (**894**)] 　　　　　　　 $|^2$ 　　　　　　　Rha	*Balanites roxburghii* (Balanitaceae)	Jain, 1987b
1010		Glc-^{4}Rha-^{4}Glc-^{3}[Diosgenin (**894**)] 　　　　　　　 $|^2$ 　　　　　　　Rha	*Allium vineale* (Liliaceae)	Chen and Snyder, 1989
1011		Glc-^{6}Glc-^{4}Glc-^{3}[Diosgenin (**894**)] 　　　 $|^3$　 $|^2$ 　　　Glc　Rha	*A. vineale* (Liliaceae)	Chen and Snyder, 1989
1012		Glc-^{6}Glc-^{4}Rha-^{4}Glc-^{3}[Diosgenin (**894**)] 　　　　　　　 $|^2$ 　　　　　　　Rha	*A. vineale* (Liliaceae)	Chen and Snyder, 1989
1013		Glc-^{6}Glc-^{4}Rha-^{4}Glc-^{3}[Diosgenin (**894**)] 　　　 $|^4$　 $|^2$ 　　　Glc　Rha	*A. vineale* (Liliaceae)	Chen and Snyder, 1989
1014	Balanitin 7	Xyl-^{3}Glc-^{4}Glc-^{3}[Diosgenin (**894**)] 　　　　　　　 $|^2$ 　　　　　　　Rha	*Balanites aegyptiaca* (Balanitaceae)	Pettit *et al.* 1991

(cont.)

Appendix 3. (Cont.)

No.	Name	Structure	Plant	Reference
1015		Ara-^{2}Glc-^{1}[Ruscogenin (**897**)]	*Ophiopogon intermedius* (Liliaceae)	Rawat et al. 1988
1016		Rha-^{2}Ara-^{1}[Ruscogenin (**897**)] $\qquad$ |^4 $\qquad$ SO$_3^-$	*O. planiscapus* (Liliaceae)	Watanabe et al. 1983
1017	Ophiopogonin-B	Rha-^{2}Fuc-^{1}[Ruscogenin (**897**)]	*O. japonicus* (Liliaceae)	Tada and Shoji, 1972
1018	Lm-2	Glc-^{2}Fuc-^{1}[Ruscogenin (**897**)]	*Liriope muscari* (Liliaceae)	Yu et al. 1990
1019	Spicatoside A	Xyl-^{3}Fuc-^{1}[Ruscogenin (**897**)] $\qquad$ |^2 $\qquad$ Glc	*L. spicata* (Liliaceae)	Lee et al. 1989
1020	Alliospiroside A	Rha-^{2}Ara-^{1}[(25S)-Ruscogenin (**898**)]	*Allium cepa* (Liliaceae)	Kravets et al. 1986b
1021	Alliospiroside B	Rha-^{2}Gal-^{1}[(25S)-Ruscogenin (**898**)]	*A. cepa* (Liliaceae)	Kravets et al. 1986a
1022	Ls-2	Fuc $\underset{\ 3}{\overset{1}{\diagup}}$[(25$S$)-Ruscogenin (**898**)] Rha $\diagdown$ Xyl	*Liriope spicata* (Liliaceae)	Yu et al. 1990
1023	Ls-3	Xyl $\underset{\ 3}{\overset{1}{\diagup}}$[(25$S$)-Ruscogenin (**898**)] Rha	*L. spicata* (Liliaceae)	Yu et al. 1990
1024	Ls-5	Xyl-^{3}Fuc-^{1}[(25S)-Ruscogenin (**898**)] $\qquad$ |^2 $\qquad$ Glc	*L. spicata* (Liliaceae)	Yu et al. 1990
1025		Ac-^{2}Rha Xyl-^{3}Glc-^{4}Gal-^{3}[Neoprazerigenin (**899**)] $\qquad$ |^2 $\qquad$ Glc	*Polygonatum sibiricum* (Liliaceae)	Son et al. 1990

No.	Name	Structure	Plant	Reference
1027		Rha-^{2}Glc-^{3}[Pennogenin (**900**)]	*P. polyphylla* (Liliaceae) *Trillium kamtschaticum* (Liliaceae)	Zhou, 1989 Nohara *et al.* 1975
1028		Glc-^{3}[Pennogenin (**900**)] $|^2$ Rha	*Paris polyphylla* (Liliaceae)	Zhou, 1989
1029		Rha-^{4}Glc-^{3}[Pennogenin (**900**)] $|^2$ Rha	*Trillium kamtschaticum* (Liliaceae) *Heloniopsis orientalis* (Liliaceae) *Paris polyphylla* (Liliaceae) *P. quadrifolia* (Liliaceae)	Nohara *et al.* 1975 Nakano *et al.* 1989a Zhou, 1989 Nohara *et al.* 1982
1030		Rha-^{3}Glc-^{3}[Pennogenin (**900**)] $|^2$ Rha	*Dracaena mannii* (Agavaceae)	Okunji *et al.* 1990, 1991
1031		Rha-^{4}Rha-^{4}Glc-^{3}[Pennogenin (**900**)] $|^2$ Rha	*Paris polyphylla* (Liliaceae) *P. quadrifolia* (Liliaceae) *Trillium grandiflorum* (Liliaceae)	Zhou, 1989 Nohara *et al.* 1982 Hufford *et al.* 1988
1032		Rha-^{2}Glc-^{3}[Isonuatigenin (**901**)]	*Allium vineale* (Liliaceae)	Chen and Snyder, 1987, 1989
1033	Alliospiroside C	Rha-^{2}Ara-^{1}[Cepagenin (**902**)]	*A. cepa* (Liliaceae)	Kravets *et al.* 1987
1034	Alliospiroside D	Rha-^{2}Gal-^{1}[Cepagenin (**902**)]	*A. cepa* (Liliaceae)	Kravets *et al.* 1987
1035		Glc-^{3}[24α-Hydroxypennogenin (**903**)]	*Paris polyphylla* (Liliaceae)	Zhou, 1989
1036		Ara-^{4}Glc-^{3}[24α-Hydroxypennogenin (**903**)] $|^2$ Rha	*P. polyphylla* (Liliaceae)	Zhou, 1989
1037		Rha-^{2}Glc-^{3}[Ophiogenin (**904**)]	*Ophiopogon japonicus* (Liliaceae)	Adinolfi *et al.* 1990

(cont.)

Appendix 3. (*Cont.*)

No.	Name	Structure	Plant	Reference
1038	Sibiricoside B	Xyl-³Glc-⁴Gal-³[Sibiricogenin (**905**)] $\overset{\displaystyle |^2}{\text{Glc}}$	*Polygonatum sibiricum* (Liliaceae)	Son *et al.* 1990
1039	Convallamarogenin triglycoside	Rha $\overset{\displaystyle 3}{\diagdown}$[Convallamarogenin (**906**)] Rha—²Qui $\diagup^1$	*Convallaria majalis* (Liliaceae)	Tschesche *et al.* 1972
1040	Ruscin	Glc-³Rha-²Ara-¹[Neoruscogenin (**907**)]	*Ruscus aculeatus* (Liliaceae) *R. ponticus* (Liliaceae)	Bombardelli *et al.* 1971 Korkashvili *et al.* 1985
1041		Rha-²Ara-¹[Neoruscogenin (**907**)]	*R. aculeatus* (Liliaceae) *R. ponticus* (Liliaceae)	Bombardelli *et al.* 1971 Korkashvili *et al.* 1985
1042		Ara-¹[Neoruscogenin (**907**)]	*R. aculeatus* (Liliaceae)	Bombardelli *et al.* 1971
1043	Agavoside A	Gal-³[Hecogenin (**908**)]	*Agave americana* (Agavaceae)	Kintia *et al.* 1975
1044	Cantalasaponin-2	Xyl-⁴Gal-³[Hecogenin (**908**)] $\overset{\displaystyle |^2}{\text{Glc-³Glc}}$	*A. cantala* (Agavaceae)	Pant *et al.* 1986
1045	Cantalasaponin-4	Xyl-⁴Gal-³[Hecogenin (**908**)] $\overset{\displaystyle |^2}{\text{Xyl-⁴Glc-³Glc}}$	*A. cantala* (Agavaceae)	Pant *et al.* 1986
1046	Chloromaloside A	Xyl-³Glc-⁴Gal-³[Neohecogenin (**909**)] $\overset{\displaystyle |^2}{\text{Glc}}$	*Chlorophytum malayense* (Anthericaceae) *Diuranthera major* (Anthericaceae)	Li *et al.* 1990a Li *et al.* 1990b

No.	Name	Structure	Source	Reference

1047	Chloromaloside B	Xyl-^{3}Glc-^{4}Gal-^{3}[Neohecogenin (**909**)]
		     |^2 |^2
		    Ara Rha
			Chlorophytum malayense (Anthericaceae)	Li *et al.* 1990a

1048	Diuranthoside B	Glc-^{3}Xyl-^{3}Glc-^{4}Gal-^{3}[Neohecogenin (**909**)]
		         |^2
		        Glc
			Diuranthera major (Anthericaceae)	Li *et al.* 1990b

1049	Diuranthoside C	Glc-^{3}Xyl-^{3}G-^{4}Gal-^{3}[Neohecogenin (**909**)]
		      |^2
		   Glc-^{3}Glc
			D. major (Anthericaceae)	Li *et al.* 1990b

1050	YS-IX	Glc-^{2}Glc-^{4}Gal-^{3}[Manogenin (**910**)]
		     |^3
		    Xyl
			Yucca gloriosa (Agavaceae)	Nakano *et al.* 1991a,b

1051	Avenacoside A	Rha-^{4}Glc-^{3}[Nuatigenin (**912**)]^{26}-Glc
		     |^2
		    Glc
			Avena sativa (Gramineae)	Tschesche *et al.* 1969d; Kesselmeier and Budzikiewicz, 1979

1052	Avenacoside B	Rha-^{4}Glc-^{3}[Nuatigenin (**912**)]^{26}-Glc
		     |^2
		   Glc-^{2}Glc
			A. sativa (Graminae)	Tschesche *et al.* 1969d; Kesselmeier and Budzikiewicz, 1979

1053		Rha-^{2}Glc-^{3}[Nuatigenin (**912**)]	*Allium vineale* (Liliaceae)	Chen and Snyder, 1987, 1989
1054		Rha-^{2}Glc-^{3}[Nuatigenin (**912**)]^{26}-Glc	*Lilium brownii* (Liliaceae)	Mimaki and Sashida, 1990a,c

1055	Aculeatiside A	Rha-^{4}Glc-^{3}[Nuatigenin (**912**)]^{26}-Glc
		     |^2
		    Rha
			Solanum aculeatissimum (Solanaceae)	Saijo *et al.* 1983

1056	Aculeatiside B	Glc-^{3}Gal-^{3}[Nuatigenin (**912**)]^{26}-Glc
		     |^2
		    Rha
			S. aculeatissimum (Solanaceae)	Saijo *et al.* 1983

1057		Rha-^{3}Rha-^{3}[Hispigenin (**913**)]	*S. hispidum* (Solanaceae)	Chakravarty *et al.* 1979
1058		Rha-^{3}Rha-^{3}[Solagenin (**914**)]	*S. hispidum* (Solanaceae)	Chakravarty *et al.* 1979

References

Abe, F. and Yamauchi, T. (1987). Glycosides of 19α-hydroxyoleanane-type triterpenoids from *Trachelospermum asiaticum* (Trachelospermum. IV). *Chem. Pharm. Bull.*, **35**, 1833–1838.

Abe, H., Odashima, S. and Arichi, S. (1978). The effects of saikosaponins on biological membranes. 2. Changes in electron spin resonance spectra from spin-labelled erythrocyte and erythrocyte ghost membranes. *Planta Medica*, **34**, 287–290.

Abe, H., Ohya, N., Yamamoto, K.F., Shibuya, T., Arichi, S. and Odashima, S. (1987). Effects of glycyrrhizin and glycyrrhetinic acid on growth and melanogenesis in cultured B16 melanoma cells. *Eur. J. Cancer Clin. Oncol.*, **23**, 1549–1555.

Abe, I., Rohmer, M. and Prestwich, G.D. (1993). Enzymatic cyclization of squalene and oxidosqualene to sterols and triterpenes. *Chem. Rev.*, **93**, 2189–2206.

Abegaz, B. and Tecle, B. (1980). A new triterpenoid glycoside from the seeds of *Glinus lotoides. Phytochemistry*, **19**, 1553–1554.

Abisch, E. and Reichstein, T. (1960). Orientierende chemische Untersuchung einiger Apocynaceen. *Helv. Chim. Acta*, **43**, 1844–1861.

Acharya, S.B., Tripathi, S.K., Tripathi, Y.C. and Pandey, V.B. (1988). Some pharmacological studies on *Zizyphus rugosa* saponins. *Indian J. Pharmacol.*, **20**, 200–202.

Achenbach, H., Hefter-Buebl, U. and Constenla, M.A. (1987). Fevicordin A and fevicordin A glucoside, novel norcucurbitacins from *Fevillea cordifolia. J. Chem. Soc., Chem. Commun.*, 441–442.

Adesina, S.K. (1985). Constituents of *Solanum dasyphyllum* fruit. *J. Nat. Prod.*, **48**, 147.

Adesina, S.K. and Gbile, Z.O. (1984). Steroidal constituents of *Solanum scabrum* subsp. *nigericum. Fitoterapia*, **55**, 362–363.

Adesina, S.K. and Reisch, J. (1985). A triterpenoid glycoside from *Tetrapleura tetraptera* fruit. *Phytochemistry*, **24**, 3003–3006.

Adesina, S.K., Adewunmi, C.O. and Marquis, V.O. (1980). Phytochemical investigations of the molluscicidal properties of *Tetrapleura tetraptera* (TAUB). *J. Afr. Med. Plants*, **3**, 7–15.

Adewunmi, C.O. (1984). Natural products as agents of schistosomiasis control in Nigeria: a review of progress. *Int. J. Crude Drug Res.*, **22**, 161–166.

Adewunmi, C.O., Awe, S.O. and Adesina, S.K. (1988). Enhanced potency of a molluscicidal glycoside isolated from *Tetrapleura tetraptera* on *Biomphalaria glabrata*. *Planta Medica*, **54**, 550–551.

Adinolfi, M., Barone, G., Corsaro, M.M., Lanzetta, R., Mangoni, L. and Parrilli, M. (1987). Glycosides from *Muscari comosum*. 7. Structures of three novel muscarosides. *Can. J. Chem.*, **65**, 2317–2326.

Adinolfi, M., Parrilli, M. and Zhu, Y. (1990). Terpenoid glycosides from *Ophiopogon japonicus* roots. *Phytochemistry*, **29**, 1696–1699.

Adinolfi, M., Corsaro, M.M., Lanzetta, R., Mancino, A., Mangoni, L. and Parrilli, M. (1993). Triterpenoid oligoglycosides from *Chionodoxa luciliae*. *Phytochemistry*, **34**, 773–778.

Adler, C and Hiller, K. (1985). Bisdesmosidische Triterpensaponine. *Pharmazie*, **40**, 676–693.

Agarwal, S.K. and Rastogi, R.P. (1974). Triterpenoid saponins and their genins. *Phytochemistry*, **13**, 2623–2645.

Ageta, M., Nonaka, G. and Nishioka, I. (1988). Tannins and related compounds. LXVII. Isolation and characterization of castanopsinins A–H, novel ellagintannins containing a triterpenoid glycoside core, from *Castanopsis cuspidata* var. *sieboldii* NAKAI. (3). *Chem. Pharm. Bull*, **36**, 1646–1663.

Agrawal, P.K. (1992). NMR spectroscopy in the structural elucidation of oligosaccharides and glycosides. *Phytochemistry*, **31**, 3307–3330.

Agrawal, P.K. and Jain, D.C. (1992). ^{13}C NMR spectroscopy of oleanane triterpenoids. *Prog. NMR Spectrosc.*, **24**, 1–90.

Agrawal, P.K., Jain, D.C., Gupta, R.K. and Thakur, R.S. (1985). Carbon-13 NMR spectroscopy of steroidal sapogenins and steroidal saponins. *Phytochemistry*, **24**, 2479–2496.

Aguwa, C.N. and Okunji, C.O. (1986). Gastrointestinal studies of *Pyrenacantha staudtii* leaf extracts. *J. Ethnopharmacol.*, **15**, 45–55.

Agzamova, M.A. and Isaev, M.I. (1991). Triterpene glycosides and their genins from *Astragalus*. XXXVIII. Cycloalpigenin D and cycloalpioside D from *Astragalus alopecurus*. *Khim. Prir. Soedin.*, 377–384.

Agzamova, M.A., Isaev, M.I., Gorovits, M.B. and Abubakirov, N.K. (1986). Triterpene glycosides of *Astragalus* and their genins. XIX. Cycloartanoic compounds and sterols of *Astragalus pamirensis* and *A. pterocephalus*. *Khim. Prir. Soedin.*, 117–118.

Agzamova, M.A., Isaev, M.I., Gorovits, M.B. and Abubakirov, N.K. (1987a). Triterpene glycosides and their genins from *Astragalus*. XXII. Cycloorbicoside A from *Astragalus orbiculatus*. *Khim. Prir. Soedin.*, 719–726.

Agzamova, M.A., Isaev, M.I., Gorovits, M.B. and Abubakirov, N.K. (1987b). Triterpene glycosides and their genins from *Astragalus*. XXIV. Cycloorbicoside G from *Astragalus orbiculatus*. *Khim. Prir. Soedin.*, 837–842.

Agzamova, M.A., Isaev, M.I., Mal'tsev, I.I., Gorovits, M.B. and Abubakirov, N.K. (1988). Triterpene glycosides and their genins from *Astragalus*. XXIX. Cycloartanes of *Astragalus kuhitangi*. *Khim. Prir. Soedin.*, 882–883.

Ahmad, V.U., Bano, N. and Bano, S. (1986a). A saponin from the stem bark of *Guaiacum officinale*. *Phytochemistry*, **25**, 951–952.

Ahmad, V.U., Bano, N., Bano, S., Fatima, A. and Kenne, L. (1986b). Guaianin, a new saponin from Guaiacum officinale. J. Nat. Prod., 49, 784–786.

Ahmad, V.U., Bano, N., Fatima, I. and Bano, S. (1988a). Isolation of two new saponins, guaianin D and E from the bark of Guaiacum officinale. Tetrahedron, 44, 247–252.

Ahmad, V.U., Sultana, V., Arif, S. and Saqib, Q.N. (1988b). Saponins from Primula denticulata. Phytochemistry, 27, 304–306.

Ahmad, V.U., Perveen, S. and Bano, S. (1989a). Guaiacin A and B from the leaves of Guaiacum officinale. Planta Medica, 55, 307–308.

Ahmad, V.U., Bano, S., Bano, N., Uddin, S., Perveen, S. and Fatima, I. (1989b). Structure of guaianin C from Guaiacum officinale. Fitoterapia, 60, 255–256.

Ahmad, V.U., Bano, S., Fatima, I., Bano, N., Riccio, R. and Minale, L. (1989c). Triterpenoid saponins of Guaiacum officinale. Gazz. Chim. Ital., 119, 31–34.

Ahmad, V.U., Uddin, S., Bano, S. and Fatima, I. (1989d). Two saponins from fruits of Guaiacum officinale. Phytochemistry, 28, 2169–2171.

Ahmad, V.U., Ghazala, Uddin, S. and Bano, S. (1990). Saponins from Zygophyllum propinquum. J. Nat. Prod., 53, 1193–1197.

Ahmad, V.U., Ahmed, W. and Usmanghani, K. (1992). Triterpenoid saponins from leaves of Castanospermum australe. Phytochemistry, 31, 2805–2807.

Ahmad, V.U., Noorwala, M., Mohammed, F.V., Sener, B., Gilani, A. and Aftab, K. (1993a). Symphytoxide A, a triterpenoid saponin from the roots of Symphytum officinale. Phytochemistry, 32, 1003–1006.

Ahmad, V.U., Shah, M.G., Mohammed, F.V. and Baqai, F.T. (1993b). Macrophyllicin, a saponin from Primula macrophylla. Phytochemistry, 32, 1543–1547.

Ahmed, E.M., Bashir, A.K. and El Khier, Y.M. (1985). Molluscicidal activity and chemical composition of Gardenia lutea. Fitoterapia, 56, 354–356.

Akai, E., Takeda, T., Kobayashi, Y. and Ogihara, Y. (1985a). Sulfated triterpenoid saponins from the leaves of Bupleurum rotundifolium L. Chem. Pharm. Bull., 33, 3715–3723.

Akai, E., Takeda, T., Kobayashi, Y., Chen, Y. and Ogihara, Y. (1985b). Minor triterpenoid saponins from the leaves of Bupleurum rotundifolium L. Chem. Pharm. Bull., 33, 4685–4690.

Akamatsu, H., Komura, J., Asada, Y. and Niwa, Y. (1991). Mechanism of anti-inflammatory action of glycyrrhizin: effect of neutrophil functions including reactive oxygen species generation. Planta Medica, 57, 119–121.

Akimaliev, A., Alimbaeva, P.K., Mzhelskaya, L.G. and Abubakirov, N.G. (1976). Triterpene glycosides of Scabiosa soongorica. II. Structure of songorosides C, G and I. Khim. Prir. Soedin., 472–476.

Alaniya, M.D. (1988). Cyclogaleginoside A from Astragalus falcatus LAM. Izv. Akad. Nauk Gruz. SSR, Ser. Khim., 14, 73–74.

Alaniya, M.D., Isaev, M.I., Gorovits, M.B., Abdullaev, N.D., Kemertelidze, E.P. and Abubakirov, N.K. (1984). Astragalus triterpene glycosides and their genins. XVI. Cyclogaleginosides A and B from Astragalus galegiformis. Khim. Prir. Soedin., 477–481.

Aleshinskaya, E.E., Aleshkina, Y.A., Berezhinskaya, V.V. and Trutneva, E.A. (1964). Pharmacology of Glycyrrhiza glabra (licorice) preparations. Farmakol. i Toksikol., 27, 217–222 (Chem. Abstr., 61, 12506).

Aliev, A.M. and Movsumov, U.S. (1976). Triterpene glycosides from *Cephalaria kotchyi* flowers. *Khim. Prir. Soedin.*, 264–265.

Alimbaeva, P.K., Akimaliev, A. and Mukhamedziev, M.M. (1977). Triterpene glycosides of some representatives of the Dipsacaceae family. *Khim. Prir. Soedin.*, 708–709.

Alzerreca, A. and Hart, G. (1982). Molluscicidal steroid glycoalkaloids possessing stereoisomeric spirosolane structures. *Toxicol. Lett.*, **12**, 151–155.

Amanmuradov, K. and Tanyurcheva, T.N. (1969). Triterpenic glycoside from *Acanthophyllum adenophorum*. *Khim. Prir. Soedin.*, 326.

Amer, M.A., Dawidar, A.M. and Fayez, M.B.E. (1974). Constituents of local plants. XVIII. The triterpenoid constituents of *Cornulaca monacantha*. *Planta Medica*, **26**, 289–292.

Amoros, M. and Girre, R.L. (1987). Structure of two antiviral triterpene saponins from *Anagallis arvensis*. *Phytochemistry*, **26**, 787–791.

Amoros, M., Fauconnier, B. and Girre, R.L. (1987). *In vitro* antiviral activity of a saponin from *Anagallis arvensis*, Primulaceae, against herpes simplex virus and poliovirus. *Antiviral Res.*, **8**, 13–25.

Amoros, M., Fauconnier, B. and Girre, R.L. (1988). Effect of saponins from *Anagallis arvensis* on experimental herpes simplex keratitis in rabbits. *Planta Medica*, **54**, 128–131.

Andersson, L. (1987). Biological screening of and chemical studies on some Swedish marine organisms. PhD. Thesis, Uppsala University, Sweden.

Andersson, L., Nasir, A. and Bohlin, L. (1987). Studies of Swedish marine organisms, IX. Polyhydroxylated steroidal glycosides from the starfish *Porania pulvillus*. *J. Nat. Prod.*, **50**, 944–947.

Andersson, L., Bohlin, L., Iorizzi, M., Riccio, R., Minale, L. and Moreno-Lopez, W. (1989). Biological activity of saponins and saponin-like compounds from starfish and brittle stars. *Toxicon*, **27**, 179–188.

Anil, H. (1979). 21-Benzoyl-barringtogenol C, a sapogenin from *Styrax officinalis*. *Phytochemistry*, **18**, 1760–1761.

Anisimov, M. and Chirva, V.J. (1980). Die biologische Bewertung von Triterpenglykoside. *Pharmazie*, **35**, 731–738.

Anisimov, M.M., Shentsova, E.B., Shcheglov, V.V. *et al.* (1978). Mechanism of cytotoxic action of some triterpene glycosides. *Toxicon*, **16**, 207–218.

Anisimov, M.M., Shcheglov, V.V., Strigina, L.I. *et al.* (1979). Chemical structure and antifungal activity of some triterpenoids. *Izv. Akad. Nauk SSSR, Ser. Biol.*, 570–575 (*Chem. Abstr.*, **91**, 187213).

Anisimov, M.M., Prokofeva, N.G., Korotkikh, L.Y., Kapustina, I.I. and Stonik, V.A. (1980). Comparative study of cytotoxic activity of triterpene glycosides from marine organisms. *Toxicon*, **18**, 221–223.

Annaev, C. and Abubakirov, N.K. (1984). Triterpene glycosides of *Climacoptera transoxana*. IV. Structure of copterosides G and H. *Khim. Prir. Soedin.*, 60–64.

Annaev, C., Isamukhamedova, M. and Abubakirov, N.K. (1983). Triterpene glycosides of *Salsola micranthera*. I. Structure of salsolosides C and D. *Khim. Prir. Soedin.*, 727–732.

Annaev, C., Isamukhamedova, M. and Abubakirov, N.K. (1984). Triterpene glycosides of *Salsola micranthera*. II. Structure of salsoloside E. *Khim. Prir. Soedin.*, 65–69.

Annoni, F., Mauri, A., Marincola, F. and Resele, L.F. (1979). Venotonic activity of escin on the human saphenous vein. *Arzneimittelforsch.*, **29**, 672–675.

Anonymous (1991). Pflanzliche Urologika: Warum man besser von 'Aquaretika' sprechen soll. *Dtsch. Apoth. Ztg.*, **131**, 838–840.

Ansamatova, R.A., Denikeeva, M.F., Sydygalieva, C.Z. and Koshoev, K.K. (1979). Triterpene glycosides of *Zygophyllum obliquum. Khim. Prir. Soedin.*, 237–238.

Ansari, A.A., Kenne, L. and Atta-ur-Rahman (1987). Isolation and characterization of two saponins from *Fagonia indica. Phytochemistry*, **26**, 1487–1490.

Ansari, A.A., Hassan, S., Kenne, L., Atta-ur-Rahman and Wehler, T. (1988). Structural studies on a saponin isolated from *Nigella sativa. Phytochemistry*, **27**, 3977–3979.

Aoki, T., Ohta, S., Aratani, S., Hirata, T. and Suga, T. (1982). The structures of four novel C_{31}-secodammarane-type triterpenoid saponins from the female flowers of *Alnus serrulatoides. J. Chem. Soc., Perkin* 1, 1399–1403.

Aoki, T., Ohta, S. and Suga, T. (1990). Triterpenoids, diarylheptanoids and their glycosides in the flowers of *Alnus* species. *Phytochemistry*, **29**, 3611–3614.

Appel, H.H. and Haleby, A. (1965). Chemical composition of espergularin. *Scientia*, **32**, 17–21 (*Chem. Abstr.*, **66**, 38202).

Applebaum, S.W., Gestetner, B. and Birk, Y. (1965). Physiological aspects of host specificity in the bruchidae – IV. Developmental incompatibility of soybeans for *Callosobruchus. J.Insect Physiol.*, **11**, 611–616.

Applebaum, S.W., Marco, S. and Birk, Y. (1969). Saponins as possible factors of resistance of legume seeds to the attack of insects. *J. Agric. Food Chem.*, 618–622.

Apsamotova, *see* Ansamotova.

Apsimon, J.W., Belanger, J., Girard, M. *et al.* (1984). Saponins from marine invertebrates. *Stud. Org. Chem.*, **17**, 273–286 (*Chem. Abstr.*, **102**, 59596).

Aquino, R., De Simone, F., Pizza, C., Cerri, R. and De Mello, J.F. (1988). Quinovic acid glycosides from *Guettarda platypoda. Phytochemistry*, **27**, 2927–2930.

Aquino, R., De Simone, F., Pizza, C. and De Mello, J.F. (1989a). Further quinovic acid glycosides from *Guettarda platypoda. Phytochemistry*, **28**, 199–201.

Aquino, R., De Simone, F., Pizza, C., Conti, C. and Stein, M.L. (1989b). Plant metabolites. Structure and *in vitro* antiviral activity of quinovic acid glycosides from *Uncaria tomentosa* and *Guettarda platypoda. J. Nat. Prod.*, **52**, 679–685.

Aquino, R., De Feo, V., De Simone, F., Pizza, C. and Cirino, G. (1991). Plant metabolites. New compounds and anti-inflammatory activity of *Uncaria tomentosa. J. Nat. Prod.*, **54**, 453–459.

Archibald, R.G. (1933). The use of the fruit of the tree *Balanites aegyptiaca* in the control of schistosomiasis in the Sudan. *Trans. R. Soc. Trop. Med. Hyg.*, **27**, 207–211.

Arneson, P.A. and Durbin, R.D. (1968). The mode of action of tomatine as a fungitoxic agent. *Plant Physiol.*, **43**, 683–686.

Arpino, P. (1990). Coupling techniques in LC/MS and SFC/MS. *J. Anal. Chem.*, **337**, 667–685.

Asada, Y., Ueoka, T. and Furuya, T. (1989a). Novel acylated saponins from montbretia (*Crocosmia crocosmiiflora*). Isolation of saponins and the structures of crocosmiosides A, B and H. *Chem. Pharm. Bull.*, **37**, 2139–2146.

Asada, Y., Ikeno, M., Ueoka, T. and Furuya, T. (1989b). Desacylsaponins, desacylmasonosides 1, 2 and 3, from the corms of *Crocosmia masonorum*. *Chem. Pharm. Bull.*, **37**, 2747–2752.

Asakawa, J., Kasai, R., Yamasaki, K. and Tanaka, O. (1977). ^{13}C NMR study of ginseng sapogenins and their related dammarane type triterpenes. *Tetrahedron*, **33**, 1935–1939.

Assa, Y., Gestetner, B., Chet, I. and Henis, Y. (1972). Fungistatic activity of lucerne saponins and digitonin as related to sterols. *Life Sci.*, **11**, 637–647 (*Chem. Abstr.*, 136133).

Aubert, L. and Anthoine, P. (1985). Cosmetic or pharmaceutical composition from plant extracts having an effect on capillary fragility. *Chem. Abstr.*, **103**, 183548.

Avilov, S.A. and Stonik, V.A. (1988). New triterpene glycosides from the holothurian *Cladolabes* sp. *Khim. Prir. Soedin.*, 764–765.

Avilov, S.A., Tishchenko, L.Y. and Stonik, V.A. (1984). Structure of cucumarioside A_2-2-triterpene glycoside from the holothurian *Cucumaria japonica*. *Khim. Prir. Soedin.*, 799–800.

Baba, M. and Shigeta, S. (1987). Antiviral activity of glycyrrhizin against varicella-zoster virus *in vitro*. *Antiviral Res.*, **7**, 99–107.

Babady-Bila, Ngalamulume, T., Kilonda, A., Toppet, S., Compernolle, F. and Hoornaert, G. (1991). An ursadienedioic acid glycoside from *Crossopteryx febrifuga*. *Phytochemistry*, **30**, 3069–3072.

Backer, R.C., Bianchi, E. and Cole, J.R. (1972). A phytochemical investigation of *Yucca schotti* (Liliaceae). *J. Pharm. Sci.*, **61**, 1665–1666.

Badalbaeva, T.A., Kondratenko, E.S. and Abubakirov, N.K. (1974). Triterpene glycosides of *Gleditschia triacanthus*. VI. Triacanthoside F. *Khim. Prir. Soedin.*, 105–106.

Bader, G. and Hiller, K. (1987). Neue Ergebnisse zur Struktur und Wirkungsweise von Triterpensaponinen. *Pharmazie*, **42**, 577–597.

Bader, G., Kulhanek, Y. and Ziegler–Böhme, H. (1990). Zur antimyzetischen Wirksamkeit von Polygalasäureglycosiden. *Pharmazie*, **45**, 618–620.

Bader, G., Zieschang, M., Wagner, K., Gründemann, E. and Hiller, K. (1991). New triterpenoid saponins from *Helianthus annuus*. *Planta Medica*, **57**, 471–474.

Baerheim Svendsen, A. and Verpoorte, R. (1983). *Chromatography of the Alkaloids. Part A: Thin-layer Chromatography*. Elsevier, Amsterdam.

Bahuguna, S. and Sati, O.P. (1990). Spirostanol saponins from *Yucca aloifolia* rhizomes. *Phytochemistry*, **29**, 342–343.

Bahuguna, R.P., Jangwan, J.S., Kaiya, T. and Sakakibara, J. (1989). Clemontanoside-A, a bisglycoside from *Clematis montana*. *Phytochemistry*, **28**, 2511–2513.

Bahuguna, S., Kishor, N., Sati, O.P., Sati, S.P., Sakakibara, J. and Kaiya, T. (1991). A new spirostanol glycoside from *Yucca aloifolia*. *J. Nat. Prod.*, **54**, 863–865.

Baker, D.C., Keeler, R. and Gaffield, W. (1988). Mechanism of death in Syrian hamster gavaged potato sprout material. *Toxicol. Pathol.*, **16**, 333–339.

Baker, D.C., Keeler, R. and Gaffield, W. (1989). Pathology in hamsters administered *Solanum* plant species that contain steroidal alkaloids. *Toxicon*, **27**, 1331–1337.

Baldwin, C.A., Anderson, L.A. and Phillipson, J.D. (1986). What pharmacists should know about ginseng. *Pharm. J.*, **238**, 583–586.

Bandara, B.M.R., Jayasinghe, U.L.B., Karunaratne, V. *et al.* (1990). Triterpenoidal constituents of *Diploclisia glaucescens. Planta Medica,* **56,** 290–292.

Banerji, N. (1977). A new saponin from stem bark of *Anthocephalus cadamba* MIQ. *Indian J. Chem.,* **15B,** 654–655.

Banerji, N. (1979). Chemical constituents of *Amaranthus spinosus* roots. *Indian J. Chem.,* **17B,** 180–181.

Banerji, R. and Nigam, S.K. (1980). Chemistry of *Acacia concinna* and *A. caesia* bark. *J. Indian Chem. Soc.,* **57,** 1043–1044.

Banerji, R., Srivastava, A.K., Misra, G., Nigam, S.K., Singh, S. and Nigam, S.C. (1979a). Steroid and triterpenoid saponins as spermicidal agents. *Indian Drugs,* **17,** 6–8.

Banerji, R., Misra, G. and Nigam, S.K. (1979b). Chemical examination of root and leaves of *Albizzia procera* BENTH. *Indian J. Pharmacol. Sci.,* **41,** 115–116.

Baranov, A.I. (1982). Medicinal uses of ginseng and related plants in the Soviet Union: recent trends in the Soviet literature. *J. Ethnopharmacol.,* **6,** 339–353.

Barber, M., Bordoli, R.S., Sedgwick, R.D. and Tyler, A.N. (1981). Fast atom bombardment of solids (F.A.B.): a new ion source for mass spectrometry. *J. Chem. Soc., Chem. Commun.,* 325–327.

Barber, M., Bordoli, R.S., Elliott, G.J., Sedgwick, R.D. and Tyler, A.N. (1982). Fast atom bombardment mass spectrometry. *Anal. Chem.,* **54,** 645A–657A.

Barton, D.H.R. and Brooks, C.J.W. (1951). Triterpenoids. Part I. Morolic acid, a new triterpenoid sapogenin. *J. Chem. Soc.,* 257–277.

Barua, A.K., Pal, S.K. and Dutta, S.P. (1972). Triterpenoids. XXXIX. Constitution of barrinic acid, a new triterpene from *Barringtonia acutangula. J. Indian Chem. Soc.,* **49,** 519–520.

Barua, A.K., Chakravarti, S., Basak, A., Ghosh, A. and Chakrabarti, P. (1976). A new triterpene glycoside from *Mollugo hirta. Phytochemistry,* **15,** 831–832.

Barua, A.K., Ray, S., Dutta, P.K. and Venkateswaran, R.V. (1986). A triterpene glycoside from *Mollugo spergula. Phytochemistry,* **25,** 1762–1764.

Basterrechea, M., Preiss, A., Coll, F., Voigt, D., Mola, J.L. and Adam, G. (1984). Havanine – a steroidal alkaloid glycoside from *Solanum havanense. Phytochemistry,* **23,** 2057–2058.

Basterrechea, M.J., Mola, J.L., Coll, F. and Verez, V. (1986). Steroidal glycosides of *Solanum havanense* JACQ. Part. II. *Rev. Cubana Quim.,* **2,** 71–73 (*Chem. Abstr.,* **106,** 153103).

Basu, N. and Rastogi, R.P. (1967). Triterpenoid saponins and sapogenins. *Phytochemistry,* **6,** 1249–1270.

Baup, M. (1826). Extrait d'une lettre de M. Baup aux rédacteurs sur plusieurs nouvelles substances. *Ann. Chim. Phys.,* **31,** 108–109.

Baxter, R.L., Price, K.R. and Fenwick, G.R. (1990). Sapogenin structure: analysis of the ^{13}C- and ^{1}H-NMR spectra of soyasapogenol B. *J. Nat. Prod.,* **53,** 298–302.

Becchi, M., Bruneteau, M., Trouilloud, M., Combier, H., Sartre, J. and Michel, G. (1979). Structure of a new saponin: chrysantellin A from *Chrysanthellum procumbens* RICH. *Eur. J. Biochem.,* **102,** 11–20.

Becchi, M., Bruneteau, M., Trouilloud, M., Combier, H., Pontanier, H. and Michel, G. (1980). Structure de la chrysantelline B: nouvelle saponine isolée de *Chrysanthellum procumbens* RICH. *Eur. J. Biochem.,* **108,** 271–277.

Beckey, H.D. and Schulten, H.-R. (1975). Field desorption mass spectrometry. *Angew. Chem.*, **87**, 425–438.

Bedoya Zurita, M., Ahond, A., Poupat, C., Potier, P. and Menou, J.L. (1986). Invertébrés marins du lagon néo-Calédonien. VII. Etude structurale d'un nouveau saponoside sulfate extrait de l'holothurie, *Neothyonidium magnum*. *J. Nat. Prod.*, **49**, 809–813.

Begley, M.J., Crombie, L., Crombie, W.M.L. and Whiting, D.A. (1986). The isolation of avenacins A-1, A-2, B-1 and B-2, chemical defences against cereal 'take-all' disease. Structure of their 'aglycones', the avenestergenins, and their anhydro dimers. *J. Chem. Soc., Perkin I*, 1905–1915.

Benidze, M.M., Dzhikiya, O.D., Pkheidze, T.A., Kemertelidze, E.P. and Shashkov, A.S. (1987). Steroidal glycosides from leaves of *Yucca aloifolia*. *Khim. Prir. Soedin.* 537–542.

Benninghoven, A. and Sichtermann, W.K. (1978). Detection, identification and structural investigation of biologically important compounds by secondary ion mass spectrometry. *Anal. Chem.*, **50**, 1180–1184.

Bep Oliver-Bever (1986). *Medicinal Plants in Tropical West Africa*. Cambridge University Press, Cambridge.

Bergsteinsson, I. and Noller, C.R. (1934). Saponins and sapogenins I. Echinocystic acid. *J. Am. Chem. Soc.*, **56**, 1403–1405.

Besso, H., Saruwatari, Y., Futamura, K., Kunihiro, K., Fuwa, T. and Tanaka, O. (1979). High performance liquid chromatographic determination of ginseng saponin by ultraviolet derivatisation. *Planta Medica*, **37**, 226–233.

Bhandari, P., Gray, A.I. and Rastogi, R.P. (1987). Triterpenoid saponins from *Caltha palustris*. *Planta Medica*, **53**, 98–100.

Bhandari, S.P.S., Agrawal, P.K. and Garg, H.S. (1990). A triterpenoid saponin from *Polycarpone loeflingiae*. *Phytochemistry*, **29**, 3889–3892.

Bhargava, S.K. (1988a). Antifertility agents from plants. *Fitoterapia*, **59**, 163–177.

Bhargava, S.K. (1988b). Antifertility effects of *Sapindus trifoliatus* L. fruit extract in rats. *Int. J. Crude Drug Res.*, **26**, 229–233.

Bhattacharya, S.K. and Mitra, S.K. (1991). Anxiolytic activity of *Panax ginseng* roots: an experimental study. *J. Ethnopharmacol.*, **34**, 87–92.

Bielenberg, J. (1992). Lakritzerzeugnisse – gesundheitlich unbedenklich? *Pharmazie in unserer Zeit*, **21**, 157–158.

Bilia, A.R., Catalano, S., Fontana, C., Morelli, I. and Palme, E. (1992). A new saponin from *Potentilla tormentilla*. *Planta Medica*, **58**, A723.

Bilia, A.R., Flammini, F., Flamini, G., Morelli, I. and Marsili, A. (1993). Flavonoids and a saponin from *Spartium junceum*. *Phytochemistry*, **34**, 847–852.

Birk, Y. (1969). Saponins. In *Toxic Constituents of Foodstuffs* ed. I.E. Liener. pp. 169–210. Academic Press, New York.

Birk, Y. and Peri, I. (1980). Saponins. In *Toxic Constituents of Plant Foodstuffs*, 2nd edn, vol. 1, ed. I.E. Liener, pp. 161–182. Academic Press, New York.

Bischof, B., Jeger, O. and Ruzicka, L. (1949). Zur Kenntnis der Triterpene. 143. Mitteilung. Uber die Lage der zweiten sekundären Hydroxyl-Gruppe in Echinocystsäure, Quillajasäure, Maniladiol und Genin A (aus *Primula officinalis* JACQUIN). Uber die Konstitution der Oleanolsäure. *Helv. Chim. Acta*, **32**, 1911–1921.

Bitter, T. and Muir, H.M. (1962). A modified uronic acid carbazole reaction. *Anal. Biochem.*, **4**, 330–334.

Björndal, H., Lindberg, B. and Svensson, S. (1967). Gas-liquid chromatography of partially methylated alditols as their acetates. *Acta Chem. Scand.*, **21**, 1801–1804.

Björndal, H., Hellerqvist, C.G., Lindberg, B. and Svensson, S. (1970). Gas–liquid chromatography and mass spectrometry in methylation analysis of polysaccharides. *Angew. Chem., Int. Ed.*, **9**, 610–619.

Blackley, C.R. and Vestal, M.L. (1983). Thermospray interface for liquid chromatography/mass spectrometry. *Anal. Chem.*, **55**, 750–754.

Blunden, G., Culling, C. and Jewers, K. (1975). Steroidal sapogenins: a review of actual and potential plant sources. *Trop. Sci.*, **17**, 139–154.

Blunt, J.W. and Munro, M.H.G. (1980). Carbon-13 NMR spectra of some tetra- and pentacyclic triterpene methyl ethers. *Org. Mag. Res.*, **13**, 26–27.

Boar, R.B. and Romer, C.R. (1975). Cycloartane triterpenoids. *Phytochemistry*, **14**, 1143–1146.

Bock, K. and Pedersen, C. (1974). A study of ^{13}CH coupling constants in hexopyranoses. *J. Chem. Soc., Perkin Trans. II*, 293–297.

Bock, K. and Pedersen, C. (1983). Carbon-13 nuclear magnetic resonance spectroscopy of monosaccharides. *Adv. Carbohydr. Chem. Biochem.*, **41**, 27–66.

Bock, K. and Thorgensen, H. (1982). Nuclear magnetic resonance spectroscopy in the study of mono- and oligosaccharides. *Annu. Rep. NMR Spectrosc.*, **13**, 1–57.

Bogacheva, N.G., Kogan, L.M. and Libizov, N.I. (1972). Terpenoid glycosides from *Chenopodium ambrosioides*. *Khim. Prir. Soedin*, 395.

Bogacheva, N.G., Kiselev, V.P. and Kogan, L.M. (1976). Isolation of 3,26-bisglycoside of yamogenin from *Trigonella foenum-graecum*. *Khim. Prir. Soedin.*, 268–269.

Boiteau, P., Pasich, B. and Rakoto Ratsimamanga, A. (1964). *Les Triterpénoïdes en Physiologie Végétale et Animale*. Gauthier-Villars, Paris.

Boll, P.M. and von Philipsborn, W. (1965). NMR studies and the absolute configuration of *Solanum* alkaloids (spiroaminoketalalkaloids). *Acta Chem. Scand.*, **19**, 1365–1370.

Bombardelli, E., Bonati, A., Gabetta, B. and Mustich, G. (1971). Glycosides from rhizomes of *Ruscus aculeatus* L. Part I. *Fitoterapia*, **42**, 127–136.

Bombardelli, E., Bonati, A., Gabetta, B. and Mustich, G. (1972). Glycosides from rhizomes of *Ruscus aculeatus* L. Part II. *Fitoterapia*, **43**, 3–10.

Bombardelli, E., Bonati, A., Gabetta, B., Martinelli, E.M., Mustich, G. and Danieli, B. (1975). Passiflorine, a new glycoside from *Passiflora edulis*. *Phytochemistry*, **14**, 2661–2665.

Bombardelli, E., Bonati, A., Gabetta, B., Martinelli, E.M. and Mustich, G. (1976). Gas–liquid chromatographic and mass spectrometric investigations on saponins in *Panax ginseng* extracts. *Fitoterapia*, **47**, 99–106.

Bombardelli, E., Gabetta, B., Martinelli, E.M. and Mustich, G. (1979). Gas chromatographic–mass spectrometric (GC–MS) analysis of medicinal plants. Part III. Quantitative evaluation of glycyrrhetic acid and GC–MS investigation on licorice triterpenoids. *Fitoterapia*, **50**, 11–24.

Bomford, R. (1988). Immunomodulators from plants and fungi. *Phytotherapy Res*, **2**, 159–164.

Borel, C. and Hostettmann, K. (1987). Molluscicidal saponins from *Swartzia madagascariensis* DESVAUX. *Helv. Chim. Acta*, **70**, 570–576.

Borel, C., Gupta, M.P. and Hostettmann, K. (1987). Molluscicidal saponins from *Swartzia simplex*. *Phytochemistry*, **26**, 2685–2689.

Bouché, R. (1955). Chemical study of the saponin of an adulterated *Polygala senega*, *Spergularia marginata* or *Spergularia media*. *J. Pharmacol. Belg.*, **10**, 169–191.

Brain, K., Hadgraft, J. and Al–Shatalebi, M. (1990). Membrane modification in activity of plant molluscicides. *Planta Medica*, **56**, 663.

Brekhman, I.I. and Dardymov, I.V. (1969a). Pharmacological investigation of glycosides from ginseng and eleutherococcus. *Lloydia*, **32**, 46–51.

Brekhman, I.I. and Dardymov, I.V. (1969b). Substances of plant origin which increase non-specific resistance. *Annu. Rev. Pharmacol.*, **9**, 419–430.

Breton, J.L. and Gonzalez, A.G. (1963). Glucosides and aglycones from Canary Scrophulariaceae. Part VI. Structure of two new triterpenes from *Scrophularia smithii* WYDLER. *J. Chem. Soc.*, 1401–1406.

Brieskorn, C.H. and Kilbinger, W. (1975). Struktur eines Saponins aus *Polygala chinensis* L. *Arch. Pharm.*, **308**, 824–832.

Brieskorn, C.H. and Wunderer, H. (1967). Uber den chemischen Aufbau der Apfelschale, IV. Pomol- und Pomonsäure. *Chem. Ber.*, **100**, 1252–1265.

British Herbal Pharmacopoeia (1983). British Herbal Medicine Association, Bournemouth, UK.

Büchi, J. and Dolder, R. (1950). Die Bestimmung des Hämolytischen Index (H.I.) offizineller Arzneidrogen. *Pharm. Acta Helv.*, **25**, 179–188.

Budzikiewicz, H., Wilson, J.M. and Djerassi, C. (1963). Mass spectrometry in structural and stereochemical problems. XXXII. Pentacyclic triterpenes. *J. Am. Chem. Soc.*, **85**, 3688–3699.

Bugorskii, P.S. and Melnikov, V.N. (1977). Triterpenoid and steroid glycosides of *Rosa gallica*. *Khim. Prir. Soedin.*, 420–421.

Bukharov, V.G. and Bukharova, I.L. (1974). Triterpene glycoside from *Viscaria*. *Khim. Prir. Soedin.*, 534.

Bukharov, V.G. and Karlin, V.V. (1970a). Triterpenic glycosides from *Patrinia scabiosofolia*. II. *Khim. Prir. Soedin.*, 211–214.

Bukharov, V.G. and Karlin, V.V. (1970b). Structure of scabioside F. *Khim. Prir. Soedin.*, 372–373.

Bukharov, V.G. and Karlin, V.V. (1970c). Structure of scabioside G. *Khim. Prir. Soedin.*, 373–374.

Bukharov, V.G. and Karlin, V.V. (1970d). *Patrinia sibirica* glycosides. I. *Khim. Prir. Soedin.*, 60–64.

Bukharov, V.G. and Karlin, V.V. (1970e). *Patrinia sibirica* glycosides. II. *Khim. Prir. Soedin.*, 64–69.

Bukharov, V.G. and Karneeva, L.N. (1970a). Triterpenoid glycosides from *Sanguisorba officinalis*. *Izv. Akad. Nauk SSSR, Ser. Khim.*, 2402–2403.

Bukharov, V.G. and Karneeva, L.N. (1970b). Glycosides of ursolic acid from *Empetrum sibiricum*. *Izv. Akad. Nauk SSSR, Ser. Khim.*, 171–172.

Bukharov, V.G. and Karneeva, L.N. (1971). Structure of nutanoside. *Khim. Prir. Soedin.*, 412–414.

Bukharov, V.G. and Shcherbak, S.P. (1969). Triterpene glycosides from *Saponaria officinalis*. *Khim. Prir. Soedin.*, 389–394.

Bukharov, V.G. and Shcherbak, S.P. (1970). Triterpenoid glycosides of *Herniaria glabra*. I. *Khim. Prir. Soedin.*, 307–311.

Bukharov, V.G. and Shcherbak, S.P. (1971). Triterpenoid glycosides from *Dianthus deltoides* (carnation). II. Structure of dianthoside C. *Khim. Prir. Soedin.*, 420–425.

Bukharov, V.G. and Shcherbak, S.P. (1973). Triterpene glycosides from *Arenaria graminifolia*. *Khim. Prir. Soedin.*, 123–124.

Bukharov, V.G., Karlin, V.V. and Talan, V.A. (1969). Triterpene glycosides of *Patrinia intermedia*. IV. Structures of the carbohydrate chains of patrinosides C and D. *Khim. Prir. Soedin.*, 89–93.

Bukharov, V.G., Karlin, V.V. and Sidorovich, T.N. (1970). Triterpene glycosides of *Patrinia scabiosofolia*. I. *Khim. Prir. Soedin.*, 69–74.

Bukharov, V.G., Shcherbak, S.P. and Beshchekova, A.P. (1971). Triterpenoid glycosides of *Dianthus deltoides*. I. *Khim. Prir. Soedin.*, 33–38.

Burnell, D.J. and Apsimon, J.W. (1983). Echinoderm saponins. In *Marine Natural Products: Chemical and Biological Perspectives*, vol. 5, ed. P.J. Scheuer, pp. 287–389. Academic Press, New York.

Burnouf-Radosevich, M. and Delfel, N.E. (1984). High-performance liquid chromatography of oleanane-type triterpenes. *J. Chromatogr.*, **292**, 403–409.

Burnouf-Radosevich, M. and Delfel, N.E. (1986). High–performance liquid chromatography of triterpene saponins. *J. Chromatogr.*, **368**, 433–438.

Burnouf-Radosevich, M., Delfel, N.E. and England, R. (1985). Gas chromatography–mass spectrometry of oleanane- and ursane-type triterpenes – application to *Chenopodium quinoa* triterpenes. *Phytochemistry*, **24**, 2063–2066.

Burrows, J.C., Price, K.R. and Fenwick, G.R. (1987). Soyasaponin IV, an additional monodesmosidic saponin isolated from soybean. *Phytochemistry*, **26**, 1214–1215.

Bushway, R.J. and Storch, R.H. (1982). Semi-preparative high performance liquid chromatographic separation of potato glycoalkaloids. *J. Liq. Chromatogr.*, **5**, 731–742.

Bushway, R.J., Barden, E.S., Bushway, A.W. and Bushway, A.A. (1979). High-performance liquid chromatographic separation of potato glycoalkaloids. *J. Chromatogr.*, **178**, 533–541.

Bushway, R.J., Bureau, J.L. and King, J. (1986). Modification of the rapid high-performance liquid chromatographic method for the determination of potato glycoalkaloids. *J. Agric. Food Chem.*, **34**, 277–279.

Calam, D.H. and Callow, R.K. (1964). Histamine protection produced by plant tumour extracts. The active principle of tomato plants infected with crowngall. *Br. J. Pharmacol.*, **22**, 486–498.

Calis, I. (1987). Saponin and sapogenol. I. Comparative study on the sapogenol constituents of three *Primula* species. *Doga Tip Eczacilik*, **11**, 206–215 (*Chem. Abstr.*, **107**, 172512).

Calis, I. (1989). Saponin and sapogenol. II. The main saponin from the roots and rhizomes of *Primula auriculata, P. megaseifolia* and *P. longipes* (Primulaceae). *Doga: Turk Saglik Bilimferi Derg.*, **13**, 111–120 (*Chem. Abstr.*, **112**, 195216).

Calis, I., Yuruker, A., Rüegger, H., Wright, A.D. and Sticher, O. (1992). Triterpene saponins from *Primula veris* subsp. *macrocalyx* and *Primula elatior* subsp. *meyeri. J. Nat. Prod.*, **55**, 1299–1306.

Cambie, R.C. and Couch, R.A.F. (1967). Extractives of some *Myrsine* species. *New Zealand J. Sci.*, **10**, 1020–1029.

Cao, Z., Yu, J., Gan, L. and Chen, Y. (1985). Structure of astramembrannins. *Huaxue Xuebao*, **43**, 581–585 (*Chem. Abstr.*, **103**, 175381).

Capon, B. and Thacker, D. (1964). The nuclear magnetic resonance spectra of some aldofuranosides and acyclic aldose acetals. *Proc. Chem. Soc. Lond.*, 369.

Capra, C. (1972). Pharmacology and toxicology of some components of *Rusculus aculeatus. Fitoterapia*, **43**, 99–113.

Carmely, S., Roll, M., Loya, Y. and Kashman, Y. (1989). The structure of eryloside A, a new antitumour and antifungal 4-methylated steroidal glycoside from the sponge *Erylus lenddenfeld. J. Nat. Prod.*, **52**, 167–170.

Carpani, G., Orsini, F., Sisti, M. and Verotta, L. (1989). Saponins from *Albizzia anthelmintica. Phytochemistry*, **28**, 863–866.

Cart, J., Reznicek, G., Korhammer, S., Haslinger, E., Jurenitsch, J. and Kubelka, W. (1992). The first spectroscopically confirmed saponin from *Herniaria glabra. Planta Medica*, **58**, A709.

Cerri, R., Aquino, R., de Simone, F. and Pizza, C. (1988). New quinovic acid glycosides from *Uncaria tomentosa. J. Nat. Prod.*, **51**, 257–261.

Chakravarty, A.K., Saha, C.R. and Pakrashi, S.C. (1979). New spirostane saponins and sapogenins from *Solanum hispidum* seeds. *Phytochemistry*, **18**, 902–903.

Cham, B.E. and Meares, H.M. (1987). Glycoalkaloids from *Solanum sodomaeum* are effective in the treatment of skin cancers in man. *Cancer Lett.*, **36**, 111–118.

Cham, B.E., Gilliver, M. and Wilson, L. (1987). Antitumour effects of glycoalkaloids isolated from *Solanum sodomaeum. Planta Medica*, **53**, 34–36.

Chandel, R.S. and Rastogi, R.P. (1980). Triterpenoid saponins and sapogenins: 1973–1978. *Phytochemistry*, **19**, 1889–1908.

Charrouf, Z., Wieruszeski, J.M., Fkih–Tetouani, S., Leroy, Y., Charrouf, M. and Fournet, B. (1992). Triterpenoid saponins from *Argania spinosa. Phyto-chemistry*, **31**, 2079–2086.

Chaturvedi, S.K. and Saxena, V.K. (1985). A new saponin, lupanol-3-*O*-β-D-glucopyranosyl(1→5)-*O*-β-D-xylofuranoside from the roots of *Streblus asper* (LOUR). *Indian J. Chem.*, **24B**, 562–563.

Chauhan, J.S. and Srivastava, S.K. (1978). Lupa-20,29-ene-3-*O*-β-D-maltoside from the roots of *Cordia obliqua. Phytochemistry*, **17**, 1005–1006.

Chausseaud, L.F., Fry, B.J., Hawkins, D.R., Lewis, J.D., Sward, I.P., Taylor, T. and Hathway, D.E. (1971). The metabolism of asiatic acid, madecassic acid and asiaticoside in the rat. *Arzneimittelforsch.*, **21**, 1379–1384.

Chavali, S.R. and Campbell, J.B. (1987). Adjuvant effects of orally administered saponins on humoral and cellular immune responses in mice. *Immunobiol.*, **174**, 347–359.

Chavali, S.R., Francis, T. and Campbell, J.B. (1987). An *in vitro* study of immunomodulatory effects of some saponins. *Int. J. Immunopharmacol.*, **9**, 675–683.

Cheban, P.L. and Chirva, V.J. (1969). Structure of helianthoside B. *Khim. Prir. Soedin.*, 327.

Cheban, P.L., Chirva, V.J. and Lazurevskii, G.V. (1969a). Structure of heliantoside A. *Khim. Prir. Soedin.*, 59.

Cheban, P.L., Chirva, V.J. and Lazurevskii, G.V. (1969b). Structure of helianthoside C, a sunflower saponin. *Khim. Prir. Soedin.*, 129–130.

Cheeke, P.R. (1971). Nutritional and physiological implications of saponins. *Can. J. Animal Sci.*, **51**, 621–632.

Chemli, R., Babadjamian, A., Faure, R., Boukef, K., Balansard, G. and Vidal, E. (1987). Arvensoside A and B, triterpenoid saponins from *Calendula arvensis*. *Phytochemistry*, **26**, 1785–1788.

Chen, H., Tanaka, T., Nonaka, G., Fujioka, T. and Mihashi, K. (1993a). Hydrolysable tannins based on a triterpenoid glycoside core, from *Castanopsis hystrix*. *Phytochemistry*, **32**, 1457–1460.

Chen, N., Chen, N., Li, H. *et al.* (1987a). Mass spectrometry of natural products. V. Negative fast atom bombardment and collisional activation mass spectrometry in the structure analysis of oligoglycosides. *Huaxue Xuebao*, **45**, 682–686 (*Chem. Abstr.*, 187104).

Chen, Q., Zheng, Z., Yang, S. and Chen, G. (1987c). Pharmacology of total saponins of the fibrous roots of *Panax notoginseng*. *Zhongyao Tongbao*, **12**, 173–175 (*Chem. Abstr.*, **107**, 32634).

Chen, R., Yu, Z., Zhang, X., Zeng, J. and Koo, T. (1982). Zhi mu (*Anemarrhena asphedoloides* BGE.) sapogenin is a powerful inhibitor of sodium, potassium ATPase. *Shengwu Huaxue Yu Shengwu Wuli Xuebao*, **14**, 159–164 (*Chem. Abstr.*, **97**, 192832).

Chen, S. and Snyder, J.K. (1987). Molluscicidal saponins from *Allium vineale*. *Tetrahedron Lett.*, **28**, 5603–5606.

Chen, S. and Snyder, J.K. (1989). Diosgenin-bearing, molluscicidal saponins from *Allium vineale*: an NMR approach for the structural assignment of oligosaccharide units. *J. Org. Chem.*, **54**, 3679–3689.

Chen, W., Lin, Q., Chen, L., Kasai, R. and Tanaka, O. (1990a). Saponin of the Chinese drug Bai-Tou-Weng. IV. Structure of anemosides B_4 and A_3. *Huaxue Xuebao* **48**, 501–505 (*Chem. Abstr.*, **113**, 158517).

Chen, W.-H. and Wu, D.-G. (1978). Study on sapogenin of *Schima argentea* PRITZ. *Hua Hsueh Pao*, **36**, 229–232 (*Chem. Abstr.*, **90**, 138052).

Chen, X.K., Zhang, R.Y., Zhang, Z.L., Jiang, T.Y. and Wang, B. (1993b). Isolation and identification of two new saponins from *Bupleurum smithii* WOLFF. *Yaoxue Xuebao*, **28**, 352–357 (*Chem. Abstr.*, **119**, 199534).

Chen, Y., Takeda, T. and Ogihara, Y (1985). Studies on the constituents of *Xanthoceras sorbifolia* BUNGE V. Major saponins from the fruits of *Xanthoceras sorbifolia* BUNGE. *Chem. Pharm. Bull.*, **33**, 1387–1394.

Chen, Y., Xie, H., Xu, S., Ma, Q., Pei, Y. and Yao, X. (1986). New minor constituents of leaves of *Panax ginseng* C.A. MEYER. *Shenyang Yaoxueyuan Xuebao*, **3**, 191 (*Chem. Abstr.*, **106**, 2889).

Chen, Y., Chen, N., Li, H., Zhao, F. and Chen, N. (1987b). Fast atom bombardment and collisional activation mass spectrometry in the structural analysis of steroidal oligoglycosides. *Biomed. Environ. Mass Spectrom.*, **14**, 9–15.

Chen, Y.-Q., Azi, G. and Luo, Y.-R. (1990b). Astrailienin A from *Astragalus iliensis*. *Phytochemistry*, **29**, 1941–1943.

Cheng, X.-J., Liu, Y.-L., Lin, G.-F. and Luo, X.-T. (1986). Effects of ginseng root saponins on central transmitters and plasma corticosterone in warm stress rats. *Acta Pharmacol. Sinica*, **7**, 6–8.

Chevillard, L., Ranson, M. and Senault, B. (1965). Anti-inflammatory activity of extracts of holly (*Ruscus aculeatus*). *Med. Pharmacol. Exp.*, **12**, 109–114.

Chirva, V.J. and Kintya, P.K. (1969). Structure of saponoside A. *Khim. Prir. Soedin.*, 188.

Chirva, V.J. and Kintya, P.K. (1970). Structure of saponaside D. *Khim. Prir. Soedin.*, 214–218.

Chirva, V.J. and Kintya, P.K. (1971). Triterpenoid glycoside from *Dianthus inacotinus*. *Khim. Prir. Soedin.*, 532.

Chirva, V.J. and Konyukhov, V.P. (1968a). Structure of clematoside A. *Khim. Prir. Soedin.*, 140–141.

Chirva, V.J. and Konyukhov, V.P. (1968b). Structure of clematoside B. *Khim. Prir. Soedin.*, 141–142.

Chirva, V.J. and Konyukhov, V.P. (1969). Structure of clematoside A'. *Khim. Prir. Soedin.*, 60–61.

Chirva, V.J., Cheban, P.L. and Lazurevskii, G.V. (1968). Sunflower saponins. *Khim. Prir. Soedin.*, 140.

Chirva, V.J., Kintya, P.K. and Sosnovskii, V.A. (1970a). Triterpenoid glycosides of *Koelreuteria paniculata*. IV. Structure of Koelreuteria saponins A and B. *Khim. Prir. Soedin.*, 328–331.

Chirva, V.J., Kintya, P.K. and Sosnovskii, V.A. (1970b). Triterpenoid glycosides of *Sapindus mukorossi*. V. Structure of sapindoside E. *Khim. Prir. Soedin.*, 431–434.

Chirva, V.J., Kintya, P.K., Sosnovskii, V.A. Krivenchuk, P.E. and Zykova, N.J. (1970c). Triterpene glycosides of *Sapindus mukorossi*. II. Structure of sapindosides A and B. *Khim. Prir. Soedin.*, 218–221.

Chirva, V.J., Kintya, P.K., Sosnovskii, V.A. and Zolotarev, B.M. (1970d). Triterpenoid glycosides of *Sapindus mukorossi*. IV. Structure of sapindoside D. *Khim. Prir. Soedin.*, 316–319.

Chirva, V.J., Cheban, P.L. and Kintya, P.K. (1970e). Meristotropic acid glycoside. *Khim. Prir. Soedin.*, 636.

Chirva, V.J., Kretsu, L.G. and Kintya, P.K. (1970f). Triterpenic glycosides from plants. I. Chemical characteristics of phaseoloside E. *Khim. Prir. Soedin.*, 377–378.

Chirva, V.J., Cheban, P.L., Kintya, P.K. and Bobeiko, V.A. (1971). Structure of triterpenoid glycosides from *Chenopodium anthelminticum* roots. *Khim. Prir. Soedin.*, 27–30.

Chirva, V.J., Kintia, P.K. and Melnikov, V.N. (1978). New data on triterpene glycosides. *Chem. Abstr.*, **88**, 148971.

Cho, Y., Higuchi, R., Marubayashi, N., Ueda, I. and Komori, T. (1992). X-ray crystallographic analysis of desulfated asterosaponin P_1 isolated from the starfish *Asterina pectinifera*. *Annalen*, 79–81.

Choi, J.S. and Woo, W.S. (1987). Triterpenoid glycosides from the roots of *Patrinia scabiosaefolia*. *Planta Medica*, **53**, 62–65.

Choi, Y.-H., Kinghorn, A.D., Shi, X., Zhang, H. and Teo, B.K. (1989). Abrusoside A: a new type of highly sweet triterpene glycoside. *J. Chem. Soc., Chem. Commun.*, 887–888.

Chong, S.K.F. and Oberholzer, V.G. (1988). Ginseng – is there a use in clinical medicine? *Postgrad. Med. J.*, **64**, 841–846.

Chou, S.C., Ramanathan, S., Matsui, A., Rogers, J. and Cutting, W.C. (1971). Isolation of saponins with antifertility activity from *Gleditschia horrida*. *Indian J. Exp. Biol.*, **9**, 503–504.

Chowdhury, A.K.A., Jahirullah, I.J., Talukder, S.A. and Khan, A.K.A. (1987). Biological activity of the alcohol extract and the glycosides of *Hydrocotyle asiatica* LINN. *J. Bangladesh Acad. Sci.*, **11**, 75–82 (*Chem. Abstr.*, **107**, 108725).

Christensen, S.B. and Omar, A.A. (1985). *Atriplex nummularia*, a source for the two molluscicide saponins: hederagenin-3-O-β-D-glucuronopyranoside and calenduloside E. *J. Nat. Prod.*, **48**, 161.

Chu, J.H. and Chou, T.Q. (1941). The saponins of the Chinese drug, sanchi, *Aralia bipinnatifida*. II. Note on arasaponin B. *Chin. J. Physiol.*, **16**, 139–141.

Ciucanu, I. and Kerek, F. (1984). A simple and rapid method for the permethylation of carbohydrates. *Carbohydr. Res*, **131**, 209–217.

Ciulei, I., Mateescu, G. and Herman, G. (1966). Chemical study of the saponin and sapogenin in the fruit of *Gleditsia australis*. *Farmacia*, **14**, 663–666.

Coates, M.L. and Wilkins, C.L. (1986). Laser desorption/Fourier transform mass spectra of glycoalkaloids and steroid glycosides. *Biomed. Environ. Mass Spectrom.*, **13**, 199–204.

Coget, J.M. (1983). *Int. Angiology*, **2** (Suppl. 3), 24.

Cole, A.R.H., Downing, D.T., Watkins, J.C. and White, D.E. (1955). The constitution of A_1-barrigenol. *Chem. Ind.*, 254–255.

Coll, F., Preiss, A., Padron, G., Basterechea, M. and Adam, G. (1983). Bahamagenin – a steroidal sapogenin from *Solanum bahamense*. *Phytochemistry*, **22**, 787–788.

Cordell, G.A., Lyon, R.L., Fong, H.H.S., Benoit, P.S. and Farnsworth, N.R. (1977). Biological and phytochemical investigations of *Dianthus barbatus* CV. 'China Doll'. *Lloydia*, **40**, 361–363.

Corsano, S., Piancatelli, G. and Panizzi (1969). Sullo acteino, principo attivo dell'*Actaea racemosa*, – Nota III. *Gazz. Chim. Ital.*, **99**, 915–932.

Crabbe, P.G. and Fryer, C. (1980). Rapid quantitative analysis of solasodine, solasodine glycosides and soladiene by high-pressure liquid chromatography. *J. Chromatogr.*, **187**, 87–100.

Crielaard, J.M. and Franchimont, P. (1988). Behandlung von Muskelverletzungen in der Sporttraumatologie. *Therapiewoche*, **38**, 2505–2509.

Crombie, L., Crombie, W.M.L. and Whiting, D.A. (1987). The chemical defences of oat roots against 'Take-all' disease. In *Biologically Active Natural Products*, ed. K. Hostettmann and P.J. Lea, pp. 244–259. Clarendon Press, Oxford.

Cui, B., Sakai, Y., Takeshita, T., Kinjo, J. and Nohara, T. (1992a). Four new oleanene derivatives from the seeds of *Astragalus complanatus*. *Chem. Pharm. Bull.*, **40**, 136–138.

Cui, B., Inoue, J., Takeshita, T., Kinjo, J. and Nohara, T. (1992b). Triterpene glycosides from the seeds of *Astragalus sinicus* L. *Chem. Pharm. Bull.*, **40**, 3330–3333.

Cui, B., Kinjo, J. and Nohara, T. (1992c). Triterpene glycosides from the bark of *Robinia pseudo-acacia* L. *Chem. Pharm. Bull.*, **40**, 2995–2999.

Cunitz, G. (1968). Zur Wirkung von Succus Liquiritiae auf die Heilung experimentaler Hautwunden. *Arzneimittelforsch.*, **18**, 434–435.

Curl, C.L., Price, K.R. and Fenwick, G.R. (1986). Isolation and structural elucidation of a triterpenoid saponin from guar, *Cyamopsis tetragonoloba*. *Phytochemistry*, **25**, 2675–2676.

Curl, C.L., Price, K.R. and Fenwick, G.R. (1988a). Isolation and structure elucidation of a new saponin ('soyasaponin V') from haricot bean (*Phaseolus vulgaris* L.). *J. Sci. Food Agric.*, **43**, 101–107.

Curl, C.L., Price, K.R. and Fenwick, G.R. (1988b). Soyasaponin A3, a new monodesmosidic saponin isolated from the seeds of *Glycine max*. *J. Nat. Prod.*, **51**, 122–124.

Dalsgaard, K. (1974). Saponin adjuvants. III. Isolation of a substance from *Quillaja saponaria* with adjuvant activity in foot-and-mouth disease vaccines. *Arch. Gesamte Virusforsch.*, **44**, 243–254.

Dang Tam, N. (1967). Sur les bokétonosides, saponosides du boket ou *Gleditschia fera* MERR. (*australis* HEMSL.; *sinensis* LAM.). *C.R. Acad. Sci.*, **264**, 121–124.

Das, M.K. and Mahato, S.B. (1983). Triterpenoids. *Phytochemistry*, **22**, 1071–1095.

D'Auria, M.V., Fontana, A., Minale, L. and Riccio, R. (1990). Starfish saponins. Part XLII. Isolation of twelve steroidal glycosides from the Pacific Ocean starfish *Henricia laeviuscola*. *Gazz. Chim. Ital.*, **120**, 155–163.

D'Auria, M.V., Paloma, L.G., Minale, L., Riccio, R. and Debitus, C. (1992). Structure characterisation by two-dimensional NMR spectroscopy of two marine triterpene oligoglycosides from a Pacific sponge of the genus *Erylus*. *Tetrahedron*, **48**, 491–498.

D'Auria, M.V., Minale, L. and Riccio, R. (1993). Polyoxygenated steroids of marine origin. *Chem. Rev.*, **93**, 1839–1895.

Davidyants, E.S., Putieva, Z.M., Bandyukova, V.A. and Abubakirov, N.K. (1984a). Triterpene glycosides from *Silphium perfoliatum* II. *Khim. Prir. Soedin.*, 120–121.

Davidyants, E.S., Putieva, Z.M., Bandyukova, V.A. and Abubakirov, N.K. (1984b). Triterpene glycosides of *Silphium perfoliatum* III. Structure of silphioside E. *Khim. Prir. Soedin.*, 750–753.

Davidyants, E.S., Putieva, Z.M., Shashkov, A.S., Bandyukova, V.A. and Abubakirov, N.K. (1985). Triterpene glycosides of *Silphium perfoliatum* IV. Structure of silphioside C. *Khim. Prir. Soedin.*, 519–522.

Davidyants, E.S., Putieva, Z.M., Bandyukova, V.A. and Abubakirov, N.K. (1986). Triterpene glycosides of *Silphium perfoliatum* V. Structure of silphioside C. *Khim. Prir. Soedin.*, 63–66.

Davis, E.A. and Morris, D.J. (1991). Medicinal uses of licorice through the millennia: the good and plenty of it. *Mol. Cell. Endocrinol.*, **78**, 1–6.

Décosterd, L.A., Dorsaz, A.-C. and Hostettmann, K. (1987). Application of semi-preparative high-performance liquid chromatography to difficult natural product separations. *J. Chromatogr.*, **406**, 367–373.

Défago, G. (1977). Rôle des saponines dans la résistance des plantes aux maladies fongiques. *Ber. Schweiz. Bot. Ges.*, **87**, 79–132.

Dekanosidze, G.E. (1973). Ursolic acid from *Rhododendron ungernii. Khim. Prir. Soedin.*, 278.

Dekanosidze, G.E., Pkheidze, T.A. and Kemertelidze, E.P. (1970a). Structure of hederacaucaside B, saponin from *Hedera caucasigena. Khim. Prir. Soedin.*, 491.

Dekanosidze, G.E., Pkheidze, T.A., Gorovits, T.T. and Kemertelidze, E.P. (1970b). Triterpene oligoside, hederacolchiside E from *Hedera colchica. Khim. Prir. Soedin.*, 484–485.

Dekanosidze, G.E., Arazashvili, A.I. and Kemertelidze, E.P. (1979). Chemical study of *Clematis orientalis. Izvest. Akad. Nauk Gruz. SSR, Ser. Khim.*, **5**, 307–310.

Dekanosidze, G.E., Dzhikiya, O.D., Vugalter, M.M. and Kemertelidze, E.P. (1985). Triterpene glycosides of *Hedera colchica.* Structure of hederacolchisides E and F. *Khim. Prir. Soedin.*, 747–750.

Delaude, C. (1973). Saponin content of the Polygala. Identification of the saponin extract of *Polygala exelliana. Bull. Soc Roy. Sci. Liège*, **42**, 631–634.

Delaude, C. (1990a). Contribution à l'étude chimique des Polygalaceae. Examen de *Polygala myrtifolia. Bull. Soc. Roy. Sci. Liège*, **59**, 453–454.

Delaude, C. (1990b). Contribution à l'étude chimique des Polygalaceae. Examen de *Muraltia ononidifolia. Bull. Soc. Roy. Sci. Liège*, **59**, 455–456.

Delaude, C. (l990c). Contribution à l'étude chimique des Polygalaceae. Examen de *Nylandtia spinosa. Bull. Soc. Roy. Sci. Liège*, **59**, 457–458.

Delaude, C. and Davreux, M. (1973). Phytochemistry of Olacaceae. Identification of a saponin of *Olax angustifolia. Bull. Soc. Roy. Sci. Liège*, **42**, 70–72.

Delaude, C., Breyne, H. and Welter, A. (1976). Contribution to the study of saponins contained in Sapindaceae. Study of the saponoside from *Chytranthus macrobotrys* (GILG.) EXELL ET MENDONCA and *Lecaniodiscus cupanoides* PLANCH. *Bull. Soc. Roy. Sci. Liège*, **45**, 458–461.

Delgado, M.C.C., Da Silva, M.S. and Braz Filho, R. (1986). 3β-O-β-D-Gluco-pyranosyl-21β-(E)-cinnamoyloxyolean-12-en-28-oic acid, a new triterpene glucoside of *Enterolobium contorstisiliquum* (VELL.) MORONG. *Quim. Nova*, **9**, 119–121 (*Chem. Abstr.* **106**, 153101).

Deneke, Th. (1936). Zur Allgemeinbehandlung der Psoriasis. *Dtsch. med. Wochen-schr.*, **62**, 337–341.

Desfosses, M. (1821). Extrait d'une lettre de M. Desfosses, pharmacien, à Besançon, à M. Robiquet. *J. Pharm.*, **6**, 374–376.

De Simone, F., Dini, A., Finamore, E. *et al.* (1981). Starfish saponins Part 5. Structure of sepositoside A, a novel steroidal cyclic glycoside from the starfish *Echinaster sepositus. J. Chem. Soc., Perkin Trans.* 1, 1885–1862.

De Tommasi, N., Conti, C., Stein, M.L. and Pizza, C. (1991). Structure and *in vitro* antiviral activity of triterpenoid saponins from *Calendula arvensis. Planta Medica*, **57**, 250–253.

De Tommasi, N., Piacente, S., De Simone, F., Pizza, C. and Zhou, Z.L. (1993). Characterization of three new triterpenoid saponins from *Ardisia japonica. J. Nat. Prod.*, **56**, 1669–1675.

Diallo, B., Vanhaelen-Fastré, R. and Vanhaelen, M. (1991). Direct coupling of high-speed counter-current chromatography to thin-layer chromatography.

Application to the separation of asiaticoside and madecassoside from *Centella asiatica*. *J. Chromatogr.*, **558**, 446–450.

Diemath, H.E. (1981). Zur Behandlung und Nachbehandlung Schädelhirnverletzter. *Wien klin. Wochenschr*, **93** (Suppl. 131), 1–8.

Dijoux, M.-G., Lavaud, C., Massiot, G., Le Men-Olivier, L. and Sheeley, D.M. (1993). A saponin from leaves of *Aphloia madagascariensis*. *Phytochemistry*, **34**, 497–499.

Dimbi, M.Z., Warin, R., Delaude, C. and Huls, R. (1987). Structure of zanhin and of medicagin, two prosapogenins from *Zanha golungensis* HIERN. *Bull. Soc. Chim. Belg.*, **96**, 207–217.

Ding, J.K., Fujino, H., Kasai, R. *et al.* (1986). Chemical evaluation of *Bupleurum* species collected in Yunnan, China. *Chem. Pharm. Bull.*, **34**, 1158–1167.

Ding, N., Yahara, S. and Nohara, T. (1992a). Structure of mimengosides A and B, new triterpenoid glycosides from Buddlejae flos produced in China. *Chem. Pharm. Bull.*, **40**, 780–782.

Ding, S. and Zhu, Z. (1993). Gycomoside I: a new dammarane saponin from *Gynostemma compressum*. *Planta Medica*, **59**, 373–375.

Ding, Y., Chen, Y.-Y., Wang, D.-Z. and Yang, C.-R. (1989). Steroidal saponins from a cultivated form of *Agave sisalana*. *Phytochemistry*, **28**, 2787–2791.

Ding, Y., Kinjo, J., Yang, C.-R. and Nohara, T. (1991a). Oleanene glycosides from *Crotalaria albida*. *Chem. Pharm. Bull.*, **39**, 496–498.

Ding, Y., Kinjo, J., Yang, C.-R. and Nohara, T. (1991b). Triterpenes from *Mucuna birdwoodiana*. *Phytochemistry*, **30**, 3703–3707.

Ding, Y., Tian, R., Takeshita, T., Kinjo, J. and Nohara, T. (1992b). Four new oleanene glycosides from Sophorae subprostatae radix. III. *Chem. Pharm. Bull.*, **40**, 1831–1834.

Dini, A., Rastrelli, L., Saturnino, P. and Schetlino, O. (1991). Minor components in food plants. Note II. Triterpenoidal saponins from *Ullucus tuberosus*. *Boll.-Soc. Ital. Biol. Sper.*, **67**, 1059–1065 (*Chem. Abstr.*, **118**, 167899).

Djerassi, C. and Lippman, A.E. (1954). Terpenoids. X. The triterpenes of the cactus *Lemaireocercus hystrix*. *J. Am. Chem. Soc.*, **76**, 5780–5781.

Djerassi, C., McDonald, R.M. and Lemin, A.J. (1953). Terpenoids. III. The isolation of erythrodiol, oleanolic acid and a new triterpene triol, longispinogenin, from the cactus *Lemaireocereus longispinus*. *J. Am. Chem. Soc.*, **75**, 5940–5942.

Djerassi, C., Farkas, E., Lemin, A.J., Collins, J.C. and Walls, F. (1954a). Terpenoids. VI. Dumortierigenin, a new triterpene lactone from the cactus *Lemaireocereus dumortieri*. *J. Am. Chem. Soc.*, **75**, 2969–2973.

Djerassi, C., Geller, L.E. and Lemin, A.J. (1954b). Terpenoids. VIII. The structures of the cactus triterpenes gummosogenin and longispinogenin. *J. Am. Chem. Soc.*, **76**, 4089–4091.

Djerassi, C., Liu, L.H., Farkas, E. *et al.* (1955). Terpenoids. XI. Investigation of nine cactus species. Isolation of two new triterpenes, stellatogenin and machaeric acid. *J. Am. Chem. Soc.*, **77**, 1200–1203.

Djerassi, C., Bowers, A., Burstein, S. *et al.* (1956a). Terpenoids. XXII. Triterpenes from some Mexican and South American plants. *J. Am. Chem. Soc.*, **78**, 2312–2315.

Djerassi, C., Henry, J.A., Lemin, A.J., Rios, T. and Thomas, G.H. (1956b). Terpenoids. XXIV. The structure of the cactus triterpene queretaroic acid. *J. Am. Chem. Soc.*, **78**, 3783–3787.

Djerassi, C., Burstein, S., Estrada, H. *et al.* (1957). Terpenoids. XXVIII. The triterpene composition of the genus *Myrtillocactus*. *J. Am. Chem. Soc.*, **79**, 3525–3528.

Djerassi, C., Budzikiewicz, H. and Wilson, J.M. (1962). Mass spectrometry in structural and stereochemical problems. Unsaturated pentacyclic triterpenoids. *Tetrahedron Lett.*, 263–270.

Doddrell, D.M., Khong, P.W. and Lewis, K.G. (1974). The stereochemical dependence of ^{13}C chemical shifts in olean-12-enes and urs-12-enes as an aid to structural assignment. *Tetrahedron Lett.*, 2381–2384.

Doll, R., Hill, I.D., Hutton, C. and Underwood, D.J. (1962). Clinical trial of a triterpenoid liquorice compound in gastric and duodenal ulcer. *Lancet*, ii, 793–796.

Dominguez, X.A., Franco, R., Pugliese, O., Escobar, N. and Jean, J.A. (1979). Medicinal plants of Mexico. XXXV. Chemical study of guanacastle or parota (*Enterolobium cyclocarpum* JACQ.), a legume, bark and fruit. *Rev. Latinoamer. Quim.*, **10**, 46–48.

Domon, B. and Hostettmann, K. (1983). Saponins with molluscicidal properties from *Lonicera nigra* L. *Helv. Chim. Acta*, **66**, 422–428.

Domon, B. and Hostettmann, K. (1984). New saponins from *Phytolacca dodecandra* L'HERIT. *Helv. Chim. Acta*, **67**, 1310–1315.

Domon, B., Dorsaz, A.-C. and Hostettmann, K. (1984). High-performance liquid chromatography of oleanane saponins. *J. Chromatogr.*, **315**, 441–446.

Dong, J. and Han, G.Y. (1991). A new active steroidal saponin from *Anemarrhena asphodeloides*. *Planta Medica*, **57**, 460–462.

Dorman, D.E. and Roberts, J.D. (1970). Nuclear magnetic resonance spectroscopy. Carbon-13 spectra of some pentose and hexose aldopyranoses. *J. Am. Chem. Soc.*, **92**, 1355–1361.

Dorsaz, A.-C. and Hostettmann, K. (1986). Further saponins from *Phytolacca dodecandra* L'HERIT. *Helv. Chim. Acta*, **69**, 2038–2047.

Dorsaz, A.-C., Hostettmann, M. and Hostettmann, K. (1988). Molluscicidal saponins from *Sesbania sesban*. *Planta Medica*, **54**, 225–227.

Dragalin, I.P. and Kintia, P.K. (1975). Steroidal saponins of *Yucca filamentosa*: yuccoside C and protoyuccoside C. *Phytochemistry*, **14**, 1817–1820.

Du, H.Q., Zhao, X., Zhao, T.Z., Wang, M.T., Zhang, Z.W., Yao, M. and Yu, S.Z. (1983). Studies on the chemical constituents of the roots of *Rosa multiflora* THUNB. *Yaoxue Xuebao*, **18**, 314–316 (*Chem. Abstr.*, **100**, 48578).

Dua, P.R., Shanker, G., Srimal, R.C. *et al.* (1989). Adaptogenic activity of Indian *Panax pseudoginseng*. *Indian J. Exp. Biol.*, **27**, 631–634.

Dubois, M.-A., Ilyas, M. and Wagner, H. (1986). Cussonosides A and B, two triterpene-saponins from *Cussonia barteri*. *Planta Medica*, **52**, 80–83.

Dubois, M.-A., Higuchi, R., Komori, T. and Sasaki, T. (1988a). Structures of two new oligoglycoside sulfates, pectiniosides E and F and biological activities of the six new pectiniosides. *Annalen*, 845–850.

Dubois, M.-A., Bauer, R., Cagiotti, M.R. and Wagner, H. (1988b). Foetidissimoside A, a new 3,28-bidesmosidic triterpenoid saponin, and cucurbitacins from *Curcurbita foetidissima. Phytochemistry*, **27**, 881–885.

Dugan, J.J. and de Mayo, P. (1965). Terpenoids. X. Presenegenin, a quite normal triterpenoid. *Can. J. Chem.*, **43**, 2033–2046.

Düker, E., Kopanski, L., Jarry, H. and Wuttke, W. (1991). Effects of extracts from *Cimicifuga racemosa* on gonadotropin release in menopausal women and ovariectomized rats. *Planta Medica*, **57**, 420–424.

Eade, R.A., Rossler, L.P., Simes, H.V. and Simes, J.J.H. (1965). Extractives of Australian timbers. VI. Ebelin lactone. *Aust. J. Chem.*, **18**, 1451–1470.

Eade, R.A., Hunt, K., Simes, J.J.H. and Stern, W. (1969). Extractives of Australian timbers. IX. The saponin of *Planchonella pohlmanianum. Aust. J. Chem.*, **22**, 2703–2707.

Eade, R.A., McDonald, F.J. and Simes, J.J.H. (1973). Extractives of Australian timbers. XIV. Triterpene glycosides of *Acacia myrtifolia. Aust. J. Chem.*, **26**, 839–844.

Eddy, C.R., Wall, M.E. and Scott, M.K. (1953). Catalog of infrared absorption. *Anal. Chem.*, **25**, 266–271.

Eggert, H. and Djerassi, C. (1975). ^{13}C NMR spectra of sapogenins. *Tetrahedron Lett.*, 3635–3638.

Ehringer, H. (1968). Objektivierbare Venentonisierung nach oraler Gabe eines Kombinationspräparates mit Rosskastanienextrakt. *Arzneimittelforsch.*, **18**, 432–435.

Elbary, A.A. and Nour, S.A. (1979). Correlation between the spermicidal activity and the haemolytic index of certain plant saponins. *Pharmazie*, **34**, 560–561.

Eldridge, A.C. and Hockridge, M.E. (1983). High-performance liquid chromatographic separation of Eastern black nightshade (*Solanum ptycanthum*). *J. Agric. Food Chem.*, **31**, 1218–1220.

Elenga, P.A., Nikolov, S. and Panova, D. (1987). Triterpene glycosides from *Astragalus glycyphyllos* L. A new natural compound of the overgrowing parts of *A. glycyphyllos* L. *Pharmazie*, **42**, 422–423.

Elias, R., De Méo, M., Vidal-Ollivier, E., Laget, M., Balansard, G. and Dumenil, G. (1990). Antimutagenic activity of some saponins isolated from *Calendula officinalis* L., *C. arvensis* L. and *Hedera helix* L. *Mutagenesis*, **5**, 327–331.

Elias, R., Diaz Lanza, A.M., Vidal-Ollivier, E., Balansard, G., Faure, R. and Babadjamian, A. (1991). Triterpenoid saponins from the leaves of *Hedera helix. J. Nat. Prod.*, **54**, 98–103.

El-Khrisy, E.A.M., Abdel-Hafez, O.M., Mahmood, K. and Abu-Mustafa, E.A. (1986). Constituents of *Solanum melongena* var. *esculentum* fruits. *Fitoterapia*, **57**, 440–441.

El-Sebakhy, N. and Waterman, P.G. (1985). 6-Oxocycloartan-3β,16β-diglucoside: a new cycloartane diglucoside from *Astragalus trigonus. Planta Medica*, **51**, 350–352.

El-Sebakhy, N.A., Harraz, F.M., Abdallah, R.M. *et al.* (1990). Cycloartane triterpene glycosides from Egyptian *Astragalus* species. *Phytochemistry*, **29**, 3271–3274.

Elujoba, A.A. and Hardman, R. (1985). Incubation conditions for fenugreek whole seed. *Planta Medica*, **51**, 113–115.

456   References

Elujoba, A.A., Fell, A.F., Linley, P.A. and Maitland, D.J. (1990). Triterpenoid saponins from fruit of *Lagenaria breviflora*. *Phytochemistry*, **29**, 3281–3285.

Elyakov, G.B., Strigina, L.I., Khorlin, A.Y., Kochetkov, N.K. (1962). Glycosides of *Panax ginseng* C.A. MEY. *Izv. Akad. Nauk SSSR, Otd. Khim. Nauk*, 1125.

Encarnacion, R., Kenne, L., Samuelsson, G. and Sandberg, F. (1981). Structural studies on some saponins from *Lecaniodiscus cupanioides*. *Phytochemistry*, **20**, 1939–1942.

Encarnacion, R., Carrasco, G., Espinoza, M., Anthoni, U., Nielsen, P.H. and Christophersen, C. (1989). Neothyoside A, proposed structure of a triterpenoid tetraglycoside from the Pacific sea cucumber *Neothyone gibbosa*. *J. Nat. Prod.*, **52**, 248–251.

Enomoto, Y., Ito, K., Kawagoe, Y., Morio, Y. and Yamasaki, Y. (1986). Positive inotropic action of saponins on isolated atrial and papillary muscles from the guinea pig. *Br. J. Pharmacol.*, **88**, 259–267.

Epstein, M.T., Espiner, E.A., Donald, R.A. and Hughes, H. (1977). Effect of eating liquorice on the renin–angiotensin aldosterone axis in normal subjects. *Br. Med. J.*, **1**(6059), 488–490.

Espada, A., Rodriguez, J., Villaverde, M.C. and Riguera, R. (1990). Hypoglucaemic triterpenoid saponins from *Boussingaultia baselloides*. *Can. J. Chem.*, **68**, 2039–2044.

Espada, A., Rodriguez, J., Villaverde, M.C. and Riguera, R. (1991). Boussingoside D1, a new triterpenoid saponin from *Boussingaultia baselloides*. *Annalen*, 291–293.

Evans, F.J. (1974). Thin-layer chromatography–mass spectrometry for the identification of saponins from *Digitalis* species. *Biomed. Mass Spectrom.*, **1**, 166–168.

Fadeev, Y.M., Isaev, M.I., Akimov, Y.A., Kintya, P.K., Gorovits, M.B. and Abubakirov, N.K. (1988). Triterpene glycosides and their genins from *Astragalus*. XXV. Cyclocanthoside D from *Astragalus tragacantha*. *Khim. Prir. Soedin.*, 73–76.

Fang, Z. and Ying, G. (1986). Structure of a new triterpene saponin B from *Polygala japonica* HOUTT. *Zhiwu Xuebao*, **28**, 196–200 (*Chem. Abstr.*, **105**, 57905).

Fang, Z., Zhou, Y. and Zeng, X. (1992). Isolation and identification of saponins from *Aralia decaisneana* HANCE. *Zhiwu Xuebao*, **34**, 461–465 (*Chem. Abstr.*, **118**, 98067).

Fardella, G. and Corsano, S. (1973). Ricerche preliminari sulla biosintesi della acteina. *Annali di Chimica*, **63**, 333–337.

Farnsworth, N.R. and Waller, D.P. (1982). Current status of plant products reported to inhibit sperm. *Res. Front. Fertil. Regul.*, **2**, 1–16.

Farnsworth, N.R., Kinghorn, A.D., Soejarto, D.D. and Waller, D.P. (1985). Siberian ginseng (*Eleutherococcus senticosus*): current status as an adaptogen. In *Economic and Medicinal Plant Research*, vol. 1, ed. H. Wagner, H. Hikino and N.R. Farnsworth, pp. 155–215. Academic Press, London.

Farooq, M.O., Varshney, I.P. and Khan, M.S.Y. (1959). Saponins and sapogenins IV. Isolation of oleanolic acid from *Sesbania aegyptica* PERS. *J. Am. Pharm. Assoc.*, **48**, 466–468.

Farooq, M.O., Varshney, I.P. and Naim, Z. (1961). Saponins and sapogenins VIII. Isolierung eines neuen Triterpenalkohols, Acaciol, aus *Acacia intsia* WILLD. *Arch. Pharmazie*, **294**, 197–200.

Faul, W.H. and Djerassi, C. (1970). Mass spectrometry in structural and stereo-chemical problems. CXCIV. Mass spectrometric fragmentations of steroid sapogenins. *Org. Mass. Spectrom.*, **3**, 1187–1213.

Felix, W. (1985). 'Venenmittel' pflanzlicher Herkunft. *Dtsch. Apoth. Ztg.*, **125**, 1333–1335.

Feng, S., Xu, R. and Han, G. (1985). Structure determination of a saponin from *Dolichos falcata* KLEIN. *Zhongcaoyao*, **16**, 47 (*Chem. Abstr.*, **102**, 225901).

Fenwick, G.R. and Hanley, A.B. (1985). The genus *Allium*. Part 2. *Crit. Rev. Food Sci. Nutr.*, **22**, 273–377.

Ferrari, F., Kiyan de Cornelio, I., Delle Monache, F. and Marini Bettolo, G.B. (1981). Quinovic acid glycosides from roots of *Macfadyena unguis-cati*. *Planta Medica*, **43**, 24–27.

Fieser, L.F. and Fieser, M. (1959). *Steroids*. Reinhold, New York.

Figurkin, B.A. and Ogurtsova, L.N. (1975). Triterpene glycosides of *Anemone ranunculoides*. *Khim. Prir. Soedin.*, 101.

Filderman, R.B. and Kovacs, B.A. (1969). Anti-inflammatory activity of the steroid alkaloid glycoside, tomatine. *Br. J. Pharmacol.*, **37**, 748–755.

Findlay, J.A. and He, Z.-Q. (1991). Polyhydroxylated steroidal glycosides from the starfish *Asterias forbesi*. *J. Nat. Prod.*, **54**, 428–435.

Findlay, J.A., Jaseja, M. and Brisson, J.-R. (1987). Forbeside C, a saponin from *Asterias forbesi*. Complete structure by nuclear magnetic resonance methods. *Can. J. Chem.*, **65**, 2605–2611.

Findlay, J.A., He, Z.-Q. and Blackwell, B. (1990). Minor saponins from the starfish *Asterias forbesi*. *Can. J. Chem.*, **68**, 1215–1217.

Fischer, L. and Goodrich, F.J. (1930). *Polypodium occidentale*. *J. Am. Pharmacol. Assoc.*, **19**, 1063–1071.

Fischer, R., Folberth, K. and Karawya, M.S. (1959). Zur Frage der Resistenz des Igels gegen Cantheridin. *Arzneimittelforsch.*, **9**, 761–763.

Fokina, G.A. (1979). Gas liquid chromatography of terpenoids. Separation of methyl esters of oleanolic and ursolic acids. *Khim. Prir. Soedin.*, 583–584.

Fontaine, T.D., Irving, G.W., Ma, R., Poole, J.B. and Doolittle, S.P. (1948). Isolation and partial characterization of crystalline tomatine, an antibiotic agent from the tomato plant. *Arch. Biochem.*, **18**, 467–475.

Foresta, P., Ghirardi, O., Gabetta, B. and Cristoni, A. (1986). Triterpene saponins having anti-inflammatory, mucolytic and antiedemic activities, process for the preparation thereof and pharmaceutical compositions containing them. Eur. Pat. 251, 197 (*Chem. Abstr.*, **109**, 236992).

Forgacs, P. and Provost, J. (1981). Olaxoside, a saponin from *Olax glabriflora* and *Olax psittacorum*. *Phytochemistry*, **20**, 1689–1691.

Forgo, I. and Kirchdorfer, A.M. (1980). Ginseng steigert die körperliche Leistung. *Arztliche Praxis*, **33**, 1784–1786.

Fraisse, D., Tabet, J.C., Becchi, M. and Raynaud, J. (1986). Fast atom bombardment mass spectrometry of triterpenic saponins. *Biomed. Environ. Mass Spectrom.*, **13**, 1–14.

Franz, C. and Jatisatienr, A. (1983). Pflanzliche Steroid-Rohstoffe. *Dtsch. Apoth. Ztg.*, **123**, 1069–1072.

Frechet, D., Christ, B., Monegier du Sorbier, B., Fischer, H. and Vuilhorgne, M. (1991). Four triterpenoid saponins from dried roots of *Gypsophila* species. *Phytochemistry*, **30**, 927–931.

Freeland, W.J., Calcott, P.H. and Anderson, L.R. (1985). Tannins and saponin: interaction in herbivore diets. *Biochem. Syst. Ecol.*, **13**, 189–195.

Friedman, M. and Dao, L. (1992). Distribution of glycoalkaloids in potato plants and commercial potato products. *J. Agric. Food Chem.*, **40**, 419–423.

Friedman, M. and Levin, C.E. (1992). Reversed-phase high-performance liquid chromatographic separation of potato glycoalkaloids and hydrolysis products on acidic columns. *J. Agric. Food Chem.*, **40**, 2157–2163.

Frolova, G.M. and Ovodov, Yu.S. (1971). Triterpenoid glycosides of *Eleutherococcus senticosus* leaves II. Structure of eleutherosides I, K, L and M. *Khim. Prir. Soedin.*, 618–622.

Frolova, G.M., Ovodov, Yu.S. and Suprunov, N.I. (1971). Triterpenoid glycosides of *Eleutherococcus senticosus* leaves I. Isolation and general characteristics. *Khim. Prir. Soedin.*, 614–618.

Fuggersberger-Heinz, R. and Franz, G. (1984). Formation of glycyrrhizinic acid in *Glycyrrhiza glabra* var. *typica*. *Planta Medica*, **50**, 409–413.

Fujii, K. and Matsukawa, T. (1936). Untersuchung über Saponine und Sterine. VIII. Mitteil. Ueber das Saponin aus der Wurzel von *Dioscorea Tokoro Makino*. 1 Abteil. *Yakugaku Zasshi*, **56**, 408–414.

Fujimoto, H., Suzuki, K., Hagiwara, H. and Yamazaki, M. (1986). New toxic metabolites from a mushroom, *Hebeloma vinosophyllum*. I. Structures of hebevinosides I, II, III, IV and V. *Chem. Pharm. Bull.*, **34**, 88–99.

Fujimoto, Y., Yamada, T., Ikekawa, N., Nishiyama, I., Matsui, T. and Hoshi, M. (1987). Structure of acrosome reaction-inducing steroidal saponins from egg jelly of the starfish *Asterias amurensis*. *Chem. Pharm. Bull.*, **35**, 1829–1832.

Fujioka, N., Kohda, H., Yamasaki, K. *et al.* (1989). Dammarane and oleanane saponins from callus tissue of *Panax japonicus*. *Phytochemistry*, **28**, 1855–1858.

Fujioka, T., Iwase, Y., Okabe, H., Mihashi, K. and Yamauchi, T., (1987). Studies on the constituents of *Actinostemma lobatum* MAXIM. II. Structures of actinostemmosides G and H, new dammarane triterpene glycosides isolated from the herb. *Chem. Pharm. Bull.*, **35**, 3870–3873.

Fujita, M. and Furuya, T. (1954). Studies on saponin-bearing drugs. II. Morphology and chemical characters of the Japanese Hai-shô, from the root of *Patrinia scabiosaefolia* LINK. *J. Pharm. Soc. Jap.*, **74**, 94–98.

Fujita, E., Fujita, T. and Suzuki, T. (1967). On the constituents of *Nauclea orientalis* L. I. Norengenin and naucleoside, a new glycoside (Terpenoids V). *Chem. Pharm. Bull.*, **15**, 1682–1686.

Fujita, M., Itokawa, H. and Kumekawa, Y. (1974). The study on the constituents of *Clematis* and *Akebia* spp. II. On the saponins isolated from the stem of *Akebia quinata* DECNE. (1). *Yakugaku Zasshi*, **94**, 194–198.

Fukuda, N., Imamura, N., Saito, E., Nohara, T. and Kawasaki, T. (1981). Steroid saponins and sapogenins of underground parts of *Trillium kamtschaticum*

PALL. IV. Additional oligoglycosides of 18-norspirostane derivatives and other steroid constituents. *Chem. Pharm. Bull.*, **29**, 325–335.

Fukuhara, K. and Kubo, I. (1991). Isolation of steroidal glycoalkaloids from *Solanum incanum* by two countercurrent chromatographic methods. *Phytochemistry*, **30**, 685–687.

Fukunaga, T., Nishiya, K., Takeya, K. and Itokawa, H. (1987). Studies on the constituents of goat's rue (*Galega officinalis* L.). *Chem. Pharm. Bull.*, **35**, 1610–1614.

Fullas, F., Kim, J., Compadre, C.M. and Kinghorn, A.D. (1989). Separation of natural product sweetening agents using overpressured layer chromatography. *J. Chromatogr.*, **464**, 213–219.

Fullas, F., Choi, Y.-H., Kinghorn. A.D. and Bunyapraphatsara, N. (1990). Sweet-tasting triterpene glycoside constituents of *Abrus fruticulosus*. *Planta Medica*, **56**, 332–333.

Furuya, T., Yoshikawa, T., Orihara, Y. and Oda, H. (1983). Saponin production in cell suspension cultures of *Panax ginseng*. *Planta Medica*, **48**, 83–87.

Furuya, T., Ueoka, T. and Asada, Y. (1988). Novel acylated saponins, crocosmiosides A and B, from Montbretia (*Crocosmia crocosmiiflora*). *Chem. Pharm. Bull.*, **36**, 444–447.

Fusetani, N., Kato, Y., Hashimoto, K., Komori, T., Itakura, Y. and Kawasaki, T. (1984). Biological activities of asterosaponins with special reference to structure–activity relationships. *J. Nat. Prod.*, **47**, 997–1002.

Gafner, F., Msonthi, J.D. and Hostettmann, K. (1985). Molluscicidal saponins from *Talinum tenuissimum* DINTER. *Helv. Chim. Acta*, **68**, 555–558.

Gafner, F., Chapuis, J.-C., Msonthi, J.D. and Hostettmann, K. (1987). Cytotoxic naphthoquinones, molluscicidal saponins and flavonols from *Diospyros zombensis*. *Phytochemistry*, **26**, 2501–2503.

Gan, H. and Chen, S. (1982). Qualitative and quantitative comparison between root and stem and leaves of *Bupleurum chinense*. *Zhongyao Tongbao*, **7**, 7–8 (*Chem. Abstr.*, **97**, 60875).

Gan, L.-X., Han, X.-B. and Chen, Y.-Q. (1986a). Astrasieversianins IX, XI and XV, cycloartane derived saponins from *Astragalus sieversianus*. *Phytochemistry*, **25**, 1437–1441.

Gan, L.-X., Han, X.-B. and Chen, Y.-Q. (1986b). The structures of thirteen astrasieversianins from *Astragalus sieversianus*. *Phytochemistry*, **25**, 2389–2393.

Ganenko, T.V., Isaev, M.I., Lutskii, V.I. *et al.* (1986a). Triterpene glycosides and their genins from *Thalictrum foetidum*. III. Structure of cyclofoetoside A. *Khim. Prir. Soedin.*, 66–71.

Ganenko, T.V., Isaev, M.I., Gromova, A.S. *et al.* (1986b). Triterpene glycosides and their genins of *Thalictrum foetidum*. V. Structure of cyclofoetoside B. *Khim. Prir. Soedin.*, 341–345.

Ganenko, T.V., Gromova, A.S., Lutskii, V.I., Semenov, A.A. and Isaev, M.M. (1988). Triterpenoids of *Thalictrum foetidum* L. and their biological activity. *Chem. Abstr.*, **111**, 39707.

Garg, H.S. and Mitra, C.R. (1967). *Blighia sapida*. I. Constituents of the fresh fruit. *Planta Medica*, **15**, 74–80.

Garg, S.C. and Oswal, V.B. (1993). Saponins from the seed bran of *Parkia biglandulosa*. *Fitoterapia*, **64**, 282.

Gariboldi, P., Verotta, L. and Gabetta, B. (1990). Saponins from *Crossopteryx febrifuga*. *Phytochemistry*, **29**, 2629–2635.

Gaunt, I.F., Grasso, P. and Gangolli, S.D. (1974). Short-term toxicity of quillaia extract in rats. *Food Cosmet. Toxicol.*, **12**, 641–650.

Gedeon, J. (1952). Ueber *Randia*-Saponine und Sapogenine. *Arch. Pharmazie*, **285**, 127–129.

Gedeon, J. and Kincl, F.A. (1956). Ueber Saponine und Sapogenine (2. Mitteilung). *Arch. Pharmazie*, **289**, 162–165.

George, A.J. (1965). Legal status and toxicity of saponins. *Food Cosmet. Toxicol.*, **3**, 85–91.

Gerlach, H. (1966). Saponins and their significance in pharmacy. I. *Pharm. Zentralh. Deut.*, **106**, 573–583.

Gessner, O. (1974). *Die Gift- und Arzneipflanzen von Mitteleuropa*, Winter-Verlag, Heidelberg, p. 160.

Gestetner, B. (1971). Lucerne saponins. V. Structure of a saponin from *Medicago sativa* (alfalfa). *Phytochemistry*, **10**, 2221–2223.

Ghosal, S., Srivastava, A.K., Srivastava, R.S. and Maitra, M. (1981). Justicisaponin-I, a new triterpenoid saponin from *Justicia simplex*. *Planta Medica*, **42**, 279–283.

Ghosh, A.C. and Dutta, N.L. (1965). Chemical investigation of *Ougeinia dalbergioides*. *J. Indian Chem. Soc.*, **42**, 831–835.

Gibson, M.R. (1978). Glycyrrhiza in old and new perspectives. *Lloydia*, **41**, 348–354.

Giral, F. and Hahn, L.R. (1960). Echinocystic acid in *Fouquieria peninsularis*. *J. Chem. Soc.*, 2373.

Glombitza, K.-W. and Kurth, H. (1987a). Saponine aus *Anagallis arvensis* L. (Primulaceae). *Arch. Pharm.*, **320**, 1083–1087.

Glombitza, K.-W. and Kurth, H. (1987b). Die Struktur einiger Triterpensaponine aus *Anagallis arvensis*. *Planta Medica*, **53**, 548–555.

Goad, L.J. (1991). Phytosterols. In *Methods in Plant Biochemistry*, vol. 7, ed. P.M. Dey and J.B. Harborne, pp. 369–434. Academic Press, London.

Goad, L.J. and Goodwin, T.W. (1972). The biosynthesis of plant sterols. In *Progress in Phytochemistry*, vol. 3, ed. L. Reinhold and Y. Liwschitz, pp. 113–198. Wiley Interscience, New York.

Godin, P. (1954). A new spray reagent for paper chromatography of polyols and cetoses. *Nature*, **174**, 134.

Goldstein, I.J., Hay, G.W., Lewis, B.A., Smith, F. (1965). Controlled degradation of polysaccharides by periodate oxidation, reduction and hydrolysis. *Methods Carbohydr. Chem.*, **5**, 361–370.

Goll, P.H., Lemma, A., Duncan, J. and Mazengia, B. (1983). Control of schistosomiasis in Adwa, Ethiopia, using the plant molluscicide endod (*Phytolacca dodecandra*). *Tropenmed. Parasitol.*, **34**, 177–183.

Gomes, A., Sharma, R.M. and Ghatak, B.J.R. (1987). Pharmacological investigation of a glycosidal fraction isolated from *Maesa chisia* D.DON var. *angustifolia* HOOK F. and TH. *Indian J. Exp. Biol.*, **25**, 826–831.

Gopalsamy, N. (1992). Studies of saponins from three Mauritian plants: *Aphloia theiformis* (Flacourtiaceae), *Polyscias dichroostachya* and *Polyscias mauritiana* (Araliaceae). PhD. Thesis, Lausanne University.

Gopalsamy, N., Vargas, D., Guého, J., Ricaud, C. and Hostettmann, K. (1988). Saponins from leaves of *Aphloia theiformis*. *Phytochemistry*, **27**, 3593–3595.

Gopalsamy, N., Guého, J., Julien, H.R., Owadally, A.W. and Hostettmann, K. (1990). Molluscicidal saponins of *Polyscias dichroostachya*. *Phytochemistry*, **29**, 793–795.

Gorin, P.A.J. and Mazurek, M. (1975). Further studies on the assignment of signals in ^{13}C magnetic resonance spectra of aldoses and derived methyl glycosides. *Can. J. Chem.*, **53**, 1212–1223.

Goryanu, G.M., Krokhmalyuk, V.V. and Kintia, P.K. (1976). Structures of glycosides of *Asparagus officinalis*. Structure of asparagosides A and B. *Khim. Prir. Soedin.*, 400–401.

Gosmann, G., Schenkel, E.P. and Seligmann, O. (1989). A new saponin from mate, *Ilex paraguariensis*. *J. Nat. Prod.*, **52**, 1367–1370.

Grishkovets, V.I., Tolkacheva, N.V., Shashkov, A.S. and Chirva, V.J. (1991). Triterpene glycosides of *Hedera taurica*. VII. Structure of taurosides A and D from Crimean ivy leaves. *Khim. Prir. Soedin*, 686–689.

Gromova, A.S., Lutskii, V.I., Semenov, A.A., Demisenko, V.A. and Isakov, V.V. (1984). Triterpene saponins from *Thalictrum minus*. IV. Structure of thalicoside A. *Khim. Prir. Soedin.*, 213–219.

Gromova, A.S., Lutskii, V.I., Semenov, A.A., Valeev, R.B., Kalabin, G.A. and Elkin, Y.N. (1985). Triterpene saponins from *Thalictrum minus*. V. Structure of thalicoside B. *Khim. Prir. Soedin.*, 670–676.

Grünweller, S., Schröder, E. and Kesselmeier, J. (1990). Biological activities of furostanol saponins from *Nicotiana tabacum*. *Phytochemistry*, **29**, 2485–2490.

Gu, W., Liu, J., Zhang, J., Liu, X., Liu, S. and Chen, Y. (1987). A study of the hypotensive action of total saponin of *Zizyphus jujuba* MILL and its mechanism. *J. Med. Coll. PLA*, **2**, 315–318 (*Chem. Abstr.*, **109**, 48104).

Guan, Y., He, H. and Chen, J. (1985). Effect of the total saponins of *Panax notoginseng* on contraction of rabbit aortic strips. *Zhongguo Yaoli Xuebao*, **6**, 267–269 (*Chem. Abstr.*, **104**, 81755).

Guan, Y.Y., Kwan, C.Y., He, H. and Daniel, E.E. (1988). Inhibition of norepinephrin-induced contractile responses of canine mesenteric artery by plant total saponins. *Blood Vessels*, **25**, 312–315.

Gubanov, I.A., Libizov, N.I. and Gladkikh, A.S. (1970). Search for saponin-containing plants among the flora of Central Asia and Southern Kazakhstan. *Farmatsiya* (*Moscow*), **19**, 23–31 (*Chem. Abstr.*, **73**, 95408).

Guédon, D., Abbe, P., Cappelaere, N. and Rames, N. (1989). Dosage des saponosides de *Panax ginseng* C.A. MEYER. Comparaison entre les méthodes officielles et la CLHP. *Ann. Pharm. Fr.*, **47**, 169–177.

Guérin, J.-C. and Réveillère, H.-P. (1984). Activité antifongique d'extraits végétaux à usage thérapeutique. I. Etude de 41 extraits sur 9 souches fongiques. *Ann. Pharm. Fr.*, **42**, 553–559.

Gunzinger, J., Msonthi, J.D. and Hostettmann, K. (1986). Molluscicidal saponins from *Cussonia spicata*. *Phytochemistry*, **25**, 2501–2503.

Gupta, D. and Singh, J. (1989). Triterpenoid saponins from *Centipeda minima*. *Phytochemistry*, **28**, 1197–1201.

Gupta, D. and Singh, J. (1990). Triterpenoid saponins from *Centipeda minima*. *Phytochemistry*, **29**, 1945–1950.

Gupta, R.K., Jain, D.C. and Thakur, R.S. (1984). Furostanol glycosides from *Trigonella foenum-graecum* seeds. *Phytochemistry*, **23**, 2605–2607.

Gupta, R.K., Jain, D.C. and Thakur, R.S. (1985). Furostanol glycosides from *Trigonella foenum-graecum* seeds. *Phytochemistry*, **24**, 2399–2401.

Gupta, R.K., Jain, D.C. and Thakur, R.S. (1986). Two furostanol saponins from *Trigonella foenum-graecum*. *Phytochemistry*, **25**, 2205–2207.

Guseinov, D.Ya. and Iskenderov, G.B. (1972). Chemical composition and biological value of saponins of some plants of Azerbaidzhan. *Biol. Nauki (Moscow)*, **15**, 85–88 (*Chem. Abstr.*, **77**, 16542).

Gutsu, E.V., Kintia, P.K., Lazur'evskii, G.V. and Balashova, N.N. (1984). Steroidal alkaloids and glycosides of *Capsicum annuum* L. *Rastit. Resur.*, **20**, 127–130 (*Chem. Abstr.*, **100**, 117851).

Habermehl, G.G. and Krebs, H.C. (1990). Toxins of echinoderms. In *Studies in Natural Products Chemistry*, vol. 7, ed. Atta-ur-Rahman, pp. 265–316. Elsevier, Amsterdam.

Habermehl, G. and Möller, H. (1974). Isolierung und Struktur von Larreagenin A. *Annalen*, 169–175.

Hahn, D.R., Kasai, R., Kim, J.H. *et al.* (1986). The glycosides of araliaceous drugs and their biological activities. *Saengyak Hakhoechi*, **17**, 78–84 (*Chem. Abstr.*, **105**, 158694).

Hahn, L.R., Sanchez, C. and Romo, J. (1965). Isolation and structure of jacquinic acid. *Tetrahedron*, **21**, 1735–1740.

Hakomori, S. (1964). A rapid permethylation of glycolipid and polysaccharide catalyzed by methylsulfinyl carbanion in dimethyl sulfoxide. *J. Biochem. (Tokyo)*, **55**, 205–208.

Hamburger, M. and Hostettmann, K. (1986). New saponins and a prosapogenin from *Polygala chamaebuxus* L. *Helv. Chim. Acta*, **69**, 221–227.

Hamburger, M., Slacanin, I., Hostettmann, K., Dyatmiko, W. and Sutarjadi (1992). Acetylated saponins with molluscicidal activity from *Sapindus rarak*: unambiguous structure determination by proton nuclear magnetic resonance and quantitative analysis. *Phytochem. Anal.*, **3**, 231–237.

Han, B.H., Park, M.H., Han, Y.N. *et al.* (1982). Degradation of ginseng saponins under mild acidic conditions. *Planta Medica*, **44**, 146–149.

Han, D., Han, J., Qiao, M., Yu, Z., Tan, W. and Xu, S. (1987). Paridiformoside. *Yaoxue Xuebao*, **22**, 746–749 (*Chem. Abstr.*, **108**, 62305).

Hanawa, T., Hirama, N., Kosoto, H., Hyun, S.J., Ohwada, S., Hasegawa, R., Haranaka, R., Nakagawa, S. (1987). Effects of Saiko-Zai on collagen metabolism in mice administered glucocorticoid. *Nihon Univ. J. Med.*, **29**, 197–205 (*Chem. Abstr.*, **108**, 31560).

Hänsel, R. (1984). Traditionelle Reizkörpertherapie, gesehen als Immunstimulation. *Dtsch. Apoth. Ztg.*, **124**, 54–59.

Hänsel, R. and Haas, H. (1983). *Therapie mit Phytopharmaka*, Springer-Verlag, Berlin.

Haraguchi, M., Motidome, M. and Gottlieb, O.R. (1988). Triterpenoid saponins and flavonol glycosides from *Phytolacca thyrsifolia*. *Phytochemistry*, **27**, 2291–2296.

Hardman, R., Kosugi, J. and Parfitt, R.T. (1980). Isolation and characterization of a furostanol glycoside from fenugreek. *Phytochemistry*, **19**, 698–700.

Hariharan, V. (1974). Structure of putranosides from the seed coats of *Putranjiva roxburghii*. *Indian J. Chem.*, **12**, 447–449.

Harkar, S., Razdan, T.K. and Waight, E.S. (1984). Further triterpenoids and ^{13}C NMR spectra of oleanane derivatives from *Phytolacca acinosa*. *Phytochemistry*, **23**, 2893–2898.

Harmatha, J., Mauchamp, B., Arnault, C. and Slama, K. (1987). Identification of a spirostane-type saponin in the flowers of leek with inhibitory effects on growth of leek-moth larvae. *Biochem. Sys. Ecol.*, **15**, 113–116.

Harnischfeger, G. and Stolze, H. (1983). *Bewährte Pflanzendrogen in Wissenschaft und Medizin*. Notamed Verlag, Bad Homburg/Melsungen.

Hasan, M., Burdi, D.K. and Ahmad, V.U. (1991). Leucasin, a triterpene saponin from *Leucas nutans*. *Phytochemistry*, **30**, 4181–4183.

Hasanean, H.A., El-Shanawany, M.A., Bishay, D.W. and Franz, G. (1992a). New triterpenoid glycosides from *Zygophyllum album* L. *Pharm. Pharmacol. Lett*, **1**, 115–117.

Hasanean, H.A., El-Shanawany, M.A., Bishay, D.W. and Franz, G. (1992b). New triterpenoid glycosides from *Taverniera aegyptiaca* BOISS. *Pharmazie*, **47**, 143–144.

Hashimoto, Y. (1979). *Marine Toxins and Other Bioactive Marine Metabolites*. Japan Scientific Societies Press, Tokyo.

Hashimoto, Y. and Yasumoto, I. (1960). Confirmation of saponin as a toxic principle of starfish. *Bull. Jpn Soc. Sci. Fisheries*, **26**, 1132.

Hashimoto, Y., Ishizone, H., Suganuma, M., Ogura, M., Nakatsu, K. and Yoshioka, H. (1983). Periandrin I, a sweet triterpene glycoside from *Periandra dulcis*. *Phytochemistry*, **22**, 259–264.

Hashizume, A. (1973). Saponin from the leaf of *Thea sinensis*. III. Component sapogenins and sugars of the saponin from the leaf of *Thea sinensis*. *Nippon Nogei Kagaku Kaishi*, **47**, 237–240 (*Chem. Abstr.*, **80**, 68381).

Haslinger, E., Korhammer, S. and Schubert-Zsilavecz, M. (1990). Sequential analysis of oligosaccharide structures by modern NMR techniques. *Annalen*, 713–714.

Hassall, C.H. and Smith, B.S.W. (1957). Hecogenin from *Agave sisalana* by microbiological hydrolysis. *Chem. Ind.*, 1570.

Hattori, M., Kawata, Y., Kakiuchi, N., Matsuura, K., Tomimori, T. and Namba, T. (1988). Application of liquid chromatography/mass spectrometry to the qualitative analysis of saponins. II. *Chem. Pharm. Bull.*, **36**, 4467–4473.

Hattori, T., Ikematsu, S., Koito, A. *et al.* (1989). Preliminary evidence for inhibitory effect of glycyrrhizin on HIV replication in patients with AIDS. *Antiviral Res.*, **11**, 255–262.

Hayashi, T. and Kubo, M. (1980). Antitumor compositions comprising saponins. *Brit. UK Pat.* 2035082 (*Chem. Abstr.*, **94**, 7771).

Hayden, A.L., Smeltzer, P.B. and Scheer, I. (1954). Detection of steroidal pseudosapogenins by infrared spectroscopy. *Anal. Chem.*, **26**, 550–552.

He, L.-Y. (1987). Studies on the method of determination of combined sugars in glycosides. *Acta Pharm. Sinica*, **22**, 70–74.

He, Z.-Q. and Findlay, J.A. (1991). Constituents of *Astragalus membranaceus*. *J. Nat. Prod.*, **54**, 810–815.

Heftmann, E. (1967). Biochemistry of steroidal saponins and glycoalkaloids. *Lloydia*, **30**, 209–230.

Heftmann, E. (1968). Biosynthesis of plant steroids. *Lloydia*, **31**, 293–317.

Heftmann, E. (1974). Recent progress in the biochemistry of plant steroids other than sterols (saponins, glycoalkaloids, pregnane derivatives, cardiac glycosides, and sex hormones). *Lipids*, **9**, 626–639.

Hemmer, R. (1985). Zur Therapie des Hirnödems beim Schädel-Hirn Trauma. *Unfallchirurg*, **88**, 93–96.

Henry, M., Pauthe Dayde, D. and Rochd, M. (1989). Extraction and high-performance liquid chromatographic determination of gypsogenin 3-*O*-glucuronide. *J. Chromatogr.*, **477**, 413–419.

Henry, M., Rochd, M. and Bennini, B. (1991). Biosynthesis and accumulation of saponins in *Gypsophila paniculata*. *Phytochemistry*, **30**, 1819–1821.

Henschler, D., Hempel, K., Schultze, B. and Maurer, W. (1971) Zur Pharmakokinetik von Aescin. *Arzneimittelforsch.*, **21**, 1682–1692.

Hensens, O.D. and Lewis, K.G. (1965). Reactions of primulagenin A – Part 2. *Tetrahedron Lett.*, 4639–4643.

Heywood, B.J. and Kon, G.A.R. (1940). Sapogenins. Part IX. The occurrence and constitution of bassic acid. *J. Chem. Soc.*, 713–720.

Heywood, B.J., Kon, G.A.R. and Ware, L.L. (1939). Sapogenins. Part V. Bassic acid. *J. Chem. Soc.*, 1124–1129.

Hiai, S. (1986). Chinese medicinal material and the secretion of ACTH and corticosteroid. *Chem. Abstr.*, **104**, 161363.

Hiai, S. and Yokoyama, H. (1986). Chronic effect of saikosaponin on adrenal and thymus growth in normal and dexamethasone-treated rats. *Chem. Pharm. Bull.*, **34**, 1195–1202.

Hiai, S., Oura, H. and Nakajima, T. (1976). Colour reaction of some sapogenins and saponins with vanillin and sulfuric acid. *Planta Medica*, **29**, 116–122.

Hiai, S., Yokoyama, H. and Oura, H. (1981). Effect of escin on adrenocorticotropin and corticosterone levels in rat plasma. *Chem. Pharm. Bull.*, **29**, 490–494.

Hiai, S., Yokoyama, H., Oura, H. and Kawashima, Y. (1983). Evaluation of corticosterone secretion-inducing activities of ginsenosides and their prosapogenins and sapogenins. *Chem. Pharm. Bull.*, **31**, 168–174.

Hiai, S., Sasayama, Y. and Oguro, C. (1987). Chronic effect of ginseng saponin, glycyrrhizin and flavin adenine dinucleotide on adrenal and thymus weight in normal and dexamethasone-treated rats. *Chem. Pharm. Bull.*, **35**, 241–248.

Hidaka, K., Ito, M., Matsuda, Y. *et al.* (1987a). New triterpene saponins from *Ilex pubescens*. *Chem. Pharm. Bull.*, **35**, 524–529.

Hidaka, K., Ito, M., Matsuda, Y., Kohda, H., Yamasaki, K. and Yamahara, J. (1987b). A triterpene and saponin from roots of *Ilex pubescens*. *Phytochemistry*, **26**, 2023–2027.

Higuchi, R. and Kawasaki, T. (1972). Seed saponins of *Akebia quinata* DECNE. II. Hederagenin 3,28-*O*-bisglycosides. *Chem. Pharm. Bull.*, **20**, 2143–2149.

Higuchi, R. and Kawasaki, T. (1976). Pericarp saponins of *Akebia quinata*

DECNE. II. Norarjunolic acids and their glycosides. *Chem. Pharm. Bull.*, **24**, 1314–1323.

Higuchi, R., Miyahara, K. and Kawasaki, T. (1972). Seed saponins of *Akebia quinata* DECNE. I. Hederagenin 3-O-glycosides. *Chem. Pharm. Bull.*, **20**, 1935–1939.

Higuchi, R., Kawasaki, T., Biswas, M., Pandey, V.B. and Dasgupta, B. (1982). Triterpenoid saponins from the bark of *Symplocos spicata*. *Phytochemistry*, **21**, 907–910.

Higuchi, R., Fujioka, T., Iwamoto, M., Komori, T., Kawasaki, T. and Lassak, E.V. (1983). Triterpenoid saponins from leaves of *Pittosporum undulatum*. *Phytochemistry*, **22**, 2565–2569.

Higuchi, R., Kubota, S., Komori, T. *et al.* (1984). Triterpenoid saponins from the bark of *Zizyphus joazeiro*. *Phytochemistry*, **23**, 2597–2600.

Higuchi, R., Tokimitsu, Y., Hamada, N., Komori, T. and Kawasaki, T. (1985). A new cleavage method for the sugar-aglycone linkage in saponin. *Annalen*, 1192–1201.

Higuchi, R., Tokimitsu, Y., Fujioka, T., Komori, T., Kawasaki, T. and Oakenful, D.G. (1987). Structure of desacylsaponins obtained from the bark of *Quillaja saponaria*. *Phytochemistry*, **26**, 229–235.

Higuchi, R., Tokimitsu, Y. and Komori, T. (1988). An acylated triterpenoid saponin from *Quillaja saponaria*. *Phytochemistry*, **27**, 1165–1168.

Hikino, H. and Kiso, Y. (1988). Natural products for liver diseases. In *Economic and Medicinal Plant Research*, vol. 2, ed. H. Wagner, H. Hikino and N.R. Farnsworth, pp. 39–72. Academic Press, London.

Hikino, H., Kiso, Y., Kinouchi, J., Sanada, S. and Shoji, J. (1985). Antihepatotoxic actions of ginsenosides from *Panax ginseng* roots. *Planta Medica*, **51**, 62–64.

Hill, R.A., Kirk, D.N., Makin, H.J. and Murphy, G.M. (1991). *Dictionary of Steroids. Chemical data, Structures and Biographies*. Chapman & Hall, London.

Hiller, K. (1987). New results on the structure and biological activity of triterpenoid saponins. In *Biological Active Natural Products*, ed. K. Hostettmann and P.J. Lea, pp. 167–184. Clarendon Press, Oxford.

Hiller, K. and Adler, C. (1982). Neue Ergebnisse über Triterpensaponine. *Pharmazie*, **37**, 619–634.

Hiller, K. and Voigt, G. (1977). Neue Ergebnisse in der Erforschung der Triterpensaponine. *Pharmazie*, **32**, 365–393.

Hiller, K., Keipert, M. and Linzer, B. (1966). Triterpensaponine. *Pharmazie*, **21**, 713–751.

Hiller, K., Thi, N., Döhnert, H. and Franke, P. (1975). Isolierung neuer Estersapogenine aus *Eryngium giganteum* M.B. *Pharmazie*, **30**, 105–109.

Hiller, K., Von Mach, B. and Franke, P. (1976). Ueber die Saponine von *Eryngium maritimum* L. *Pharmazie*, **31**, 53.

Hiller, K., Nguyen, K.Q.C. and Franke, P. (1978a). Isolierung von 3-O-D-glucopyranosyloleanolsäure-28-O-D-xylopyranosid aus *Eryngium bromeliifolium* DELAR. *Pharmazie*, **33**, 78–80.

Hiller, K., Nguyen, K.Q.C. and Franke, P. (1978b). Isolation of betulinic acid 3-O-β-D-glucopyranosyl(1→6)-β-D-glucopyranoside from *Eryngium bromeliifolium* DELAR. *Z. Chem.*, **18**, 260–261.

Hiller, K., Paulick, A., Döhnert, H. and Franke, P. (1979). Ueber die Saponine von *Polemonium caeruleum* L. *Pharmazie*, **34**, 565–566.

Hiller, K., Voigt, G. and Döhnert, H. (1981). Zur Struktur des Hauptsaponins aus *Hydrocotyle vulgaris* L. *Pharmazie*, **36**, 844–846.

Hiller, K., Bader, G. and Schulten, H.-R. (1987a). Virgaurea saponin I, ein neues bisdesmosidesches Triterpensaponin aus *Solidago virgaurea* L. *Pharmazie*, **42**, 541–543.

Hiller, K., Bardella, H. and Schulten, H.-R. (1987b). Beiträge zur Struktur der Saponine von *Solidago canadensis* L. *Pharmazie*, **42**, 622–625.

Hiller, K., Schöpke, T. and Franke, P. (1987c). Zur Artefaktbildung von Oleanolsäure und Hederagenin unter säurehydrolytischen Einflüssen. *Pharmazie*, **42**, 67.

Hiller, K., Schöpke, T., Wray, V. and Schulten, H.-R. (1988). Zur Struktur der Hauptsaponine von *Bellis perennis* L. *Pharmazie*, **43**, 850–852.

Hiller, K., Bader, G. and Schöpke, T. (1990a). Antifungal effects of glycosides of polygalacic acid. *Planta Medica*, **56**, 644.

Hiller, K., Leska, M., Gründemann, E., Dube, G., Karwatzki, A. and Franke, P. (1990b). Zur Struktur der bisdesmosidischen Saponine von *Astrantia major* L. *Pharmazie*, **45**, 615–617.

Hiller, K., Bader, G., Reznicek, G., Jurenitsch, J. and Kubelka, W. (1991). Die Hauptsaponine der arzneilich genutzten Arten der Gattung *Solidago*. *Pharmazie*, **46**, 405–408.

Himi, T., Saito, H. and Nishiyama, N. (1989). Effect of ginseng saponins on the survival of cerebral cortex neurons in cell cultures. *Chem. Pharm. Bull.*, **37**, 481–484.

Hirabayashi, K., Iwata, S., Matsumoto, H. *et al.* (1991). Antiviral activities of glycyrrhizin and its modified compounds against human immunodeficiency virus type 1 (HIV-1) and herpes simplex virus type 1 (HSV-1) *in vitro*. *Chem. Pharm. Bull.*, **39**, 112–115.

Hiraga, Y., Endo, H., Takahashi, K. and Shibata, S. (1984). High-performance liquid chromatographic analysis of licorice extracts. *J. Chromatogr.*, **292**, 451–453.

Hirai, Y., Konishi, T., Sanada, S., Ida, Y. and Shoji, J. (1982). Studies on the constituents of *Aspidistra elatior* BLUME. I. On the steroids of the underground part. *Chem. Pharm. Bull.*, **30**, 3476–3484.

Hirai, Y., Sanada, S., Ida, Y. and Shoji, J. (1984) Studies on the constituents of Palmae plants. I. The constituents of *Trachycarpus fortunei* (HOOK.) H. WENDL. (1) *Chem. Pharm. Bull.*, **32**, 295–301.

Hirai, Y., Sanada, S., Ida, Y. and Shoji, J. (1986). Studies on the constituents of Palmae plants. III. The constituents of *Chamaerops humilis* L. and *Trachycarpus wagnerianus* BECC. *Chem. Pharm. Bull.*, **34**, 82–87.

Hirschmann, H. and Hirschmann, F.B. (1958). C-22 Ketals related to the sapogenins. *Tetrahedron*, **3**, 243–254.

Homans, A.L. and Fuchs, A. (1970). Direct bioautography on thin-layer chromatograms as a method for detecting fungitoxic substances. *J. Chromatogr.*, **51**, 325–327.

Honda, T., Murae, T., Tsuyuki, T., Takahashi, T. and Sawai, M. (1976). Arjungenin, arjunglucoside I and arjunglucoside II. A new triterpene and new triterpene glucosides from *Terminalia arjuna*. *Bull. Chem. Soc. Jp.*, **49**, 3213–3218.

Hosny, M., Khalifa, T., Calis, I., Wright, A.D. and Sticher, O. (1992). Balanitoside, a furostanol glycoside, and 6-methyl diosgenin from *Balanites*. *Phytochemistry*, **31**, 3565–3569.

Hostettmann, K. (1980). Saponins with molluscicidal activity from *Hedera helix* L. *Helv. Chim Acta*, **63**, 606–609.

Hostettmann, K. (1989). Plant-derived molluscicides of current importance. In *Economic and Medicinal Plant Research*, vol. 2, ed. H. Wagner, H. Hikino and N.R. Farnsworth, pp. 73–102. Academic Press, London.

Hostettmann, K. and Marston, A. (1987). Plant molluscicide research – an update. In *Plant Molluscicides*, ed. K.E. Mott, pp. 299–320. John Wiley, Chichester.

Hostettmann, K., Hostettmann-Kaldas, M. and Nakanishi, K. (1978). Molluscicidal saponins from *Cornus florida* L. *Helv. Chim. Acta*, **61**, 1990–1995.

Hostettmann, K., Hostettmann-Kaldas, M. and Nakanishi, K. (1979). Droplet counter-current chromatography for the preparative isolation of various glycosides. *J. Chromatogr.*, **170**, 355–361.

Hostettmann, K., Hostettmann-Kaldas, M. and Sticher, O. (1980). Rapid preparative separation of natural products by centrifugal thin-layer chromatography. *J. Chromatogr.*, **202**, 154–156.

Hostettmann, K., Doumas, J. and Hardy, M. (1981). Desorption/chemical ionization mass spectrometry of naturally occurring glycosides. *Helv. Chim. Acta*, **64**, 297–303.

Hostettmann, K., Kizu, H. and Tomimori, T. (1982). Molluscicidal properties of various saponins. *Planta Medica*, **44**, 34–35.

Hostettmann, K., Hostettmann, M. and Marston, A. (1984). Isolation of natural products by droplet counter-current chromatography and related methods. *Nat. Prod. Rep.*, **1**, 471–481.

Hostettmann, K., Hostettmann, M. and Marston, A. (1986). *Preparative Chromatography Techniques – Applications in Natural Product Isolation*. Springer-Verlag, Berlin.

Hostettmann, K., Hostettmann, M. and Marston, A. (1991). Saponins. In *Methods in Plant Biochemistry*, vol. 7, ed. B.V. Charlwood and D.V. Banthorpe, pp. 435–471. Academic Press, London.

Houghton, P.J. and Lian, L.M. (1986a). Iridoids, iridoid-triterpenoid congeners and lignans from *Desfontainia spinosa*. *Phytochemistry*, **25**, 1907–1912.

Houghton, P.J. and Lian, L.M. (1986b). Triterpenoids from *Desfontainia spinosa*. *Phytochemistry*, **25**, 1939–1944.

Houseman, P.A. (1944). *Licorice*. Royal Institute of Chemistry Monograph, London.

Howes, F.N. (1930). *Fish-poison plants*. *Bull. Misc. Inform.* No. 4, pp. 129–153. Royal Botanical Gardens, Kew, UK.

Hu, S.Y. (1976). The genus *Panax* (ginseng) in Chinese medicine. *Econ. Bot.*, **30**, 11–28.

Hu, S.-Y. (1977). A contribution to our knowledge of ginseng. *Am. J. Chin. Med.*, **5**, 1–23.

Huang, W.-Y., Chen, W.-K., Chou, Y.-L. and Chu, J.-H. (1962). Saponin of Chinese drug from *Anemone chinensis*. II. Composition of anemosapogenin. *Acta Chim. Sinica*, **28**, 126–131 (*Chem. Abstr.*, **59**, 1692).

Huang, H., Chen, C., Lin, W., Yang, G., Song, J. and Ren, G. (1982a). Antitumour activity of total saponins from *Dolichos falcatus* KLEIN. *Zhongguo Yaoli Xuebao*, **3**, 286–288 (*Chem. Abstr.*, **98**, 100885).

Huang, H., Huang, N. and Li, S. (1982b). Analgesic effect of saponin from *Dolichos falcatus*. *Yaoxue Tongbao*, **17**, 22 (*Chem. Abstr.*, **97**, 49567).

Huffman, V.L., Lee. C.K. and Burns, E.E. (1975). Selected functional properties of sunflower meal (*Helianthus annuus*) *J. Food Sci.*, **40**, 70–74.

Hufford, C.D., Liu, S. and Clark, A.M. (1988). Antifungal activity of *Trillium grandiflorum* constituents. *J. Nat. Prod.*, **51**, 94–98.

Hui, W.H. and Ho, C.T. (1968). The Rubiaceae of Hong Kong. VII. The occurrence of triterpenoids, steroids, and triterpenoid saponins. *Aust. J. Chem.*, **21**, 547–549.

Hui, W.H. and Szeto, S.K. (1967). Rubiaceae of Hong Kong. IV. The occurrence of triterpenoids and triterpenoid saponins from *Antirrhoea chinensis*. *Phytochemistry*, **6**, 443–444.

Hui, W.H. and Yee, C.W. (1968). The Rubiaceae of Hong Kong. VI. The occurrence of triterpenoids, steroids and triterpenoid saponins in *Adina pilulifera*. *Aust. J. Chem.*, **21**, 543–546.

Hui, W.H., Lam, W.K. and Tye, S.M. (1971). Triterpenoid and steroid constituents of *Aster baccharoides*. *Phytochemistry*, **10**, 903–904.

Hwang, C.S. and Lee, S.W. (1983). Studies on glucoalkaloids in plants. I. High performance liquid chromatographic analysis of glucoalkaloids in periderm and cortex of potato (*Solanum tuberosum* var. May Queen). *Chem. Abstr.*, **98**, 122864.

Hylands, P.J. and Kosugi, J. (1982). Bryonoside and bryoside – new triterpene glycosides from *Bryonia dioica*. *Phytochemistry*, **21**, 1379–1384.

Hylands, P.J. and Salama, A.M. (1979). Maragenins I, II and III, new pentacyclic triterpenes from *Marah macrocarpus*. *Tetrahedron*, **35**, 417–420.

Ibragimov, A.Y. and Khazanovich, R.L. (1972). Ursolic acid from *Agrimonia asiatica*. *Khim. Prir. Soedin.*, 394–395.

Ichikawa, T., Ishida, S., Sakiya, Y., Sawada, Y. and Hanano, M. (1986). Biliary excretion and enterohepatic cycling of glycyrrhizin in rats. *J. Pharm. Sci.*, **75**, 672–675.

Ikeda, Y., Sugiura, M., Fukaya, C. *et al.* (1990). A potent inhibitor of cyclic nucleotide phosphodiesterase from *Periandra dulcis*. *Planta Medica*, **56**, 554–555.

Ikeda, Y., Sugiura, M., Fukaya, C. *et al.* (1991). Periandradulcins A, B and C: phosphodiesterase inhibitors from *Periandra dulcis* MART. *Chem. Pharm. Bull.*, **39**, 566–571.

Ikegami, S., Hirose, Y., Kamiya, A. and Tamura, S. (1972). Asterosaponins. III. Structure of carbohydrate moiety of asterosaponin A. *Agric. Biol. Chem.*, **36**, 2453–2457.

Ikegami, S., Okano, K. and Muragaki, H. (1979). Structure of glycoside B2, a steroidal saponin in the ovary of the starfish *Asterias amurensis*. *Tetrahedron Lett.*, 1769–1772.

Ikram, M., Ogihara, Y. and Yamasaki, K. (1981). Structure of a new saponin from *Zizyphus vulgaris*. *J. Nat. Prod.*, **44**, 91–93.

Ikramov, M.T., Khazanovich, R.L. and Khalmatov, K.K. (1971). Saponins from two species of *Eryngium*. *Khim. Prir. Soedin.*, 843.

Ikramov, M.T., Kharlamov, I.A., Khazanovich, R.L. and Khalmatov, K.K. (1976). Sapogenins of *Eryngium macrocalyx*. *Khim. Prir Soedin.*, 401.

Imomnazarov, B.A. and Isaev, M.I. (1992). Three new triterpene glycosides from the epigeal part of *Astragalus coluteocarpus*. *Chem. Nat. Comp.*, **28**, 195–198.

Ina, H., Ohta, Y. and Iida, H. (1985). Aquilegifolin: a triterpenoid glycoside from *Thalictrum aquilegifolium*. *Phytochemistry*, **24**, 2655–2657.

Inada, A., Murata, H., Kobayashi, M. and Nakanishi, T. (1987a). Pseudo-ginsenoside-RP_1 and zingibroside-R_1 from seeds of *Strongylodon macrobotrys* A. GRAY. *Shoyakugaku Zasshi*, **41**, 153–156.

Inada, A., Kobayashi, M., Murata, H. and Nakanishi, T. (1987b). Two new triterpenoid glycosides from leaves of *Ilex chinensis* SIMS. *Chem. Pharm. Bull.*, **35**, 841–845.

Inada, A., Yamada, M., Murata, H. *et al.* (1988). Phytochemical studies of seeds of medicinal plants. I. Two sulfated triterpenoid glycosides, sulfapatrinosides I and II, from seeds of *Patrinia scabiosaefolia* FISCHER. *Chem. Pharm. Bull.*, **36**, 4269–4274.

Indian Pat. 165,214 Isolation of a nonsteroidal antiinflammatory, analgesic, antipyretic and tranquilosedative triterpene glycoside drug from *Maesa chisia*, and biological activity and pharmaceutical compositions containing the drug. (*Chem. Abstr.*, **116**, 158896).

Inose, Y., Miyase, T. and Ueno, A. (1991). Studies on the constituents of *Solidago virga-aurea* L. I. Structural elucidation of saponins in the herb. *Chem. Pharm. Bull.*, **39**, 2037–2042.

Inoue, O., Takeda, T. and Ogihara, Y. (1978). Carbohydrate structures of three new saponins from the root bark of *Hovenia dulcis* (Rhamnaceae). *J. Chem. Soc., Perkin Trans. I*, 1289–1293.

Inoue, T., Sakurai, N., Nagai, M. and Peigen, X. (1985). Beeioside III, a cyclolanostanol xyloside from the rhizomes of *Beesia calthaefolia* and *Souliea vaginata*. *Phytochemistry*, **24**, 1329–1331.

Inoue, H., Mori, T., Shibata, S. and Koshihara, Y. (1989). Modulation by glycyrrhetinic acid derivatives of TPA-induced mouse ear oedema. *Br. J. Pharmacol.*, **96**, 204–210.

Iorizzi, M., Minale, L., Riccio, R., Debray, M. and Menou, J.L (1986). Starfish saponins, Part 23. Steroidal glycosides from the starfish *Halityle regularis*. *J. Nat. Prod.*, **49**, 67–78.

Iorizzi, M., Minale, L. and Riccio, R. (1990). Starfish saponins, Part 39. Steroidal oligoglycoside sulphates and polyhydroxysteroids from the starfish *Asterina pectinifera*. *Gazz. Chim. Ital.*, **120**, 147–153.

Ireland, P.A. and Dziedzic, S.Z. (1986a). Effect of hydrolysis on sapogenin release in soya. *J. Agric. Food Chem.*, **34**, 1037–1041.

Ireland, P.A. and Dziedzic, S.Z. (1986b). High-performance liquid chromatography of soyasaponins on silica phase with evaporative light-scattering detection. *J. Chromatogr.*, **361**, 410–416.

Ireland, P.A. and Dziedzic, S.Z. (1987). Saponins and sapogenins of chickpea, haricot bean and red kidney bean. *Food Chem.*, **23**, 105–116.

Ireland, P.A., Dziedzic, S.Z., Drew, M.G.B. and Forsyth, G.A. (1987). Structure of soyasapogenol B. *J. Agric. Food Chem.*, **35**, 971–973.

Isaac, O. (1992). *Die Ringelblume.* Wissenschaftliche Verlagsgesellschaft mbH, Stuttgart.

Isaev, M.I. and Imomnazarov, B.A. (1991). Triterpene glycosides and their genins from *Astragalus.* XXXVII. Cycloaraloside F from *Astragalus amarus* and *Astragalus villosissimus. Khim. Prir. Soedin.,* 374–377.

Isaev, M.I., Gorovits, M.B., Gorovits, T.T., Abdullaev, N.D. and Abubakirov, N.K. (1983a). *Astragalus* triterpenoid glycosides and their genins. VIII. Askendoside C from *Astragalus taschkendicus. Khim. Prir. Soedin.,* 173–180.

Isaev, M.I., Gorovits, M.B., Abdullaev, N.D. and Abubakirov, N.K. (1983b). *Astragalus* triterpenoid glycosides and their genins. IX. Askenoside D from *Astragalus taschkendicus. Khim. Prir. Soedin.,* 180–185.

Isaev, M.I., Gorovits, M.B., Abdullaev, N.D. and Abubakirov, N.K. (1983c). Triterpene glycosides of *Astragalus* and their genins. XII. Askendoside B from *Astragalus taschkendicus. Khim. Prir. Soedin.,* 457–460.

Isaev, M.I., Gorovits, M.B., Abdullaev, N.D. and Abubakirov, N.K. (1983d). Triterpene glycosides of *Astragalus* and their genins. XIV. Askendoside A from *Astragalus taschkendicus. Khim. Prir. Soedin.,* 587–592.

Isaev, M.I., Gorovits, M.B. and Abubakirov, N.K. (1986). Triterpene glycosides from *Astragalus.* XXII. Cycloorbicoside A from *Astragalus orbiculatus. Khim. Prir. Soedin.,* 719.

Isaev, M.I., Gorovits, M.B. and Abubakirov, N.K. (1988). Triterpene glycosides and their genins from *Astragalus.* XXVI. Cycloartanes and sterols of *Astragalus schachirudensis. Khim. Prir. Soedin.,* 136–137.

Isaev, M.I., Gorovits, M.B. and Abubakirov, N.K. (1989). Triterpene glycosides and their genins from *Astragalus.* XXX. Cycloaraloside A from *Astragalus amarus. Khim. Prir. Soedin.,* 806–809.

Ishida, S., Sakiya, Y., Ichikawa, T. and Awazu, S. (1989). Pharmacokinetics of glycyrrhetic acid, a major metabolite of glycyrrhizin, in rats. *Chem. Pharm. Bull.,* **37**, 2509–2513.

Ishii, H., Tori, K., Tozyo, T. and Yoshimura, Y. (1978). Structure of polygalacin-D and -D_2, platycodin-D and -D_2 and their monoacetates, saponins isolated from *Platycodon grandiflorum* A. DC., determined by carbon-13 nuclear magnetic resonance spectroscopy. *Chem. Pharm. Bull.,* **26**, 674–677.

Ishii, H., Kitagawa, I., Matsushita, K. *et al.* (1981). The configuration and conformation of the arabinose moiety in platycodins, saponins isolated from *Platycodon grandiflorum,* and Mi-saponins from *Madhuca longiflora* based on carbon-13 and hydrogen-1 NMR spectroscopic evidence: the total structures of the saponins. *Tetrahedron Lett.,* **22**, 1529–1532.

Ishii, H., Tori, K., Tozyo, T. and Yoshimura, Y. (1984). Saponins from roots of *Platycodon grandiflorum.* Part 2. Isolation and structure of new triterpene glycosides. *J. Chem. Soc., Perkin Trans.* 1, 661–668.

Iskenderov, G.B. (1970). Triterpenic glycosides from *Hedera pastuchovii.* II. Structure of pastuchoside C. *Khim. Prir. Soedin.,* 376–377.

Iskenderov, G.B. (1971). Triterpenoid glycosides from *Hedera pastuchovii* (ivy). IV. Structure of pastuchoside B. *Khim. Prir. Soedin.,* 425–429.

Iskenderov, G.B. and Isaev, M.I. (1974). Sterol and triterpene glycosides of *Crataegus pentagyna. Khim. Prir. Soedin.,* 103–104.

Itakura, Y. and Komori, T. (1986a). Structure elucidation of two new oligoglycoside sulphates, versicoside B and versicoside C. *Annalen*, 359–373.

Itakura, Y. and Komori, T. (1986b). Structure of four new oligoglycoside sulfates. *Annalen*, 499–508.

Ito, M., Nakashima, H., Baba, M. *et al.* (1987). Inhibitory effect of glycyrrhizin on the *in vitro* infectivity and cytopathic activity of the human immunodeficiency virus [HIV(HTLV-III/LAV)]. *Antiviral Res.*, 7, 127–137.

Ito, S., Ogino, T., Sugiyama, H. and Kodama, M. (1967). Structures of A_1 barrigenol and R_1-barrigenol. *Tetrahedron Lett.*, 2289–2294.

Ivanov, A.S., Zakharova, T.S., Odinokova, L.E. and Uvarova, N.I. (1987). Hypolipidemic properties of betulin glycosides. *Khim-Farm. Zh.*, 21, 1091–1094 (*Chem. Abstr.*, 108, 31926).

Ivanova, N.S. and Kuznetsova, T.A. (1985). Holothurin A – basic triterpene glycoside of the cuvierian organs of the sea cucumber *Bohadschia graeffei*. Structure of native aglycone and progenins. *Khim. Prir. Soedin.*, 123–124.

Iwamoto, M., Fujioka, T., Okabe, H., Mihashi, K. and Yamauchi, T. (1987). Studies on the constituents of *Actinostemma lobatum* MAXIM. I. Structures of actinostemmosides A, B, C and D, dammarane triterpene glycosides isolated from the herb. *Chem. Pharm. Bull.*, 35, 553–561.

Jacker, H.-J. and Hiller, K. (1976). Zur antiexsudativen Wirksamkeit der Saponine aus *Eryngium planum* L. und *Sanicula europaea* L. *Pharmazie*, 31, 747–748.

Jacker, H.-J., Voigt, G. and Hiller, K. (1982). Zum antiexsudativen Verhalten einiger Triterpensaponine. *Pharmazie*, 37, 380–382.

Jacobi, W. (1906). *Arch. Exp. Pathol. Pharmakol.*, 54, 283.

Jadhav, S.J., Sharma, R.P. and Salunkhe, D.K. (1981). Naturally occurring toxic alkaloids in food. *CRC Crit. Rev. Toxicol.*, 9, 21–104.

Jain, D.C. (1987a). Gitogenin-3-O-β-D-laminaribioside from the aerial part of *Agave cantala*. *Phytochemistry*, 26, 1789–1790.

Jain, D.C. (1987b). Antifeedant active saponin from *Balanites roxburghii* stem bark. *Phytochemistry*, 26, 2223–2225.

Jain, D.C. and Tripathi, A.K. (1991). Insect feeding-deterrent activity of some saponin glycosides. *Phytother. Res.*, 5, 139–141.

Jain, D.C., Thakur, R.S., Bajpai, A. and Sood, A.R. (1988). A soyasapogenol-B glucoside from the seeds of *Phaseolus vulgaris*. *Phytochemistry*, 27, 1216–1217.

Jain, D.C., Agrawal, P.K. and Thakur, R.S. (1991). Phaseoluside-A, a new soyasapogenol B triglucoside from *Phaseolus vulgaris* seeds. *Planta Medica*, 57, 94–95.

Jain, G.K. and Khanna, N.M. (1982). Capitogenic acid: a new triterpene acid from *Schefflera capitata* HARMS. *Indian J. Chem.*, 21B, 622–625.

Jain, G.K., Pal, R. and Khanna, N.M. (1980). Spermicidal saponins from *Pittosporum nilghrense* WIGHT ET APNOTT. *Indian J. Pharm. Sci.*, 42, 12–13.

Jain, M.P., Sethi, V.K. and Thakur, R.S. (1978). Constituents of *Viburnum nervosum* HOOK. *Indian J. Pharm. Sci.*, 40, 63–64.

Jain, S.A. and Srivastava, S.K. (1984). Betulin 3β-O-β-D-xylopyranoside from the roots of *Amoora rohituka*. *Indian J. Pharm. Sci.*, 46, 161–162.

James, S.L. and Pearce, E.J. (1988). The influence of adjuvant on induction of protective immunity by a non-living vaccine against schistosomiasis. *J. Immunol.*, **140**, 2753–2759.

Janeczko, Z., Jansson, P.E. and Sendra, J. (1987). A new steroidal saponin from *Polygonatum officinale. Planta Medica*, **53**, 52–54.

Janeczko, Z., Sendra, J., Kmiec, K. and Brieskorn, C.H. (1990). A triterpenoid glycoside from *Menyanthes trifoliata. Phytochemistry*, **29**, 3885–3887.

Janeczko, Z., Jagiello, K., Jansson, P.E. and Piskornik, M. (1992). Triterpene saponins from *Anemone coronaria. Fitoterapia*, **63**, 191.

Jansakul, C., Baumann, H., Kenne, L. and Samuelsson, G. (1987). Ardisiacrispin A and B, two utero-contracting saponins from *Ardisia crispa. Planta Medica*, **53**, 405–409.

Jansson, P.E., Kenne, L., Liedgren, H., Lindberg, B. and Lönngren, J. (1976). A practical guide to the methylation analysis of carbohydrates. *Chem. Commun.*, *Univ. Stockholm*, 1–75.

Jap. Pat. 57 165,400 Saponins from Astragali root. (*Chem. Abstr.*, **98**, 95652).

Jap. Pat. 58 59,921 Antitumor saponins from *Gynostemma pentaphyllum.* (*Chem. Abstr.*, **99**, 47915).

Jap. Pat. 58 72,523 Anticancer saponins from soybean. (*Chem. Abstr.*, **99**, 28005).

Jap. Pat. 59 20,298 Sodium 3-(α-L-arabinopyranosyl)-16,23-dihydroxyolean-12-en-28-oate. (*Chem. Abstr.*, **101**, 43572).

Jap. Pat. 60 48,995 Lucynoside saponins from a gourd. (*Chem. Abstr.*, **103**, 166146).

Jap. Pat. 61 282,395 Isolation of saikosaponins from *Bupleurum kunmingense.* (*Chem. Abstr.*, **107**, 195155).

Jap. Pat. 62 240,696 Saikosaponins from *Bupleurum kunmingense* as anticancer agents. (*Chem. Abstr.*, **109**, 11715).

Jap. Pat. 81 46,817 Saponins as neoplasm inhibitors. (*Chem. Abstr.*, **95**, 103307).

Jap. Pat. 82 163,315 Antithrombic pharmaceuticals containing ginsenoside Ro. (*Chem. Abstr.*, **98**, 78132).

Jap. Pat. 92, 05,235 Ginsenoside Ro for the treatment of hepatitis. (*Chem. Abstr.*, **116**, 158902).

Jaretzky, R. (1951). Die Wirkung von Radix Sarsaparillae, Lignum Guajaci und Esberisan auf Diurese und Ausscheidung harnpflichtiger Stoffe. *Pharmazie*, **6**, 115–117.

Javier Arriaga, F., Rumbero, A. and Vazquez, P. (1990). Two triterpene glycosides from *Isertia haenkeana. Phytochemistry*, **29**, 209–213.

Jennings, J.P., Klyne, W. and Scopes, P.M. (1964). Optical rotatory dispersion of lactones. *Proc. Chem. Soc.*, 412–413.

Jentsch, K., Spiegl, P. and Fuchs, L. (1961). Vergleichende Untersuchung der anthelminthischen Wirksamkeit von Saponinen *in vitro. Arzneimittelforsch.*, **11**, 413–414.

Jia, Z.-J., Liu, X.-Q. and Liu, Z.-M. (1993). Triterpenoids from *Sanguisorba alpina. Phytochemistry*, **32**, 155–159.

Jiang, Y., Massiot, G., Lavaud, C. *et al.* (1991). Triterpenoid glycosides from the bark of *Mimosa tenuiflora. Phytochemistry*, **30**, 2357–2360.

Jiang, Y., Weniger, B., Béji, N., Italiano, L., Beck, J.P. and Anton, R. (1993). Cytotoxic activity of a new saponin from *Sideroxylon cubense. Planta Medica*, **59**, A680.

Jie, Y.H., Cammisuli, S. and Baggiolini, M. (1984). Immunomodulatory effects of *Panax ginseng* C.A. MEYER in the mouse. *Agents Actions*, **15**, 386–391.

Jimenez, C., Villaverde, M.C., Riguera, R., Castedo, L. and Stermitz, F. (1989). Triterpene glycosides from *Mussatia* species. *Phytochemistry*, **28**, 2773–2776.

Jin, J., Wen, Y., Cheng, G., Gan, L. and Chen, Y. (1987). Constituent of *Picria fel-tarrae* LOUR. VIII. Structure of picfeltarraenin II. *Huaxue Xuebao*, **45**, 1133–1134 (*Chem. Abstr.*, **108**, 109540).

Jisaka, M., Ohigashi, H., Takagaki, T. *et al.* (1992). Bitter steroid glucosides, vernoniosides A1, A2 and A3, and related B1 from a possible medicinal plant, *Vernonia amygdalis*, used by wild chimpanzees. *Tetrahedron*, **48**, 625–632.

Jisaka, M., Ohigashi, H., Takegawa, K. *et al.* (1993). Steroid glucosides from *Vernonia amygdalina*, a possible chimpanzee medicinal plant. *Phytochemistry*, **34**, 409–413.

Jizba, J., Dolejs, L., Herout, V. *et al.* (1971a). Polypodosaponin, ein neuer Saponintyp aus *Polypodium vulgare* L. *Chem. Ber.*, **104**, 837–846.

Jizba, J., Dolejs, L., Herout, V. and Sorm, F. (1971b). The structure of osladin – the sweet principle of the rhizomes of *Polypodium vulgare* L. *Tetrahedron Lett.*, 1329–1332.

Johnson, A.L. and Shimizu, Y. (1974). Phytolaccinic acid, a new triterpene from *Phytolacca americana*. *Tetrahedron*, **30**, 2033–2036.

Jones, P.G. and Fenwick, G.R. (1981). The glycoalkaloid content of some edible solanaceous fruits and potato products. *J. Sci. Food Agric.*, **32**, 419–421.

Jones, R.N., Katzenellenbogen, E. and Dobriner, K. (1953). The infrared absorption spectrum of the steroid sapogenins. *J. Am. Chem. Soc.*, **75**, 158–166.

Joshi, B.S., Moore, K.M., Pelletier, S.W., Puar, M.S. and Pramanik, B.N. (1992). Saponin from *Collinsonia canadensis J. Nat. Prod.*, **55**, 1468–1476.

Joshi, J. and Dev, S. (1988). Chemistry of ayurvedic crude drugs. Part VIII. Shatavari. Part 2. Structure elucidation of bioactive shatavarin-1 and other glycosides. *Indian J. Chem.*, **27B**, 12–16.

Julien, J., Gasquet, M., Maillard, C., Balansard, G. and Timon-David, P. (1985). Extracts of the ivy plant, *Hedera helix*, and their anthelmintic activity on liver fluke. *Planta Medica*, **50**, 205–208.

Jurenitsch, J., Haslinger, E. and Kubelka, W. (1979). Zur Struktur der Sapogenine aus *Polemonium reptans* L. *Pharmazie*, **34**, 445–446.

Jurzysta, M. and Jurzysta, A. (1978). Gas-liquid chromatography of trimethylsilyl ethers of soyasapogenols and medicagenic acid. *J. Chromatogr.*, **148**, 517–520.

Juvik, J.A., Stevens, M.A. and Rick, C.M. (1982). Survey of the genus *Lycopersicon* for variability in α-tomatine content. *Hort. Sci.*, **17**, 764–766.

Kaizuka, H. and Takahashi, K. (1983). High-performance liquid chromatographic system for a wide range of naturally occurring glycosides. *J. Chromatogr.*, **258**, 135–146.

Kajiwara, N., Ninomiya, T., Kawai, H. and Hosogai, Y. (1984). Determination of solanine in potatoes by high performance liquid chromatography. *J. Food Hyg. Soc. Jap.*, **25**, 256–260 (*Chem. Abstr.*, **102**, 22947).

Kaku, T., Miyato, T., Uruno, T., Sako, I. and Kinoshita, A. (1975). Chemico-pharmacological studies on saponins of *Panax ginseng* C.A. MEYER. *Arznei-mittelforsch*, **25**, 343–347, 539–547.

Kakuno, T., Yoshikawa, K. and Arihara, S. (1991a). Ilexosides A, B, C and D, anti-allergic 18,19-seco-ursane glycosides from fruit of *Ilex crenata. Tetrahedron Lett.*, **32**, 3535–3538.

Kakuno, T., Yoshikawa, K., Arihara, S., Takei, M. and Endo, K. (1991b). Ilexosides E, F, G, H and I, novel 18,19-seco-ursane glycosides from fruit of *Ilex crenata. Tetrahedron*, **47**, 7219–7226.

Kalinin, V.I., Malyutin, A.N. and Stonik, V.A. (1986). Caudinoside A – a novel triterpene glycoside from the holothurian *Paracaudina ransonetii. Khim. Prir. Soedin.*, 378–379.

Kamel, M.S., Ohtani, K., Kurokawa, T. *et al.* (1991). Studies on *Balanites aegyptiaca* fruits, an antidiabetic Egyptian folk medicine. *Chem. Pharm. Bull.*, **39**, 1229–1233.

Kanaoka, M., Kato, H., Shimada, F. and Yano, S. (1992). Studies on the enzyme immunoassay of bioactive constituents in oriental medicinal drugs. VI. Enzyme immunoassay of ginsenoside Rb_1 from *Panax ginseng. Chem. Pharm. Bull.*, **40**, 314–317.

Kanazawa, H., Nagata, Y., Matsushima, Y., Tomoda, M. and Takai, N. (1987). High-performance liquid chromatographic analysis for ginsenosides in *Panax ginseng* extracts using glass-ODS column. *Chromatographia*, **24**, 517–519.

Kanazawa, H., Nagata, Y., Matsushima, Y., Tomoda, M. and Takai, N. (1990a). Simultaneous determination of ginsenosides and saikosaponins by high-performance liquid chromatography. *J. Chromatogr.*, **507**, 327–332.

Kanazawa, H., Nagata, Y., Matsushima, Y., Tomoda, M. and Takai, N. (1990b). Preparative high-performance liquid chromatography on chemically modified porous glass. Isolation of saponins from ginseng. *Chem. Pharm. Bull.*, **38**, 1630–1632.

Kanazawa, H., Nagata, Y., Matsushima, Y., Tomoda, M. and Takai, N. (1991). Preparative high-performance liquid chromatography on chemically modified porous glass. Isolation of acidic saponins from ginseng. *J. Chromatogr.*, **537**, 469–474.

Kanazawa, H., Nagata, Y., Kurosaki, E., Matsushima, Y. and Takai, N. (1993). Comparison of columns of chemically modified porous glass and silica in reversed-phase high-performance liquid chromatography of ginsenosides. *J. Chromatogr.*, **632**, 79–85.

Kaneko, H., Nakanishi, K., Murakami, M. and Kuwajima, K. (1985). Vasodilation by ginseng saponins. *Wakan Iyaku Gakkaishi*, **2**, 122–123 (*Chem. Abstr.*, **104**, 45498).

Kaneko, K., Tanaka, M.W. and Mitsuhashi, H. (1976). Origin of nitrogen in the biosynthesis of solanidine by *Veratrum grandiflorum. Phytochemistry*, **15**, 1391–1393.

Kang, B.N. and Koh, I.S. (1974). Effect of saponin on sodium-potassium activated ATPase in rabbit red cell membrane. *Taehan Saengri Hakhoe Chi*, **8**, 67–76 (*Chem. Abstr.*, **82**, 697555).

Kang, S.S. and Woo, W.S. (1987). Two new saponins from *Phytolacca americana. Planta Medica*, **53**, 338–340.

Kang, S.S., Lee, Y.S. and Lee, E.B. (1987). Isolation of azukisaponin V possessing leucocyte migration inhibitory activity from *Melilotus officinalis. Saengyak Hakhoechia*, **18**, 89–93 (*Chem. Abstr.*, **107**, 242495).

Kapundu, M., Warin, R., Delaude, C. and Huls, R. (1978). Chemical study of saponins from *Milletia laurentii* WILD. *Bull. Soc. Roy. Sci. Liège*, **47**, 84–91.

Kapundu, M., Lami, N., Dimbi, M.Z., Warin, R., Delaude, C. and Huls, R. (1984). Isolation and structure determination of majideagenin, a prosapogenin from *Majidea fosteri* RADLK. *Bull. Soc. Chim. Belg.*, **93**, 497–503.

Kapundu, M., Penders, A., Warin, R., Delaude, C. and Huls, R. (1988). Structure of kotschyioside, a novel saponoside from *Aspilia kotschyi* (SCH. BIP.) OLIV. *Bull. Soc. Chim. Belg.*, **97**, 329–342.

Karikura, M., Miyase, T., Tanizawa, H., Takino, Y., Taniyama, T. and Hayashi, T. (1990). Studies on absorption, distribution, excretion, and metabolism of ginseng saponins. V. The decomposition products of ginsenoside Rb$_2$ in the large intestine of rats. *Chem. Pharm. Bull.*, **38**, 2859–2861.

Karikura, M., Miyase, T., Tanizawa, H., Taniyama, T. and Takino, Y. (1991). Studies on absorption, distribution, excretion, and metabolism of ginseng saponins. VI. The decomposition products of ginsenoside Rb$_2$ in the stomach of rats. *Chem. Pharm. Bull.*, **39**, 400–404.

Karliner, J. and Djerassi, C. (1966). Terpenoids. LVII. Mass spectral and nuclear magnetic resonance studies of pentacyclic triterpene hydrocarbons. *J. Org. Chem.*, **31**, 1945–1956.

Karpova, V.I., Kintya, P.K. and Chirva, V.Y. (1975). Structure of saponin A from *Naumburgia thyrsiflora*. *Khim. Prir. Soedin.*, 364–366.

Kartnig, T. and Wegschaider, O. (1971). Eine Möglichkeit zur Identifizierung von Zuckern aus kleinsten Mengen von Glykosiden oder aus Zuckergemischen. *J. Chromatogr.*, **61**, 375–377.

Kartnig, T., Ri, C.Y. and Wegschaider, O. (1972). Untersuchungen über die Aufbereitung und die Stabilität von Saponindrogen. 1 Mitteilung. *Planta Medica*, **22**, 127–135.

Kasai, R., Suzuo, M., Asakawa, J. and Tanaka, O. (1977). Carbon-13 chemical shifts of isoprenoid-β-D-glucopyranosides and β-D-mannopyranosides. Stereochemical influences of aglycone alcohols. *Tetrahedron Lett.*, 175–178.

Kasai, R., Okihara, M., Asakawa, J., Mizutani, K. and Tanaka, O. (1979). ^{13}C NMR study of α- and β-anomeric pairs of D-mannopyranosides and L-rhamnopyranosides. *Tetrahedron*, **35**, 1427–1432.

Kasai, R., Miyakoshi, M., Matsumoto, K. *et al.* (1986a) Tubeimoside I, a new cyclic bisdesmoside from Chinese cucurbitaceous folk medicine 'Tu bei mu', a tuber of *Bulbostemma paniculatum*. *Chem. Pharm. Bull.*, **34**, 3974–3977.

Kasai, R., Matsumoto, K., Taniyasu, S., Tanaka, O., Kim. J.-H. and Hahn, D.-R. (1986b). 3,4-seco-lupane type triterpene glycosyl esters from a Korean medicinal plant, *Acanthopanax chiisanensis* (Araliaceae). *Chem. Pharm. Bull.*, **34**, 3284–3289.

Kasai, R., Oinaka, T., Yang, C.-R., Zhou, J. and Tanaka, O. (1987a). Saponins from Chinese folk medicine, 'Liang wang cha', leaves and stems of *Nothopanax delavayi*, Araliaceae. *Chem. Pharm. Bull.*, **35**, 1486–1490.

Kasai, R., Yamaguchi, H. and Tanaka, O. (1987b). High-performance liquid chromatography of glycosides on a new type of hydroxyapatite column. *J. Chromatogr.*, **407**, 205–210.

Kasai, R., Matsumoto, K., Nie, R. *et al.* (1987c). Sweet and bitter cucurbitane glycosides from *Hemsleya carnosiflora*. *Phytochemistry*, **26**, 1371–1376.

Kasai, R., Miyakoshi, M., Nie, R.-L. *et al.* (1988a). Saponins from *Bolbostemma paniculatum*: cyclic bisdesmosides, tubeimosides II and III. *Phytochemistry*, **27**, 1439–1446.

Kasai, R., Nishi, M., Mizutani, K. *et al.* (1988b). Trifolioside II, an acyclic sesquiterpene oligoglycoside from pericarps of *Sapindus trifoliatus*. *Phytochemistry*, **27**, 2209–2211.

Kasai, R., Tanaka, T., Nie, R., Miyakoshi, M., Zhou, J. and Tanaka, O. (1990). Saponins from Chinese medicinal plant, *Hemsleya graciliflora* (Cucurbitaceae). *Chem. Pharm. Bull.*, **38**, 1320–1322.

Kashiwada, Y., Fujioka, T., Chang, J.-J., Chen, I., Mihashi, K. and Lee, K.-H. (1992). Antitumor agents. 136. Cumingianosides A–F, potent antileukemic new triterpene glycosides, and cumindysosides A and B, trisnor- and tetranortriterpene glycosides with a 14,18-cycloapocuphane-type skeleton from *Dysoxylum cumingianum*. *J. Org. Chem.*, **57**, 6946–6953.

Kasprzyk, Z. and Wojciechowski, Z. (1967). The structure of triterpene glycosides from the flowers of *Calendula officinalis*. *Phytochemistry*, **6**, 69–75.

Kawai, H., Kuroyanagi, M., Umehara, K., Ueno, A. and Satake, M. (1988). Studies on the saponins of *Lonicera japonica* THUNB. *Chem. Pharm. Bull.*, **36**, 4769–4775.

Kawai, H., Nishida, M., Tashiro, Y., Kuroyanagi, M., Ueno, A. and Satake, M. (1989). Studies on the structures of udosaponins A, B, C, D, E and F from *Aralia cordata* THUNB. *Chem. Pharm. Bull.*, **37**, 2318–2321.

Kawai, K.-I. and Shibata, S. (1978). Pseudojujubogenin, a new sapogenin from *Bacopa monniera*. *Phytochemistry*, **17**, 287–289.

Kawai, K., Iitaka, Y. and Shibata, S. (1974). Jujubogenin *p*-bromobenzoate. *Acta Cryst. (B)*, **30**, 2886–2888.

Kawamura, N., Watanabe, H. and Oshio, H. (1988). Saponins from roots of *Momordica cochinchinensis*. *Phytochemistry*, **27**, 3585–3591.

Kawano, K., Sakai, K., Sato, H. and Sakamura, S. (1975). Bitter principle of asparagus. Isolation and structure of furostanol saponin in asparagus storage root. *Agric. Biol. Chem.*, **39**, 1999–2002.

Kawano, K., Sato, H. and Sakamura, S. (1977). A bitter principle of asparagus. Part. II. Isolation and structure of furostanol saponin in asparagus edible shoots. *Agric. Biol. Chem.*, **41**, 1–8.

Kawasaki, T. (1981). Detection and isolation of steroid saponins. In *Advances in Natural Product Chemistry*, ed. S. Natori, N. Ikekawa and M. Suzuki, pp. 292–309. John Wiley, New York.

Kawasaki, T. and Miyahara, K. (1963). Thin layer chromatography of steroid saponins and their derivatives. *Chem. Pharm. Bull.*, **11**, 1546–1550.

Kawasaki, T. and Miyahara, K. (1965). Structure of yononin. A novel type of spirostanol glycoside. *Tetrahedron*, **21**, 3633–3639.

Kawasaki, T. and Yamauchi, T. (1962). Structures of dioscin, gracillin and kikuba-saponin. (2). (Saponins of Japanese Dioscoreaceae. XI.). *Chem. Pharm. Bull.*, **10**, 703–708.

Kawasaki, T., Nishioka, I., Komori, T., Yamauchi, T. and Miyahara, K. (1965). Digitalis saponins – IV. Structure of F-Gitonin. *Tetrahedron*, **21**, 299–307.

Kawashima, K., Mimaki, Y. and Sashida, Y. (1991). Steroidal saponins from *Allium giganteum* and *A. aflatunense*. *Phytochemistry*, **30**, 3063–3067.

Keeler, R.F., Baker, D.C. and Gaffield, W. (1990). Spirosolane-containing *Solanum* species and induction of congenital craniofacial malformations. *Toxicon*, **28**, 873–884.

Keeler, R.F., Baker, D.C. and Gaffield, W. (1991). Teratogenic *Solanum* species and the responsible teratogens. In *Handbook of Natural Toxins*, vol. 6, ed. R.F. Keeler and A.T. Tu, pp. 83–99. Marcel Dekker, New York.

Kemertelidze, E.P., Gvazava, L.N., Alania, M.D., Kikoladze, V.S. and Tsitsishvili, V.G. (1992). Digitoside, a novel triterpene glycoside from *Digitalis ciliata*. *J. Nat. Prod.*, **55**, 217–220.

Kemoklidze, Z.S., Dekanosidze, G.E., Dzhikiya, O.D., Vugalter, M.M. and Kemertelidze, E.P. (1982). Triterpenoid glycosides of *Fatsia japonica* leaves. *Khim. Prir. Soedin.*, 788–789.

Kemoklidze, Z.S., Dekanosidze, G.E., Dzhikiya, O.D., Vugulter, M.M. and Kemertelidze, E.P. (1987). Triterpene saponins from *Fatsia japonica* growing in the Georgian SSR. *Soobshch. Akad. Nauk Gruz. SSR*, **125**, 569–572 (*Chem. Abstr.*, **108**, 72032).

Kennelly, E.J., Lewis, W.H., Winter, R.E.K., Johnson, S., Elvin-Lewis, M. and Gossling, J. (1993). Triterpenoid saponins from *Gouania lupuloides*. *J. Nat. Prod.*, **56**, 402–410.

Kensil, C.R., Patel, U., Lennick, M. and Marciani, D. (1991). Separation and characterization of saponins with adjuvant activity from *Quillaja saponaria* MOLINA cortex. *J. Immunol.*, **146**, 431–437.

Kerimova, K.M., Pkheidze, T.A. and Gorovits, T.T. (1971). Triterpene saponins of *Zygophyllum fabago*. *Khim. Prir. Soedin.*, 521.

Kesselmeier, J. and Budzikiewicz, H. (1979). Identification of saponins as structural building units in isolated prolamellar bodies from etioplasts of *Avena sativa* L. *Z. Pflanzenphysiol.*, **91**, 333–344.

Khamidullina, E.A., Gromova, A.S., Lutskii, V.I., Vereshchagin, A.L., Semenov, A.A. and Larin, M.F. (1989). Triterpene glycosides of *Thalictrum squarrosum*. IV. Structures of squarrosides A1, A2, B1 and B2. *Khim. Prir. Soedin.*, 516–523.

Khan, M.N.I., Nizami, S.S., Khan, M.A. and Ahmed, Z. (1993). New saponins from *Mangifera indica*. *J. Nat. Prod.*, **56**, 767–770.

Khong, P.W. and Lewis, K.G. (1975). New triterpenoid extractives from *Lemaireocereus chichipe*. *Aust. J. Chem.*, **28**, 165–172.

Khorana, M.L. and Raisinghani, K.H. (1961). Studies of *Luffa echinata*. III. The oil and the saponin. *J. Pharm. Sci.*, **50**, 687–689.

Khorlin, A.J., Venjaminova, A.G. and Kochetkov, N.K. (1964). The structure of *Kalopanax septemlobus* saponin A. *Doklady Akad. Nauk SSSR*, **115**, 619–622 (*Chem. Abstr.*, **60**, 15964).

Khorlin, A.J., Chirva, V.J. and Kochetkov, N.K. (1965). Triterpene saponins. XV. Clematoside C, a triterpene oligoside from the roots of *Clematis manshurica*. *Izv. Akad. Nauk SSSR, Ser. Khim.*, 811–818 (*Chem. Abstr.*, **63**, 5730).

Khorlin, A.J., Chirva, V.J. and Kochetkov, N.K. (1967). Triterpene saponins. XXI. Structure of clematoside C. *Izv. Akad. Nauk SSSR, Ser. Khim.* 1306–1311 (*Chem. Abstr.*, **67**, 117188).

Kikuchi, J., Nakamura, K., Nakata, O. and Morikawa, Y. (1987). Determination of monosaccharides constituting the glycosides in saponins by high-performance liquid chromatography. *J. Chromatogr.*, **403**, 319–323.

Kim, C., Kim, C.C., Kim, M.S., Hu, C.Y. and Rhe, J.S. (1970). Influence of ginseng on the stress mechanism. *Lloydia*, **33**, 43–48.

Kim, E. and Kim, Y.C. (1987). A triterpene glycoside in berries of *Rubus coreanus*. *Saengyak Hakhoechi*, **18**, 188–190 (*Chem. Abstr.*, **108**, 118802).

Kim, J. and Kinghorn, A.D. (1989). Further steroid and flavonoid constituents of the sweet plant *Polypodium glycyrrhiza*. *Phytochemistry*, **28**, 1225–1228.

Kim, J., Pezzuto, J.M., Soejarto, D.D., Lang, F.A. and Kinghorn, A.D. (1988). Polypodoside A, an intensely sweet constituent of the rhizomes of *Polypodium glycyrrhiza*. *J. Nat. Prod.*, **51**, 1166–1172.

Kim, Y.C., Higuchi, R. and Komori, T. (1992). Hydrothermolysis of triterpenoid and steroid glycosides. *Annalen*, 453–459.

Kimata, H., Hiyama, C., Yahara, S., Tanaka, O., Ishikawa, O. and Auira, M. (1979). Application of high performance liquid chromatography to the analysis of crude drugs: separatory determination of saponins in Bupleuri radix. *Chem. Pharm. Bull.*, **27**, 1836–1841.

Kimata, H., Kasai, R. and Tanaka, O. (1982). Saponins of Juk-siho and roots of *Bupleurum longeradiatum* TURCZ. *Chem. Pharm. Bull.*, **30**, 4373–4377.

Kimata, H., Nakashima, T., Kokubun, S. *et al.* (1983). Saponins of pericarps of *Sapindus mukurossi* GAERTN. and solubilization of monodesmosides by bis-desmosides. *Chem. Pharm. Bull.*, **31**, 1998–2005.

Kimata, H., Sumida, N., Matsufuji, N. *et al.* (1985). Interaction of saponin of Bupleuri radix with ginseng saponin: solubilization of saikosaponin-a with chikusetsusaponin V (=ginsenoside R_0). *Chem. Pharm. Bull.*, **33**, 2849–2853.

Kimura, M., Tohma, M. and Yoshizawa, I. (1966). Constituents of *Convallaria*. V. On the structure of convallasaponin-C. *Chem. Pharm. Bull.*, **14**, 55–61

Kimura, M., Tohma, M., Yoshizawa, I. and Fujino, A. (1968a). Constituents of *Convallaria*. XII. Convallasaponin-E: diosgenin triarabinoside. *Chem. Pharm. Bull.*, **16**, 2191–2194.

Kimura, M., Tohma, M., Yoshizawa, I. and Akiyama, H. (1968b). Constituents of *Convallaria*. X. Structures of convallasaponin-A, -B, and their glucosides. *Chem. Pharm. Bull.*, **16**, 25–33.

Kimura, O., Sakurai, N., Nagai, M. and Inoue, T. (1982). Studies on the Chinese crude drug 'Shoma'. VI. Shengmanol xyloside, a new genuine natural product of some *Cimicifuga* glycosides. *Yakugaku Zasshi*, **102**, 538–545.

Kimura, Y., Kobayashi, Y., Takeda, T. and Ogihara, Y. (1981). Three new saponins from the leaves of *Hovenia dulcis* (Rhamnaceae). *J. Chem. Soc., Perkin Trans.* 1, 1923–1927.

King, F.E., Baker, J.A. and King, T.J. (1955). The chemistry of extractives from hardwoods. Part XXIV. A saponin constituent of Makoré (*Mimusops heckelii*). *J. Chem. Soc.*, 1338–1342.

Kinjo, J., Takeshita, T., Abe, Y. *et al.* (1988). Studies on the constituents of *Pueraria lobata*. IV. Chemical constituents in the flowers and the leaves. *Chem. Pharm. Bull.*, **36**, 1174–1179.

Kintia, P.K. and Pirozhkova, N.M. (1982). *Androsace septentrionalis* triterpenoid glycosides. Structure of androseptosides A, B, C and D. *Khim. Prir. Soedin.,* 526–527.

Kintia, P.K. and Shvets, S.A. (1985a). Melongoside L and Melongoside M, steroidal saponins from *Solanum melongena* seeds. *Phytochemistry,* **24,** 197–198.

Kintia, P.K. and Shvets, S.A. (1985b). Melangosides N, O and P: steroidal saponins from seeds of *Solanum melongena. Phytochemistry,* **24,** 1567–1569.

Kintia, P.K., Dragalin, J.P. and Chirva, V.J. (1972). Steroid saponins. III. Glycosides A and B from *Yucca filamentosa. Khim. Prir. Soedin.,* 615–617.

Kintia, P.K., Wojciechowski, Z. and Kasprzyk, Z. (1974). Biosynthesis of oleanolic acid glycosides from mevalonate-2-^{14}C in germinating seeds of *Calendula officinalis. Bull. Acad. Pol. Sci., Ser. Sci. Biol.,* **22,** 73–76 (*Chem. Abstr.,* **81,** 23113).

Kintia, P.K., Bobeiko, V.A., Krokhmalyuk, V.V. and Chirva, V.J. (1975). Saponins in leaves of *Agave americana. Pharmazie,* **30,** 396–397.

Kircher, H.W. (1977). Triterpene glycosides and queretaroic acid in organ pipe cactus. *Phytochemistry,* **16,** 1078–1080.

Kiryalov, N.P. (1969). Structure of macedonic acid. *Khim. Prir. Soedin.,* 448–449.

Kiryalov, N.P. and Bogatkina, V.F. (1969). Structure of echinatic acid. *Khim. Prir. Soedin.,* 447–448.

Kiryalov, N.P., Bogatkina, V.F. and Nadezhina, T.P. (1973). Glabrolide from *Glycyrrhiza aspera* roots. *Khim. Prir. Soedin.,* 277.

Kishor, N. (1990). A new molluscicidal spirostanol glycoside from *Agava cantala. Fitoterapia,* **61,** 456–457.

Kishor, N. and Sati, O.P. (1990). A new molluscicidal spirostanol glycoside of *Yucca aloifolia. J. Nat. Prod.,* **53,** 1557–1559.

Kishor, N., Bahuguna, S., Sati, O.P., Sakakibara, J. and Kaiya, T. (1991). A new molluscicidal spirostanol glycoside from *Yucca aloifolia. Fitoterapia,* **62,** 266–269.

Kitagawa, I. (1981). Cleavage glycoside linkages. In *Advances in Natural Product Chemistry,* eds. S. Natori, N. Ikekawa and M. Suzuki, pp. 310–326. John Wiley, New York.

Kitagawa, I. (1984). Isolation of soybean plant saponins. West German Patent DE 3400258 (*Chem. Abstr.,* **102,** 75306).

Kitagawa, I. (1988). Bioactive marine natural products. *Yakugaku Zasshi,* **108,** 398–416.

Kitagawa, I. and Kobayashi, M. (1978). Saponin and sapogenol. XXVI. Steroidal saponins from the starfish *Acanthaster planci* L. (crown of thorns). (2). Structure of the major saponin thornasteroside A. *Chem. Pharm. Bull.,* **26,** 1864–1873.

Kitagawa, I., Inada, A., Yosioka, I., Somanathan, R., Sultanbawa, M.U.S. (1972a). Protobassic acid, a genuine sapogenol of seed kernels of *Madhuca longifolia* L. *Chem. Pharm. Bull.,* **20,** 630–632.

Kitagawa, I., Matsuda, A. and Yosioka, I. (1972b). Saponin and sapogenol. VII. Sapogenol constituents of five Primulaceous plants. *Chem. Pharm. Bull.,* **20,** 2226–2234.

Kitagawa, I., Yoshikawa, M. and Yosioka, I. (1973). Photolysis of uronide linkage in saponin. *Tetrahedron Lett.,* 3997–3998.

Kitagawa, I., Yoshikawa, M., Imakura, Y. and Yosioka, I. (1974). Saponin and sapogenol. VII. Photochemical cleavage of glycoside linkage in saponin. (1). Photolysis of some saponins and their structural features. *Chem. Pharm. Bull.*, **22**, 1339–1347.

Kitagawa, I., Inada, A. and Yosioka, I. (1975). Saponin and sapogenol. XII. Misaponin A and Mi-saponin B, two major bisdesmosides from the seed kernels of *Madhuca longifolia* (L.) MACBRIDE. *Chem. Pharm. Bull.*, **23**, 2268– 2278.

Kitagawa, I., Sugawara, T. and Yosioka, I. (1976a). Saponin and sapogenol. XV. Antifungal glycosides from the sea cucumber *Stichopus japonicus* SELENKA. (2) Structures of holotoxin A and holotoxin B. *Chem. Pharm. Bull.*, **24**, 275–284.

Kitagawa, I., Yoshikawa, M. and Yosioka, I. (1976b). Sapogenin and sapogenol. XIII. Structures of three soybean saponins: soyasaponin I, soyasaponin II, and soyasaponin III. *Chem. Pharm. Bull.*, **24**, 121–129.

Kitagawa, I., Yoshikawa, M., Ikenishi, Y., In, K.S. and Yosioka, I. (1976c). A new selective cleavage method of glucuronide linkage in oligoglycoside. Lead tetraacetate oxidation followed by alkali treatment. *Tetrahedron Lett.*, 549–552.

Kitagawa, I., Im, K.-S. and Yosioka, I. (1976d). Saponin and sapogenol. XVI. Structure of desacylboninsaponin A obtained from the bark of *Schima mertensiana* KOIDZ. *Chem. Pharm. Bull.*, **24**, 1260–1267.

Kitagawa, I., Ikenishi, Y., Yoshikawa, M. and Im, K.S. (1977). Saponin and sapogenol. XX. Selective cleavage of the glucuronide linkage in saponin by acetic anhydride and pyridine. *Chem. Pharm. Bull.*, **25**, 1408–1416.

Kitagawa, I., Yamanaka, H., Kobayashi, M., Nishino, T., Yosioka, I. and Sugawara, T. (1978a). Saponin and sapogenol. XXVII. Revised structure of holotoxin A and holotoxin B, two antifungal oligoglycosides from the sea cucumber *Stichopus japonicus* SELENKA. *Chem. Pharm. Bull.*, **26**, 3722–3731.

Kitagawa, I., Shirakawa, K. and Yoshikawa, M. (1978b). Saponin and sapogenol. XXIV. The structure of mi-saponin C, a bisdesmoside of protobassic acid from the seed kernels of *Madhuca longifolia* (L.) MACBRIDE. *Chem. Pharm. Bull.*, **26**, 1100–1110.

Kitagawa, I., Kamigauchi, T., Ohmori, H. and Yoshikawa, M. (1980a). Saponin and sapogenol. XXIX. Selective cleavage of the glucuronide linkage in oligoglycosides by anodic oxidation. *Chem. Pharm. Bull.*, **28**, 3078–3086.

Kitagawa, I., Yoshikawa, M., Kobayashi, K., Imakura, Y., Im, K.S. and Ikenishi, Y. (1980b). Saponin and sapogenol. XXVII. Reinvestigation of the branching positions in the glucuronide moieties of three glucuronoid saponins: desacyl-jegosaponin, desacyl-boninsaponin A, and sakuraso-saponin. *Chem. Pharm. Bull.*, **28**, 296–300.

Kitagawa, I., Yoshikawa, M., Wang, H.K. *et al.* (1982). Revised structures of soyasapogenols A, B and E, oleanene-sapogenols from soybean. Structures of soyasaponins I, II, III. *Chem. Pharm. Bull.*, **30**, 2294–2297.

Kitagawa, I., Wang, H.K., Takagi, A., Fuchida, M., Miura, I. and Yoshikawa, M. (1983a). Saponin and sapogenol. XXXIV. Chemical constituents of Astragali radix, the root of *Astragalus membranaceus* BUNGE. (1). Cyclo-astragenol, the 9,19-cyclolanostane-type aglycone of astragalosides, and the artifact aglycone astragenol. *Chem. Pharm. Bull.*, **31**, 689–697.

Kitagawa, I., Taniyama, T., Hayashi, T. and Yoshikawa, M. (1983b). Malonylginsenosides Rb$_1$, Rb$_2$, Rc and Rd, four new malonylated dammarane-type triterpene oligoglycosides from ginseng radix. *Chem. Pharm. Bull.*, **31**, 3353–3356.

Kitagawa, I., Wang, H.K., Saito, M. and Yoshikawa, M. (1983c). Saponin and sapogenol. XXXII. Chemical constituents of the seeds of *Vigna angularis* (WILLD.) OHWI ET OHASHI. (2). Azukisaponins I, II, III, and IV. *Chem. Pharm. Bull.*, **31**, 674–682.

Kitagawa, I., Wang, H.K., Saito, M. and Yoshikawa, M. (1983d). Saponin and sapogenol. XXXIII. Chemical constituents of the seeds of *Vigna angularis* (WILLD.) OHWI ET OHASHI. (3). Azukisaponins V and VI. *Chem. Pharm. Bull.*, **31**, 683–688.

Kitagawa, I., Wang, H.K. and Yoshikawa, M. (1983e). Saponin and sapogenol. XXXVII. Chemical constituents of Astragali radix, the root of *Astragalus membranaceus* BUNGE. (4). Astragalosides VII and VIII. *Chem. Pharm. Bull.*, **31**, 716–722.

Kitagawa, I., Wang, H.K., Saito, M. and Yoshikawa, M. (1983f). Saponin and sapogenol. XXXV. Chemical constituents of Astragali radix, the root of *Astragalus membranaceus* BUNGE. (2). Astragalosides I, II and IV, acetyl-astragaloside I and isoastragalosides I and II. *Chem. Pharm. Bull.*, **31**, 698–708.

Kitagawa, I., Yoshikawa, M., Hayashi, T. and Taniyama, T. (1984a). Quantitative determination of soyasaponins in soybeans of various origins and soybean products by means of high-performance liquid chromatography. *Yakugaku Zasshi*, **104**, 275–279.

Kitagawa, I., Yoshikawa, M., Hayashi, T. and Taniyama, T. (1984b). Characterization of saponin constituents in soybeans of various origins and quantitative analysis of soyasaponins by gas-liquid chromatography. *Yakugaku Zasshi*, **104**, 162–168.

Kitagawa, I., Saito, M., Taniyama, T. and Yoshikawa, M. (1985a). Saponin and sapogenol. XXXIX. Structure of soyasaponin A$_1$, a bisdesmoside of soyasapogenol A, from soybean, the seeds of *Glycine max* MERRILL. *Chem. Pharm. Bull.*, **33**, 1069–1076.

Kitagawa, I., Kobayashi, M., Inamoto, T., Fuchida, M. and Kyogoku, Y. (1985b). Marine natural products. XIV. Structures of echinosides A and B, antifungal lanostane-oligosides from the sea cucumber *Actinopyga echinites* (JAEGER). *Chem. Pharm. Bull.*, **33**, 5214–5224.

Kitagawa, I., Saito, M., Taniyama, T. and Yoshikawa, M. (1985c). Saponin and sapogenin. XXXVIII. Structure of soyasaponin A$_2$, a bisdesmoside of soyasapogenol A, from soybean, the seeds of *Glycine max* MERRILL. *Chem. Pharm. Bull*, **33**, 598–608.

Kitagawa, I., Kobayashi, M., Okamoto, Y., Yoshikawa, M. and Hanamoto, Y. (1987a). Structures of sarasinosides A$_1$ B$_1$ and C$_1$; new norlanostane-triterpenoid oligoglycosides from the Palauan marine sponge *Asteropus sarasinosum*. *Chem. Pharm. Bull.*, **35**, 5036–5039.

Kitagawa, I., Taniyama, T., Shibuya, H., Noda, T. and Yoshikawa, M. (1987b). Chemical studies on crude drug processing. V. On the constituents of ginseng radix rubra (2): comparison of the constituents of white ginseng and red ginseng prepared from the same *Panax ginseng* root. *Yakugaku Zasshi*, **107**, 495–505.

Kitagawa, I., Wang, H.K., Taniyama, T. and Yoshikawa, M. (1988a). Saponin and sapogenol. XLI. Reinvestigation of the structures of soyasapogenols A, B and E, oleanene-sapogenols from soybean. Structures of soyasaponins I, II and III. *Chem. Pharm. Bull.*, **36**, 153–161.

Kitagawa, I., Taniyama, T., Nagahama, Y., Okubo, K., Yamauchi, F. and Yoshikawa, M. (1988b). Saponin and sapogenol. XLII. Structures of acetyl-soyasaponins A_1, A_2 and A_3, astringent, partially acetylated bisdesmosides of soyasapogenol A, from American soybean, the seeds of *Glycine max* MERRILL. *Chem. Pharm. Bull.*, **36**, 2819–2828.

Kitagawa, I., Zhou, J.L., Sakagami, M., Taniyama, T. and Yoshikawa, M. (1988c). Licorice-saponins A3, B2, C2, D3 and E2, five new oleanene-type triterpene oligoglycosides from Chinese Glycyrrhizae radix. *Chem. Pharm. Bull.*, **36**, 3710–3713.

Kitagawa, I., Taniyama, T., Murakami, T., Yoshihara, M. and Yoshikawa, M. (1988d). Saponin and sapogenol. XLVI. On the constituents in aerial part of American alfalfa, *Medicago sativa* L. The structure of dehydrosoyasaponin I. *Yakugaku Zasshi*, **108**, 547–554.

Kitagawa, I., Taniyama, T., Hong, W.W., Hori, K. and Yoshikawa, M. (1988e). Saponin and sapogenol. XLV. Structure of kaikasaponins I, II and III from Sophorae flos, the buds of *Sophora japonica* L. *Yakugaku Zasshi*, **108**, 538–546.

Kitagawa, I., Kobayashi, M., Hori, M. and Kyogoku, Y. (1989a). Marine natural products. XVIII. Four lanostane-type triterpene oligoglycosides, bivittosides A, B, C and D, from the Okinawan sea cucumber *Bohadschia bivittata* MITSUKURI. *Chem. Pharm. Bull.*, **37**, 61–67.

Kitagawa, I., Sakagami, M., Hashiuchi, F., Zhou, J.L., Yoshikawa, H. and Ren, J. (1989b). Apioglycyrrhizin and araboglycyrrhizin, two new sweet oleanene-type triterpene oligoglycosides from the root of *Glycyrrhiza inflata*. *Chem. Pharm. Bull.*, **37**, 551–553.

Kitagawa, I., Kobayashi, M., Son, B.W., Suzuki, S. and Kyogoku, Y. (1989c). Marine natural products. XIX. Pervicosides A, B and C, lanostane-type triterpene-oligoglycoside sulphates from the sea cucumber *Holothuria pervica*. *Chem. Pharm. Bull.*, **37**, 1230–1234.

Kitagawa, I., Taniyama, T., Yoshikawa, M., Ikenishi, Y. and Nakagawa, Y. (1989d). Chemical studies on crude drug processing. VI. Chemical structures of malonyl-ginsenosides Rb_1, Rb_2, Rc, and Rd isolated from the root of *Panax ginseng* C.A. MEYER. *Chem. Pharm. Bull.*, **37**, 2961–2970.

Kitagawa, I., Zhou, J.L., Sakagami, M., Uchida, E. and Yoshikawa, M. (1991). Licorice saponins F3, G2, H2, J2 and K2, five new oleanene-triterpene oligoglycosides from the root of *Glycyrrhiza uralensis*. *Chem. Pharm. Bull.*, **39**, 244–246.

Kitajima, J. and Tanaka, Y. (1989). Two new triterpenoid glycosides from the leaves of *Schefflera octophylla*. *Chem. Pharm. Bull.*, **37**, 2727–2730.

Kitajima, J., Komori, T., Kawasaki, T. and Schulten, H.-R. (1982). Basic steroid saponins from the aerial parts of *Fritillaria thunbergii*. *Phytochemistry*, **21**, 187–192.

Kitasato, Z. and Sone, C. (1932). Ueber die Konstitution des Hederagenins und der Oleanolsäure. I. *Acta Phytochim.* (Tokyo), **6**, 179–222.

Kiuchi, F., Liu, H. and Tsuda, Y. (1990). Two new gymnemic acid congeners containing a hexulopyranoside moiety. *Chem. Pharm. Bull.*, **38**, 2326–2328.

Kiyosawa, S., Hutoh, M., Komori, T., Nohara, T., Hosokawa, I. and Kawasaki, T. (1968). Detection of prototype compounds of diosgenin and other spirostanol-glycosides. *Chem. Pharm. Bull.*, **16**, 1162–1163.

Kizu, H. and Tomimori, T. (1979). Studies on the constituents of *Clematis* species. I. On the saponins of the root of *Clematis chinensis* OSBECK. *Chem. Pharm. Bull.*, **27**, 2388–2393.

Kizu, H. and Tomimori, T. (1982). Studies on the constituents of *Clematis* species. V. On the saponins of the root of *Clematis chinensis* OSBECK. *Chem. Pharm. Bull.*, **30**, 3340–3346.

Kizu, H., Koshijima, M. and Tomimori, T. (1985a). Studies on the constituents of *Hedera rhombea* BEAN. III. On the dammarane triterpene glycosides. (2). *Chem. Pharm. Bull.*, **33**, 3176–3181.

Kizu, H., Hirabayashi, S., Suzuki, M. and Tomimori, T. (1985b). Studies on the constituents of *Hedera rhombea* BEAN. IV. On the hederagenin glycosides. (2). *Chem. Pharm. Bull.*, **33**, 3473–3478.

Kizu, H., Kitayama, S., Nakatani, F., Tomimori, T. and Namba, T. (1985c). Studies on Nepalese crude drugs. III. On the saponins of *Hedera nepalensis* K. KOCH. *Chem. Pharm. Bull.*, **33**, 3324–3329.

Klimek, B., Lavaud, C. and Massiot, G. (1992). Saponins from *Verbascum nigrum*. *Phytochemistry*, **31**, 4368–4370.

Klyne, W. (1950). The configuration of the anomeric carbon atoms in some cardiac glycosides. *Biochem. J.*, **47**, xli–xlii.

Klyne, W. (1960). The use of optical rotation for the study of the structure of organic substances. *Chemie*, **10**, 293–300.

Knight, J.O. and White, D.E. (1961). Triterpenoid compounds. 7β-Hydroxy-A_1-barrigenol. *Tetrahedron Lett.*, 100–104.

Kobayashi, M., Okamoto, Y. and Kitagawa, I. (1991). Marine natural products. XXVIII. The structures of sarasinosides A_1, A_2, A_3, B_1, B_2, B_3, C_1, C_2 and C_3. Nine new norlanostane-triterpenoidal oligoglycosides from the Palauan marine sponge *Asteropus sarasinosum*. *Chem. Pharm. Bull.*, **39**, 2867–2877.

Kobert, R. (1887). Ueber Quillajasäure. Ein Beitrag zur Kenntnis der Saponingruppe. *Arch. Exper. Pathol. Pharmakol.*, **23**, 233–272.

Koch, H.P. (1993). Saponine in Knoblauch und Küchenzwiebel, *Dtsch. Apoth. Ztg.*, **133**, 63–75.

Kochetkov, N.K. and Khorlin, A.J. (1966). Oligoside, ein neuer Typ von Pflanzenglykosiden. *Arzneimittelforsch.*, **16**, 101–109.

Kochetkov, N.K., Khorlin, A.J. and Vaskovsky, V.E. (1962). The structure of aralosides A and B. *Tetrahedron Lett.*, 713–716.

Kochetkov, N.K., Khorlin, A.J. and Ovodov, J.S. (1963). The structure of gypsoside – triterpenic saponin from *Gypsophila pacifica* KOM. *Tetrahedron Lett.*, 477–482.

Kochetkov, N.K., Khorlin, A.J. and Ovodov, J.S. (1964a). Triterpene saponins. IX. The structure of gypsoside. *Izv. Akad. Nauk SSSR, Ser. Khim.*, 1436–1446 (*Chem. Abstr.*, **65**, 7223).

Kochetkov, N.K., Khorlin, A.J. and Snjatkova, V.I. (1964b). Triterpene saponins. XIII. Halolysis of glucosides of the triterpene series and synthesis of glucosides of oleanolic acid. *Izv. Akad. Nauk SSSR, Ser. Khim.*, 2028–2036 (*Chem. Abstr.*, **63**, 667).

Kofler, L. (1927). *Die Saponine*. Julius Springer Verlag, Vienna.

Kohda, H. and Tanaka, O. (1975). Enzymatic hydrolysis of ginseng saponins and their related glycosides. *Yakugaku Zasshi*, **95**, 246–249.

Kohda, H., Takeda, O. and Tanaka, S. (1989). Molluscicidal triterpenoidal saponin from *Lysimachia sikokiana*. *Chem. Pharm. Bull.*, **37**, 3304–3305.

Kohda, H., Tanaka, S. and Yamaoka, Y. (1990). Saponins from leaves of *Acanthopanax hypoleucus* MAKINO. *Chem. Pharm. Bull.*, **38**, 3380–3383.

Kohda, H., Tanaka, S., Yamaoka, Y. and Ohhara, Y. (1991). Saponins from *Amaranthus hypochondriacus*. *Chem. Pharm. Bull.*, **39**, 2609–2612.

Kohda, H., Yamaoka, Y., Morinaga, S., Ishak, M. and Darise, M. (1992). Saponins from *Talinum triangulare*. *Chem. Pharm. Bull.*, **40**, 2557–2558.

Koizumi, H., Sanada, S., Ida, Y. and Shoji, J. (1982). Studies on the saponins of ginseng. IV. On the structure and enzymatic hydrolysis of ginsenoside-Ra$_1$. *Chem. Pharm. Bull.*, **30**, 2393–2398.

Kojima, H. and Ogura, H. (1989). Configurational studies on hydroxy groups at C-2, 3 and 23 or 24 of oleanene and ursene-type triterpenes by NMR spectroscopy. *Phytochemistry*, **28**, 1703–1710.

Komori, T., Ida, Y., Mutou, Y., Miyahara, K., Nohara, T. and Kawasaki, T. (1975). Mass spectra of spirostanol and furostanol glycosides. *Biomed. Mass Spectrom.*, **2**, 65–77.

Komori, T., Krebs, H.C., Itakura, Y. *et al.* (1983). Steroid-Oligoglycoside aus dem Seestern *Luidia maculata* MUELLER ET TROSCHEL, 1. Die Strukturen eines neuen Aglyconsulfats und von zwei neuen Oligoglycosidsulfaten. *Annalen*, 2092–2113.

Komori, T., Kawasaki, T. and Schulten, H.-R. (1985). Field desorption mass spectrometry of natural products. XII. Field desorption and fast atom bombardment mass spectrometry of biologically active natural oligoglycosides. *Mass Spectrom. Rev.*, **4**, 255–293.

Kondo, N. and Shoji, S. (1975). Studies on the constituents of Himalayan ginseng, *Panax pseudoginseng*. II. The structures of the saponins (2). *Chem. Pharm. Bull.*, **23**, 3282–3285.

Kondo, N., Shoji, J., Nagumo, N. and Komatsu, N. (1969). Studies on the constituents of Panacis japonici rhizoma. II. The structure of chikusetsusaponin IV and some observations on the structural relationship with araloside A. *Yakugaku Zasshi*, **89**, 846–850.

Kondo, N., Marumoto, Y. and Shoji, J. (1971). Studies on the constituents of Panacis japonici rhizoma. IV. The structure of chikusetsusaponin V. *Chem. Pharm. Bull.*, **19**, 1103–1107.

Kondo, T., Kurotori, S., Teshima, M. and Sumimoto, M. (1963). Termiticidal substance from *Kalopanax septemlobus* wood. *Nippon Mokuzai Gakkaishi*, **9**, 125–129 (*Chem. Abstr.*, **60**, 2273).

Kondratenko, E.S., Putieva, Z.M. and Abubakirov, N.K. (1981). Triterpenoid glycosides from plants of the Caryophyllaceae family. *Khim. Prir. Soedin.*, 417–433.

Kong, F., Zhu, D., Xu, R. *et al.* (1986). Structural study of tubeimoside I, a constituent of Tu-bei-mu. *Tetrahedron Lett.*, **27**, 5765–5768.

Kong, J., Li, X.-C., Wei, B.-Y. and Yang, C.-R. (1993). Triterpene glycosides from *Decaisnea fargesii*. *Phytochemistry*, **33**, 427–430.

Kong, L., Shao, C. and Xu, J. (1988). Chemical constituents of the root of *Acanthopanax sessiliflorus*. *Zhongcaoyao*, **19**, 482–486 (*Chem. Abstr.*, **110**, 132187).

Konishi, T. and Shoji, J. (1979). Studies on the constituents of Asparagi radix. 1. On the structures of furostanol oligoglycosides of *Asparagus cochinchinensis* (LOUREIO) MERRILL. *Chem. Pharm. Bull.*, **27**, 3086–3094.

Konishi, T., Tada, A., Shoji, J., Kasai, R. and Tanaka, O. (1978). The structures of platycodin A and C, monoacetylated saponins of the roots of *Platycodon grandiflorum* A.DC. *Chem. Pharm. Bull.*, **26**, 668–670.

Konishi, T., Kiyosawa, S. and Shoji, J. (1984). Studies on the constituents of *Aspidistra elatior* BLUME. II. On the steroidal glycosides of the leaves. (1). *Chem. Pharm. Bull.*, **32**, 1451–1460.

Kono, H. and Odajima, T. (1988). Antitumor pharmaceuticals containing saiko-saponins of parsley. Jap. Pat. 62 187,408 (*Chem. Abstr.*, **108**, 62472).

Konoshima, T. and Sawada, T. (1982). Legume saponins of *Gleditsia japonica* MIQUEL V. ^{13}C Nuclear magnetic resonance spectral studies on the structures of gleditsia saponins D_2, G and I. *Chem. Pharm. Bull.*, **30**, 4082–4087.

Konoshima, T., Inui, H., Sato, K., Yonezawa, M. and Sawada, T. (1980). Legume saponins of *Gleditsia japonica* MIQUEL. II. Desmonoterpenyl-glycoside of echinocystic acid. *Chem. Pharm. Bull.*, **28**, 3473–3478.

Konoshima, T., Sawada, T. and Kimura, T. (1985). Studies on the constituents of leguminous plants. VIII. The structure of a new triterpenoid saponin from the fruits of *Gymnocladus chinensis* BAILLON. *Chem. Pharm. Bull.*, **33**, 4732–4739.

Konoshima, T., Kozuka, M., Haruna, M. and Ito, K. (1991). Constituents of leguminous plants, XIII. New triterpenoid saponins from *Wisteria brachybotrys*. *J. Nat. Prod.*, **54**, 830–836.

Konoshima, T., Kokumai, M., Kozuka, M., Tokuda, H., Nishino, H. and Iwashima, A. (1992). Anti-tumor-promoting activities of afromosin and soyasaponin I isolated from *Wistaria brachybotrys*. *J. Nat. Prod.*, **55**, 1776–1778.

Koo, J.H. (1983). The effect of ginseng saponin on the development of experimental atherosclerosis. *Hanyang Uidae Haksulchi*, **3**, 273–286 (*Chem Abstr.*, **101**, 17145).

Korkashvili, T.Sh., Dzhikiya, O.D., Vugalter, M.M., Pkheidze, T.A. and Kemert-elidze, E.P. (1985). Steroid glycosides of *Ruscus ponticus*. *Soobshch. Akad. Nauk Gruz. SSR*, **120**, 561–564 (*Chem. Abstr.*, **105**, 3499).

Kotake, M. and Taguchi, K. (1932). *Sci. Pap. Inst. Phys. Chem. Res.*, **18**, 5.

Kouam, J., Nkengfack, A.E., Fomum, Z.T., Ubillas, R., Tempesta, M.S. and Meyer, M. (1991). Two new triterpenoid saponins from *Erythrina sigmoidea*. *J. Nat. Prod.*, **54**, 1288–1295.

Kouno, I., Tsuboi, A., Nanri, M. and Kawano, N. (1990). Acetylated triterpene glycoside from roots of *Dipsacus asper*. *Phytochemistry*, **29**, 338–339.

Kravets, S.D., Vollerner, Y.S., Gorovits, M.B., Shashkov, A.S. and Abubakirov, N.K. (1986a). Spirostan and furostan type steroids from plants of the *Allium* genus. XXI. Structure of alliospiroside A and alliofuroside A from *Allium cepa. Khim. Prir. Soedin.*, 188–196.

Kravets, S.D., Vollerner, Y.S., Gorovits, M.B., Shashkov, A.S. and Abubakirov, N.K. (1986b). Steroids of the spirostane and furostane series from plants of the genus *Allium*. XXII. The structure of alliospiroside B from *Allium cepa. Khim. Prir. Soedin.*, 589–592.

Kravets, S.D., Vollerner, Y.S., Shashkov, A.S., Gorovits, M.B. and Abubakirov, N.K. (1987). Steroids of the spirostan and furostan series from plants of the genus *Allium*. XXIII. Structure of cepagenin and alliospirosides C and D from *Allium cepa. Khim. Prir. Soedin.*, 843–849.

Kravets, S.D., Vollerner, Y.S., Gorovits, M.B. and Abubakirov, N.K. (1990). Steroids of the spirostan and furostan series from plants of the genus *Allium. Chem. Nat. Compd.*, **26**, 359–373.

Kretsu, L.G. and Lazurevskii, G.V. (1970). Triterpenic glycosides of *Beta vulgaris. Khim. Prir. Soedin.*, 272–273.

Krokhmalyuk, V.V., Kintya, P.K. and Chirva, V.J. (1975). Structure of saponin B from *Clematis songarica. Khim. Prir. Soedin.*, 600–603.

Kubo, T., Hamada, S., Nohara, T. *et al.* (1989). Study on the constituents of *Desmodium styracifolium. Chem. Pharm. Bull.*, **37**, 2229–2231.

Kubo, S., Mimaki, Y., Sashida, Y., Nikaido, T. and Ohmoto, T. (1992). Steroidal saponins from the rhizomes of *Smilax sieboldii. Phytochemistry*, **31**, 2445–2450.

Kubota, T. and Hinoh, H. (1968). The constitution of saponins isolated from *Bupleurum falcatum* L. *Tetrahedron Lett.*, 303–306.

Kubota, T. and Matsushima, T. (1928). *J. Pharm. Soc. Jap.*, **48**, 25.

Kubota, T., Kitatani, H. and Hinoh, H. (1969). Isomerisation of quillaic acid and echinocystic acid with hydrochloric acid. *Tetrahedron Lett.*, 771–774.

Kudou, S., Tonomura, M., Tsukamoto, C., Shimoyamada, M., Uchida, T. and Okubo, K. (1992). Isolation and structure elucidation of the major genuine soybean saponin. *Biosci. Biotech. Biochem.*, **56**, 142–143.

Kudou, S., Tonomura, M., Tsukamoto, C. *et al.* (1993). Isolation and structural elucidation of DDMP-conjugated soyasaponins as genuine saponins from soybean seeds. *Biosci. Biotech. Biochem.*, **57**, 546–550.

Kuhn, R. and Löw, I. (1961). Zur Konstitution der Leptine. *Chem. Ber.*, **94**, 1088–1095.

Kuhn, R., Löw, I. and Trischmann, H. (1955a). Die Konstitution des Solanins. *Chem. Ber.*, **88**, 1492–1507.

Kuhn, R., Löw, I. and Trischmann, H. (1955b). Die Konstitution des α-Chaconins. *Chem. Ber.*, **88**, 1690–1693.

Kühner, K., Voigt, G., Hiller, K. *et al.* (1985). Glycosidstrukturen der Saponine von *Sanicula europaea*. 39. Mitteilung: Zur Kenntnis der Inhaltstoffe einiger Saniculoidea. *Pharmazie*, **40**, 576–578.

Kulshreshtha, M.J. and Rastogi, R.P. (1972). Chemical constituents of *Zizyphus rugosa. Indian J. Chem.*, **10**, 152–154.

Kumagai, A. (1967). Effects of glycyrrhizin on steroid hormone activities. *Minophagen Med. Rev.*, **12**, 14–23.

Kundu, A.P. and Mahato, S.B. (1993). Triterpenoids and their glycosides from *Terminalia chebula. Phytochemistry*, **32**, 999–1002.

Kunz, K., Schaffler, K., Biber, A. and Wauschkuhn, C.H. (1991). Bioverfügbarkeit von β-Aescin nach oraler Gabe zweier Aesculus-Extrakt enthaltender Darreichsformen an gesunden Probanden. *Pharmazie*, **46**, 145.

Kupchan, S.M., Barboutis, S.J., Knox, J.R. and Cam, C.A.L. (1965). β-Solamarine: tumor inhibitor isolated from *Solanum dulcamara*. *Science*, **150**, 1827–1828.

Kupchan, S.M., Hemingway, R.J., Knox, J.R., Barboutis, S.J., Werner, D. and Barboutis, M.A. (1967). Tumor inhibitors XXI. Active principles of *Acer negundo* and *Cyclamen persicum*. *J. Pharm. Sci.*, **56**, 603–608.

Kupchan, S.M., Takasugi, M., Smith, R.M. and Steyn, P.S. (1970). Acerotin and acerocin, novel triterpene ester aglycones from the tumor-inhibitory saponins of *Acer negundo*. *J. Chem. Soc., Chem. Commun.*, 969–970.

Kurihara, Y., Ookubo, K., Tasaki, H. *et al.* (1988). Studies on the taste modifiers. I. Purification and structure determination of sweetness inhibiting substance in leaves of *Ziziphus jujuba*. *Tetrahedron*, **44**, 61–66.

Kurilchenko, V.A., Zemtsova, G.N. and Bandyukova, V.Y. (1971). Chemical study of *Scabiosa bipinnata*. *Khim. Prir. Soedin.*, 534–535.

Kurilenko, M.I. (1968). *Pulsatilla nigricans* root glucoside. *Farm. Zh. (Kiev)*, **23**, 75–80 (*Chem. Abstr.*, **70**, 99583).

Kusumoto, K., Nagao, T., Okabe, H. and Yamauchi, T. (1989). Studies on the constituents of *Luffa operculata* COGN. I. Isolation and structures of luperosides A–H, dammarane-type triterpene glycosides in the herb. *Chem. Pharm. Bull.*, **37**, 18–22.

Kutney, J.P. (1963). Nuclear magnetic resonance (N.M.R.) study in the steroidal sapogenin series. Stereochemistry of the spiro ketal system. *Steroids*, **2**, 225–235.

Kuwada, S. (1931). *J. Pharm. Soc. Jap.*, **51**, 57.

Kuwada, S. and Matsukawa, T. (1934). *J. Pharm. Soc. Jap.*, **54**, 8.

Kuwahara, M., Kawanishi, F., Komiya, T. and Oshio, H. (1989). Dammarane saponins of *Gynostemma pentaphyllum* MAKINO and isolation of malonylginsenosides-Rb_1, -Rd and malonylgypenoside V. *Chem. Pharm. Bull.*, **37**, 135–139.

Kuznetsova, T.A., Anisimov, M.M., Popov, A.M. *et al.* (1982). A comparative study *in vitro* of physiological activity of triterpene glycosides of marine invertebrates of echinoderm type. *Comp. Biochem. Physiol.*, **73C**, 41–43 (*Chem. Abstr.*, **98**, 86542).

Lacaille-Dubois, M.-A., Hanguet, B., Rustaiyan, A. and Wagner, H. (1993). Squarroside A, a biologically active triterpene saponin from *Acanthophyllum squarrosum*. *Phytochemistry*, **34**, 489–495.

Laidlaw, R.A. (1954). Wood saponins. Part I. A preliminary investigation of the saponins from Morabukea (*Mora gonggrijpii* (KLEINH.) SANDWITH). *J. Chem. Soc.*, 752–757.

Lalitha, T., Seshadri, R. and Venkataraman, L.V. (1987). Isolation and properties of saponins from *Madhuca butyracea* seeds. *J. Agric. Food Chem.*, **35**, 744–748.

Lang, W. (1977). Studies on the percutaneous absorption of tritium-labeled aescin in pigs. *Res. Exp. Med.*, **169**, 175–187 (*Chem. Abstr.*, **86**, 165049).

Lavaud, C., Massiot, G., Le Men-Olivier, L., Viari, A., Vigny, P. and Delaude, C. (1992). Saponins from *Steganotaenia araliacea. Phytochemistry*, **31**, 3177–3181.

Lazurevskii, G.V., Chirva, V.J., Kintia, P.K. and Kretsu, L.G. (1971). Triterpenoid glycosides of beans. *Dokl. Akad. Nauk SSSR*, **199**, 226–227 (*Chem. Abstr.*, **75**, 115903).

Lee, D.Y., Son, K.H., Do, J.C. and Kang, S.S. (1989). Two new steroid saponins from the tubers of *Liriope spicata. Arch. Pharmacal Res.*, **12**, 295–299 (*Chem. Abstr.*, **113**, 55831).

Lee, S.-S., Su, W.-C. and Liu, K.C. (1991). Two new triterpene glycosides from *Paliurus ramosissimus. J. Nat. Prod.*, **54**, 615–618.

Lee, T.M. and der Marderosian, A.H. (1981). Two-dimensional TLC analysis of ginsenosides from root of dwarf ginseng (*Panax trifolius* L.), Araliaceae. *J. Pharm. Sci.*, **70**, 89–91.

Lee, T.M. and der Marderosian, A.H. (1988). Studies on the constituents of dwarf ginseng. *Phytother. Res.*, **2**, 165–170.

Lee, Y.B., Yoo, S.J., Kim, J.S. and Kang, S.S. (1992). Triterpenoid saponins from the roots of *Caragana sinica. Arch. Pharmacal Res.*, **15**, 62–68 (*Chem. Abstr.*, **117**, 208892).

Lehtola, T. and Huhtikangas, A. (1990). Radio-immunoassay of aescine, a mixture of triterpene glycosides. *J. Immunoassay*, **11**, 17–19.

Lei, W., Shi, Q. and Yu, S. (1984). Analgesic and CNS inhibitory effects of total saponins from the leaves of *Panax notoginseng. Zhongao Tongbao*, **9**, 134–137 (*Chem. Abstr.*, **101**, 103905).

Lemieux, R.U., Kullnig, R.K., Bernstein, H.J. and Schneider, W.G. (1958). Configurational effects on the proton magnetic resonance spectra of six-membered ring compounds. *J. Am. Chem. Soc.*, **80**, 6098–6105.

Lemma, A. (1970). Laboratory and field evaluation of the molluscicidal properties of *Phytolacca dodecandra. Bull. W.H.O.*, **42**, 597–612.

Leonart, R., Moreira, E.A., Loureiro, M.M. and Miguel, O.G. (1982). Solasodine in fruits of *Solanum sanctae-catharinae* DUNAL. *Trib. Farm.*, **49–50**, 61–63 (*Chem. Abstr.*, **100**, 48612).

Leung, A.Y. (1980). *Encyclopedia of Common Natural Ingredients Used in Food Drugs and Cosmetics*. John Wiley, New York.

Levy, M., Zehavi, U., Naim, M. and Polacheck, I. (1986). An improved procedure for the isolation of medicagenic acid 3-*O*-β-D-glucopyranoside from alfalfa roots and its antifungal activities on plant pathogens. *J. Agric. Food Chem.*, **34**, 960–963.

Levy, M., Zehavi, U., Naim, M. and Polacheck, I. (1989). Isolation, structure determination, synthesis, and antifungal activity of a new native alfalfa-root saponin. *Carbohydr. Res.*, **193**, 115–123.

Lewis, J.R. (1974). Carbenoxolone sodium in the treatment of peptic ulcer: a review. *J. Am. Med. Assoc.*, **229**, 460–462.

Li, H.-Y., Koike, K., Ohmoto, T. and Ikeda, K. (1993). Dianchinenosides A and B, two new saponins from *Dianthus chinensis. J. Nat. Prod.*, **56**, 1065–1070.

Li, X.-C., Wang, D.-Z. and Yang, C.-R. (1990a). Steroidal saponins from *Chlorophytum malayense. Phytochemistry*, **29**, 3893–3898.

Li, X.-C., Wang, Y.-F., Wang, D.-Z. and Yang, C.-R. (1990b). Steroidal saponins from *Diuanthera major*. *Phytochemistry*, **29**, 3899–3901.

Li, X.-C., Wang, D.-Z., Wu, S.-G. and Yang, C.-R. (1990c). Triterpenoid saponins from *Pulsatilla campanella*. *Phytochemistry*, **29**, 595–599.

Li, X.J. and Zhang, B.H. (1988). Studies on the antiarrhythmic effects of panaxatriol saponins (PTS) isolated from *Panax notoginseng*. *Yaoxue Xuebao*, **23**, 168–173 (*Chem. Abstr.*, **109**, 539).

Lin, C.-N. and Gan, K.-H. (1989). Antihepatotoxic principles of *Solanum capsicastrum*. *Planta Medica*, **55**, 48–50.

Lin, C.N., Lin, S.Y., Chung, M.I., Gan, K.H. and Lin, C.C. (1986). Studies on the constituents of Formosan *Solanum* species. Part III. Isolation of steroidal alkaloids from *Solanum incanum* L. *T'ai-wan Yao Hsueh Tsa Chih*, **38**, 166–171 (*Chem. Abstr.*, **107**, 172434).

Lin, C.-N., Chung, M.-I. and Lin, S.-Y. (1987). Steroidal alkaloids from *Solanum capsicastrum*. *Phytochemistry*, **26**, 305–307.

Lin, C.-N., Chung, M.-I. and Gan, K.-H. (1988). Novel hepatotoxic principles of *Solanum incanum*. *Planta Medica*, **54**, 222.

Lin, C.-N., Lu, C.-M., Cheng, M.-K., Gan, K.-H. and Won, S.-J. (1990). The cytotoxic principles of *Solanum incanum*. *J. Nat. Prod.*, **53**, 513–516.

Lin, J.-T., Nes, W.D. and Heftmann, E. (1981). High-performance liquid chromatography of triterpenoids. *J. Chromatogr.*, **207**, 457–463.

Lin, T.D., Kondo, N. and Shoji, J. (1976). Studies on the constituents of Panacis japonici rhizoma. V. The structures of chikusetsu saponin I, Ia, Ib, IVa and glycoside P_1. *Chem. Pharm. Bull.*, **24**, 253–261.

Lindberg, B. (ed.) (1972). Methylation analysis of polysaccharides. In *Methods in Enzymology*, vol. 28, pp. 178–195. Academic Press, New York.

Linde, H. (1964). Die Inhaltsstoffe von *Cimicifuga racemosa*. 1. Mitteilung: Ueber die Isolierung von zwei Glykosiden. *Arzneimittelforsch.*, **14**, 1037–1040.

Lins Brandao, M.G., Lacaille-Dubois, M.-A., Teixera, M.A. and Wagner, H. (1992). Triterpene saponins from the roots of *Ampelozizyphus amazonicus*. *Phytochemistry*, **31**, 352–354.

Liu, C.-X. and Xiao, P.-G. (1992). Recent advances on ginseng research in China. *J. Ethnopharmacol.*, **36**, 27–38.

Liu, H.-W. and Nakanishi, K. (1981). A micromethod for determining the branching points in oligosaccharides based on circular dichroism. *J. Am. Chem. Soc.*, **103**, 7005–7006.

Liu, H.-W. and Nakanishi, K. (1982). The structures of balanitins, potent molluscicides isolated from *Balanites aegyptica*. *Tetrahedron*, **38**, 513–519.

Loloiko, A.A., Grishkovets, V.I., Shashkov, A.S. and Chirva, V.J. (1988). Triterpene glycosides of *Hedera taurica*. II. Structure of taurosides B and C in leaves of Crimean ivy. *Khim. Prir. Soedin.*, 379–382.

Lontsi, D., Sondengam, B.L., Ayafor, J.F., Tsoupras, M.G. and Tabacchi, R. (1990). Further triterpenoids of *Musanga cecropioides*: the structure of cecropic acid. *Planta Medica*, **56**, 287–289.

Lorenz, D. and Marek, M.L. (1960). Das therapeutisch wirksame Prinzip der Rosskastanie (*Aesculus hippocastanum*). 1. Mitteilung: Aufklärung des Wirkstoffes. *Arzneimittelforsch.*, **10**, 263–272.

Lourens, W.A. and O'Donovan, M.B. (1961). The lucerne saponins. II. The isolation of triterpene genins from lucerne saponin. *S. Afr. J. Agric. Sci.*, **4**, 293–300.

Lower, E.S. (1984). Activity of the saponins. *Drug Cosmet. Ind.*, **135**, 39–44.

Luchanskaya, V.N., Kondratenko, E.S., Gorovits, T.T. and Abubakirov, N.K. (1971a). Triterpenoid glycosides from *Gypsophylla trichotoma*. II. Structure of trichoside A. *Khim. Prir. Soedin.*, 151–153.

Luchanskaya, V.N., Kondratenko, E.S. and Abubakirov, N.K. (1971b). Triterpenoid glycosides from *Gypsophila trichotoma*. Structure of trichoside B. *Khim. Prir. Soedin.*, 431–434.

Luchanskaya, V.N., Kondratenko, E.S. and Abubakirov, N.K. (1971c). Triterpenoid glycosides of *Gypsophila trichotoma*. IV. Structure of trichoside C. *Khim. Prir. Soedin.*, 608–610.

Luckner, M. (1990). *Secondary Metabolism in Microorganisms, Plants and Animals*, 3rd edn. V.E.B. Gustav Fischer Verlag, Jena.

Luo, S.-Q., Jin, H.-F., Kawai, H., Seto, H. and Otake, N. (1987). Isolation of new saponins from the aerial part of *Bupleurum kunmingense* Y. LI ET S.L. PAN. *Agric. Biol. Chem.*, **51**, 1515–1519.

Luo, S.-Q., Lin, L.-Z. and Cordell, G.A. (1993). Saikosaponin derivatives from *Bupleurum wenchuanense*. *Phytochemistry*, **33**, 1197–1205.

Lutomski, J. (1983). Neues über die biologischen Eigenschaften einiger Triterpensaponinen. *Pharmazie in unserer Zeit*, **12**, 149–153.

Lythgoe, B. and Trippett, S. (1950). The constitution of the disaccharide of glycyrrhinic acid. *J. Chem. Soc.*, 1983–1990.

Ma, J.C.N. and Lau, F.W. (1985). Structure characterization of haemostatic diosgenin glycosides from *Paris polyphylla*. *Phytochemistry*, **24**, 1561–1565.

Ma, W.-G., Wang, D.-Z., Zeng, Y.-L. and Yang, C.-R. (1991). Three triterpenoid saponins from *Triplostegia grandiflora*. *Phytochemistry*, **30**, 3401–3404.

Ma, W.-W., Heinstein, P.F. and McLaughlin, J.L. (1989). Additional toxic, bitter saponins from the seeds of *Chenopodium quinoa*. *J. Nat. Prod.*, **52**, 1132–1135.

Mackie, A.M., Singh, H.T. and Owen, J.M. (1977). Studies on the distribution, biosynthesis and formation of steroidal saponins in echinoderms. *Comp. Biochem. Physiol.*, **56B**, 9–14.

Maeda, M., Iwashita, T. and Kurihara, Y. (1989). Studies on taste modifiers. II. Purification and structure determination of gymnemic acids, antisweet active principles from *Gymnema sylvestre* leaves. *Tetrahedron Lett.*, **30**, 1547–1550.

Maga, J.A. (1980). Potato glycoalkaloids. *Crit. Rev. Food Sci. Nutr.*, **15**, 371–404.

Maharaj, I., Froh, K.J. and Campbell, J.B. (1986). Immune responses of mice to inactivated rabies vaccine administered orally: potentiation by quillaja saponin. *Can. J. Microbiol.*, **32**, 414–420.

Mahato, S.B. (1991). Triterpenoid saponins from *Medicago hispida*. *Phytochemistry*, **30**, 3389–3393.

Mahato, S.B. and Nandy, A.K. (1991). Triterpenoid saponins discovered between 1987 and 1989. *Phytochemistry*, **30**, 1357–1390.

Mahato, S.B. and Pal, B.C. (1987). Triterpenoid glycosides of *Corchorus acutangulus* LAM. *J. Chem. Soc., Perkin Trans 1*, 629–634.

Mahato, S.B., Dutta, N.L. and Chakravarti, R.N. (1973). Triterpenes from *Careya arborea*. Structure of careyagenol D. *J. Indian Chem. Soc.*, **50**, 254–259.

Mahato, S.B., Sahu, N.P. and Pal, B.C. (1978). New steroidal saponins from *Dioscorea floribunda*: structure of floribunda saponins C, D, E and F. *Indian J. Chem.*, **16B**, 350–354.

Mahato, S.B., Sahu, N.P., Ganguly, A.N., Kasai, R. and Tanaka, O. (1980a). Steroidal alkaloids from *Solanum khasianum*: application of ^{13}C NMR spectroscopy to their structure elucidation. *Phytochemistry*, **19**, 2017–2020.

Mahato, S.B., Sahu, N.P. and Ganguly, A.N. (1980b). Carbon-13 NMR spectra of dioscin and gracillin isolated from *Costus speciosus*. *Indian J. Chem.*, **19B**, 817–819.

Mahato, S.B., Ganguly, A.N. and Sahu, N.P. (1982). Steroid saponins. *Phytochemistry*, **21**, 959–978.

Mahato, S.B., Sahu, N.P., Luger, P. and Müller, E. (1987). Stereochemistry of a triterpenoid trisaccharide from *Centella asiatica*. X-ray determination of the structure of asiaticoside. *J. Chem. Soc., Perkin Trans.* 2, 1509–1515.

Mahato, S.B., Sarkar, S.K. and Poddar, G. (1988). Triterpenoid saponins. *Phytochemistry*, **27**, 3037–3067.

Mahato, S.B., Pal, B.C. and Price, K.R. (1989). Structure of acaciaside, a triterpenoid trisaccharide from *Acacia auriculiformis*. *Phytochemistry*, **28**, 207–210.

Mahato, S.B., Nandy, A.K., Luger, P. and Weber, M. (1990). Determination of structure and stereochemistry of tomentosic acid by X-ray crystallography. A novel mechanism for transformation of arjungenin to tomentosic acid. *J. Chem. Soc., Perkin Trans.* 2, 1445–1450.

Mahato, S.B., Sahu, N.P., Roy, S.K. and Sen, S. (1991). Structure elucidation of four new triterpenoid oligoglycosides from *Anagallis arvensis*. *Tetrahedron*, **47**, 5215–5230.

Mahato, S.B., Nandy, A.K. and Roy, G. (1992). Triterpenoids. *Phytochemistry*, **31**, 2199–2249.

Maheas, M.R. (1961). Structure of the triterpene alcohols isolated from *Jacquinia armillaris* JACQ. *C.R. Hebd. Scéances Acad. Sci.*, **252**, 805–807.

Maillard, M. and Hostettmann, K. (1993). Analysis of saponins in crude plant extracts by thermospray LC-MS. *J. Chromatogr.*, **647**, 137–146.

Maillard, M., Adewunmi, C.O. and Hostettmann, K. (1989). New triterpenoid *N*-acetyl glycosides with molluscicidal activity from *Tetrapleura tetraptera* TAUB. *Helv. Chim. Acta*, **72**, 668–674.

Maiti, P.C., Roy, S. and Roy, A. (1968). Chemical investigation of Indian soapnut, *Sapindus laurifolius*. *Experientia*, **24**, 1091.

Majester-Savornin, B., Elias, R., Diaz-Lanza, A.M., Balansard, G., Gasquet, M. and Delmas, F. (1991). Saponins of the ivy plant, *Hedera helix*, and their leishmanicidic activity. *Planta Medica*, **57**, 260–262.

Malaviya, N., Pal, R. and Khanna, N.M. (1989). Chemical examination of *Ardisia neriifolia* WALL. *Indian J. Chem.*, **28B**, 522–523.

Malaviya, N., Pal, R. and Khanna, N.M. (1991). Saponins from *Deutzia corymbosa*. *Phytochemistry*, **30**, 2798–2800.

Malinow, M.R., McLaughlin, P. and Stafford, C. (1978). Prevention of hypercholesterolemia in monkeys (*Macaca fascicularis*) by digitonin. *Am. J. Clin. Nutr.*, **31**, 814–818.

Malinow, M.R., McNulty, W.P., McLaughlin, P. *et al.* (1981). The toxicity of alfalfa saponins in rats. *Food Cosmet. Toxicol.*, **19**, 443–445.

Mallavarapu, G.R. (1990). Recent advances in oleanane triterpenes. In *Studies in Natural Products Chemistry*, vol. 7, ed. Atta-ur-Rahman, pp. 131–174. Elsevier, Amsterdam.

Maltsev, I.I., Stekhova, S.I., Shentsova, E.B., Anisimov, M.M. and Stonik, V.A. (1985). Antimicrobial activity of glycosides from holothurians of the Stichopodidae family. *Khim. Farm. Zh.*, **19**, 54–56 (*Chem. Abstr.*, **102**, 146003).

Mancera, O., Ringold, H.J., Djerassi, C., Rosenkranz, G. and Sondheimer, F. (1953). Steroids. XLIII. A ten step conversion of progesterone to cortisone. The differential reduction of pregnane-3,20-diones with sodium borohydride. *J. Am. Chem. Soc.*, **75**, 1286–1290.

Mangan, J.L. (1958). Bloat in cattle. VII. The measurement of foaming properties of surface-active compounds. *New Zealand J. Agric. Res.*, **1**, 140–147.

Margineau, C., Cucu, V., Grecu, L. and Parvu, C. (1976). Anticandida action of a saponin from *Primula*. *Planta Medica*, **30**, 35–38.

Marhuenda, E., Martin, M.J. and Alarcon de la Lastra, C. (1993). Antiulcerogenic activity of aescine in different experimental models. *Phytother. Res.*, **7**, 13–16.

Marker, R.E. and Lopez, J. (1947). Steroidal sapogenins. No. 163. The biogenesis of steroidal sapogenins in plants. *J. Am. Chem. Soc.*, **69**, 2383–2385.

Marker, R.E. and Rohrmann, E. (1939). Sterols. LIII. The structure of the side chain of sarsasapogenin. *J. Am. Chem. Soc.*, **61**, 846–851.

Marker, R.E., Wagner, R.B., Ulshafer, P.R., Wittbecker, E.L., Goldsmith, D.P.J. and Ruof, C.H. (1947). Steroidal sapogenins. *J. Am. Chem. Soc.*, **69**, 2167–2230.

Marsh, C.A. and Levvy, G.A. (1956). Glucuronide metabolism in plants. III. Triterpene glucuronides. *Biochem. J.*, **63**, 9–14.

Marston, A. and Hostettmann, K. (1985). Plant molluscicides. *Phytochemistry*, **24**, 639–652.

Marston, A. and Hostettmann, K. (1991a). Assays for molluscicidal, cercaricidal, schistosomicidal and piscicidal activities. In *Methods in Plant Biochemistry*, vol. 6, ed. K. Hostettmann, pp. 153–178. Academic Press, London.

Marston, A. and Hostettmann, K. (1991b). Modern separation methods. *Nat. Prod. Rep.*, **8**, 391–413.

Marston, A., Gafner, F., Dossaji, S.F. and Hostettmann, K. (1988a). Fungicidal and molluscicidal saponins from *Dolichos kilimandscharicus*. *Phytochemistry*, **27**, 1325–1326.

Marston, A., Borel, C. and Hostettmann, K. (1988b). Separation of natural products by centrifugal partition chromatography. *J. Chromatogr.*, **450**, 91–99.

Marston, A., Slacanin, I. and Hostettmann, K. (1990). Centrifugal partition chromatography in the separation of natural products. *Phytochem. Anal.*, **1**, 3–17.

Marston, A., Slacanin, I. and Hostettmann, K. (1991). Tackling analytical problems of saponins by HPLC. *HPLC '91*, Basle, Switzerland. Book of abstracts, p. 142.

Martindale (1982). *The Extra Pharmacopoeia*, 28th edn. Pharmaceutical Press, London.

Masood, M., Minocha, P.K., Tiwari, K.P. and Srivastava, K.C. (1981). Narcissiflorine, narcissiflormine and narcissifloridine, three triterpene saponins from *Anemone narcissiflora*. *Phytochemistry*, **20**, 1675–1679.

Massiot, G. (1991). How far can one go in the field of structural elucidation of natural products? *J. Ethnopharmacol.*, **32**, 103–110.

Massiot, G., Lavaud, C., Guillaume, D., Le Men-Olivier, L. and van Binst, G. (1986). Identification and sequencing of sugars in saponins using 2D ^{1}H NMR spectroscopy. *J. Chem. Soc., Chem Commun.* 1485–1487.

Massiot, G., Lavaud, C., Guillaume, D. and Le Men-Olivier, L. (1988a). Reinvestigation of the sapogenins and prosapogenins from alfalfa (*Medicago sativa*). *J. Agric. Food Chem.*, **36**, 902–909.

Massiot, G., Lavaud, C., Le Men-Olivier, L., van Binst, G., Miller, S.P.F. and Fales, H.M. (1988b). Structural elucidation of alfalfa root saponins by mass spectrometry and nuclear magnetic resonance analysis. *J. Chem. Soc., Perkin Trans 1*, 3071–3079.

Massiot, G., Lavaud, C., Delaude, C., van Binst, G., Miller, S.P.F. and Fales, H.M. (1990). Saponins from *Tridesmostemon claessenssi*. *Phytochemistry*, **29**, 3291–3298.

Massiot, G., Lavaud, C. and Nuzillard, J.M. (1991a). Révision des structures des chrysantéllines par résonance magnétique nucléaire. *Bull. Soc. Chim. Fr.*, **127**, 100–107.

Massiot, G., Lavaud, C., Besson, V., Le Men-Olivier, L. and van Binst, G. (1991b). Saponins from aerial parts of alfalfa (*Medicago sativa*). *J. Agric. Food Chem.*, **39**, 78–82.

Massiot, G., Chen, X.-F., Lavaud, C. *et al.* (1992a). Saponins from stem bark of *Petersianthus macrocarpus*. *Phytochemistry*, **31**, 3571–3576.

Massiot, G., Lavaud, C., Benkhaled, M. and Le Men-Olivier, L. (1992b). Soyasaponin VI, a new maltol conjugate from alfalfa and soybean. *J. Nat. Prod.*, **55**, 1339–1342.

Matos, M.E.O. and Tomassini, T.C.B. (1983). Wedelin, a saponin from *Wedelia scaberrima*. *J. Nat. Prod.*, **46**, 836–840.

Matos, M.E.O., Sousa, M.P., Machado, M.I.L. and Braz Filho, R. (1986). Quinovic acid glycosides from *Guettarda angelica*. *Phytochemistry*, **25**, 1419–1422.

Matsuda, H., Namba, K., Fukuda, S., Tani, T. and Kubo, M. (1986a). Pharmacological study on *Panax ginseng* C.A. MEYER. III. Effects of red ginseng on experimental disseminated intravascular coagulation. (2) Effects of ginsenoside on blood coagulative and fibrinolytic systems. *Chem. Pharm. Bull.*, **34**, 1153–1157.

Matsuda, H., Namba, K., Fukuda, S., Tani, T. and Kubo, M. (1986b). Pharmacological study on *Panax ginseng* C.A. MEYER. IV. Effects of red ginseng on experimental disseminated intravascular coagulation. (3) Effect of ginsenoside-Ro on the blood coagulative and fibrinolytic system. *Chem. Pharm. Bull.*, **34**, 2100–2104.

Matsumoto, K., Kasai, R., Kanamaru, F., Kohda, H. and Tanaka, O. (1987). 3,4-Secolupane-type triterpene glycosyl esters from leaves of *Acanthopanax divaricatus* SEEM. *Chem. Pharm. Bull.*, **35**, 413–415.

Matsuura, H., Ushiroguchi, T., Itakura, Y., Hayashi, N. and Fuwa, T. (1988). A furostanol glycoside from garlic bulbs of *Allium sativum* L. *Chem. Pharm. Bull.*, **36**, 3659–3663.

Matsuura, H., Ushiroguchi, T., Itakura, Y. and Fuwa, T. (1989a). Further studies on steroidal glycosides from bulbs, roots and leaves of *Allium sativum* L. *Chem. Pharm. Bull.*, **37**, 2741–2743.

Matsuura, H., Ushiroguchi, T., Itakura, Y. and Fuwa, T. (1989b). A furostanol glycoside from *Allium chinense* G. DON. *Chem. Pharm. Bull.*, **37**, 1390–1391.

McIlroy, R.J. (1951). *The Plant Glycosides*, Ch. IX. Edward Arnold, London.

McMillan, M. and Thompson, J.C. (1979). An outbreak of suspected solanine poisoning in schoolboys: examination of criteria of solanine poisoning. *Quart. J. Med.*, **48**, 227–243.

McShefferty, J. and Stenlake, J.B. (1956). Caulosapogenin and its identity with hederagenin. *J. Chem. Soc.*, 2314–2316.

Menu, P., Bouquet, A., Cavé, A. and Pousset, J.L. (1976). Etude préliminaire des saponosides des fruits du *Brenania brieyi* DE WILD. (13) (Rubiacées). *Ann. Pharmacol. Fr.*, **34**, 427–438.

Meyer, B.N., Heinstein, P.F., Burnouf-Radosevich, M., Delfel, N.E. and McLaughlin, J.L. (1990). Bioactivity-directed isolation and characterization of quinoside A: one of the toxic/bitter principles of quinoa seeds (*Chenopodium quinoa* WILLD.). *J. Agric. Food Chem.*, **38**, 205–208.

Mimaki, Y. and Sashida, Y. (1990a). Steroidal saponins from the bulbs of *Lilium brownii. Phytochemistry*, **29**, 2267–2271.

Mimaki, Y. and Sashida, Y. (1990b). Studies on the chemical constituents of the bulbs of *Fritillaria camtschatcensis. Chem. Pharm. Bull.*, **38**, 1090–1092.

Mimaki, Y. and Sashida, Y. (1990c). Steroidal saponins and alkaloids from the bulbs of *Lilium brownii* var. *colchesteri. Chem. Pharm. Bull.*, **38**, 3055–3059.

Mimaki, Y. and Sashida, Y. (1991). Steroidal and phenolic constituents of *Lilium speciosum. Phytochemistry*, **30**, 937–940.

Mimaki, Y., Sashida, Y. and Shimomura, H. (1989). Lipid and steroidal constituents of *Lilium auratum* var. *platyphyllum* and *L. tenuifolium. Phytochemistry*, **28**, 3453–3458.

Mimaki, Y., Sashida, Y. and Kawashima, K. (1991). Steroidal saponins from the bulbs of *Camassia cusickii. Phytochemistry*, **30**, 3721–3727.

Mimaki, Y., Ori, K., Kubo, S. *et al.* (1992). Scillasaponins A, B and C, new triterpenoid oligosaccharides from the plants of the subfamily Scilloideae. *Chem. Lett.*, 1863–1866.

Min, Z. and Qin, K. (1984). A new triterpene glucoside of *Ilex hainanensis* MERR. *Yaozue Xuebao*, **19**, 691–696 (*Chem. Abstr.*, **102**, 59302).

Minale, L., Pizza, C., Riccio, R. and Zollo, F. (1982). Steroidal oligoglycosides from starfishes. *Pure Appl. Chem.*, **54**, 1935–1950.

Minale, L., Riccio, R. and Zollo, F. (1993). Steroidal glycosides and polyhydroxysteroids from echinoderms. *Prog. Chem. Org. Nat. Prod.*, **62**, 75–308.

Minghua, C., Ruilin, N., Nagasawa, H., Isogai, A., Jun, Z. and Suzuki, A. (1992). A dammarane saponin from *Neoalsomitra integrifoliola. Phytochemistry*, **31**, 2451–2453.

Minocha, P.K. and Tandon, R.N. (1980). A new triterpene glycoside from the stem of *Ichnocarpus frutescens. Phytochemistry*, **19**, 2053–2055.

Minocha, P.K. and Tiwari, K.P. (1981). A triterpenoidal saponin from roots of *Acanthus illicifolius*. *Phytochemistry*, **20**, 135–137.

Mishra, M. and Srivastava, S.K. (1985). Betulinic acid-3-*O*-β-D-maltoside from *Acacia leucophloea* WILD. *Indian J. Pharm. Sci.*, **47**, 154–155.

Mitra, A.K. and Karrer, P. (1953). Ueber ein neues Vorkommen von Hederagenin in *Holboellia latifolia* WALL. *Helv. Chim. Acta*, **36**, 1401.

Miyahara, K. and Kawasaki, T. (1969). Structure of tokoronin. *Chem. Pharm. Bull.*, **17**, 1369–1376.

Miyahara, K., Isozaki, F. and Kawasaki, T. (1969). Isolation and structure of a new tokorogenin glycoside. *Chem. Pharm. Bull.*, **17**, 1735–1739.

Miyahara, K., Kudo, K. and Kawasaki, T. (1983). Co-occurrence and high-performance liquid chromatographic separation of the glycosides of rhodea-sapogenin and its analogs which differ in the F-ring structure. *Chem. Pharm. Bull.*, **31**, 348–351.

Miyakoshi, M., Kasai, R., Nishioka, M., Ochiai, H. and Tanaka, O. (1990). Solubilising effect and inclusion reaction of cyclic bisdesmosides from tubers of *Bolbostemma paniculatum*. *Yakugaku Zasshi*, **110**, 943–949.

Miyakoshi, M., Ida, Y., Isoda, S. and Shoji, J. (1993). 3-*Epi*-oleanene type triterpene glycosyl esters from leaves of *Acanthopanax spinosus*. *Phytochemistry*, **33**, 891–895.

Miyamoto, T., Togawa, K., Higuchi, R., Komori, T. and Sasaki, T. (1990). Six newly identified biologically active triterpenoid glycoside sulphates from the sea cucumber *Cucumaria echinata*. *Annalen*, 453–460.

Miyase, T., Kohsaka, H. and Ueno, A. (1992). Tragopogonosides A-I, oleanane saponins from *Tragopogon pratensis*. *Phytochemistry*, **31**, 2087–2091.

Mizoguchi, Y., Ikemoto, Y., Arai, T., Yamamoto, S. and Morisawa, S. (1984). Effect of glycyrrhizin on the antibody production of pokeweed mitogen-stimulated lymphocytes. *Arerugi*, **33**, 328–335 (*Chem. Abstr.*, **101**, 183642).

Mizui, F., Kasai, R., Ohtani, K. and Tanaka, O. (1988). Saponins from brans of quinoa, *Chenopodium quinoa* WILLD. I. *Chem. Pharm. Bull.*, **36**, 1415–1418.

Mizui, F., Kasai, R., Ohtani, K. and Tanaka, O. (1990). Saponins from brans of quinoa, *Chenopodium quinoa* WILLD. II. *Chem. Pharm. Bull.*, **38**, 375–377.

Mizuno, M., Tan, R.-X., Zhen, P., Min, Z.-D., Iinuma, M. and Tanaka, T. (1990). Two steroidal alkaloid glycosides from *Veratrum taliense*. *Phytochemistry*, **29**, 359–361.

Mizutani, K., Ohtani, K., Wei, J.-X., Kasai, R. and Tanaka, O. (1984). Saponins from *Anemone rivularis*. *Planta Medica*, **50**, 327–331.

Monder, C., Stewart, P.M., Lakshmi, V., Valentino, R., Burt, D. and Edwards, C.R.W. (1989). Licorice inhibits corticosteroid 11β-dehydrogenase of rat kidney and liver: *in vivo* and *in vitro* studies. *Endocrinology*, **125**, 1046–1053.

Moon, C.K., Kang, N.Y., Yun, Y.P., Lee, S.H., Lee, H.A. and Kang, T.L. (1984). Effect of red ginseng crude saponin on plasma lipid levels in rats fed on a diet high in cholesterol and triglyceride. *Arch. Pharmacal Res.*, **7**, 41–45 (*Chem. Abstr.*, **101**, 204131).

Morein, B. (1988). The iscom antigen-presenting system. *Nature*, **332**, 287–288.

Moreira, E.A., Cecy, C., Nakashima, T., Cavazzani, J.R., Miguel, O.G. and Krambeck, R. (1981). Solasodine in *Solanum erianthum* D. DON. *Cienc. Cult.* (Sao Paolo), **33** (Suppl.) 114–128 (*Chem. Abstr.*, **97**, 195831).

Morgan, M.R.A., Coxon, D.T., Bramham, S., Chan, H.W., van Gelder, W.M.J. and Allison, M.J. (1985). Determination of the glycoalkaloid content of potato tubers by three methods including enzyme-linked immunosorbent assay. *J. Sci. Food Agric.*, **36**, 282–288.

Morisset, R., Côté, N.G., Panisset, J.C., Jemni, L., Camirand, P. and Brodeur, A. (1987). Evaluation of the healing activity of hydrocotyle tincture in the treatment of wounds. *Phytother. Res.*, **1**, 117–121.

Morita, T., Kasai, R., Tanaka, O., Zhou, J., Yang, T.-R. and Shoji, J. (1982). Saponins of Zu-tziseng, rhizomes of *Panax japonicus* C.A. MEYER var. *major* (BURK.) C.Y. WU ET K.M. FENG, collected in Yunnan, China. *Chem. Pharm. Bull.*, **30**, 4341–4346.

Morita, T., Tanaka, O. and Kohda, H. (1985). Saponin composition of rhizomes of *Panax japonicus* collected in South Kyushu, Japan, and its significance in oriental traditional medicine. *Chem. Pharm. Bull.*, **33**, 3852–3858.

Morita, T., Zhou, J. and Tanaka, O. (1986a). Saponins from *Panax pseudo-ginseng* WALL. subsp. *pseudo-ginseng* HARA collected at Nielamu, Tibet, China. *Chem. Pharm. Bull.*, **34**, 4833–4835.

Morita, T., Nie, R.-L., Fujino, H. *et al.* (1986b). Saponins from Chinese cucurbit-aceous plants: solubilization of saikosaponin-a with hemslosides Ma2 and Ma3 and structure of hemsloside H_1 from *Hemsleya chinensis*. *Chem. Pharm. Bull.*, **34**, 401–405.

Morita, T., Ushiroguchi, T., Hayashi, N., Matsuura, H., Itakura, Y. and Fuwa, T. (1988). Steroidal saponins from elephant garlic, bulbs of *Allium ampeloprasum* L. *Chem. Pharm. Bull.*, **36**, 3480–3486.

Morris, S.C. and Lee, T.H. (1984). The toxicity and teratogenicity of Solanaceae glycoalkaloids, particularly those of the potato (*Solanum tuberosum*): a review. *Food Technol. Aust.*, **36**, 118–124.

Mott, K.E. (1987). *Plant Molluscicides.* John Wiley, Chichester.

Mukhamedziev, M.M., Alimbaeva, P.K., Gorovits, T.T. and Abubakirov, N.K. (1971). Structure of the triterpenoid glycoside from *Dipsacus azureus*, dipsa-coside B. *Khim. Prir. Soedin.*, 153–158.

Mukharya, D. (1985). A saponin 'lupeoside' from the seeds of *Canavalia ensiformis*. *Acta Cienc. Indica, Chem.*, **11**, 24–25 (*Chem. Abstr.*, **107**, 214787).

Mukharya, D.K. and Ansari, A.H. (1987). Olean-12-ene-3-*O*-β-D-galactopyranosyl (1→4)-*O*-α-L-rhamnopyranoside: a new triterpenoidal saponin from the roots of *Spilanthes acmella* (MURR.). *Indian J. Chem.*, **26B**, 87.

Mukherjee, K. and Bose, L. (1975). Chemical investigation of *Canthium diococcum*. *J. Indian Chem. Soc.*, **52**, 1112.

Murakami, K., Nohara, T. and Tomimatsu, T. (1984). Studies on the constituents of *Solanum* plants. IV. On the constituents of *Solanum japonense* NAKAI. *Yakugaku Zasshi*, **104**, 195–198.

Murakami, K., Ezima, H., Takaishi, Y. *et al.* (1985). Studies on the constituents of *Solanum* plants. V. The constituents of *S. lyratum* THUNB. II. *Chem. Pharm. Bull.*, **33**, 67–73.

Murakami, T., Nagasawa, M., Urayama, S. and Satake, T. (1986). New triter-penoid saponins in the rhizome and roots of *Caulophyllum robustum* MAXIM. *Yakugaku Zasshi*, **88**, 321–324.

Murata, T., Imai, S., Imanishi, M. and Goto, M. (1970). Anti-microbial glycosides of *Euptelea polyandra* SIEB. ET ZUCC. II. The structure of eupteleogenin. *Yakugaku Zasshi*, **90**, 744–751.

Mustich, G. (1975). Terpenes. *Ger. Offen.*, 2,503,135 (*Chem. Abstr.*, **84**, 59791).

Muto, Y., Takagi, K., Kitagawa, O. and Kumagai, K. (1987). Triterpenoid saponins as antiulcer agents. Jap. Pat. 62 205,025 (*Chem. Abstr.*, **108**, 173547).

Mzhelskaya, L.G. and Abubakirov, N.K. (1967a). Triterpene glycosides of *Leontice eversmanni*. II. The structure of leontoside A and leontoside B. *Khim. Prir. Soedin.*, 101–105.

Mzhelskaya, L.G. and Abubakirov, N.K. (1967b). The structure of leontoside E. *Khim. Prir. Soedin.*, 218–219.

Mzhelskaya, L.G. and Abubakirov, N.K. (1968a). Triterpenic glycosides of *Leontice eversmanni*. IV. Structure of leontoside D. *Khim. Prir. Soedin.*, 153–158.

Mzhelskaya, L.G. and Abubakirov, N.K. (1968b). Triterpenic glycosides of *Leontice eversmanni*. V. Structure of leontoside C. *Khim. Prir. Soedin.*, 216–221.

Nabata, H., Saito, H. and Takagi, K. (1973). Pharmacological studies of neutral saponins (GNS) of *Panax ginseng* root. *Jpn J. Pharmacol.*, **23**, 29–41.

Nagamoto, N., Noguchi, H., Itokawa, A. *et al.* (1988). Antitumour constituents from bulbs of *Crocosmia crocosmiiflora*. *Planta Medica*, **54**, 305–307.

Nagao, T. and Okabe, H. (1992). Studies on the constituents of *Aster scaber* THUNB. III. Structures of scaberosides B_7, B_8 and B_9, minor oleanolic acid glycosides isolated from the root. *Chem. Pharm. Bull.*, **40**, 886–888.

Nagao, T., Okabe, H., Mihashi, K. and Yamauchi, T. (1989a). Studies on the constituents of *Thladiantha dubia* BUNGE. I. The structure of dubiosides A, B and C, the quillaic acid glucuronide saponins isolated from the tuber. *Chem. Pharm. Bull.*, **37**, 925–929.

Nagao, T., Hachiyama, S., Oka, H. and Yamauchi, T. (1989b). Studies on the constituents of *Aster tataricus* L. f. II. Structures of aster saponins isolated from the root. *Chem. Pharm. Bull.*, **37**, 1977–1983.

Nagao, T., Okabe, H. and Yamauchi, T. (1990). Studies on the constituents of *Aster tataricus* L. f. II. Structures of aster saponins E and F, isolated from the root. *Chem. Pharm. Bull.*, **38**, 783–785.

Nagao, T., Tanaka, R., Iwase, Y., Hanazono, H. and Okabe, H. (1991). Studies on the constituents of *Luffa acutangula* ROXB. I. Structures of acutosides A–G, oleanane-type triterpene saponins isolated from the herb. *Chem. Pharm. Bull.*, **39**, 599–606.

Nagata, T. (1986). Useful components of tea in leaves of the genus *Camellia*. *Chagyo Shikenjo Kenkyu Hokoku*, 59–120 (*Chem. Abstr.*, **107**, 93436).

Nagata, T., Tsushida, T., Hamaya, E., Enoki, N., Manabe, S. and Nishino, C. (1985). Camellidins, antifungal saponins isolated from *Camellia japonica*. *Agric. Biol. Chem.*, **49**, 1181–1186.

Nagumo, S., Kishi, S., Inoue, T. and Nagai, M. (1991). Saponins of Anemarrhenae rhizoma. *Yakugaku Zasshi*, **111**, 306–310.

Nakai, S., Takagi, N., Miichi, H. *et al.* (1984). Pfaffosides, nortriterpenoid saponins from *Pfaffia paniculata*. *Phytochemistry*, **23**, 1703–1705.

Nakanishi, K., Goto, T., Ito, S., Natori, S. and Nozoe, S. (1983). ^{13}C Nuclear magnetic resonance data for triterpenoids. In *Natural Products Chemistry*, vol. 3, pp. 179–184. Oxford University Press, Oxford.

Nakanishi, T., Terai, H., Nasu, M., Miura, I. and Yoneda, K. (1982). Two triterpenoid glycosides from leaves of *Ilex cornuta*. *Phytochemistry*, **21**, 1373–1377.

Nakano, K., Matsuda, E., Tsurumi, K. *et al.* (1988). The steroidal glycosides of the flowers of *Yucca gloriosa*. *Phytochemistry*, **27**, 3235–3239.

Nakano, K., Murakami, K., Takaishi, Y., Tomimatsu, T. and Nohara, T. (1989a). Studies on the constituents of *Heloniopsis orientalis* (THUNB.) C. TANAKA. *Chem. Pharm. Bull.*, **37**, 116–118.

Nakano, K., Yamasaki, T., Imamura, Y., Murakami, K., Takaishi, Y. and Tomimatsu, T. (1989b). The steroidal glycosides from the caudex of *Yucca gloriosa*. *Phytochemistry*, **28**, 1215–1217.

Nakano, K., Midzuta, Y., Hara, Y., Murakami, K., Takaishi, Y. and Tomimatsu, T. (1991a). 12-Keto steroidal glycosides from the caudex of *Yucca gloriosa*. *Phytochemistry*, **30**, 633–636.

Nakano, K., Hara, Y., Murakami, K., Takaishi, Y. and Tomimatsu, T. (1991b). 12-Hydroxy steroidal glycosides from the caudex of *Yucca gloriosa*. *Phytochemistry*, **30**, 1993–1995.

Nakano, T., De Azcunes, B.C. and Spinelli, A.C. (1976). Chemical studies on the constituents of *Lupinus verbasciformis*. *Planta Medica*, **29**, 241–246.

Nakashima, H., Okubo, K., Honda, Y., Tamura, T., Matsuda, S. and Yamamoto, N. (1989). Inhibitory effect of glycosides like saponin from soybean on the infectivity of HIV *in vitro*. *AIDS*, **3**, 655–658.

Nakatani, M., Hatanaka, S., Komura, H., Kubota, T. and Hase, T. (1989). The structure of rotungenoside, a new bitter triterpene glucoside from *Ilex rotunda*. *Bull. Chem. Soc. Jpn*, **62**, 469–473.

Nakayama, K., Fujino, H., Kasai, R., Tanaka, O. and Zhou, J. (1986a). Saponins of pericarps of Chinese *Sapindus delavayi* (Pyi-shiau-tzu), a source of natural surfactants. *Chem. Pharm. Bull.*, **34**, 2209–2213.

Nakayama, K., Fujino, H., Kasai, R., Mitoma, Y., Yota, N. and Tanaka, O. (1986b). Solubilizing properties of saponins from *Sapindus mukurossi* GAERTN. *Chem. Pharm. Bull.*, **34**, 3279–3283.

Namba, T., Matsushige, K., Morita, T. and Tanaka, O. (1986). Saponins of plants of *Panax* species collected in central Nepal and their chemotaxonomical significance. I. *Chem. Pharm. Bull.*, **34**, 730–788.

Namba, T., Huang, X.-L., Shu, Y.-Z. *et al.* (1989). Chronotropic effect of the methanolic extracts of the plants of the *Paris* species and steroidal glycosides isolated from *P. vietnamensis* on spontaneous beating of myocardial cells. *Planta Medica*, **55**, 501–505.

Nandy, A.K., Podder, G., Sahu, N.P. and Mahato, S.B. (1989). Triterpenoids and their glucosides from *Terminalia bellerica*. *Phytochemistry*, **28**, 2769–2772.

Naplatanova, D. (1984). Effect of some factors on the production of tribestan tablets. *Farmatsiya* (Sofia), **34**, 29–32 (*Chem. Abstr.*, **101**, 216342).

Neszmelyi, A., Tori, K., Lukacs, G. (1977). Use of carbon-13 spin-lattice relaxation times for sugar sequence determination in steroidal oligosaccharides. *J. Chem. Soc., Chem. Commun.*, 613–614.

Ng, T.B. and Yeung, H.W. (1986). Scientific basis of the therapeutic effects of ginseng. In *Folk Medicine* ed. R.P. Steiner, pp. 139–151. American Chemical Society, Washington.

Nicollier, G. and Thompson, A.C. (1983). A new triterpenoid saponin from the flowers of *Melilotus alba*, white sweet clover. *J. Nat. Prod.*, **46**, 183–186.

Nie, R.-L., Morita, T., Kasai, R., Zhou, J., Wu, C.-Y. and Tanaka, O. (1984). Saponins from Chinese medicinal plants. (1). Isolation and structures of hemslosides. *Planta Medica*, **50**, 322–327.

Nie, R.-L., Tanaka, T., Miyakoshi, M. *et al.* (1989). A triterpenoid saponin from *Thladiantha hookeri* var. *pentadactyla*. *Phytochemistry*, **28**, 1711–1715.

Nigam, S.K., Li, X.-C., Wang, D.-Z., Misra, G. and Yang, C.-R. (1992). Triterpenoidal saponins from *Madhuca butyracea*. *Phytochemistry*, **31**, 3169–3172.

Nishie, K., Gumbmann, M.R. and Keyl, A.C. (1971). Pharmacology of solanine. *Toxicol. Appl. Pharmacol.*, **19**, 81–92.

Nishimoto, N., Nakai, S., Takagi, N. *et al.* (1984). Pfaffosides and nortriterpenoid saponins from *Pfaffia paniculata*. *Phytochemistry*, **23**, 139–142.

Nishimoto, N., Shiobara, Y., Fujino, M. *et al.* (1987). Ecdysteroids from *Pfaffia iresinoides* and reassignment of some ^{13}C NMR chemical shifts. *Phytochemistry*, **26**, 2505–2507.

Nishino, C., Manabe, S., Enoki, N., Nagata, T., Tsushida, T. and Hamaya, E. (1986). The structure of the tetrasaccharide unit of camellidins, saponins possessing antifungal activity. *J. Chem. Soc., Chem. Commun.*, 720–723.

Nishizawa, M., Nishide, H., Kosela, S. and Hayashi, Y. (1983). Structure of lansiosides: biologically active new triterpene glycosides from *Lansium domesticum*. *J. Org. Chem.*, **48**, 4462–4466.

Noguchi, Y., Higuchi, R., Marubayashi, N. and Komori, T. (1987). Steroid oligoglycosides from the starfish *Asterina pectinifera* MULLER ET TROSCHEL, 1. Structures of two new sapogenins and two new oligoglycoside sulphates: pectinioside A and pectinioside B. *Annalen*, 341–348.

Nohara, T., Miyahara, K. and Kawasaki, T. (1975). Steroid saponins and sapogenins of underground parts of *Trillium kamtschaticum* PALL. II. Pennogenin- and kryptogenin 3-O-glycosides and related compounds. *Chem. Pharm. Bull.*, **23**, 872–885.

Nohara, T., Komori, T. and Kawasaki, T. (1980). Steroid saponins and sapogenins of underground parts of *Trillium kamtschaticum* PALL. III. On the structure of a novel type of steroid glycoside, trillenoside A, an 18-norspirostanol oligoside. *Chem. Pharm. Bull.*, **28**, 1437–1448.

Nohara, T., Ito, Y., Seike, H. *et al.* (1982). Studies on the constituents of *Paris quadrifolia* L. *Chem. Pharm. Bull.*, **30**, 1851–1856.

Nose, M., Ametani, S. and Ogiwara, Y. (1988a). Preparation of saponin derivatives from saikosaponin c and antiinflammatory agents containing them. Jap. Pat. 63 179, 886 (*Chem. Abstr.*, **110**, 8587).

Nose, M., Ametani, S. and Ogiwara, Y. (1988b). Preparation of saponin derivatives from saikosaponin c and antiinflammatory agents containing them. Jap. Pat. 63 179, 887 (*Chem. Abstr.*, **110**, 8588).

Nose, M., Amagaya, S. and Ogihara, Y. (1989a). Corticosterone secretion-inducing activity of saikosaponin metabolites formed in the alimentary tract. *Chem. Pharm. Bull.*, **37**, 2736–2740.

Nose, M., Amagaya, S. and Ogihara, Y. (1989b). Effects of saikosaponin metabolism on the hemolysis of red blood cells and their adsorbability on the cell membrane. *Chem. Pharm. Bull.*, **37**, 3306–3310.

Nowacka, J. and Oleszek, W. (1992). High performance liquid chromatography of zanhic acid glycoside in alfalfa (*Medicago sativa*). *Phytochem. Anal.*, **3**, 227–230.

Numata, A., Kitajima, A., Katsuno, T. *et al.* (1987). An antifeedant for the yellow butterfly larvae in *Camellia japonica*: a revised structure of camellidin II. *Chem. Pharm. Bull.*, **35**, 3948–3951.

Oakenfull, D. (1981). Saponins in food. *Food Chem.*, **6**, 19–40.

Oakenfull, D. (1986). Aggregation of saponins and bile acids in aqueous solution. *Aust. J. Chem.*, **39**, 1671–1683.

Oakenfull, D. and Topping, D.L. (1983). *Atherosclerosis*, **48**, 301.

Odani, T., Tanizawa, H. and Takino, Y. (1983). Studies on the absorption, distribution, excretion and metabolism of ginseng saponins. II. The absorption, distribution and excretion of ginsenoside Rg_1 in the rat. *Chem. Pharm. Bull.*, **31**, 292–298.

Odani, T., Ushio, Y. and Arichi, S. (1986). The effect of ginsenosides on adrenocorticotropin secretion in primary culture of rat pituitary cells. *Planta Medica*, **52**, 177–179.

O'Dell, B.L., Regam, W.O. and Beach, T.J. (1959). Toxic principle in red clover. *Missouri Univ. Agric. Exp. Sta. Res. Bull.*, **702**, 12.

Ogihara, Y. and Nose, M. (1986). Novel cleavage of the glycosidic bond of saponins in alcoholic alkali metal solution containing a trace of water. *J. Chem. Soc., Chem. Commun.*, 1417.

Ogihara, Y., Chen, Y. and Kobayashi, Y. (1987). A new prosapogenin from hovenia saponin D by mild alkaline degradation. *Chem. Pharm. Bull.*, **35**, 2574–2575.

Ogino, T., Hayasaka, T., Ito, S. and Takahashi, T. (1968). Triterpenic constituents of *Stewartia monadelpha* SIEB. ET ZUCC. Camelliagenin A 16-cinnamate as a sapogenin. *Chem. Pharm. Bull.*, **16**, 1846–1847.

Ogunkoya, L. (1977). Application of the method of molecular rotation differences in structural and stereochemical problems in triterpenes. *Tetrahedron*, **33**, 3321–3330.

Ohigashi, H., Jisaka, M., Takagaki, T. *et al.* (1991). Bitter principle and a related steroid glucoside from *Vernonia amygdalina*, a possible medicinal plant for wild chimpanzees. *Agric. Biol. Chem.*, **55**, 1201–1203.

Ohtani, K. and Hostettmann, K. (1992). Molluscicidal and antifungal triterpenoid saponins from the leaves of *Rapanea melanophloeos*. *Planta Medica*, **58**, A708.

Ohtani, K., Mizutani, K., Kasai, R. and Tanaka, O. (1984). Selective cleavage of ester type glycoside-linkages and its application to structure determination of natural oligoglycosides. *Tetrahedron Lett.*, 4537–4540.

Ohtani, K., Miyajima, C., Takahasi, T. *et al.* (1990). A dimeric triterpene-glycoside from *Rubus coreanus*. *Phytochemistry*, **29**, 3275–3280.

Ohtani, K., Aikawa, Y., Kasai, R., Chou, W.-H., Yamasaki, H. and Tanaka, O. (1992a). Minor diterpene glycosides from sweet leaves of *Rubus suavissimus*. *Phytochemistry*, **31**, 1553–1559.

Ohtani, K., Ogawa, K., Kasai, R. *et al.* (1992b). Oleanane glycosides from *Glycyrrhiza yunnanensis* roots. *Phytochemistry*, **31**, 1747–1752.

Ohtani, K., Mavi, S. and Hostettmann, K. (1993). Molluscicidal and antifungal triterpenoid saponins from *Rapanea melanophloeos* leaves. *Phytochemistry*, **33**, 83–86.

Okabe, H., Miyahara, Y., Yamauchi, T., Miyahara, K. and Kawasaki, T. (1980). Studies on the constituents of *Momordica charantia* L. I. Isolation and characterization of momordicosides A and B, glycosides of a pentahydroxy-cucurbitane triterpene. *Chem. Pharm. Bull.*, **28**, 2753–2762.

Okabe, H., Nagao, T., Hachiyama, S. and Yamauchi, T. (1989). Studies on the constituents of *Luffa operculata* COGN. II. Isolation and structure elucidation of saponins in the herb. *Chem. Pharm. Bull.*, **37**, 895–900.

Okada, Y., Koyama, K., Takahashi, K., Okuyama, T. and Shibata, S. (1980). Gleditsia saponins I. Structures of monoterpene moieties of gleditsia saponin C. *Planta Medica*, **40**, 185–192.

Okada, Y., Shibata, S., Ikekawa, T., Javellana, A.M.J. and Kamo, O. (1987). Entada saponin-III, a saponin isolated from the bark of *Entada phaseoloides*. *Phytochemistry*, **26**, 2789–2796.

Okada, Y., Shibata, S., Javellana, A.M.J. and Kamo, O. (1988a). Entada saponins (ES) II and IV from the bark of *Entada phaseoloides*. *Chem. Pharm. Bull.*, **36**, 1264–1269.

Okada, Y., Shibata, S., Kamo, O. and Okuyama, T. (1988b). ^{13}C NMR spectral studies of entagenic acid to establish its structure. *Chem. Pharm. Bull.*, **36**, 5028–5030.

Okamura, N., Nohara, T., Yagi, A. and Nishioka, I. (1981). Studies on constituents of Zizyphi fructus. III. Structures of dammarane-type saponins. *Chem. Pharm. Bull.*, **29**, 676–683.

Okanishi, T., Akahori, A. and Yasuda, F. (1965). Studies on the steroidal components of domestic plants. XLVII. Constituents of the stem of *Smilax sieboldi* MIQ. (1). The structure of laxogenin. *Chem. Pharm. Bull.*, **13**, 545–550.

Okano, K., Ohkawa, N. and Ikegami, S. (1985). Structure of ovarian astero-saponin-4, an inhibitor of spontaneous oocyte maturation from the starfish *Asterias amurensis*. *Agric. Biol. Chem.*, **49**, 2823–2826.

Okuda, H. and Yoshida, R. (1980). Studies on the effects of ginseng components on diabetes mellitus. *Proceedings of the Third International Ginseng Symposium*, pp. 53–57. Korea Ginseng Research Institute, Seoul, Korea.

Okunji, C.O., Okeke, C.N., Gugnani, H.C. and Iwu, M.M. (1990). An antifungal spirostanol saponin from fruit pulp of *Dracaena mannii*. *Int. J. Crude Drug Res.*, **28**, 193–199.

Okunji, C.O., Iwu, M.M. and Hostettmann, K. (1991). Molluscicidal saponins from the fruit pulp of *Dracaena mannii*. *Int. J. Pharmacognosy*, **29**, 66–70.

Okuyama, T., Takata, M., Takahashi, K., Ishikawa, T., Miyasaka, K. and Kaneyama, N. (1989). High-performance liquid chromatographic analysis of naturally occurring glycosides and saponins. *J. Chromatogr.*, **466**, 390–398.

Oleszek, W., Shannon, S. and Robinson, R.W. (1986). Steroidal alkaloids of *Solanum lycopersicoides*. *Acta Soc. Bot. Pol.*, **55**, 653–657 (*Chem. Abstr.*, **107**, 4309).

Oleszek, W., Jurzysta, M., Price, K.R. and Fenwick, G.R. (1990). High-performance liquid chromatography of alfalfa root saponins. *J. Chromatogr.*, **519**, 109–116.

Oleszek, W., Jurzysta, M., Ploszynski, M., Colquhoun, I.J., Price, K.R. and Fenwick, G.R. (1992). Zanhic acid tridesmoside and other dominant saponins from alfalfa (*Medicago sativa* L.) aerial parts. *J. Agric. Food Chem.*, **40**, 191–196.

Oliveira, M.M. and Andrade, S.O. (1961). Toxic saponin from *Elvira biflora*. *J. Pharm. Sci.*, **50**, 780–782.

Onken, D. and Onken, D. (1980). Zur Rohstoffproblematik der Steroidhormonsynthesen. *Pharmazie*, **35**, 193–198.

Ono, M., Hamada, T. and Nohara, T. (1986). An 18-norspirostanol glycoside from *Trillium tschonoskii*. *Phytochemistry*, **25**, 544–545.

Oobayashi, K., Yoshikawa, K. and Arihara, S. (1992). Structural revision of bryonoside and structure elucidation of minor saponins from *Bryonia dioica*. *Phytochemistry*, **31**, 943–946.

Ori, K., Mimaki, Y., Sashida, Y., Nikaido, T. and Ohmoto, T. (1992). Steroidal alkaloids from the bulbs of *Fritillaria persica*. *Phytochemistry*, **31**, 4337–4341.

Orsini, F. and Verotta, L. (1985). Separation of natural polar substances by reversed-phase high-performance liquid chromatography, centrifugal thin-layer chromatography and droplet counter-current chromatography. *J. Chromatogr.*, **349**, 69–75.

Orsini, F., Pelizzoni, F. and Verotta, L. (1986). Quadranguloside, a cycloartane triterpene glycoside from *Passiflora quadrangularis*. *Phytochemistry*, **25**, 191–193.

Orsini, F., Pelizzoni, F., Ricca, G. and Verotta, L. (1987). Triterpene glycosides related to quadranguloside from *Passiflora quadrangularis*. *Phytochemistry*, **26**, 1101–1105.

Orsini, F., Pelizzoni, F. and Verotta, L. (1991). Saponins from *Albizzia lucida*. *Phytochemistry*, **30**, 4111–4115.

Osborne, R. and Pegel, K.H. (1984). Jessic acid and related acid triterpenoids from *Combretum elaeagnoides*. *Phytochemistry*, **23**, 635–637.

Oshima, Y., Ohsawa, T., Oikawa, K., Konno, C. and Hikino, H. (1984a). Structures of dianosides A and B, analgesic principles of *Dianthus superbus* var. *longicalycinus* herbs. *Planta Medica*, **50**, 40–47.

Oshima, Y., Ohsawa, T. and Hikino, H. (1984b). Structure of dianosides C, D, E and F, triterpenoid saponins of *Dianthus superbus* var. *longicalycinus* herb. *Planta Medica*, **50**, 43–47.

Oshima, Y., Ohsawa, T. and Hikino, H. (1984c). Structures of dianosides G, H and I, triterpenoid saponins of *Dianthus superbus* var. *longicalycinus* herbs. *Planta Medica*, **50**, 254–258.

Osman, S.F. (1984). Steroidal glycoalkaloid biosynthesis and function in *Solanum* spp. In *Isopentenoids in Plants*, ed. W.D. Nes, G. Fuller and L.-S. Tsai, pp. 519–530. Marcel Dekker, New York.

Osman, S.F. and Sinden, S.L. (1982). The glycoalkaloids of *Solanum demissum*. *Phytochemistry*, **21**, 2763–2764.

Osman, S.F., Herb, S.F., Fitzpatrick, T.J. and Sinden, S.L. (1976). Commersonine, a new glycoalkaloid from two *Solanum* species. *Phytochemistry*, **15**, 1065–1067.

Otsuka, H., Morita, Y., Ogihara, Y. and Shibata, S. (1977). The evaluation of ginseng and its congeners by droplet counter-current chromatography (DCC). *Planta Medica*, **32**, 9–17.

Otsuka, H., Akiyama, T., Kawai, K., Shibata, S., Inoue, O. and Ogihara, Y. (1978a). The structure of jujubosides A and B, the saponins isolated from the seeds of *Zizyphus jujuba*. *Phytochemistry*, **17**, 1349–1352.

Otsuka, H., Kobayashi, S. and Shibata, S. (1978b). Separation and determination of saponins of *Bupleuri radix* by droplet counter current chromatography (DCC). *Planta Medica*, **33**, 152–159.

Pailer, M. and Berner, H. (1967). Constitution of tormentol (tormentoside). *Monatsh. Chem.*, **98**, 2082–2088.

Pakrashi, A., Ray, H., Pal, B.C. and Mahato, S.B. (1991). Sperm immobilizing effect of triterpene saponins from *Acacia auriculiformis*. *Contraception*, **43**, 475–483.

Pal, B.C. and Mahato, S.B. (1987). New triterpenoid pentasaccharides from *Androsace saxifragifolia*. *J. Chem. Soc., Perkin Trans.* 1, 1963–1967.

Pal, B.C., Achari, B. and Price, K.R. (1991). A triterpenoid glucoside from *Barringtonia acutangula*. *Phytochemistry*, **30**, 4177–4179.

Pambou Tchivounda, H., Koudogbo, B., Tabet, J.C. and Casadevall, E. (1990). A triterpene saponin from *Cylicodiscus gabunensis*. *Phytochemistry*, **29**, 2723–2725.

Pambou Tchivounda, H., Koudogbo, B., Besace, Y. and Casadevall, E. (1991). Triterpene saponins from *Cylicodiscus gabunensis*. *Phytochemistry*, **30**, 2711–2716.

Pang, P.K.T., Wang, L.C.H., Benishin, C.G. and Liu, H.J. (1991). Ginsenosides Rb_1 and Rg_1 in composition and method for treatment of Alzheimer-type senile dementia, and purification and enrichment of ginsenoside Rb_1. *Chem. Abstr.*, **114**, 17584.

Pant, G. (1988). On the spermicidal activity of the constituents of rhizomes of *Agave cantala*. *Fitoterapia*, **59**, 340–341.

Pant, G., Sati, O.P., Miyahara, K. and Kawasaki, T. (1986). Spirostanol glycosides from *Agave cantala*. *Phytochemistry*, **25**, 1491–1494.

Pant, G., Sati, O.P., Miyahara, K. and Kawasaki, T. (1987). Search for molluscicidal agents: saponins from *Agave cantala* leaves. *Int. J. Crude Drug Res.*, **25**, 35–38.

Pant, G., Panwar, M.S., Negi, D.S., Rawat, M.S.M. and Morris, G.A. (1988a). Spirostanol glycoside from fruits of *Asparagus officinalis*. *Phytochemistry*, **27**, 3324–3325.

Pant, G., Panwar, M.S., Rawat, M.S.M. and Negi, D.S. (1988b). Spermicidal glycosides from *Hedera nepalensis* K. KOCH (inflorescence). *Pharmazie*, **43**, 294.

Pant, G., Panwar, M.S., Negi, D.S. and Rawat, M.S.M. (1988c). Spermicidal activity and chemical analysis of *Pentapanax leschenaultii*. *Planta Medica*, **54**, 477.

Pant, G., Panwar, M.S., Negi, D.S., Rawat, M.S.M., Morris, G.A. and Thompson, R.I.G. (1988d). Structure elucidation of a spirostanol glycoside from

Asparagus officinalis fruits by concerted use of two-dimensional NMR techniques. *Mag. Reson. Chem.*, **26**, 911–918.

Paphassarang, S., Raynaud, J. and Lussignol, M. (1988). La polysciasaponine P₇ de *Polyscias scutellaria* (BURM. F.) FOSB. (Araliaceae). *Pharmazie*, **43**, 296–297.

Paphassarang, S., Raynaud, J. and Lussignol, M. (1989a). Triterpenoid saponins from *Polyscias scutellaria*. *J. Nat. Prod.*, **52**, 239–242.

Paphassarang, S., Raynaud, J., Lussignol, M. and Becchi, M. (1989b). Triterpenic glycosides from *Polyscias scutellaria*. *Phytochemistry*, **28**, 1539–1541.

Paphassarang, S., Raynaud, J., Lussignol, M. and Cabalion, P. (1989c). Sur une nouvelle saponine des feuilles de *Polyscias scutellaria* (BURM. F. FOSB.) (Araliaceae). *Pharmazie*, **44**, 580–581.

Paphassarang, S., Raynaud, J. and Lussignol, M. (1990). A new oleanolic glycoside from *Polyscias scutellaria*. *J. Nat. Prod.*, **53**, 163–166.

Parente, J.P. and Mors, W.B. (1980). Derrissaponin, a new hydrophilic constituent of Timbo-urucu. *An. Acad. Brasil. Cienc.*, **52**, 503–514 (*Chem. Abstr.*, **94**, 99764).

Parkhurst, R.M., Thomas, D.W., Skinner, W.A. and Cary, L.W. (1973a). Molluscicidal saponins of *Phytolacca dodecandra*. Oleanoglycotoxin-A. *Phytochemistry*, **12**, 1437–1442.

Parkhurst, R.M., Thomas, D.W., Skinner, W.A. and Cary, L.W. (1973b). Molluscicidal saponins of *Phytolacca dodecandra*, lemmatoxin C. *Indian J. Chem.*, **11**, 1192–1195.

Parkhurst, R.M., Thomas, D.W., Skinner, W.A. and Cary, L.W. (1974). Molluscicidal saponins of *Phytolacca dodecandra*. Lemmatoxin. *Can. J. Chem.*, **52**, 702–705.

Parkhurst, R.M., Thomas, D.W., Cary, L.W. and Reist, E.J. (1980). A new triterpene lactone from *Gymnocladus dioica*. *Phytochemistry*, **19**, 273–275.

Parnell, A., Bhuva, V.S. and Bintcliffe, E.J.B. (1984). The glycoalkaloid content of potato varieties. *J. Natl. Inst. Agric. Bot.*, **16**, 535–541 (*Chem. Abstr.*, **102**, 201214).

Parrilli, M., Adinolfi, M. and Mangoni, L. (1979). Carbon-13 NMR study of two spirocyclic furanoid nortriterpenes from *Muscari comosum*. *Gazz. Chim. Ital.*, **109**, 611–613.

Patel, A.V., Blunden, G., Crabb, T.A., Sauvaire, Y. and Baccou Y.C. (1987). A review of naturally-occurring steroidal sapogenins. *Fitoterapia*, **58**, 67–107.

Patkhullaeva, M., Mzhelskaya, L.G. and Abubakirov, N.K. (1972). Triterpenoid glycosides of *Ladyginia bucharica*. II. Structure of ladyginosides A and B. *Khim. Prir. Soedin.*, 466–471.

Patra, A., Mitra, A.K., Ghosh, S., Ghosh, A. and Barua, A.K. (1981). Carbon-13 NMR spectroscopy of some pentacyclic triterpenoids. *Org. Magn. Reson.*, **15**, 399–400.

Patterson, M.J., Bland, J. and Lindgren, E.W. (1978). Physiological response of symbiotic polychaetes to host saponins. *J. Exp. Mar. Biol. Ecol.*, **33**, 51–56 (*Chem. Abstr.*, **89**, 194311).

Pedersen, M.W. (1975). Relative quantity and biological activity of saponins in germinated seeds, roots and foliage of alfalfa. *Crop Sci.*, **15**, 541–543.

Pegel, K.H. and Rogers, C.B. (1976). Mollic acid 3β-D-glucoside, a novel

1α-hydroxycycloartane saponin from *Combretum molle* (Combretaceae). *Tetrahedron Lett.*, 4299–4302.

Pegel, K.H. and Rogers, C.B. (1985). The characterisation of mollic acid 3β-D-xyloside and its genuine aglycone mollic acid, two novel 1α-hydroxycycloartenoids. *J. Chem. Soc., Perkin Trans.* 1, 1711–1715.

Pelletier, S.W. and Nakamura, S. (1967). A prosapogenin from *Polygala senega* and *Polygala tenuifolia*. *Tetrahedron Lett.*, 5303–5306.

Penders, A., Delaude, C., Pepermans, H. and van Binst, G. (1989). Identification and sequencing of sugars in an acetylated saponin of *Blighia welwitschii* by N.M.R. spectroscopy. *Carbohyd. Res.*, **190**, 109–120.

Penders, A., Kapundu, M. and Delaude, C. (1992). Structure du principal saponoside de *Alternanthera sessilis*. *Bull. Soc. Chim. Belg.*, **101**, 227–232.

Perera, P., Andersson, R., Bohlin, L. *et al.* (1993). Structure determination of a new saponin from the plant *Alphitonia zizyphoides* by NMR spectroscopy. *Mag. Res. Chem.*, **31**, 472–480.

Peskar, B.M. (1980). Effect of carbenoxolone on prostaglandin synthesizing and metabolizing enzymes and correlation with gastric mucosal carbenoxolone concentrations. *Scand. J. Gastroenterol.*, Suppl. **15**, 109–114 (*Chem. Abstr.*, **95**, 654).

Petersen, T.G. and Palmqvist, B. (1990). Utilizing column selectivity in developing a high-performance liquid chromatographic method for ginsenoside assay. *J. Chromatogr.*, **504**, 139–149.

Peterson, D.H., Murray, H.C., Eppstein, S.H. *et al.* (1952). Microbiological transformations of steroids. I. Introduction of oxygen at carbon-11 of progesterone. *J. Am. Chem. Soc.*, **74**, 5933–5936.

Petricic, J. and Radosevic, A. (1969). Asperosid, a new bisdesmoside 22-hydroxy-furostanol saponin from *Smilax aspera*. *Farmaceutski Glasnik*, **25**, 91–95 (*Chem. Abstr.*, **71**, 64049).

Pettit, G.R., Herald, C.L. and Herald, D.L. (1976). Antineoplastic agents XLV: sea cucumber cytotoxic saponins. *J. Pharm. Sci.*, **65**, 1558–1559.

Pettit, G.R., Doubek, D.L., Herald, D.L. *et al.* (1991). Isolation and structure of cytostatic steroidal saponins from the African medicinal plant *Balanites aegyptiaca*. *J. Nat. Prod.*, **54**, 1491–1502.

Phillips, J.C., Butterworth, K.R., Gaunt, I.F., Evans, J.G. and Grasso, P. (1979). Long-term toxicity study of quillaja extract in mice. *Food Cosmet. Toxicol.*, **17**, 23–27.

Phillipson, J.D. and Anderson, L.A. (1984). Ginseng – quality, safety and efficacy? *Pharmaceut. J.*, 161–165.

Pietta, P., Mauri, P. and Rava, A. (1986). Improved high-performance liquid chromatographic method for the analysis of ginsenosides in *Panax ginseng* extracts and products. *J. Chromatogr.*, **356**, 212–219.

Pietta, P., Mauri, P., Facino, R.M. and Carini, M. (1989). High-performance liquid chromatographic analysis of β-escin. *J. Chromatogr.*, **478**, 259–263.

Pinhas, H. and Bondiou, J.-C. (1967). Sur la constitution chimique de la partie glucidique du madécassoside. *Bull. Soc. Chim. Fr.*, 1888–1895.

Pistelli, L., Bilia, A.R., Marsili, A., De Tommasi, N. and Manunta, A. (1993). Triterpenoid saponins from *Bupleurum fruticosum*. *J. Nat. Prod.*, **56**, 240–244.

Pizza, C., Zhong-Liang, Z. and De Tommasi, N. (1987). Plant metabolites. Triterpenoid saponins from *Calendula arvensis*. *J. Nat. Prod.*, **50**, 927–931.

506   References

Plhak, L.C. and Sporns, P. (1992). Enzyme immunoassay for potato glycoalkaloids. *J. Agric. Food Chem.*, **40**, 2533–2540.

Polacheck, I., Zehavi, U., Naim, M., Levy, M. and Evron, R. (1986). The susceptibility of *Cryptococcus neoformans* to an antimycotic agent (G2) from alfalfa. *Chem. Abstr.*, **105**, 206033.

Polonsky, J. (1968). Bitter principles of the simaroubaceous plants and some constituents of other medicinal plants. *Cienc. Cult.* (Sao Paulo), **20**, 19–31 (*Chem. Abstr.*, **70**, 830).

Polonsky, J. and Zylber, M.J. (1961). Determination of the configuration of the primary hydroxyl of asiatic acid. *Bull. Soc. Chim. Fr.*, 1586–1591.

Polonsky, J., Pourrat, H. and Seiligmann, J. (1960). Study of the constituents of the roots of *Polygala paenea*. The structure of a new sapogenin, polygalacic acid. *C.R. Hebd. Scéances Acad. Sci.*, **251**, 2374–2376.

Pompei, R., Flore, O., Marccialis, M.A., Pani, A. and Loddo, B. (1979). Glycyrrhizic acid inhibits virus growth and inactivates virus particles. *Nature*, **281**, 689–690.

Potter, D.A. and Kimmerer, T.W. (1989). Inhibition of herbivory on young holly leaves: evidence for the defensive role of saponins. *Oecologia*, **78**, 322–329.

Potterat, O., Hostettmann, K., Stoeckli-Evans, H. and Saadou, M. (1992). Saponins with an unusual secoursene skeleton from *Sesamum alatum* THONN. *Helv. Chim. Acta*, **75**, 833–841.

Pourrat, A. and Pourrat, H. (1961). Isolement et structure du tormentoside. *Bull. Soc. Chim. Fr.*, 884.

Pracros, P. (1988). Mesure de l'activité des saponines de la luzerne par les larves du ver de farine: *Tenebrio molitor* L. (Coléoptère, Tenebrionidae). II. Recherche des fractions de saponines responsables des effets antinutritionnels observés. *Agronomie*, **8**, 793–799.

Price, K.R. and Fenwick, G.R. (1984). Soyasaponin I, a compound possessing undesirable taste characteristics isolated from the dried pea (*Pisum sativum* L.). *J. Sci. Food Agric.*, **35**, 887–892.

Price, K.R., Mellon, F.A., Self, R., Fenwick, G.R. and Osman, S.F. (1985). Fast atom bombardment mass spectrometry of *Solanum* glycoalkaloids and its potential for mixture analysis. *Biomed. Mass Spectrom.*, **12**, 79–85.

Price, K.R., Curl, C.L. and Fenwick, G.R. (1986). The saponin content and sapogenol composition of the seed of 13 varieties of legume. *J. Sci. Food Agric.*, **37**, 1185–1191.

Price, K.R., Johnson, I.T. and Fenwick, G.R. (1987). The chemistry and biological significance of saponins in foods and feedingstuffs. *C.R. Food Sci. Nutr.*, **26**, 27–135.

Price, K.R., Eagles, J. and Fenwick, G.R. (1988). Saponin composition of 13 varieties of legume seed using fast atom bombardment mass spectrometry. *J. Sci. Food Agric.*, **42**, 183–193.

Primorac, M., Sekulovic, D. and Antonic, S. (1985). *In vitro* determination of the spermicidal activity of plant saponins. *Pharmazie*, **40**, 585.

Proserpio, G., Gatti, S. and Genesi, P. (1980). Cosmetic uses of horse-chestnut (*Aesculus hippocastanum*) extracts, of escin and of the cholesterol/escin complex. *Fitoterapia*, **51**, 113–128.

Puri, R., Wong, T.C. and Puri, R.K. (1993). Solasodine and diosgenin: ¹H and ¹³C assignments by two-dimensional NMR spectroscopy. *Mag. Res. Chem.*, **31**, 278–282.

Puri, R., Wong, T.C. and Puri, R.K. (1994). ¹H- and ¹³C-NMR assignments and structural determination of a novel glycoalkaloid from *Solanum platanifolium*. *J. Nat. Prod.*, **57**, 587–596.

Purohit, M.C., Pant, G. and Rawat, M.S.M. (1991). A betulinic acid glycoside from *Schefflera venulosa*. *Phytochemistry*, **30**, 2419.

Putieva, Z.M., Kondratenko, E.S. and Abubakirov, N.K. (1974). Triterpene glycosides of *Acanthophyllum paniculata*. II. Structure of paniculatoside C. *Khim. Prir. Soedin.*, 104–105.

Qin, G., Chen, Z., Xu, R., Jiang, Z. and Liang, J. (1987). Chemical constituents of *Ilex pubescens* HOOK ET AM. II. The structure of ilex saponin A. *Huaxue Xuebao*, **45**, 249–255 (*Chem. Abstr.*, **107**, 4283).

Qin, W., Wu, X., Zhao, J., Fukuyama, Y., Yamada, T. and Nakagawa, K. (1986). Triterpenoid glycosides from leaves of *Ilex cornuta*. *Phytochemistry*, **25**, 913–916.

Qiu, F., Chen, C., Chang, X. and Zhou, J. (1982). Steroidal glycoalkaloids of *Notholirion bulbuliferum*. I. *Yunnan Zhiwu Yanjiu*, **4**, 419–423 (*Chem. Abstr.*, **98**, 104286).

Quader, M.A., Gray, A.I., Waterman, P.G., Lavaud, C., Massiot, G., Hasan, C.M. and Ahmed, M.-D. (1990). Capsugenin-25,30-O-β-diglucopyranose: a new glycoside from the leaves of *Corchorus capsularis*. *J. Nat. Prod.*, **53**, 527–530.

Quershi, M.A. and Thakur, R.S. (1977). Chemical constituents of *Randia tetrasperma*. *Planta Medica*, **32**, 229–232.

Racz-Kotilla, E., Petre, M. and Racz, G. (1982). Anti-nociceptive effect of *Platycodon grandiflorum* extracts. *Rev. Med.*, **28**, 180–182 (*Chem. Abstr.*, **100**, 132512).

Radeglia, R., Adam, G. and Ripperger, H. (1977). ¹³C NMR spectroscopy of *Solanum* steroid alkaloids. *Tetrahedron Lett.*, 903–906.

Rakhimov, K.D., Vermenichev, S.M., Lutskii, V.I., Gromova, A.S., Ganenko, T.V. and Semenov, A.A. (1987). Triterpene glycosides from *Thalictrum foetidum* and *T. minus* (Ranunculaceae) and their antineoplastic activity. *Khim.-Farm. Zh.*, **21**, 1434–1436 (*Chem. Abstr.*, **108**, 124139).

Rangaswami, S. and Sarangan, S. (1969). Sapogenins of *Clerodendron serratum*. Constitution of a new pentacyclic triterpene acid, serratogenic acid. *Tetrahedron*, **25**, 3701–3705.

Rangaswami, S. and Seshadri, T.R. (1971). Constitution of *Putranjiva* saponins A, B, C and D. *Indian J. Chem.*, **9**, 189.

Rao, A.V. and Kendall, C.W. (1986). Dietary saponins and serum lipids. *Food Chem. Toxicol.*, **24**, 441.

Rao, G.S., Sinsheimer, J.E. and Cochran, K.W. (1974). Antiviral activity of triterpenoid saponins containing acylated β-amyrin aglycones. *J. Pharm. Sci.*, **63**, 471–473.

Rao, M.G., Row, L.R. and Rukmini, C. (1969). Chemistry of saponins. IV. New triterpene sapogenin, castanogenol, from the bark of *Castanospermum australe*. *Indian J. Chem.*, **7**, 1203–1205.

Rauwald, H.W. and Janssen, B. (1988). Desglucoruscin und Desglucoruscosid als Leitstoffe des *Ruscus aculeatus*-Wurzelstocks. *Pharm. Ztg. Wiss.*, **133**, 61–68.

Rawat, M.S.M., Negi, D.S., Panwar, M.S. and Pant, G. (1988). A spirostanol glycoside from the rhizomes of *Ophiopogon intermedius*. *Phytochemistry*, **27**, 3326–3327.

Rawat, M.S.M., Negi, D.S., Pant, G. and Panwar, M.S. (1989). Spermicidal activity and chemical investigation of *Albizzia chinensis*. *Fitoterapia*, **60**, 168–169.

Razafintsalama, J.M., Adriantsiferana, R. and Rasoanaivo, P. (1987). Ilexosides, processes for their extraction from *Ilex mitis*, and their use in pharmaceutical and cosmetic compositions. *Chem. Abstr.*, **109**, 115855.

Reddy, G.C.S., Ayengar, K.N.N. and Rangaswami, S. (1975). Triterpenoids of *Gardenia latifolia*. *Phytochemistry*, **14**, 307.

Reddy, K.S., Shekhani, M.S., Berry, D.E., Lynn, D.G. and Hecht, S.M. (1984). Afromontoside. A new cytotoxic principle from *Dracaena afromontana*. *J. Chem Soc., Perkin Trans.* 1, 987–992.

Rees, H.H., Goad, L.J. and Goodwin, T.W. (1968). Cyclization of 2,3-oxido-squalene to cycloartenol in a cell-free system from higher plants. *Tetrahedron Lett.*, 723–725.

Reher, G. and Budesinsky, M. (1992). Triterpenoids from plants of the Sanguis-orbeae. *Phytochemistry*, **31**, 3909–3914.

Reher, G., Reznicek, G. and Baumann, A. (1991). Triterpenoids from *Sarco-poterium spinosum* and *Sanguisorba minor*. *Planta Medica*, **57**, 506.

Reichert, R.D., Tatarynovich, J.T. and Tyler, R.T. (1986). Abrasive dehulling of quinoa (*Chenopodium quinoa*): effect on saponin content as determined by an adapted hemolytic assay. *Cereal Chem.*, **63**, 471–475.

Reshef, G., Gestetner, B., Birk, Y. and Bondi, A. (1976). Effect of alfalfa saponins on the growth and some aspects of lipid metabolism of mice and quails. *J. Sci. Food Agric.*, **27**, 63–72.

Revers, F.E. (1946). Has licorice juice (succus liquiritiae) a healing action in gastric ulcer? *Nederland Tijdschr. Geneeskunde*, **90**, 135–137.

Reznicek, G., Jurenitsch, J., Robien, W. and Kubelka, W. (1989a). Saponins in *Cyclamen* species: configuration of cyclamiretin C and structure of iso-cyclamin. *Phytochemistry*, **28**, 825–828.

Reznicek, G., Jurenitsch, J., Michl, G. and Haslinger, E. (1989b). The first structurally confirmed saponin from *Solidago gigantea*: structure elucidation by modern NMR techniques. *Tetrahedron Lett.*, **30**, 4097–4100.

Reznicek, G., Jurenitsch, J., Kubelka, W., Michl, G., Korhammer, S. and Haslinger, E. (1990a). Isolierung und Struktur der vier Hauptsaponine aus *Solidago gigantea* var. *serotina*. *Annalen*, 989–994.

Reznicek, G., Jurenitsch, J., Kubelka, W., Korhammer, S., Haslinger, E. and Hiller, K. (1990b). The first spectroscopically confirmed saponins from *Solidago canadensis*. *Planta Medica*, **56**, 554.

Reznicek, G., Jurenitsch, J., Plasun, M. *et al.* (1991). Four major saponins from *Solidago canadensis*. *Phytochemistry*, **30**, 1629–1633.

Rezu-Ul-Jalil, Jabbar, A. and Hasan, C.M. (1986). Hypoglycemic activities of the glycosides of *Momordica cochinchinensis*. *J. Bangladesh Acad. Sci.*, **10**, 25–30 (*Chem. Abstr.*, **105**, 75994)

Riccio, R., Iorizzi, M., Squillace-Greco, O., Minale, L., Debray, M. and Menou,

J.L. (1985a). Starfish saponins, part 22. Asterosaponins from the starfish *Halityle regularis*: a novel 22,23-epoxy-steroidal glycoside sulphate. *J. Nat. Prod.*, **48**, 756–765.

Riccio, R., Iorizzi, M., Squillace-Greco, O., Minale, L., Laurent, D. and Barbin, Y. (1985b). Starfish saponins. XXVI. Steroidal glycosides from the starfish *Poraster superbus*. *Gazz. Chim. Ital.*, **115**, 505–509.

Riccio, R., Squillace-Greco, O., Minale, L. *et al.* (1986a). Starfish saponins, part 28. Steroidal glycosides from Pacific starfishes of the genus *Nardoa*. *J. Nat. Prod.*, **49**, 1141–1143.

Riccio, R., Valeria D'Auria, M. and Minale, L. (1986b). Two new steroidal glycoside sulphates, longicaudoside-A and -B from the Mediterranean ophiuroid *Ophiderma longicaudum*. *J. Org. Chem.*, **51**, 533–536.

Riccio, R., Iorizzi, M. and Minale, L. (1986c). Starfish saponins. XXX. Isolation of sixteen steroidal glycosides and three polyhydroxy steroids from the Mediterranean starfish *Coscinasterias tenuispina*. *Bull. Soc. Chim. Belg.*, **95**, 869–893.

Riccio, R., Iorizzi, M. and Minale, L. (1987a). Recent advances in the chemistry of echinoderms. In *Biologically Active Natural Products*, ed. K. Hostettmann and P.J. Lea, pp. 153–165. Clarendon Press, Oxford.

Riccio, R., Minale, L., Bano, S. and Ahmad, V.U. (1987b). Starfish saponins part 32. Structure of a novel steroidal 5-*O*-methyl galactofuranoside from the starfish *Astropecten indicus*. *Tetrahedron Lett.*, **28**, 2291–2294.

Riccio, R., Iorizzi, M., Minale, L., Oshima, Y. and Yasumoto, T. (1988). Starfish saponins. Part 34. Novel steroidal glycoside sulphates from the starfish *Asterias amurensis*. *J. Chem. Soc., Perkin Trans. 1*, 1337–1347.

Richou, R., Jensen, R., Belin, C. and Richou, H. (1964). Inflammatory effect of saponin. *Rev. Immunol.*, **28**, 49–62.

Ridout, C.L., Wharf, S.G., Price, K.R., Johnson, I.T. and Fenwick, G.R. (1988). UK mean daily intakes of saponins – intestine-permeabilizing factors in legumes. *Food Sci. Nutr.*, **42F**, 111–116.

Ridout, C.L., Price, K.R., Coxon, D.T. and Fenwick, G.R. (1989). Glycoalkaloids from *Solanum nigrum* L., α-solamargine and α-solasonine. *Pharmazie*, **44**, 732–733.

Rimpler, H. and Rizk, A.M. (1969). Isolation of oleanolic acid from *Fagonia cretica*. *Phytochemistry*, **8**, 2269.

Ripperger, H. and Schreiber, K. (1968). Struktur von Paniculonin A und B, zwei neuen Spirostanglykosiden aus *Solanum paniculatum* L. *Chem. Ber.*, **101**, 2450–2458.

Ripperger, H. and Schreiber, K. (1981). *Solanum* steroid alkaloids. In *The Alkaloids*, vol. XIX, ed. R.G.A. Rodrigo, pp. 81–192. Academic Press, New York.

Ripperger, H., Preiss, A. and Schmidt, J. (1981). *O*(3)-(2-Acetylamino-2-deoxy-β-D-glucopyranosyl)-oleanolic acid, a novel triterpenoid glycoside from two *Pithecellobium* species. *Phytochemistry*, **20**, 2434–2435.

Roddick, J.G. (1974). Steroidal glycoalkaloid α-tomatine. *Phytochemistry*, **13**, 9–25.

Roddick, J.G. (1989). The acetylcholinesterase-inhibiting activity of steroidal glycoalkaloids and their aglycones. *Phytochemistry*, **28**, 2631–2634.

Roddick, J.G. and Rijnenberg, A.L. (1986). Effect of steroidal glycoalkaloids on the permeability of liposome membranes. *Physiol. Plantarum*, **68**, 436–440.

Roddick, J.G., Rijnenberg, A.L. and Osman, S.F. (1988). Synergistic interaction between potato glycoalkaloids α-solanine and α-chaconine in relation to destabilization of cell membranes: ecological implications. *J. Chem. Ecol.*, **14**, 889–902.

Roddick, J.G., Rijnenberg, A.L. and Weissenberg, M. (1990). Membrane-disrupting properties of the steroidal glycoalkaloids solasonine and solamargine. *Phytochemistry*, **29**, 1513–1518.

Rodriguez, J. and Riguera, R. (1989). Lefevreiosides: four novel triterpenoid glycosides from the sea cucumber *Cucumaria lefevrei*. *J. Chem. Res.*, **(S)**, 342–343.

Rodriguez, J., Castro, R. and Riguera, R. (1991). Holothurinosides: new antitumour non sulphated triterpenoid glycosides from the sea cucumber *Holothuria forskalii*. *Tetrahedron*, **47**, 4753–4762.

Rogers, C.B. (1988). Pentacyclic triterpenoid rhamnosides from *Combretum imberbe* leaves. *Phytochemistry*, **27**, 3217–3220.

Rogers, C.B. (1989). Isolation of the 1α-hydroxycycloartenoid mollic acid α-L-arabinoside from *Combretum edwardsii* leaves. *Phytochemistry*, **28**, 279–281.

Rogers, C.B. and Thevan, I. (1986). Identification of mollic acid α-L-arabinoside, a 1α-hydroxycycloartenoid from *Combretum molle* leaves. *Phytochemistry*, **25**, 1759–1761.

Rollier, R. (1951). Treatment of leprosy by a *Smilax* species. *Maroc Méd.*, **30**, 776–780.

Rollier, R., Rollier, M. and Sekkat, A. (1979). Comparative study of five groups of patients: dapsone with another drug. *Int. J. Leprosy*, **47**, 2(Suppl.), 441.

Romussi, G. and De Tommasi, N. (1993). Revised structures of triterpene saponins from *Anchusa officinalis* L. *Pharmazie*, **48**, 70.

Romussi, G., Cafaggi, S. and Bignardi, G. (1980). Hemolytic action and surface activity of triterpene saponins from *Anchusa officinalis* L. Part 2: on the constituents of Boraginaceae. *Pharmazie*, **35**, 498–499.

Romussi, G., Bignardi, G., Sancassan, F. and Falsone, G. (1983). Inhaltsstoffe von Cupuliferae, 6. Ein neues Triterpensaponin aus *Quercus Ilex*. *Annalen*, 1448–1450.

Romussi, G., Bignardi, G., Falsone, G. and Wendisch, D. (1987). Inhaltsstoffe von Cupuliferae, 10. Mitt. Triterpensaponine aus *Fagus sylvatica* L. *Arch. Pharm*, **320**, 153–158.

Romussi, G., Cafaggi, S. and Pizza, C. (1988a). Anchusosid-11, ein neues Nebensaponin aus *Anchusa officinalis* L. *Arch. Pharm.*, **321**, 753–754.

Romussi, G., Ciarallo, G. and Parodi, B. (1988b). Glycoside aus *Quercus cerris* L. 11. Mitteilung: Ueber Inhaltsstoffe von Cupuliferae. *Pharmazie*, **43**, 294–295.

Romussi, G., Bignardi, G., Pizza, C. and De Tommasi, N. (1991). Constituents of Cupuliferae, XIII. New and revised structures of acylated flavonoids from *Quercus suber* L. *Arch. Pharm.*, **324**, 519–524.

Romussi, G., Caviglioli, G., Pizza, C. and De Tommasi, N. (1993). Constituents of Cupuliferae, XVII. Triterpene saponins and flavonoids from *Quercus laurifolia* MICHX. *Arch. Pharm.*, **326**, 525–528.

Roques, R., Druet, D. and Comeau, L.C. (1978). Structure cristalline et moléculaire de l'hédéragénine. *Acta Cryst. B*, **34**, 1634–1639.

Rose, M.E., Veares, M.P., Lewis, I.A.S. and Goad, J. (1983). Analysis of steroid conjugates by fast atom bombardment mass spectrometry. *Biochem. Soc. Trans.*, **11**, 602–603.

Rothkopf, V.M. and Vogel, G. (1976). Neue Befunde zur Wirksamkeit und zum Wirkungsmechanismus des Rosskastanien-saponins Aescin. *Arzneimittelforsch.* **26**, 225–235.

Rothman, E.S., Wall, M.E. and Eddy, C.R. (1952a). Steroidal sapogenins. III. Structure of steroidal saponins. *J. Am. Chem. Soc.*, **74**, 4013–4016.

Rothman, E.S., Wall, M.E. and Walens, H.A. (1952b). Steroidal saponins. IV. Hydrolysis of steroidal saponins. *J. Am. Chem. Soc.*, **74**, 5791–5792.

Row, L.R. and Rukmini, C. (1966). Saponins of *Sapindus emarginatus*. *Indian J. Chem.*, **4**, 149–150.

Rowan, D.D. and Newman, R.H. (1984). Noroleanane saponins from *Celmisia petriei*. *Phytochemistry*, **23**, 639–644.

Rücker, G., Mayer, R. and Shin-Kim, J.-S. (1991). Triterpensaponine aus der chinesischen Droge 'Daxueteng' (*Caulis sargentodoxae*). *Planta Medica*, **57**, 468–470.

Rüegg, A. and Waser, P.G. (1993). Glycyrrhizingehalt von Lakritze-Erzeugnissen. *Schweiz. Apotheker-Zeitung*, **131**, 70.

Ruggieri, G.D. and Nigrelli, R.F. (1974). In *Bioactive Compounds from the Sea*, ed. H. Humm and C. Lane, pp. 183–195. Marcel Dekker, New York.

Rumbero-Sanchez, A. and Vazquez, P. (1991). A nor-triterpene glycoside from *Isertia haenkeana* and a ^{13}C NMR study of cincholic acid. *Phytochemistry*, **30**, 623–626.

Russo, G. (1967). Triterpenes of *Glycyrrhiza glabra*. *Fitoterapia*, **38**, 98–109.

Saeki, I., Sumimoto, M. and Kondo, T. (1968). Antitermitic substance of *Ternstroemia japonica* wood. III. Biological tests of the antitermitic substance. *Mokuzai Gakkaishi*, **14**, 110–114 (*Chem. Abstr.*, **70**, 30214).

Sagara, K., Ito, Y., Oshima, T., Kawaura, M. and Misaki, T. (1985). Application of ion-pair high-performance liquid chromatography to the analysis of glycyrrhizin in *Glycyrrhizae radix*. *Chem. Pharm. Bull.*, **33**, 5364–5368.

Saharia, G.S. and Seshadri, V. (1980). Chemical investigation on *Randia* saponins; isolation and characterization of randioside A. *Indian J. For*, **3**, 6–8 (*Chem. Abstr.*, **94**, 153460).

Sahu, N.P., Banerji, N. and Chakravarti, R.N. (1974). New saponin of oleanolic acid from *Pereskia grandiflora*. *Phytochemistry*, **13**, 529–530.

Sahu, N.P., Roy, S.K. and Mahato, S.B. (1989). Spectroscopic determination of structures of triterpenoid trisaccharides from *Centella asiatica*. *Phytochemistry*, **28**, 2852–2854.

Saijo, R., Fuke, C., Murakami, K., Nohara, T. and Tomimatsu, T. (1983). Two steroidal glycosides, aculeatiside A and B from *Solanum aculeatissimum*. *Phytochemistry*, **22**, 733–736.

Saito, H. (1983). *Metabolism*, **10**, 556.

Saito, K., Horie, M., Hoshino, Y., Nose, N. and Nakazawa, H. (1990a). High-performance liquid chromatographic determination of glycoalkaloids in potato products. *J. Chromatogr*, **508**, 141–147.

Saito, S., Sumita, S., Tamura, N. *et al.* (1990b). Saponins from the leaves of *Aralia elata* SEEM. (Araliaceae). *Chem. Pharm. Bul.*, **38**, 411–414.

Saitoh, T., Kinoshita, T. and Shibata, S. (1976). Flavonols of licorice root. *Chem. Pharm. Bull.*, **24**, 1242–1245.

Sakai, Y., Takeshita, T., Kinjo, J., Ito, Y. and Nohara, T. (1990). Two new triterpenoid sapogenols and a new saponin from *Abrus cantoniensis* (II). *Chem. Pharm. Bull.*, **38**, 824–826.

Sakakibara, J., Hotta, Y. and Yasue, M. (1974). Studies on the constituents of *Lyonia ovalifolia* DRUDE var. *elliptica* HAND.-MAZZ. XVII. Structure of a triterpene arabinoside, ovalifolioside. (2). *Yakugaku Zasshi*, **94**, 170–175.

Sakamoto, S., Kofuji, S., Kuroyanagi, M., Ueno, A. and Sekita, S. (1992). Saponins from *Trifolium repens*. *Phytochemistry*, **31**, 1773–1777.

Sakuma, S. and Motomura, H. (1987). Purification of saikosaponins a, c and d: application of large-scale reversed-phase high performance liquid chromatography. *J. Chromatogr.*, **400**, 293–295.

Sakuma, S. and Shoji, J. (1982). Studies on the constituents of the root of *Polygala tenuifolia* WILLDENOW. II. On the structures of onjisaponins A, B and E. *Chem. Pharm. Bull.*, **30**, 810–821.

Sakurai, N., Inoue, T. and Nagai, M. (1972). Studies on the Chinese crude drug. 'Shôma'. II. Triterpenes of *Cimicifuga dahurica* MAXIM. *Yakugaku Zasshi*, **92**, 724–728.

Sakurai, N., Nagai, M., Nagase, H., Kawai, K.-I., Inoue, T. and Peigen, X. (1986). Studies on the constituents of *Beesia calthaefolia* and *Souliea vaginata*. II. Beeioside II, a cyclolanostanol xyloside from rhizomes of *Beesia calthaefolia*. *Chem. Pharm. Bull.*, **34**, 582–589.

Saluja, A.K. and Santani, D.D. (1986). A saponin from pulps of *Xeromphis spinosa*. *Planta Medica*, **52**, 72–73.

Sandberg, F. (1973). Two glycoside-containing genera of the Araliaceae family *Panax* and *Eleutherococcus*. *Planta Medica*, **24**, 392–396.

Sandberg, F. and Michel, K.H. (1962). Phytochemische Studien über die Flora Aegyptens. VI. Ueber die Saponine und Prosapogenine von *Anabasis articulata*. *Lloydia*, **25**, 142–150.

Sandberg, F. and Shalaby, A.F. (1960). Phytochemical studies on the flora of Egypt. IV. The saponins of *Anabasis articulata* and *Anabasis setifera*. *Svensk Farmac. Tidskr.*, **64**, 677–690.

Sannié, C., Heitz, S. and Lapin, H. (1951). Paper partition chromatography of steroid sapogenins. *C.R. Acad. Sci.*, **223**, 1670–1672.

Sano, S., Sanada, S., Ida, Y. and Shoji, J. (1991). Studies on the constituents of the bark of *Kalopanax pictus* NAKAI. *Chem. Pharm. Bull.*, **39**, 865–870.

Sapeika, N. (1969). *Food Pharmacology*, pp. 67–68. Thomas, Springfield, IL.

Sasaki, Y., Mizutani, K., Kasai, R. and Tanaka, O. (1988). Solubilizing properties of glycyrrhizin and its derivatives: solubilization of saikosaponin-a, the saponin of *Bupleuri radix*. *Chem. Pharm. Bull.*, **36**, 3491–3495.

Sashida, Y., Kawashima, K. and Mimaki, Y. (1991a). Novel polyhydroxylated steroidal saponins from *Allium giganteum*. *Chem. Pharm. Bull.*, **39**, 698–703.

Sashida, Y., Ori, K. and Mimaki, Y. (1991b). Studies on the chemical constituents of the bulbs of *Lilium mackliniae*. *Chem. Pharm. Bull.*, **39**, 2362–2368.

Sati, O.P. and Rana, U. (1987). A new molluscicidal triterpenic glycoside from *Aesculus indica. Int. J. Crude Drug Res.*, **25**, 158–160.

Sati, O.P. and Sharma, S.C. (1985). New steroidal glycosides from *Asparagus curillus* (roots). *Pharmazie*, **40**, 417–418.

Sati, O.P., Pant, G. and Hostettmann, K. (1984). Potent molluscicides from asparagus. *Pharmazie*, **39**, 581

Sati, O.P., Chaukiyal, D.C. and Rana, U. (1986). Molluscicidal saponins of *Xeromphis spinosa. Planta Medica*, **52**, 381–383.

Sati, O.P., Rana, U., Chaukiyal, D.C., Madhusudanan, K.P. and Bhakuni, D.S. (1987). Molluscicidal triterpenoidal glycosides of *Xeromphis spinosa. Planta Medica*, **53**, 530–532.

Sati, O.P., Bahuguna, S., Uniyal, S. and Bhakuni, D.S. (1989). Triterpenoid saponins from *Randia uliginosa* fruits. *Phytochemistry*, **28**, 575–577.

Sati, O.P., Bahuguna, S., Uniyal, S., Sakakibara, J., Kaiya, T. and Nakamura, A. (1990). A new saponin from *Deeringia amaranthoides. J. Nat. Prod.*, **53**, 466–469.

Sato, H. and Sakamura, S. (1973). Bitter principle of tomato seeds. Isolation and structure of a new furostanol saponin. *Agric. Biol. Chem.*, **37**, 225–231.

Sauvaire, Y., Ribes, G., Baccou, J.-C. and Loubatières-Mariani, M.-M. (1991). Implication of steroid saponins and sapogenins in the hypocholesterolemic effect of fenugreek. *Lipids*, **26**, 191–197.

Sawada, H., Miyakoshi, M., Isoda, S., Ida, Y. and Shoji, J. (1993). Saponins from leaves of *Acanthopanax sieboldianus. Phytochemistry*, **34**, 1117–1121.

Sawardeker, J.S. and Sloneker, J.H. (1965). Quantitative determination of mono-saccharides by gas liquid chromatography. *Anal. Chem.*, **37**, 945–947.

Saxena, R.C. and Bhalla, T.N. (1968). Antipyretic effect of glycyrrhetic acid and imipramine. *Jap. J. Pharmacol.*, **18**, 353–355.

Schantz, von M. (1966). Die Saponindrogen *Dansk Tidsskr. Farm.*, **40**, 140–155.

Scheidegger, J.J. and Cherbuliez, E. (1955). L'hédéracoside A, un nouvel hétéroside extrait du lierre (*Hedera helix* L.). *Helv. Chim. Acta*, **38**, 547–556.

Schenkel, E.P., Werner, W. and Schulte, K.E. (1991). Die Saponine aus *Thinouia coriacea. Planta Medica*, **57**, 463–467.

Scheuer, P.J. (1987). *Bioorganic Marine Chemistry*, vol. 1. Springer-Verlag, Berlin.

Schilcher, H. (1990). Pharmakologie und Klinik 'pflanzlicher Diuretika'. *Schweiz. Z. Ganzheits-Med.*, **3**, 128–137.

Schilcher, H. and Emmrich, D. (1992). Pflanzliche Urologika zur Durchspülungs-therapie. *Dtsch. Apoth. Ztg.*, **132**, 2549–2555.

Schlemmer, W. and Bosse, J. (1963). Zur quantitativen Bestimmung von Triterpen-verbindungen mit Hilfe der Eisenchlorid-Reaktion. *Arch. Pharm.*, **296**, 785–791.

Schlösser, E. (1973). Rôle of saponins in antifungal resistance. II. The hedera-saponins in leaves of English ivy. *Z. Pflanzenkr. Pflanzenschutz*, **80**, 704–710.

Schlösser, E. and Wulff, G. (1969). Ueber die Strukturspezifität der Saponin-hämolyse I. Triterpensaponine und -aglykone. *Z. Naturforsch.*, **24b**, 1284–1290.

Schmiedeberg, O. (1875). Untersuchung über die pharmakologisch wirksamen Bestandteile der *Digitalis purpurea* L. *Arch. Exp. Path, Pharmakol.*, **3**, 16–43.

Schmitz, F.J., Ksebati, M.B., Gunasekera, S.P. and Agarwal, S. (1988). Sarasinoside A$_1$, a saponin containing amino sugars isolated from a sponge. *J. Org. Chem.*, **53**, 5941–5947.

Schöpfer, H.J. (1976). Neuestes aus der Ginseng-Forschung. *Panax ginseng* C.A. MEYER, Chemie und Wirkung – eine Uebersicht. *Dtsch. Apoth. Ztg.*, **116**, 1–7.

Schöpke, T. and Hiller, K. (1990). Triterpenoid saponins. Part 6. *Pharmazie*, **45**, 313–342.

Schöpke, T., Wray, V., Rzazewska, B. and Hiller, K. (1991). Bellissaponins BA$_1$ and BA$_2$, acylated saponins from *Bellis perennis*. *Phytochemistry*, **30**, 627–631.

Schreiber, K. (1968). Steroid alkaloids: the *Solanum* group. In *The Alkaloids. Chemistry and Physiology*, vol. X, ed. R.H.F. Manske, pp. 1–192. Academic Press, New York.

Schreiber, K. and Ripperger, H. (1966). Jurubine, a novel type of steroidal saponin with (25S)-3β-amino-5α-furostane-22α,26-diol O(26)-β-D-glucopyranoside structure from *Solanum paniculatum* L. *Tetrahedron Lett.*, 5997–6002.

Schteingart, C.D. and Pomilio, A.B. (1984). Two saponins from *Zexmania buphthalmiflora*. *Phytochemistry*, **23**, 2907–2910.

Schulten, H.-R. and Games, D.E. (1974). High resolution field desorption mass spectrometry. II. Glycosides. *Biomed. Mass Spectrometry*, **1**, 120–123.

Schulten, H.-R., Komori, T., Nohara, T., Higuchi, R. and Kawasaki, T. (1978). Field desorption mass spectrometry of natural products-II. Physiologically active pennogenin- and hederagenin-glycosides. *Tetrahedron*, **34**, 1003–1010.

Schulten, H.-R., Komori, T., Kawasaki, T., Okuyama, T. and Shibata, S. (1982). Confirmation of new, high-mass saponins from *Gleditsia japonica* by field desorption mass spectrometry. *Planta Medica*, **46**, 67–73.

Schulten, H.-R., Singh, S.B. and Thakur, R.S. (1984). Field desorption mass spectrometry of natural products. Part XIV. Field desorption and fast atom bombardment mass spectrometry of spirostanol and furostanol saponins from *Paris polyphylla*. *Z. Naturforsch.*, **39C**, 201–211.

Scott, M.T., Goss-Sampson, M. and Bomford, R. (1985). Adjuvant activity of saponin: antigen localization studies. *Int. Arch. Allergy Appl. Immunol.*, **77**, 409–412.

Seaforth, C.E., Mohammed, S., Maxwell, A., Tinto, W.F. and Reynolds, W.F. (1992). Mabioside A, a new saponin from *Colubrina elliptica*. *Tetrahedron Lett.*, **33**, 4111–4114.

Segal, R. and Schlösser, E. (1975). Role of glycosidases in the membranolytic, antifungal action of saponins. *Arch. Microbiol.*, **104**, 147–150.

Segal, R., Milo-Goldzweig, I. and Zaitschek, D.V. (1969). Sapogenin content of *Anabasis articulata*. *Phytochemistry*, **8**, 521.

Segal, R., Shatkovsky, P. and Milo-Goldzweig, I. (1974). On the mechanism of saponin hemolysis – I. Hydrolysis of the glycosidic bond. *Biochem. Pharmacol.*, **23**, 973–981.

Segiet-Kujawa, E. and Kaloga, M. (1991). Triterpenoid saponins of *Eleutherococcus senticosus* roots. *J. Nat. Prod.*, **54**, 1044–1048.

Segura de Correa, R., Riccio, R., Minale, L. and Duque, C. (1985). Starfish saponins, part 21. Steroidal glycosides from the starfish *Oreaster reticulatus*. *J. Nat Prod.*, **48**, 751–755.

Sehgal, S.L., Bhakuni, D.S., Sharma, V.N. and Kaul, K.N. (1961). Chemical constituents of the seed of *Luffa graveolens*. *J. Sci. Indian Res.*, **20B**, 461–462.

Seifert, K., Preiss, A., Johne, S. *et al.* (1991). Triterpene saponins from *Verbascum songaricum. Phytochemistry*, **30**, 3395–3400.

Semenchenko, V.F. and Muravev, I.A. (1968). The triterpenic saponins from *Glycyrrhiza echinata* roots. *Rast. Resur.*, **4**, 62–67 (*Chem. Abstr.*, **69**, 21864).

Sendra, J. (1963). Phytochemical studies on *Prunella vulgaris* and *Prunella grandiflora*. I. Saponin and triterpene compounds. *Dissertat. Pharm.* (Warszawa), **15**, 333.

Seo, S., Tomita, Y., Tori, K. and Yoshimura, Y. (1978). Determination of the configuration of a secondary hydroxy group in a chiral secondary alcohol using glycosidation shifts in carbon-13 nuclear magnetic resonance spectroscopy. *J. Am. Chem. Soc.*, **100**, 3331–3339.

Sergienko, T.V., Mogilevtseva, T.B. and Chirva, V.J. (1977). Chemical study of *Albizzia julibrissin* beans. *Khim. Prir. Soedin.*, 708.

Serova, N.A. (1961). Sapogenin from *Eryngium incognitum. Med. Prom SSSR*, **15**, 26–27.

Seshadri, V. and Rangaswami, S. (1975). Constitution of putranjiva saponins A, B, C and D. *Indian J. Chem.*, **13**, 447–452.

Seshadri, V., Batta, A.K. and Rangaswami, S. (1981). Structure of two new saponins from *Achyranthes aspera. Indian J. Chem.*, **20B**, 773–775.

Seto, H., Otake, N., Kawai, H., Luo, S.-Q., Qian, F.-G. and Pan, S.-L. (1986). Studies on the chemical constituents of *Bupleurum genus*. Part III. Isolation of triterpenoid glycosides (saikosaponins) from *Bupleurum polyclonum* Y. LI ET S.L. PAN and their chemical structures. *Agric. Biol. Chem.*, **50**, 1607–1611.

Seto, T., Tanaka, T., Tanaka, O. and Naruhashi, N. (1984). β-Glucosyl esters of 19α-hydroxyursolic acid derivatives in leaves of *Rubus* species. *Phytochemistry*, **23**, 2829–2834.

Setty, B.S., Kamboj, V.P., Garg, H.S. and Khanna, N.M. (1976). Spermicidal potential of saponins isolated from Indian medicinal plants. *Contraception*, **14**, 571–578.

Sezik, E. and Yesilada, E. (1985). A new saponin from the roots of *Polygala pruinosa. Fitoterapia*, **56**, 159–163.

Shakirov, R., Nuriddinov, R.N. and Yunusov, S.Yu. (1965). Alkaloids of *Petilium eduardi. Khim. Prir. Soedin.*, 384–392.

Shao, C.-J., Kasai, R., Xu, J.-D. and Tanaka, O. (1988). Saponins from leaves of *Acanthopanax senticosus* HARMS., Ciwujia: structures of ciwujianosides B, C_1, C_2, C_3, C_4, D_1, D_2 and E. *Chem. Pharm. Bull.*, **36**, 601–608.

Shao, C.-J., Kasai, R., Xu, J.-D. and Tanaka, O. (1989a). Saponins from leaves of *Acanthopanax senticosus* HARMS., Ciwujia. II. Structures of ciwujianosides A_1, A_2, A_3, A_4 and D_3. *Chem. Pharm. Bull.*, **37**, 42–45.

Shao, C.-J., Kasai, R., Xu, J.-D. and Tanaka, O. (1989b). Saponins from roots of *Kalopanax septemlobus*. (THUNB.) KOIDZ., Ciqiu: structures of kalopanax-saponins C, D, E and F. *Chem. Pharm. Bull.*, **37**, 311–314.

Sharma, S.C. and Kumar, R. (1982). Constituents from leaves of *Zizyphus mauritiana* LAMK. *Pharmazie*, **37**, 809–810.

Sharma, S.C. and Kumar, R. (1983). Zizynummin, a dammarane saponin from *Zizyphus nummularia. Phytochemistry*, **22**, 1469–1471.

Sharma, S.C. and Walia, S. (1983). Structures of sonunin I and sonunin II: two new saponins from *Acacia concinna* D.C. *Pharmazie*, **38**, 632–633.

Sharma, S.C., Chand, R. and Sati, O.P. (1982). Steroidal saponins of *Asparagus adscendens*. *Phytochemistry*, **21**, 2075–2078.

Shashkov, A.S. and Kemertelidze, E.P. (1988). Triterpene saponins of *Caltha polypetala*. *Glycosides* G and I. *Khim. Prir. Soedin.*, 229–236.

Shashkov, A.S., Grishkovets, V.I., Loloiko, A.A. and Chirva, V.J. (1987). Triterpenoid glycosides of *Hedera taurica*. I. Structure of tauroside E from *Hedera taurica* leaves. *Khim. Prir. Soedin.*, 363.

Shatilo, V.V. (1971). Ursolic acid from *Oxycoccus quadripetalus* berries. *Khim. Prir. Soedin.*, 534.

Shen, Y.-L., Ye, W.-C., Zhao, S.-X. *et al.* (1988). Isolation of two saponins from *Anemone flaccida* and their effects on reverse transcriptase. *Shoyakugaku Zasshi*, **42**, 35–40 (*Chem. Abstr.*, **109**, 222012).

Shibata, S. (1977). Saponins with biological and pharmacological activity. In *New Natural Products and Plant Drugs with Pharmacological, Biological or Therapeutical Activity*, ed. H. Wagner and P. Wolff, pp. 177–196. Springer, Berlin.

Shibata, S. (1986). Pharmacology and chemical study of dammarane-type triterpenoids. In *Advances in Medicinal Phytochemistry*, ed. D. Barton and W.D. Ollis, pp. 159–172. John Libbey, London.

Shibata, S., Tanaka, O., Shoji, J. and Saito, H. (1985). Chemistry and pharmacology of *Panax*. In *Economic and Medicinal Plant Research*, vol. 1, ed. H. Wagner, H. Hikino and N.R. Farnsworth, pp. 217–284. Academic Press, London.

Shibata, S., Takahashi, K., Yano, S. *et al.* (1987). Chemical modification of glycyrrhetinic acid in relation to the biological activities. *Chem. Pharm. Bull.*, **35**, 1910–1918.

Shigenaga, S., Kouno, I. and Kawano, N. (1985). Triterpenoids and glycosides from *Geum japonicum*. *Phytochemistry*, **24**, 115–118.

Shihata, I.M., El-Gendi, A.Y.I. and Abd El-Malik, M.M. (1977). Pharmacological studies on saponin fraction of *Opilia celtidifolia*. *Planta Medica*, **31**, 60–67.

Shimada, S. (1969). Antifungal steroid glycoside from sea cucumber. *Science*, **163**, 1462.

Shimizu, K., Amagaya, S. and Ogihara, Y. (1983). Separation and quantitative analysis of saikosaponins by high-performance liquid chromatography. *J. Chromatogr.*, **268**, 85–91.

Shimizu, K., Amagaya, S. and Ogihara, Y. (1985). New derivatives of saikosaponins. *Chem. Pharm. Bull.*, **33**, 3349–3355.

Shimizu, M. and Takemoto, T. (1967). Studies on the constituents of caryophyllaceous plants. I. On the saponin of *Dianthus superbus* var. *longicalycinus*. (1). *Yakugaku Zasshi*, **87**, 250–254.

Shimizu, M., Shingyouchi, K.-I., Morita, N., Kizu, H. and Tomimori, T. (1978). Triterpenoid saponins from *Pulsatilla cernua* SPRENG. I. *Chem. Pharm. Bull.*, **26**, 1666–1671.

Shimizu, S., Ishihara, N., Umehara, K., Miyase, T. and Ueno, A. (1988). Sesquiterpene glycosides and saponins from *Cynara cardunculus* L. *Chem. Pharm. Bull.*, **36**, 2466–2474.

Shimizu, Y. (1971). Antiviral substances in starfish. *Experientia*, **27**, 1188–1189.

Shimomura, H., Sashida, Y. and Mimaki, Y. (1988). 26-*O*-acylated furostanol saponins pardarinoside A and B from the bulbs of *Lilium pardarum*. *Chem. Pharm. Bull.*, **36**, 3226–3229.

Shimomura, H., Sashida, Y. and Mimaki, Y. (1989). Steroidal saponins, pardarinoside A–G from the bulbs of *Lilium pardarinum*. *Phytochemistry*, **28**, 3163–3170.

Shimoyamada, M., Suzuki, M., Sonta, H., Maruyama, M. and Okubo, K. (1990). Antifungal activity of the saponin fraction obtained from *Asparagus officinalis* L. and its active principle. *Agric. Biol. Chem.*, **54**, 2553–2557.

Shiobara, Y., Inoue, S., Nishiguchi, Y. *et al.* (1992). Pfaffane-type nortriterpenoids from *Pfaffia pulverulenta*. *Phytochemistry*, **31**, 1737–1740.

Shiojima, K., Arai, Y., Masuda, K., Takase, Y., Ageta, T. and Ageta, H. (1992). Mass spectra of pentacyclic triterpenoids. *Chem. Pharm. Bull.*, **40**, 1683–1690.

Shiraiwa, M., Kudo, S., Shimoyamada, M., Harada, K. and Okubo, K. (1991). Composition and structure of 'group A saponin' in soybean seed. *Agric. Biol. Chem.*, **55**, 315–322.

Shnulin, A.N., Amirova, G.S. and Mustafaev, N.M. (1972). X-ray study of a crystal of methyl meristotropate. *Khim. Prir. Soedin.*, 661–662.

Shoji, J. (1981). Isolation of triterpenoidal saponins. In *Advances in Natural Products Chemistry*, ed. S. Natori, N. Ikekawa and M. Suzuki, pp. 275–291. Wiley, New York.

Shoji, S., Kawanishi, S. and Tsukitani, Y. (1971). On the structure of senegin-II of *Senegae radix*. *Chem. Pharm. Bull.*, **19**, 1740–1742.

Sholichin, M., Yamasaki, K., Kasai, R. and Tanaka, O. (1980). ^{13}C Nuclear magnetic resonance of lupane-type triterpenes, lupeol, betulin and betulinic acid. *Chem. Pharm. Bull.*, **28**, 1006–1008.

Shrivastava, K. and Saxena, V.K. (1988). A new saponin from the roots of *Albizzia lebbek*. *Fitoterapia*, **59**, 479–480.

Shukla, Y.N. and Thakur, R.S. (1988). An acetylated saponin from *Panax pseudo-ginseng* subsp. *himalaicus* var. *angustifolius*. *Phytochemistry*, **27**, 3012–3104.

Shukyurov, D.Z., Putieva, Z.M., Kondratenko, E.S. and Abubakirov, N.K. (1974). Triterpenoid glycosides of *Poterium polygamum* and *Poterium lasiocarpum*. *Khim. Prir. Soedin.*, 531–532.

Siddiqui, S., Faizi, S. and Siddiqui, B.S. (1983). Studies in the chemical constituents of the fresh berries of *Solanum xanthocarpum* SCHRAD. AND WENDLE. *J. Chem. Soc. Pak.*, **5**, 99–102.

Sidhu, G.S., Upson, B. and Malinow, M.R. (1987). Effects of soy saponins and tigogenin cellobioside on intestinal uptake of cholesterol, cholate and glucose. *Nutr. Rep. Int.*, **35**, 615–623 (*Chem. Abstr.*, **107**, 22075).

Silva, M. and Mancinelli, P. (1961). Oleanolic acid in *Lardizabala biternata*. *J. Pharm. Sci.*, **50**, 975.

Silva, M. and Stück, R. (1962). Ueber das Vorkommen von Oleanolsäure in *Boquilla trifoliata* (D.C.) DCNE. *Arch. Pharm.*, **295**, 58–59.

Silva, M., Balocchi, M. and Sammes, P.G. (1968). Triterpenoid constituents of *Griselinia scandens*. *Phytochemistry*, **7**, 333–334.

Simes, J.J.H., Tracey, J.G., Webb, L.J. and Dunstan, W.J. (1959). An Australian phytochemical survey. III. Saponins in Eastern Australian flowering plants.

Australia, Commonwealth Sci. Ind. Res. Organisation, Bull. No. 281 (Chem. Abstr., 53, 22277).

Simoes, C.M.O., Amoros, M., Schenkel, E.P., Shin-Kim, J.-S., Rücker, G. and Girre, L. (1990). Preliminary studies of antiviral activity of triterpenoid saponins: relationships between their chemical structure and antiviral activity. Planta Medica, 56, 652–653.

Sinden, S.L., Sanford, L.L. and Deahl, K.L. (1986). Segregation of leptine glycoalkaloids in Solanum chacoense bitter. J. Agric. Food Chem., 34, 372–377.

Singh, B., Agrawal, P.K. and Thakur, R.S. (1986). Aesculuside-A, a new triterpene glycoside from Aesculus indica. Planta Medica, 52, 409–410.

Singh, B., Agrawal, P.K. and Thakur, R.S. (1987). Aesculuside-B, a new triterpene glycoside from Aesculus indica. J. Nat. Prod., 50, 781–783.

Singh, P.N. and Singh, S.B. (1980). A new saponin from mature tubers of Cyperus rotundus. Phytochemistry, 19, 2056.

Singh, S., Khanna, N.M. and Dhar, M.M. (1974). Solaplumbine, a new anticancer glycoside from Nicotiana plumbaginifolia. Phytochemistry, 13, 2020–2022.

Singh, S.B., Thakur, R.S. and Schulten, H.-R. (1982a). Furostanol saponins from Paris polyphylla: structures of polyphyllin G and H. Phytochemistry, 21, 2079–2082.

Singh, S.B., Thakur, R.S. and Schulten, H.-R. (1982b). Spirostanol saponins from Paris polyphylla, structures of polyphyllin C, D, E and F. Phytochemistry, 21, 2925–2929.

Singh, V.K., Agarwal, S.S. and Gupta, B.M. (1984). Immunomodulatory activity of Panax ginseng extract. Planta Medica, 50, 462–465.

Skinner, F.A. (1955). Antibiotics. In Moderne Methoden der Pflanzenanalyse, vol. III, ed. K. Peach and M.V. Tracey, pp. 626–725. Springer-Verlag, Berlin.

Slacanin, I., Marston, A. and Hostettmann, K. (1988). High-performance liquid chromatographic determination of molluscicidal saponins from Phytolacca dodecandra (Phytolaccaceae). J. Chromatogr., 448, 265–274.

Slacanin, I., Marston, A., Hostettmann, K., Guédon, D. and Abbe, P. (1991). The isolation of Eleutherococcus senticosus constituents by centrifugal partition chromatography and their quantitative determination by high performance liquid chromatography. Phytochem. Anal., 2, 137–142.

Slanina, P. (1990). Solanine (glycoalkaloids) in potato – toxicological evaluation. Food Chem. Toxicol., 28, 759–761 (Chem. Abstr., 114, 141714).

Soboleva, V.A. and Boguslavskaya, L.I. (1987). Triterpene glycosides from Euphorbia sequieriana. Farm. Zh. (Kiev), 76–77 (Chem. Abstr., 107, 55752).

Sokolov, S.Y. (1986). Comparative pharmacological studies of the neurologic properties of a group of natural products – the triterpene glycosides. In Advances in Medicinal Phytochemistry, ed. D. Barton and W.D. Ollis, pp. 173–178. Libbey, London.

Sokolskii, I.N., Bankovskii, A.I. and Zinkevich, E.P. (1975). Triterpenoid glycosides from Camellia oleifera and Camellia sasanqua. Khim. Prir. Soedin., 102–103.

Soldati, F. and Sticher, O. (1980). HPLC separation and quantitative determination of ginsenosides from Panax ginseng, Panax quinquefolia and from ginseng drug preparations. Planta Medica, 39, 348–357.

Sollorz, G. (1985). Qualitätsbeurteilung von Ginsengwurzeln. Quantitative HPLC-Bestimmung der Ginsenoside. *Dtsch. Apoth. Ztg.*, **125**, 2052–2055.

Son, K.H., Do, J.C. and Kang, S.S. (1990). Steroidal saponins from the rhizomes of *Polygonatum sibiricum*. *J. Nat. Prod.*, **53**, 333–339.

Sonnenborn, U. (1987). Ginseng. *Dtsch. Apoth. Ztg.*, **127**, 433–441.

Sonnenborn, U. (1989). Ginseng-Nebenwirkungen: Fakten oder Vermutungen? *Med. Monatsschrift Pharm.*, **12**, 46–53.

Sonnenborn, U. and Hänsel, R. (1992). *Panax ginseng*. In *Adverse Effects of Herbal Drugs*, vol. 1, ed. P.A.G.M. de Smet, K. Keller, R. Hänsel and R.F. Chandler, pp. 179–192. Springer-Verlag, Berlin.

Sood, A.R., Bajpai, A. and Dixit, M. (1985). Pharmacological and biological studies on saponins. *Indian J. Pharmacol.*, **17**, 178–179.

Sosa, A. (1962). Some heterosides of *Mollugo nudicaulis*. *Ann. Pharmac. Fr.*, **20**, 257–279.

Sotheeswaran, S., Bokel, M. and Kraus, W. (1989). A hemolytic saponin, randianin, from *Randia dumetorum*. *Phytochemistry*, **28**, 1544–1546.

Spinks, E.A. and Fenwick, G.R. (1990). The determination of glycyrrhizin in selected UK liquorice products. *Food Additives and Contaminants*, 7, 769–778.

Srivastava, R. and Kulshreshtha, D.K. (1986). Glochidioside, a triterpene glycoside from *Glochidion heyneanum*. *Phytochemistry*, **25**, 2672–2674.

Srivastava, R. and Kulshreshtha, D.K. (1988). Triterpenoids from *Glochidion heyneanum*. *Phytochemistry*, **27**, 3575–3578.

Srivastava, S.K. (1989). An acetylated saponin from *Schefflera impressa*. *J. Nat. Prod.*, **52**, 1342–1344.

Srivastava, S.K. and Agnihotri, V.K. (1983). Chemical investigation of the stem bark of *Aphanamixis polystachya*. *Natl. Acad. Sci., Lett.* (India) 7, 213–214 (*Chem. Abstr.*, **102**, 163680).

Srivastava, S.K. and Jain, D.C. (1989). Triterpenoid saponins from plants of Araliaceae. *Phytochemistry*, **28**, 644–647.

Srivastava, S.K., Srivastava, S.D. and Nigam, S.S. (1983). Lupa-20(29)-ene-3-*O*-α-L-rhamnopyranoside from the roots of *Cordia obliqua*. *Indian Chem. Soc.*, **60**, 202.

Stahl, E. (1969). *Thin-layer Chromatography*, 2nd edn. Springer-Verlag, Berlin.

Stahl, E. and Glatz, A. (1982). Optimization of aldehyde-sulphuric acid reagents for the detection in thin-layer chromatography. *J. Chromatogr.*, **243**, 139–143.

Stecka, L. (1968). Identification of *Polemonium* saponosides. *Acta Polon. Pharmac.*, **25**, 463–469.

Steinegger, E. and Hänsel, R. (1988). *Lehrbuch der Pharmakognosie und Phytopharmazie*, 4th edn. Springer-Verlag, Berlin.

Steinegger, E. and Marty, S.T. (1976). Glycyrrhizic acid determination in *Radix liquiritiae*. Comparison of existing methods. 1. *Pharm. Helv. Acta*, **51**, 374.

Steiner, M. and Holtzem, H. (1955). Triterpene und Triterpen-Saponine. In *Moderne Methoden der Pflanzenanalyse*, vol. III ed. K. Peach and M.V. Tracey, pp. 58–140. Springer-Verlag, Berlin.

Sticher, O. and Soldati, F. (1979). HPLC Trennung und quantitative Bestimmung der Ginsenoside von *Panax ginseng*, *Panax quinquefolium* und von Ginseng-spezialitäten. *Planta Medica*, **36**, 30–42.

Still, W.C., Kahn, M. and Mitra, A. (1978). Rapid chromatographic techniques for preparative separations with moderate resolution. *J. Org. Chem.*, **43**, 2923–2925.

Stoeckli-Evans, H. (1989). Structure of medicagenic acid: a triterpene. *Acta Cryst.*, *C*, **45**, 341–343.

Stolzenberg, S.J. and Parkhurst, R.M. (1974). Spermicidal action of extracts and compounds from *Phytolacca dodecandra*. *Contraception*, **10**, 135–143.

Stonik, V.A. (1986). Some terpenoid and steroid derivatives from echinoderms and sponges. *Pure Appl. Chem.*, **58**, 423–436.

Stonik, V.A., Mal'tsev, I.I. and Elyakov, G.B. (1982). Structure of thelentosides A and B from the sea cucumber *Thelenota ananas*. *Khim. Prir. Soedin.*, 624–627.

Strigina, L.I., Remennikova, T.M., Shcherdrin, A.P. and Elyakov, G.B. (1972). Triterpenic glycosides of *Caltha silvestris*. *Khim. Prir. Soedin.*, 303–306.

Strömbom, J., Sandberg, F. and Dencker, L. (1985). Studies on absorption and distribution of ginsenoside Rg_1 by whole-body autoradiography and chromatography. *Acta Pharm. Suecia*, **22**, 113–122.

Su, H. and Guo, R. (1986). Experimental studies on bolbostemmosaponins used as intravaginal spermatocidal agents. *Xi'an Yike Daxue Xuebao*, **7**, 225–229 (*Chem. Abstr.*, **108**, 49459).

Suga, Y., Maruyama, Y., Kawanishi, S. and Shoji, J. (1978). Studies on the constituents of phytolaccaceous plants. I. On the structures of phytolaccasaponin B, E and G from the roots of *Phytolacca americana* L. *Chem. Pharm. Bull.*, **26**, 520–525.

Sugishita, E., Amagaya, S. and Ogihara, Y. (1982). Structure–activity studies of some oleanane triterpenoid glycosides and their related compounds from the leaves of *Tetrapanax papyriferum* on antiinflammatory activities. *J. Pharmacobio-Dyn.*, **5**, 379–387.

Sun, R.-Q. and Jia, Z.-J. (1990). Saponins from *Oxytropis glabra*. *Phytochemistry*, **29**, 2032–2034.

Sun, R.-Q., Jia, Z.-J. and Cheng, D.-L. (1991). Three saponins from *Oxytropis* species. *Phytochemistry*, **30**, 2707–2709.

Sun, R., Cheng, D., Jia, Z. and Zhu, Z. (1987). Chemical components of *Oxytropis ochrocephala* BUNGE. II. Structures of two triterpenoid saponins. *Huaxue Xuebao*, **45**, 145–149. (*Chem. Abstr.*, **107**, 36602).

Sung, T.V. and Adam, G. (1991). A sulphated triterpenoid saponin from *Schefflera octophylla*. *Phytochemistry*, **30**, 2717–2720.

Sung, T.V., Steglich, W. and Adam, G. (1991). Triterpene glycosides from *Schefflera octophylla*. *Phytochemistry*, **30**, 2349–2356.

Sung, T.V., Lavaud, C., Porzel, A., Steglich, W. and Adam, G. (1992). Triterpenoids and their glycosides from the bark of *Schefflera octophylla*. *Phytochemistry*, **31**, 227–231.

Suter, R., Tanner, M., Borel, C., Hostettmann, K. and Freyvogel, T.A. (1986). Laboratory and field trials at Ifakara (Kilombero District, Tanzania) on the plant molluscicide *Swartzia madagascariensis*. *Acta Tropica*, **43**, 69–83.

Svechnikova, A.N., Umarova, R.U., Abdullaev, N.D., Gorovits, M.B. and Abubakirov, N.K. (1982). Triterpenoid glycosides of *Astragalus* and their genins VII. Structure of cyclosiversiosides A and C. *Khim. Prir. Soedin.*, 629–632.

Svechnikova, A.N., Umarova, R.U., Gorovits, M.B., Abdullaev, N.D. and Abubakirov, N.K. (1983a). Triterpene glycosides of *Astragalus* and their genins.

XI. Cyclosiversioside G – a triglycoside from *Astragalus sieversianus*. *Khim. Prir. Soedin.*, 312–315.

Svechnikova, A.N., Umarova, R.U., Abdullaev, N.D., Gorovits, M.B., Gorovits, T.T. and Abubakirov, N.K. (1983b). Triterpene glycosides of *Astragalus* and their genins. XIII. Structure of cyclosiversioside H – a triglycoside from *Astragalus sieversianus*. *Khim. Prir. Soedin.*, 460–463.

Sweeley, C.C., Bentley, R., Makita, M. and Wells, W.W. (1963). Gas–liquid chromatography of trimethylsilyl derivatives of sugars and related substances. *J. Am. Chem. Soc.*, **85**, 2497–2507.

Tachibana, K. and Gruber, S.H. (1988). Shark repellant lipophilic constituents in the defense secretion of the Moses sole (*Pardachirus marmoratus*). *Toxicon*, **26**, 839–853.

Tachibana, K., Sakaitanai, M. and Nakanishi, K. (1984). Pavoninins: shark-repelling ichthyotoxins from the defense secretion of the Pacific sole. *Science*, **226**, 703–705.

Tachibana, K., Nakanishi, K. and Gruber, S.H. (1985a). Shark repellents of the Red Sea Moses sole. *Tennen Yuki Kagobutsu Toronkai Koen Yoshishu*, 545 (*Chem. Abstr.*, **104**, 165710).

Tachibana, K., Sakaitani, M. and Nakanishi, K. (1985b). Pavoninins, shark-repelling and ichthyotoxic steroid *N*-acetylglucosaminides from the defense secretion of the sole *Pardachirus pavoninus* (Soleidae). *Tetrahedron*, **41**, 1027–1037.

Tada, A. and Shoji, J. (1972). Studies on the constituents of Ophiopogonis tuber. II. On the structure of ophiopogonin B. *Chem. Pharm. Bull.*, **20**, 1729–1734.

Tada, A., Kaneiwa, Y., Shoji, J. and Shibata, S. (1975). Studies on the saponins of the root of *Platycodon grandiflorum* A. DECANDOLLE. I. Isolation and the structure of platycodin-D. *Chem. Pharm. Bull.*, **23**, 2965–2972.

Tagiev, S.A. and Ismailov, A.I. (1978). Study of sapogenins of *Gypsophila bicolor* roots. *Khim. Prir. Soedin.*, 138–139.

Takabe, S., Takeda, T., Ogihara, Y. and Yamasaki, K. (1981). Triterpenoid glycosides from the roots of *Tetrapanax papyriferum* K. KOCH. Part 2. Structures of new glycosides. *J. Chem. Res. (S)*, 16.

Takabe, S., Takeda, T., Chen, Y. and Ogihara, Y. (1985). Triterpenoid glycosides from the roots of *Tetrapanax papyriferum* K. KOCH. III. Structures of four new saponins. *Chem. Pharm. Bull.*, **33**, 4701–4706.

Takagi, K., Park, E.-H. and Kato, H. (1980). Anti-inflammatory activities of hederagenin and crude saponin isolated from *Sapindus mukorossi* GAERTN. *Chem. Pharm. Bull.*, **28**, 1183–1188.

Takata, M., Shibata, S. and Okuyama, T. (1990). Structures of glycosides of Bai-hua qian-hu. *Planta Medica*, **56**, 133.

Takechi, M. and Tanaka, Y. (1990). Structure–activity relationships of the saponin α-hederin. *Phytochemistry*, **29**, 451–452.

Takechi, M. and Tanaka, Y. (1991). Structure–activity relationships of synthetic diosgenyl monoglycosides. *Phytochemistry*, **30**, 2557–2558.

Takechi, M. and Tanaka, Y. (1992). Structure–activity relationships of synthetic methyl oleanolate glycosides. *Phytochemistry*, **31**, 3789–3791.

Takechi, M., Shimada, S. and Tanaka, Y. (1991). Structure–activity relationships of the saponins dioscin and dioscinin. *Phytochemistry*, **30**, 3943–3944.

Takechi, M., Shimada, S. and Tanaka, Y. (1992). Time course and inhibition of saponin-induced hemolysis. *Planta Medica*, **58**, 128–130.

Takeda, K. (1972). The steroidal sapogenins of the Dioscoreaceae. In *Progress in Phytochemistry*, vol. 3, ed. L. Reinhold and Y. Liwschitz, pp. 287–333. Wiley Interscience, New York.

Takeda, K., Hara, S., Wada, A. and Matsumoto, N. (1963). A systematic simultaneous analysis of steroid sapogenins by thin-layer chromatography. *J. Chromatogr.*, **11**, 562–564.

Takeda, O., Tanaka, S., Yamasaki, K., Kohda, H., Iwanaga, Y. and Tsuji, M. (1989). Screening for molluscicidal activity in crude drugs. *Chem. Pharm. Bull.*, **37**, 1090–1091.

Takemoto, T. (1983). Saponins from *Gynostemma pentaphyllum*. Jap. Pat. 58 57,398 (*Chem. Abstr.*, **99**, 76853).

Takemoto, T. and Kometani, K. (1965). Triterpenglykoside (Mubenine) aus Samen von *Stauntonia hexaphylla*. *Annalen*, **685**, 237–246.

Takemoto, T., Kusano, G. and Yamamoto, N. (1970). Studies on the constituents of *Cimicifuga* spp. VII. Structure of cimicifugenol. *Yakugaku Zasshi*, **90**, 68–72.

Takemoto, T., Nishimoto, N., Nakai, S. *et al.* (1983a). Pfaffic acid, a novel nor-triterpene from *Pfaffia paniculata* KUNTZE. *Tetrahedron Lett.*, **24**, 1057–1060.

Takemoto, T., Arihara, S., Nakajima, T. and Okuhira, M. (1983b). Studies on the constituents of *Gynostemma pentaphyllum* MAKINO. I. Structures of gypenoside I–XIV. *Yakugaku Zasshi*, **103**, 173–185.

Takemoto, T., Arihara, S., Nakajima, T. and Okuhira, M. (1983c). Studies on the constituents of Fructus momordicae. III. Structure of mogrosides. *Yakugaku Zasshi*, **103**, 1167–1173.

Takemoto, T., Arihara, S., Yoshikawa, K., Kusumoto, K., Yano, I. and Hayashi, T. (1984). Studies on the constituents of Cucurbitaceae plants. VI. On the saponin constituents of *Luffa cylindrica* ROEM. (1). *Yakugaku Zasshi*, **104**, 246–255.

Takemoto, T., Arihara, S., Yoshikawa, K., Tanaka, R. and Hayashi, T. (1985). Studies on the constituents of Cucurbitaceae plants. XIII. On the saponin constituents of *Luffa cylindrica* ROEM. (2). *Yakugaku Zasshi*, **105**, 834–839.

Takemoto, T., Arihara, S. and Yoshikawa, K. (1986). Studies on the constituents of Cucurbitaceae plants. XIV. On the saponin constituents of *Gynostemma pentaphyllum* MAKINO. (9). *Yakugaku Zasshi*, **106**, 664–670.

Tamura, K., Mizutani, K. and Yamamoto, S. (1990). Isolation of monodesmoside saponins from rind of *Sapindus trifoliatus* or *-mukorosii* fruits as skin fungicides. Jap. Pat. 02 160,798 (*Chem. Abstr.*, **113**, 218226).

Tanaka, M., Mano, N., Akazai, E., Narni, Y., Kato, F. and Koyama, Y. (1987). Inhibition of mutagenicity by glycyrrhiza extract and glycyrrhizin. *J. Pharmacobio-Dyn.*, **10**, 685–688.

Tanaka, O. (1985). Application of ^{13}C-nuclear magnetic resonance spectrometry to structural studies on glycosides: saponins of *Panax* spp. and natural sweet glycosides. *Yakugaku Zasshi*, **105**, 323–351.

Tanaka, O. (1987). Solubilizing properties of ginseng saponins. *Korean J. Ginseng Sci.*, **11**, 211–218.

Tanaka, O. and Kasai, R. (1984). Saponins of ginseng and related plants. *Prog. Chem. Org. Nat. Prods.*, **46**, 1–76.

Tanaka, R., Nagao, T., Okabe, H. and Yamauchi, T. (1990). Studies on the constituents of *Aster tataricus* L.f. IV. Structures of aster saponins isolated from the herb. *Chem. Pharm. Bull.*, **38**, 1153–1157.

Tani, T., Katsuki, T., Kubo, M., Arichi, S. and Kitagawa, I. (1985). Histochemistry. V. Soyasaponins in soybeans (*Glycine max* MERRILL, seeds). *Chem. Pharm. Bull.*, **33**, 3829–3833.

Taniyama, T., Nagahama, Y., Yoshikawa, M. and Kitagawa, I. (1988a). Saponin and sapogenol. XLIII. Acetylsoyasaponins A_4, A_5 and A_6, new astringent bisdesmosides of soyasapogenol A, from Japanese soybean, the seeds of *Glycine max* MERRILL. *Chem. Pharm. Bull.*, **36**, 2829–2839.

Taniyama, T., Yoshikawa, M. and Kitagawa, I. (1988b). Saponin and sapogenol. XLIV. Soyasaponin composition in soybeans of various origins and soyasaponin content in various organs of soybean. Structure of soyasaponin V from soybean hypocotyl. *Yakugaku Zasshi*, **108**, 562–571.

Tewari, N.C., Ayengar, K.N.N. and Rangaswami, S. (1970). Structure of sapogenins of *Caccinia glauca*. Caccigenin, caccigenin lactone and 23-deoxycaccigenin. *Indian J. Chem.*, **8**, 593–597.

Texeira, J.R.M., Lapa, A.J., Souccar, C. and Valle, J.R. (1984). Timbós: ichthyotoxic plants used by Brazilian Indians. *J. Ethnopharmacol.*, **10**, 311–318.

Texier, O., Ahond, A., Regerat, F. and Pourrat, H. (1984). A triterpenoid saponin from *Ficaria ranunculoides* tubers. *Phytochemistry*, **23**, 2903–2905.

Tian, J., Wu, F., Qiu, M. and Nie, R. (1993a). Two triterpenoid saponins from *Pterocephalus bretschneidri*. *Phytochemistry*, **32**, 1539–1542.

Tian, J., Wu, F., Qiu, M. and Nie, R. (1993b). Triterpenoid saponins from *Pterocephalus hookeri*. *Phytochemistry*, **32**, 1535–1538.

Tillekeratne, L.M.V., Liyanage, G.K., Ratnasooriya, W.D., Ksebati, M.B. and Schmitz, F.J. (1989). A new spermatostatic glycoside from the soft coral *Sinularia crispa*. *J. Nat. Prod.*, **52**, 1143–1145.

Timbekova, A.E. and Abubakirov, N.K. (1984). Triterpene glycosides of alfalfa. I. Medicoside G – a novel bisdesmoside from *Medicago sativa*. *Khim. Prir. Soedin.*, 451–458.

Timbekova, A.E. and Abubakirov, N.K. (1985). Triterpene glycosides of alfalfa. II. Medicoside C. *Khim. Prir. Soedin.*, 805–808.

Timbekova, A.E. and Abubakirov, N.K. (1986). Triterpene glycosides of alfalfa. III. Medicoside I. *Khim. Prir. Soedin.*, 607–610.

Timon-David, P., Julien, J., Gasquet, M., Balansard, G. and Bernard, P. (1980). Recherche d'une activité antifongique de plusieurs principes actifs. Extraits du lierre grimpant: *Hedera helix* L. *Ann. Pharmac. Fr.*, **38**, 545–552.

Tingey, W.M. (1984). Glycoalkaloids as pest resistance factors. *Am. Potato J.*, **61**, 157–167.

Tiwari, K.P. and Choudhary, R.N. (1979). A new saponin from *Wisteria sinensis*. *Indian J. Chem.*, **18B**, 389–390.

Tiwari, K.P. and Masood, M. (1980). Obtusilobicinin, a new saponin from *Anemone obtusiloba*. *Phytochemistry*, **19**, 1244–1247.

Tiwari, K.P. and Minocha, P.K. (1980). Pavophylline, a new saponin from the stem of *Pavonia zeylanica*. *Phytochemistry*, **19**, 701–704.

Tiwari, K.P., Srivastava, S.D. and Srivastava, S.K. (1980). α-L-Rhamnopyranosyl-3β-hydroxy-lup-20(29)-en-28-oic acid from the stem of *Dillenia pentagyna*. *Phytochemistry*, **19**, 980–981.

Tokuda, H., Konoshima, T., Kozuka, M. and Kimura, T. (1988). Inhibitory effects of 12-*O*-tetradecanoylphorbol-13-acetate and teleocidin B induced Epstein–Barr virus by saponin and its related compounds. *Cancer Lett.*, **40**, 309–317.

Tomas-Barbaren, F.A. and Hostettmann, K. (1988). A cytotoxic triterpenoid and flavonoids from *Crossopteryx febrifuga*. *Planta Medica*, **54**, 266–267.

Tomita, Y. and Uomori, A. (1971). Biosynthesis of sapogenins in tissue cultures of *Dioscorea tokoro* MAKINO. *J. Chem. Soc., Chem. Commun.*, 284.

Tomowa, M.P., Panowa, D. and Wulfson, N.S. (1974). Steroid saponins and sapogenins. IV. Saponins from *Tribulus terrestris*. *Planta Medica*, **25**, 231–237.

Tomowa, M., Gyulemetova, R., Zarkova, S., Peeva, S., Pangarova, T. and Simova, M. (1981). Steroidal saponins from *Tribulus terrestris* L. with a stimulating action on the sexual functions. *Int. Conf. Chem. Biotechnol. Biol. Act. Nat. Prod.*, **3**, 298–302.

Topping, D.L., Hood, R.L., Illman, R.J., Storer, G.B. and Oakenfull, D.G. (1978). Effects of dietary saponin on bile acid secretion and plasma cholesterol in the rat. *Proc. Nutr. Soc. Aust.*, **3**, 68 (*Chem. Abstr.*, **90**, 85661).

Topping, D.L., Illman, R.J., Fenwick, D.E. and Oakenfull, D.G. (1980). *Proc. Nutr. Soc. Aust.*, **5**, 195.

Tori, K. and Aono, K. (1964). *Ann. Rept. Shionogi Res. Lab.*, **14**, 136.

Tori, K., Seo, S., Shimaoka, A. and Tomita, Y. (1974). Carbon-13 NMR spectra of olean-12-enes. Full signal assignments including quaternary carbon signals assigned by use of direct ^{13}C, ^{1}H spin couplings. *Tetrahedron Lett.*, 4227–4230.

Tori, K., Yoshimura, Y., Seo, S., Sakurawi, K., Tomita, Y. and Ishii, H. (1976a). Carbon-13 NMR spectra of saikogenins. Stereochemical dependence in hydroxylation effects upon carbon-13 chemical shifts of oleanene-type triterpenoids. *Tetrahedron Lett.*, 4163–4166.

Tori, K., Seo, S., Yoshimura, Y., Nakamura, M., Tomita, Y. and Ishii, H. (1976b). Carbon-13 NMR spectra of saikosaponins A, C, D and F isolated from *Bupleurum falcatum* L. *Tetrahedron Lett.*, 4167–4170.

Tori, K., Seo, S., Terui, Y., Nishikawa, J. and Yasuda, F. (1981). Carbon-13 NMR spectra of 5β-steroidal sapogenins. Reassignment of the F-ring carbon signals of (25*S*)-spirostans. *Tetrahedron Lett.*, **22**, 2405–2408.

Toyota, M., Msonthi, J.D. and Hostettmann, K. (1990). A molluscicidal and antifungal triterpenoid saponin from the roots of *Clerodendrum wildii*. *Phytochemistry*, **29**, 2849–2851.

Tripathi, R.D. and Tiwari, K.P. (1980). Geniculatin, a triterpenoid saponin from *Euphorbia geniculata*. *Phytochemistry*, **19**, 2163–2166.

Trofimova, N.A., Romanova, G.A., Benson, N.A. and Dmitruk, S.E. (1985). Chemical composition of isolated tissue cultures and the intact *Orthosiphon stamineus* BENTH. plant. *Rastit. Resur.*, **21**, 90–95 (*Chem. Abstr.*, **102**, 128861).

Tsai, T.-H. and Chen, C.-F. (1991). Determination of three active principles in licorice extract by reversed-phase high-performance liquid chromatography. *J. Chromatogr.*, **542**, 521–525.

Tschesche, R. (1971). Advances in the chemistry of antibiotic substances from higher plants. In *Pharmacognosy and Phytochemistry*, ed. H. Wagner and L. Hörhammer, pp. 274–289. Springer, Berlin.

Tschesche, R. and Axen, U. (1964). Triterpenes. XV. The cryptoescin question. *Naturwissenschaften*, 51, 359–360.

Tschesche, R. and Balle, G. (1963). Ueber Saponine der Spirostanol-reihe – X. Zur Konstitution der Samen Saponine von *Digitalis lanata* EHRH. *Tetrahedron*, 19, 2323–2332.

Tschesche, R. and Forstmann, D. (1957). Ueber Triterpene, IV. Musennin, ein Wurmwirksames Saponin aus der Rinde von *Albizzia anthelmintica*. *Chem. Ber.*, 90, 2383–2394.

Tschesche, R. and Gutwinski, H. (1975). Steroidsaponine mit mehr als einer Zuckerkette, X. Capsicosid, ein bisdesmosidisches 22-Hydroxyfurostanol-Glycosid aus dem Samen von *Capsicum annuum* L. *Chem. Ber.*, 108, 265–272.

Tschesche, R. and Heesch, A. (1952). Ueber Triterpene, II. Mitteil.: Gratiosid, ein Triterpenglykosid aus *Gratiola officinalis* L. *Chem. Ber.*, 85, 1067–1077.

Tschesche, R. and Hulpke, H. (1966). Biosynthesis of plant steroids. III. Biosynthesis of spirostanols from cholesterol glucoside. *Z. Naturforsch.*, 21B, 494–495.

Tschesche, R. and Kämmerer, F.-J. (1969). Ueber Triterpene, XXVII. Die Struktur von Musennin und Desglucomusennin. *Annalen*, 724, 183–193.

Tschesche, R. and Richert, K.-H. (1964). Ueber Saponine der Spirostanolreihe – XI. Nuatigenin, ein Cholegenin-Analogon des Pflanzenreiches. *Tetrahedron*, 20, 387–398.

Tschesche, R. and Schultz, H. (1974). Ueber Triterpene, XXX. Ueber das Hauptsaponin der Kornrade (*Agrostemma githago* L.). *Chem. Ber.*, 107, 2710–2719.

Tschesche, R. and Striegler, H. (1965). Triterpenes. XVI. Identity of tenuifolic acid with senegenin. *Naturwissenschaften*, 52, 303.

Tschesche, R. and Wulff, G. (1963). Ueber Saponine der Spirostanolreich – IX. Die Konstitution des Digitonins. *Tetrahedron*, 19, 621–634.

Tschesche, R. and Wulff, G. (1964). Konstitution und Eigenschaften der Saponine. *Planta Medica*, 12, 272–292.

Tschesche, R. and Wulff, G. (1965). Ueber die antimikrobielle Wirksamkeit von Saponinen. *Z. Naturforsch.*, 20B, 543–546.

Tschesche, R. and Wulff, G. (1972). Chemie und Biologie der Saponine. *Prog. Chem. Org. Nat. Prods.*, 30, 462–606.

Tschesche, R. and Ziegler, F. (1964). Ueber Triterpene, XII. Ueber die Saponine der Wurzeln von *Primula elatior* L. SCHREBER. *Annalen*, 674, 185–195.

Tschesche, R., Duphorn, J. and Snatzke, G. (1963). Ueber Triterpene, IX. Ueber die Konstitution der Cincholsäure und die sauren Triterpenglykoside der Chinarinde. *Annalen*, 667, 151–163.

Tschesche, R., Inchaurrondo, F. and Wulff, G. (1964). Ueber Triterpene, XVII. Das Aglykon des Cyclamins. *Annalen*, 680, 107–118.

Tschesche, R., Schmidt, W. and Wulff, G. (1965). Triterpenes. XIX. Isolation and structure elucidation of ivy saponins (*Hedera helix*). *Z. Naturforsch.*, 20B, 708–709.

Tschesche, R., Striegler, H. and Fehlhaber, H.-W. (1966a). Ueber Triterpene. XIX. Die Struktur des Cyclamiretins A. *Annalen*, **691**, 165–171.

Tschesche, R., Kottler, R. and Wulff, G. (1966b). Ueber Saponine der Spirostanolreihe, XII. Ueber Parillin, ein Saponin mit stark verzweigter Zuckerkette. *Annalen*, **699**, 212–222.

Tschesche, R., Tjoa, B.T. and Wulff, G. (1966c). Ueber Triterpene, XXI. Ueber die Sapogenine aus den Wurzeln von *Primula veris* L. *Annalen*, **696**, 160–179.

Tschesche, R., Tjoa, B.T., Wulff, G. and Noronha, R.V. (1968). Steroidsaponine mit mehr als einer Zuckerkette, III. Convallamarosid, ein weiteres 22-Hydroxy-Furostanolsaponin. *Tetrahedron Lett.*, 5141–5144.

Tschesche, R., Weber, A. and Wulff, G. (1969a). Ueber Triterpene, XXV. Ueber die Struktur des 'Theasaponins', eines Gemisches von Saponinen aus *Thea sinensis* L. mit stark antiexsudativer Wirksamkeit. *Annalen*, **721**, 209–219.

Tschesche, R., Lüdke, G. and Wulff, G. (1969b). Steroidsaponine mit mehr als einer Zuckerkette, II. Sarsaparillosid, ein bidesmosidisches 22-Hydroxyfurostanol-saponin. *Chem. Ber.*, **102**, 1253–1269.

Tschesche, R., Mercker, H.-J. and Wulff, G. (1969c). Ueber Triterpene, XXIV. Die Konstitution der Zuckerkette des Cyclamins. *Annalen*, **721**, 194–208.

Tschesche, R., Tauscher, M., Fehlhaber, H.-W. and Wulff, G. (1969d). Steroidsaponine mit mehr als einer Zuckerkette, IV. Avenacosid A, ein bisdesmosidisches Steroidsaponin aus *Avena sativa*. *Chem. Ber.*, **102**, 2072–2082.

Tschesche, R., Rehkämper, H. and Wulff, G. (1969e). Ueber Triterpene, XXVIII. Ueber die Saponine des Spinats (*Spinacia oleracea* L.). *Annalen*, **726**, 125–135.

Tschesche, R., Wulff, G., Weber, A. and Schmidt, H. (1970). Frasshemmende Wirkung von Saponinen auf Termiten (*Isoptera, Reticulitermes*). *Z. Naturforsch.*, **25B**, 999–1000.

Tschesche, R., Seidel, L., Sharma, S.C. and Wulff, G. (1972). Steroidsaponine mit mehr als einer Zuckerkette, VI. Ueber Lanatigosid und Lanagitosid, zwei bisdesmosidische 22-Hydroxy-Furostanol-Glykoside aus den Blättern von *Digitalis lanata* EHRH. *Chem Ber.*, **105**, 3397–3406.

Tschesche, R., Hermann, K.-H., Langlais, R., Tjoa, B.T. and Wulff, G. (1973). Steroidsaponine mit mehr als einer Zuckerkette, VII. Convallomarosid, ein trisdesmosidisches 22-Hydroxyfurostanol-Glykosid aus den Wurzeln von *Convallaria majalis* L. *Chem. Ber.*, **106**, 3010–3019.

Tschesche, R., Goossens, B. and Töpfer, A. (1976). Zur Einführung des Stickstoffs und zum gemeinsamen Vorkommen von 25(*R*)- und 25(*S*)-Steroidalkaloiden in Solanaceen. *Phytochemistry*, **15**, 1387–1389.

Tschesche, R., Sepulveda, S. and Braun, T.M. (1980). Ueber Triterpene, XXXIII. Ueber das Saponin der Blüten von *Verbascum phlomoides* L. *Chem. Ber.*, **113**, 1754–1760.

Tschesche, R., Wagner, R. and Widera, W. (1983). Saponine aus den Wurzeln von *Primula elatior* (L.) SCHREBER. Konstitution eines Nebensaponins und Revision der Zuckerkette des Hauptsaponins. *Annalen*, 993–1000.

Tsuda, Y., Kaneda, M., Yasufuku, N. and Shimizu, Y. (1981). *Lycopodium* triterpenoids. (11). The structures of inundoside-A, -B, -C, -D_1, -D_2, -E, -F and -G, triterpenoid-glycosides occurring in *Lycopodium inundatum* L. *Chem. Pharm. Bull.*, **29**, 2123–2134.

Tsuda, Y., Kiuchi, F. and Liu, H.-M. (1989). Establishment of the structure of gymnemagenin by X-ray analysis and the structure of deacylgymnemic acid. *Tetrahedron Lett.*, **30**, 361–362.

Tsuji, H., Osaka, S. and Kiwada, H. (1991). Targeting of liposomes surface-modified with glycyrrhizin to the liver. I. Preparation and biological disposition. *Chem. Pharm. Bull.*, **39**, 1004–1008.

Tsukamoto, T. and Ueno, Y. (1936). Glucosides of *Dioscorea tokoro* MAKINO. I. Dioscin, dioscoreasapotoxin and diosgenin. *Yakugaku Zasshi*, **56**, 802.

Tsukamoto, T., Ueno, Y. and Ohta, Z. (1936). Constitution of diosgenin. I. Glucoside of *Dioscorea tokoro* MAKINO. *J. Pharm. Soc. Jpn.*, **56**, 931–940.

Tsukamoto, T., Kawasaki, T., Naraki, A. and Yamauchi, T. (1954). Saponins of Japanese Dioscoreaceae. II. Water-insoluble saponins from *Dioscorea nipponica* MAKINO, *D. gracillima* MIQ., and *D. tenuipes* FRANCH. ET SAV. *Yakugaku Zasshi*, **74**, 984–987.

Tsukitani, Y. and Shoji, J. (1973). Studies on the constituents of Senegae radix. III. The structures of senegin-III and -IV, saponins from *Polygala senega* LINNE var. *latifolia* TORRY ET GRAY. *Chem, Pharm. Bull.*, **21**, 1564–1574.

Tsukitani, Y., Kawanishi, S. and Shoji, J. (1973). Studies on the constituents of Senegae radix. II. The structure of senegin-II, a saponin from *Polygala senega* LINNE var. *latifolia* TORRY ET GRAY. *Chem. Pharm. Bull.*, **21**, 791–799.

Tsurumi, S., Takagi, T. and Hashimoto, T. (1992). A γ-pyronyl-triterpenoid saponin from *Pisum sativum*. *Phytochemistry*, **31**, 2435–2438.

Tursch, B., Tursch, E., Harrison, I.T. *et al.* (1963). Terpenoids. LIII. Demonstration of ring conformational changes in triterpenes of the β-amyrin class isolated from *Stryphnodendron coriaceum*. *J. Org. Chem.*, **28**, 2390–2394.

Tursch, B., Leclercq, J. and Chiurdoglu, G. (1965). Structure·de l'acide mesembryanthémoidigénique, triterpène nouveau des cactacées. *Tetrahedron Lett.*, 4161–4166.

Ujikawa, K. and Purchio, A. (1989). Substâncias antifungicas, inhibidoras de *Aspergillus flavus* e de outras fungicas, isoladas de *Agave sisalana* (sisal). *Cienc. Cult.* (Sao Paulo), **41**, 1218–1224.

Uniyal, G.C., Agrawal, P.K., Thakur, R.S. and Sati, O.P. (1990). Steroidal glycosides from *Agave cantala*. *Phytochemistry*, **29**, 937–940.

Uniyal, G.C., Agrawal, P.K., Sati, O.P. and Thakur, R.S. (1991). Agaveside C, a steroidal glycoside from *Agave cantala*. *Phytochemistry*, **30**, 1336–1339.

Uniyal, S.K. and Sati, O.P. (1992). Triterpenoid saponins from roots of *Clematis grata*. *Phytochemistry*, **31**, 1427–1428.

Ushio, Y. and Abe, H. (1991). The effects of saikosaponin on macrophage functions and lymphocyte proliferation. *Planta Medica*, **57**, 511–514.

Ushio, Y. and Abe, H. (1992). Inactivation of measles virus and herpes simplex virus by saikosaponin a. *Planta Medica*, **58**, 171–173.

Usmanghani, K. (1977). Isolation of sapogenin structures from the roots of *Lysimachia mauritiana* LAM. *Pak. J. Sci. Ind. Res.*, **20**, 393–395 (*Chem. Abstr.*, **91**, 27215).

Uvarova, N.I., Gorshkova, R.P. and Elyakov, G. B. (1963). Preparation of the total glycoside preparation from ginseng by using sephadex. *Chem. Abstr.*, **60**, 13099.

Vagujfalvi, D. (1965). Biogenesis of steroid glycoalkaloids in plants. *Herba Hung.*, **4**, 170–184.

Valent. B.S., Darvill, A.G., McNeil, M., Robertsen, B.K. and Albersheim, P. (1980). A general and sensitive chemical method for sequencing the glycosyl residues of complex carbohydrates. *Carbohydr. Res.*, **79**, 165–192.

van Antwerp, C.L., Eggert, H., Meakins, G.D., Miners, J.O. and Djerassi, C. (1977). Additivity relationships in carbon-13 nuclear magnetic resonance spectra of dihydroxy steroids. *J. Org. Chem.*, **42**, 789–793.

van Gelder, W.M.J. (1984). A new hydrolysis technique for steroid glycoalkaloids with unstable aglycones from *Solanum* species. *J. Sci. Food Agric.*, **35**, 487–494.

Vanhaelen, M. (1972). New saponoside from *Bersama yangambiensis*. *Phytochemistry*, **11**, 1111–1116.

Vanhaelen, M. and Vanhaelen-Fastré, R. (1984). Quantitative determination of biologically active constituents in crude extracts of medicinal plants by thin-layer chromatography-densitometry. II. *Eleutherococcus senticosus* MAXIM., *Panax ginseng* MEYER and *Picrorrhiza kurroa* ROYLE. *J. Chromatogr.*, **312**, 497–503.

Vanhoutte, P.M. (1986). Alpha-adrenergic effects of *Ruscus aculeatus* on cutaneous veins. Modulation by local temperature. In *Advances in Medicinal Phytochemistry*, ed. D. Barton and W.D. Ollis, pp. 187–194. Libby, London.

van Kampen, E.J. and Zijlstra, W.G. (1961). Standardisation of hemoglobinometry. II. The hemiglobin cyanide method. *Clin. Chim. Acta*, **6**, 538–544 (*Chem. Abstr.*, **55**, 27503).

Varshney, I.P. and Badhwar, G. (1971). Saponins and sapogenins of *Bassia longifolia* and *Mimusops elengi*. *Proc. Natl. Acad. Sci. India, Sect. A*, **41**, 21-23 (*Chem. Abstr.*, **77**, 123778).

Varshney, I.P. and Beg. M.F.A. (1977) saponins from the seeds of *Luffa aegyptiaca* MILL.: isolation of aegyptinin A and aegyptinin B. *Indian J. Chem.*, **15B**, 394.

Varshney, I.P. and Khan, M.S.Y. (1961). Saponins and sapogenins. VII. Acid sapogenins isolated from Maharashtrian *Albizzia odoratissima* seeds. *J. Pharm. Sci.*, **50**, 923–925.

Varshney, I.P. and Khan, M.S.Y. (1962). Saponins and sapogenins. XII. Isolation of odoratissimin, a new saponin from the seeds of *Albizzia odoratissima*. *J. Sci. Indian Res.*, **21B**, 30–33 (*Chem. Abstr.*, **57**, 9938).

Varshney, I.P. and Khanna, N. (1978). Partial structure of a new saponin samanin-D from the flowers of *Pithecolobium saman* BENTH *Indian J. Pharmac. Sci.*, **40**, 60–61.

Varshney, I.P. and Shamsuddin, K.M. (1962). Saponins and sapogenins. XVI. Echinocystic acid from the seeds of *Albizzia amara*. *J. Sci. Indian Res.*, **21B**, 347.

Varshney, I.P. and Sharma, S.C. (1968). Saponins and sapogenins. XXII. Chemical investigation of seeds of *Myristica fragrans*. *Indian J. Chem.*, **6**, 474.

Varshney, I.P., Jain, D.C., Beg, M.F.A. and Srivastava, H.C. (1985). Samanin-C and samanin-E, new saponins from *Pithecolobium saman*. *Fitoterapia*, **56**, 281–283.

Vasanth, S., Kundu, A.B., Panda, S.K. and Patra, A. (1991). Vicoside A, a 28-nortriterpenoid glucoside from *Vicoa indica*. *Phytochemistry*, **30**, 3053–3055.

Vasyukova, N.I., Paseshnichenko, V.A., Davydova, M.A. and Chalenko, G.I. (1977). Study of the fungitoxic properties of steroid saponins from *Dioscorea deltoidea* rhizomes. *Prikl. Biokhim. Mikrobiol*, **13**, 172–176 (*Chem. Abstr.*, **87**, 33923).

Vazquez, M.J., Quinoa, E., Riguera, R., San Martin, A. and Darias, J. (1992). Santiagoside, the first asterosaponin from an Antarctic starfish (*Neosmilaster georgianus*). *Tetrahedron*, **48**, 6739–6746.

Vecherko, L.P., Kabanov, V.S. and Zinkevich, E.P. (1971). Structure of calenduloside B from *Calendula officinalis* roots. *Khim. Prir. Soedin.*, 533.

Vecherko, L.P., Zinkevich, E.P., Kogan, L.M. (1973). Structure of calenduloside F from *Calendula officinalis* roots. *Khim. Prir. Soedin*, 561–562.

Vecherko, L.P., Sviridov, A.F., Zinkevich, E.P. and Kogan, L.M. (1975). Structure of calendulosides C and D from *Calendula officinalis* roots. *Khim. Prir. Soedin.*, 366–373.

Velieva, M.N., Iskenderov, G.B., Amiralieva, U.A. and Gazhieva, G.Ya (1988). Structure and hypocholesteremic properties of saponins from *Gypsophila capitata*. *Chem. Abstr.*, **109**, 70410.

Vereskovskii, V.V. and Chekalinskaya, I.I. (1980). Polyphenols and saponins of *Rhaponticum carthamoides* (WILLD) JLJIN. *Chem. Abstr.*, **92**, 3212

Vidal-Ollivier, E., Babadjamian, A., Maillard, C., Elias, R. and Balansard, G. (1989a). Identification et dosage par chromatographie liquide haute performance de six saponosides dans les fleurs de *Calendula officinalis*. *Pharm. Acta Helv.*, **64**, 156–159.

Vidal-Ollivier, E., Balansard, G., Faure, R. and Babadjamian, A. (1989b). Revised structures of triterpenoid saponins from the flowers of *Calendula officinalis*. *J. Nat. Prod.*, **52**, 1156–1159.

Vidal-Ollivier, E., Babadjamian, A., Faure, R. *et al.* (1989c). Two-dimensional NMR studies of triterpenoid glycosides. II. 1H NMR assignment of arvensoside A and B, calenduloside C and D. *Spectrosc. Lett.*, **22**, 579–584.

Vogel, G. (1963). Zur Pharmakologie von Saponinen. *Planta Medica*, **11**, 362–376.

Vogel, G. and Marek, M.-L. (1962). Zur Pharmakologie einiger Saponine. *Arzneimittelforsch.*, **12**, 815–825.

Vogel, G., Marek, M.-L. and Oertner, R. (1968a). Zur Frage der Wertbestimmung antiexsudativ wirkender Pharmaka. *Arzneimittelforsch.*, **18**, 426–429.

Vogel, G., Marek, M.-L. and Oertner, R. (1968b). Zur Pharmakologie des Teesamen-Saponins, eines Saponingemisches aus *Thea sinensis* L. *Arzneimittelforsch.*, **18**, 1466–1467.

Vogel, G., Marek, M.-L. and Oertner, R. (1970). Untersuchungen zum Mechanismus der therapeutischen und toxischen Wirkung des Rosskastanien-Saponins Aescin. *Arzneimittelforsch.*, **20**, 699–703.

Voigt, G. and Hiller, K. (1987). Neuere Ergebnisse zur Chemie und Biologie der Steroidsaponine. *Sci. Pharm.*, **55**, 201–227.

Voigt, G., Thiel, P., Hiller, K., Franke, P. and Habisch, D. (1985). Zur Struktur des Hauptsaponins der Wurzeln von *Eryngium planum* L. *Pharmazie*, **40**, 656.

Vora, P.S. (1982). High-pressure liquid chromatographic determination of glycyrrhizic acid or glycyrhizic acid salts in various licorice products: collaborative study. *J. Assoc. Off. Anal. Chem.*, **65**, 572–574.

Vugalter, M.M., Dekanosidze, G.E., Dzhikiya, O.D. and Kemertelidze, E.P. (1986). Triterpene glycosides of *Caltha polypetala*. Glycosides A, B, D, E and F. *Khim. Prir. Soedin.*, 712–716.

Vugalter, M.M., Dekanosidze, G.E., Dzhikiya, O.D. and Kemertelidze, E.P. (1987). Triterpene glycosides from *Caltha polypetala*. *Izv. Akad. Nauk Gruz. SSR, Ser. Khim.*, 13, 314–315.

Wagner, H. and Reger, H. (1986a). Folium Hederae-Extrakte: HPLC-Analyse. *Dtsch. Apoth. Ztg.*, 126, 2613–2617.

Wagner, H. and Reger, H. (1986b). Radix Primulae-Extrakte: HPLC-Analyse. *Dtsch. Apoth. Ztg.*, 126, 1489–1493.

Wagner, H., Ott, S., Jurcic, K., Morton, J. and Neszmelyi, A. (1983). Chemistry, ^{13}C-NMR study and pharmacology of two saponins from *Colubrina asiatica*. *Planta Medica*, 48, 136–141.

Wagner, H., Bladt, S. and Zgainski, E.M. (1984a). *Plant Drug Analysis*. Springer-Verlag, Berlin.

Wagner, H., Nickl, H. and Aynehchi, Y (1984b). Molluscicidal saponins from *Gundelia tournefortii*. *Phytochemistry*, 23, 2505–2508.

Wagner, H., Reger, H. and Bauer, R. (1985). Saponinhaltige Drogen und Fertigarzneimittel: HPLC-Analyse am Beispiel von Rosskastaniensamen. *Dtsch. Apoth. Ztg.*, 125, 1513–1518.

Wagner, H., Ludwig, C., Grotjahn, L. and Khan, M.S.Y. (1987). Biologically active saponins from *Dodonaea viscosa*. *Phytochemistry*, 26, 697–701.

Wagner, J. and Hoffmann, H. (1967). Ueber Inhaltsstoffe des Rosskastanien-samens, V. Untersuchung an Triterpenglykosiden. *Z. Physiol. Chem.*, 348, 1697–1698.

Wagner, J. and Sternkopf, G. (1958). Chemische und physiologische Unter-suchungen über das Saponin der Zuckerrübe. *Nahrung*, 2, 338–357.

Wall, M.E. and Serota, S. (1959). Conversion of steroidal sapogenins to pseudo-sapogenins. US Patent 2 870 143 (*Chem. Abstr.*, 53, 16211).

Wall, M.E., Eddy, C.R., McClennan, M.L. and Klumpp. M.E. (1952). Detection and estimation of steroidal sapogenins in plant tissue. *Anal. Chem.*, 24, 1337–1341.

Wallenfels, K. (1950). Ueber einen neuen Nachweis reduzierender Zucker im Papierchromatogramm und dessen quantitative Auswertung. *Naturwissen-schaften*, 37, 491–492.

Wallenfels, K., Bechtler, G., Kuhn, R., Trischmann, H. and Egge, H. (1963). Permethylation of oligomeric and polymeric hydrocarbons and quantitative analysis of the cleavage products. *Angew. Chem.* , 75, 1014–1022.

Waller, D.P., Zaneveld, L.J.D. and Fong, H.H.S. (1980). *In vitro* spermicidal activity of gossypol. *Contraception*, 22, 183–187.

Walter, E.D. (1960). Note on saponins and their sapogenins from strawberry clover. *J. Am. Pharm. Assoc.*, 49, 735–736.

Walter, E.D. (1961). Isolation of oleanolic acid saponin from trefoil (*Lotus corniculatus* var. *viking*). *J. Pharm. Sci.*, 50, 173.

Waltho, J.P., Williams, D.H., Mahato, S.B., Pal, B.C. and Barna, J.C.J. (1986). Structure elucidation of two triterpenoid tetrasaccharides from *Androsace saxifragifolia*. *J. Chem. Soc., Perkin 1*, 1527–1531.

Wang, B.Z., Wang, G.Z. and Liu, A.J. (1981). Antiinflammatory effect of saikosaponins. *Chung-Kuo Yaoli Hsueh Pao*, **2**, 60–63 (*Chem. Abstr.*, **94**, 150447).

Wang, H., He, K., Ling, L. and Li, H. (1989a). Chemical study of *Astragalus* plant. II. Structures of asernestioside A and B, isolated from *Astragalus ernestii* COMB. *Huaxue Xuebao*, **47**, 583–587 (*Chem. Abstr.*, **111**, 228965).

Wang, H.K., He, K., Ji, L., Tezuka, Y., Kikuchi, T. and Kitagawa, I. (1989b). Asernestioside C, a new minor saponin from the roots of *Astragalus ernestii* COMB.; first example of negative nuclear Overhauser effect in the saponins. *Chem. Pharm. Bull.*, **37**, 2041–2046.

Wang, H., Yu, D., Liang, X., Watanabe, N., Tamai, M. and Omura, S. (1989c). Yemuoside YM_7, YM_{11}, YM_{13} and YM_{14}: four nortriterpenoid saponins from *Stauntonia chinensis. Planta Medica*, **55**, 303–306.

Wang, M., Guan, X., Han, X. and Hong, S. (1992). A new triterpenoid saponin from *Ardisia crenata. Planta Medica*, **58**, 205–207.

Wang, M.-K., Wu, F.-E. and Chen, Y.-Z. (1993). Triterpenoid saponins from *Anemone hupehensis. Phytochemistry*, **34**, 1395–1397.

Warashina, T., Miyase, T. and Ueno, A. (1991). Novel acylated saponins from *Tragopogon porrifolius* L. Isolation and the structures of tragopogonsaponins A–R. *Chem. Pharm. Bull.*, **39**, 388–396.

Wasicky, R. and Wasicky, M. (1961). *Polygala brasiliensis* L., eine saponinreiche *Polygala*-Art Brasiliens. *Qualitas Plant Mat. Vegetabiles*, **8**, 65–79.

Watanabe, K., Fujino, H., Morita, T., Kasai, R. and Tanaka, O. (1988). Solubilization of saponins of *Bupleuri radix* with ginseng saponins: cooperative effect of dammarane saponins. *Planta Medica*, **54**, 405–409.

Watanabe, N., Saeki, J., Sumimoto, M., Kondo, T. and Kurotori, S. (1966). Antitermitic substance of *Ternstroemia japonica* wood. I. *Nippon Mokuzai Gakkaishi*, **12**, 236–238 (*Chem. Abstr.*, **66**, 102477).

Watanabe, Y., Sanada, S., Ida, Y. and Shoji, J. (1983). Comparative studies on the constituents of Ophiopogonis tuber and its congeners. II. Studies on the constituents of the subterranean part of *Ophiopogon planiscapus* NAKAI. *Chem. Pharm. Bull.*, **31**, 3486–3495.

Weil, L. (1901). Beiträge zur Kenntnis der Saponinsubstanzen und ihrer Verbreitung. *Arch. Pharm.*, **239**, 363–373.

Weissenberg, M., Klein, M., Meisner, J. and Ascher, K.R.S. (1986). Larval growth inhibition of the spiny bollworm, *Earias insulana*, by some steroidal secondary plant compounds. *Entomol. Exp. Appl.*, **42**, 213–217.

Wenkert, E., Baddeley, G.V., Burfitt, I.R. and Moreno, L.N. (1978). Carbon-13 nuclear magnetic resonance spectroscopy of naturally-occurring substances. LVII. Triterpenes related to lupane and hopane. *Org. Magn. Reson.*, **11**, 337–343.

Weston, R.J., Gottlieb, H.E., Hagaman, E.W. and Wenkert, E. (1977). Carbon-13 nuclear magnetic resonance spectroscopy of naturally occurring substances. *Aust. J. Chem.*, **30**, 917–921.

Wierenga, J.M. and Hollingworth, R.M. (1992). Inhibition of insect acetylcholinesterase by the potato glycoalkaloid α-chaconine. *Natural Toxins*, **1**, 96–99.

Wilkins, A.L., Ronaldson, K.J., Jager, P.M. and Bird, P.W. (1987). A carbon-13 NMR study of some oxygenated hopane triterpenes. *Aust. J. Chem.*, **40**, 1713–1721.

Wilkomirski, B. and Kasprzyk, Z. (1979). Free and ester-bound triterpene alcohols and sterols in cellular subfractions of Calendula officinalis flowers. Phyto-chemistry, 18, 253–255.

Willker, W. and Leibfritz, D. (1992). Complete assignment and conformational studies of tomatine and tomatidine. Mag. Res. Chem., 30, 645–650.

Willuhn, G. and Köthe, U. (1983). Das bittere Prinzip des Bittersüssen Nacht-schattens, Solanum dulcamara L. – Isolierung und Struktur neuer Furos-tanolglykoside. Arch. Pharm., 316, 678–687.

Wilson, R.H., Poley, G.W. and de Eds, F. (1961). Some pharmacologic and toxicologic properties of tomatine and its derivatives. Toxicol. Appl. Phar-macol. 3, 39–48.

Windaus, A. (1909). Ueber die Entgiftung der Saponine durch Cholesterin. Ber., 42, 238–246.

Windaus, A. and Schneckenburger, A. (1913). Ueber Gitonin, ein neues Digitalis-Glykosid. Ber., 46, 2628–2633.

Winterstein, A. and Stein, G. (1932). Untersuchungen in der Saponinreihe. XI. Mitteilung. Ueber die Oxy-triterpensäure-glykoside der Araliaceen. Z. Physiol. Chem., 211, 5–18.

Woitke, H.-D., Kayser, J.-P. and Hiller, K. (1970). Fortschritte in der Erforschung der Triterpensaponine. Pharmazie, 25, 133–143, 213–241.

Wojciechowski, Z.A. (1975). Biosynthesis of oleanolic acid glycosides by sub-cellular fractions of Calendula officinalis seedlings. Phytochemistry, 14, 1749–1753.

Wojciechowski, Z., Jelonkiewicz-Konador, A., Tomaszewski, M., Jankowski, J. and Kasprzyk, Z. (1970). Structure of glycosides of oleanolic acid isolated from the roots of Calendula officinalis. Phytochemistry, 10, 1121–1124.

Wolfender, J.-L., Maillard, M., Marston, A. and Hostettmann, K. (1992). Mass spectrometry of underivatized naturally occurring glycosides. Phytochem. Anal., 3, 193–214.

Wolters, B. (1965). Der Anteil der Steroidsaponine an der antibiotischen Wirkung von Solanum dulcamara. Planta Medica, 13, 189–193.

Wolters, B. (1966a). Zur antimikrobiellen Wirksamkeit pflanzlicher Steroide und Triterpene. Planta Medica, 14, 392–401.

Wolters, B. (1966b). The action of various triterpene saponins on fungi Natur-wissenschaften, 53, 253.

Wolters, B. (1968a). The antibiotic action of saponins. III. Saponins as plant fungistatic compounds. Planta (Berlin), 79, 77–83.

Wolters, B. (1968b). Ueber die antibiotische Wirkung neutraler Steroidglykoside mit und ohne Saponincharacter. Planta Medica, 16, 114–119.

Woo, E.H., Woo, W.S., Chmurny, G.N. and Hilton, B.D. (1992). Melandrioside A, a saponin from Melandrium firmum. J. Nat. Prod., 55, 786–794.

Woo, W.S. and Kang, S.S. (1977). The structure of phytolaccoside G. Yakhak Hoe Chi, 21, 159-162 (Chem. Abstr., 88, 191324).

Woo, W.S., Kang, S.S. and Yoon, M.M. (1976). Phytochemical study on Clematis apiifolia. Soul Taehakkyo Saengyak Yonguso Opjukjip, 15, 1–4 (Chem. Abstr., 88, 94727).

Woo, W.S., Kang, S.S., Wagner, H., Seligmann, O. and Chari, V.M. (1978).

Triterpenoid saponins from the roots of *Phytolacca americana*. *Planta Medica*, **34**, 87–92.

Wu, F., Koike, K., Ohmoto, T. and Chen, W.-X. (1989). Saponins from Chinese folk medicine, 'Zhu jie xiang fu', *Anemone raddeana* REGEL. *Chem. Pharm. Bull.*, **37**, 2445.

Wu, J. and Chen, J. (1988). Depressant actions of *Panax notoginseng* saponins on vascular smooth muscle. *Zhongguo Yaoli Xuebao*, **9**, 147–152 (*Chem. Abstr.*, **108**, 179841).

Wulff, G. (1965). Determination of sugars in glycosides and oligosaccharides by gas chromatography. *J. Chromatogr.*, **18**, 285–296.

Wulff, G. (1968). Neuere Entwicklungen auf dem Saponingebiet. *Dtsch. Apoth. Ztg.*, **108**, 797–808.

Wulff, G. and Tschesche, R. (1969). Ueber Triterpene–XXVI. Ueber die Struktur des Rosskastaniensaponine (Aescin) und die Aglykone verwandter Glykoside. *Tetrahedron*, **25**, 415–436.

Xu, C.-J. and Lin, J.-T. (1985). Comparison of silica-, C_{18}-, and NH_2-HPLC columns for the separation of neutral steroid saponins from *Dioscorea* plants. *J. Liq. Chromatogr.*, **8**, 361–368.

Xu, D., Zhang, B., Qi, Y. and Ma, J. (1983). Studies on alkaloids of Ping Bei Mu (*Fritillaria ussuriensis* MAXIM.). II. Isolation and identification of sipeimine glycoside. *Zhongcaoyao*, **14**, 55–56 (*Chem. Abstr.*, **98**, 212853).

Xu, D.-M., Xu, M.-L., Wang, S.-Q. *et al.* (1990). Two new steroidal alkaloids from *Fritillaria ussuriensis*. *J. Nat. Prod.*, **53**, 549-552.

Xu, J.-P., Xu, R.-S., Luo, Z. and Dong, J.-Y. (1991). New triterpenoidal saponins from folk contraceptive medicine, *Mussaenda pubescens*. *Huaxue Xuebao*, **49**, 621–624 (*Chem. Abstr.*, **115**, 131997).

Xu, J.-P., Xu, R.-S. and Li, X.-Y. (1992a). Four new cycloartane saponins from *Curculigo orchioides*. *Planta Medica*, **58**, 208–210.

Xu, J.-P., Xu, R.-S., Luo, Z., Dong, J.-Y. and Hu, H.-M. (1992b). Mussaendosides M and N, new saponins from *Mussaenda pubescens*. *J. Nat. Prod.*, **55**, 1124–1128.

Xu, S., Chen, Y., Cai, Z. and Yao, X. (1987). Chemical constituents of *Panax quinquefolius* LINN. *Yaoxue Xuebao*, **27**, 750–755 (*Chem. Abstr.*, **108**, 72123).

Xu, W.-H. and Xue, Z. (1986). Chemical investigation of glycosidal steroidal alkaloids of *Northoliron hyacinthinum*. *Yaoxue Xuebao*, **21**, 177–182 (*Chem. Abstr.*, **105**, 21663).

Xu, W.H. and Xue, Z. (1988). The chemical structure of neohyacinthoside. *Yaoxue Xuebao*, **23**, 61–63 (*Chem. Abstr.*, **109**, 51701).

Xu, X. and Zhong, C. (1988a). Chemical constituents of Chinese paris (*Paris podophylla* var. *chinensis*). II. Structural determination of sapogenin C. *Zhongcaoyao*, **19**, 242–243 (*Chem. Abstr.*, **109**, 226726).

Xu, X. and Zhong, C. (1988b). The chemical constituents of Chinese paris (*Paris polyphylla* var. *chinensis*). I. Isolation and structural determination of sapogenins A, B and D. *Zhongcaoyao*, **19**, 194–198 (*Chem. Abstr.*, **109**, 134875).

Xue, S.-R., Liu, J.-Q., Wang, G., Shi, J.-Q., Wu, Q.-J. and Hu, S.-Z. (1992a). Triterpenoid saponin of *Clinopodium chinense* (BENTH.) O. KUNTZE. *Yaoxue Xuebao*, **27**, 207–212 (*Chem. Abstr.*, **117**, 86700).

Xue, S.-R., Liu, J.-Q. and Wang, G. (1992b). Triterpenoid saponins from *Clinopodium polycephalum*. *Phytochemistry*, **31**, 1049–1050.

Yadav, N., Misra, G. and Nigam, S.K. (1976). Triterpenoids of *Adenanthera pavonina* bark. *Planta Medica*, **29**, 176–178.

Yahara, S., Murakami, N., Yamasaki, M., Hamada, T., Kinjo, J. and Nohara, T. (1985). A furostanol glucuronide from *Solanum lyratum*. *Phytochemistry*, **24**, 2748–2750.

Yahara, S., Emura, S., Feng, H. and Nohara, T. (1989a). Studies on the constituents of the bark of *Dalbergia hupeana*. *Chem. Pharm. Bull.*, **37**, 2136–2138.

Yahara, S., Ohtsuka, M., Nakano, K. and Nohara, T. (1989b). Studies on the constituents of solanaceous plants. XIII. A new steroidal glucuronide from Chinese *Solanum lyratum*. *Chem. Pharm. Bull.*, **37**, 1802–1804.

Yamada, H., Nishizawa, M. and Katayama, C. (1992). Osladin, a sweet principle of *Polypodium vulgare*. Structure revision. *Tetrahedron Lett.*, **33**, 4009–4010.

Yamaguchi, H., Matsuura, H., Kasai, R. *et al.* (1986). Application of borate ion-exchange mode high-performance liquid chromatography to separation of glycosides: saponins of ginseng, *Sapindus mukurossi* GAERTN, and *Anemone rivularis* BUCH.-HAM. *Chem. Pharm. Bull.*, **34**, 2859–2867.

Yamaguchi, H., Kasai, R., Matsuura, H., Tanaka, O. and Fuwa, T. (1988a). High-performance liquid chromatographic analysis of acidic saponins of ginseng and related plants. *Chem. Pharm. Bull.*, **36**, 3468–3473.

Yamaguchi, H., Matsuura, H., Kasai, R. *et al.* (1988b). Analysis of saponins of wild *Panax ginseng*. *Chem. Pharm. Bull.*, **36**, 4177–4181.

Yamahara, J., Mibu, H., Sawada, T. *et al.* (1979). Anti-inflammatory effects of Mi-saponin. *Yakugaku Zasshi*, **99**, 612–617.

Yamahara, J., Kubomura, Y., Miki, K. and Fujimura, H. (1987a). Anti-ulcer action of *Panax japonicus* rhizome. *J. Ethnopharmacol.*, **19**, 95–101.

Yamahara, J., Kubomura, Y., Miki, K. and Fujimura, H. (1987b). Anti-ulcer action of chikusetsusaponin III. *Yakugaku Zasshi.*, **107**, 135–139.

Yamamoto, A., Miyase, T., Ueno, A. and Maeda, T. (1991). Buddlejasaponins I-IV, four new oleanane-triterpene saponins from the aerial parts of *Buddleja japonica* HEMSL. *Chem. Pharm. Bull.*, **39**, 2764–2766.

Yamamoto, A., Suzuki, H., Miyase, T., Ueno, A. and Maeda, T. (1993). Clinoposaponins VI and VIII, two oleanane-triterpene saponins from *Clinopodium micranthum*. *Phytochemistry*, **34**, 485–488.

Yamamoto, M. and Uemura, T. (1980). Endocrinological and metabolic actions of ginseng principles. *Proceedings of the Third International Ginseng Symposium*, pp. 115–119. Korea Ginseng Research Institute, Seoul, Korea.

Yamamoto, M., Kumagai, A. and Yamamura, Y. (1975). Structure and actions of saikosaponins isolated from *Bupleurum falcatum* L. I. Anti-inflammatory action of saikosaponins. *Arzneimittelforsch.*, **25**, 1021–1023.

Yamanouchi, T. (1943). *Zool. Mag.* (Tokyo), **55**, 87.

Yamasaki, Y., Ito, K., Enomoto, Y. and Sutko, J.L. (1987). Alterations by saponins of passive calcium permeability and sodium-calcium exchange activity of canine cardiac sarcolemmal vesicles. *Biochim. Biophys. Acta*, **897**, 481–487.

Yamashita, T., Fujimura, N., Yahara, S., Nohara, T., Kawanobu, S. and Fujieda, K. (1990). Structures of three new steroidal alkaloid glycosides, solaverines

I, II and III from *Solanum toxicarum* and *S. verbascifolium*. *Chem. Pharm. Bull.*, **38**, 827–829.

Yamashita, T., Matsumoto, T., Yahara, S., Yoshida, N. and Nohara, T. (1991). Structures of two new steroidal glycosides, soladulcosides A and B from *Solanum dulcamara*. *Chem. Pharm. Bull.*, **39**, 1626–1628.

Yamauchi, T. (1959). Saponins of Japanese Dioscoreaceae. IX. Hydrolysis of diosgenin glycosides. *Chem. Pharm. Bull.*, **7**, 343–348.

Yang, C., Jiang, Z., Wu, M., Zhou, J. and Tanaka, O. (1984). Studies on saponins of rhizomes of *Panax zingiberensis* WU ET FENG. *Yaoxue Xuebao*, **19**, 232–236 (*Chem. Abstr.*, **103**, 3686).

Yang, C., Jiang, Z., Zhou, J., Kasai, R. and Tanaka, O. (1985). Two new oleanolic acid-type saponins from *Panax stipuleanatus*. *Yunnan Zhiwu Yanjiu*, **7**, 103–108 (*Chem. Abstr.*, **102**, 218350).

Yang, D., Zhong, Z.C. and Xie, Z.M. (1992). Sweet principles from the leaves of *Cyclocarya paliurus* (BATAL.) ILJINSKAYA. *Yaoxue Xuebao*, **27**, 841–844 (*Chem. Abstr.*, **118**, 100698).

Yano, I., Nishiizumi, C., Yoshikawa, K. and Arihara, S. (1993). Triterpenoid saponins from *Ilex integra*. *Phytochemistry*, **32**, 417–420.

Yasukawa, K., Takido, M., Takeuchi, M. and Nakagawa, S., (1988). Inhibitory effect of glycyrrhizin and caffeine on two-stage carcinogenesis in mice. *Yakugaku Zasshi*, **108**, 794–796.

Yasumoto, T., Tanaka, M. and Hashimoto, Y. (1966). Distribution of saponin in echinoderms. *Nippon Suisan Gakkaishi*, **32**, 673–676 (*Chem. Abstr.*, **69**, 94029).

Yata, N., Sugihara, N., Yamajo, R. *et al.* (1985). Enhanced rectal absorption of β-lactam antibiotics in rat by monodesmosides isolated from pericarps of *Sapindus mukurossi* (Enmei-hi). *J. Pharmacobio-Dyn.*, **8**, 1041–1047.

Yatsyno, A.I., Belova, L.F., Lipkina, G.S., Sokolov, S.Ya. and Trutneva, E.A. (1978). Pharmacology of calenduloside B – a new triterpene glycoside from rhizomes of *Calendula officinalis*. *Farmakol. Toksikol.* (Moscow), **41**, 556–560 (*Chem. Abstr.*, **89**, 209164).

Yepez, A.M., Lock de Ugaz, O., Alvarez, C.M. *et al.* (1991). Quinovic acid glycosides from *Uncaria guianensis*. *Phytochemistry*, **30**, 1635–1636.

Yi, Y.-H. (1990). Esculentoside L and K: two new saponins from *Phytolacca esculenta*. *Planta Medica*, **56**, 301–303.

Yi, Y. and Wang, Z. (1984). Active principles of *Phytolacca esculenta*. I. Isolation and identification of the triterpene saponins. *Zhongcaoyao*, **15**, 55–59 (*Chem. Abstr.*, **100**, 180001).

Yi, Y.-H. and Wang, C.-L. (1989). A new active saponin from *Phytolacca esculenta*. *Planta Medica*, **55**, 551–552.

Yi, Y.-H. and Dai, F.-B. (1991). A new triterpenoid and its glycoside from *Phytolacca esculenta*. *Planta Medica*, **57**, 162–164.

Yi, Y. and Wu, Z. (1991). A new triterpene glycoside of *Thalictrum foeniculaceum*. *Zhongguo Yaoke Daxue Xuebao*, **22**, 270–274 (*Chem. Abstr.*, **117**, 66557).

Yokoyama, H., Hiai, S., Oura, H. and Hayashi, T. (1982a). Effects of total saponins extracted from several crude drugs on rat adrenocortical hormone secretion. *Yakugaku Zasshi*, **102**, 555–559.

Yokoyama, H., Hiai, S. and Oura, H. (1982b). Rat plasma corticosterone secretion-inducing activities of total saponin and prosapogenin methyl esters from the roots of *Platycodon grandiflorum* A.DC. *Yakugaku Zasshi*, **102**, 1191–1194.

Yokoyama, H., Hiai, S. and Oura, H. (1984). Effect of saikosaponins on dexamethasone suppression of the pituitary-adrenocortical system. *Chem. Pharm. Bull.*, **32**, 1224–1227.

Yoshikawa, K., Takemoto, T. and Arihara, S. (1987a). Studies on the constituents of Cucurbitaceae plants. XVI. On the saponin constituents of *Gynostemma pentaphyllum* MAKINO (11). *Yakugaku Zasshi*, **107**, 262–267.

Yoshikawa, K., Mitake, M., Takemoto, T. and Arihara, S. (1987b). Studies on the constituents of Cucurbitaceae plants. XVII. On the saponin constituents of *Gynostemma pentaphyllum* MAKINO (12). *Yakugaku Zasshi*, **107**, 355–360.

Yoshikawa, K., Amimoto, K., Arihara, S. and Matsuura, K. (1989a). Structure studies of new antisweet constituents from *Gymnema sylvestre*. *Tetrahedron Lett.*, **30**, 1103–1106.

Yoshikawa, K., Amimoto, K., Arihara, S. and Matsuura, K. (1989b). Gymnemic acid V, VI and VII from Gur-ma, the leaves of *Gymnema sylvestre* R. BR. *Chem. Pharm. Bull.*, **37**, 852–854.

Yoshikawa, K., Arihara, S. and Matsuura, K. (1991a). A new type of antisweet principle occurring in *Gymnema sylvestre*. *Tetrahedron Lett.*, **32**, 789–792.

Yoshikawa, K., Shimono, N. and Arihara, S. (1991b). Antisweet substances, jujubasaponins I–III from *Zizyphus jujuba*. Revised structure of ziziphin. *Tetrahedron Lett.*, **32**, 7059–7062.

Yoshikawa, K., Arihara, S., Wang, J.-D., Narui, T. and Okuyama, T. (1991c). Studies of two new fibrinolytic saponins from the seed of *Luffa cylindrica* ROEM. *Chem. Pharm. Bull.*, **39**, 1185–1188.

Yoshikawa, M., Wang, H.K., Kayakiri, H., Taniyama, T. and Kitagawa, I. (1985). Saponin and sapogenol. XL. Structure of sophoraflavoside I, a bisdesmoside of soyasapogenol B, from Sophorae radix, the root of *Sophora flavescens* AITON. *Chem. Pharm. Bull.*, **33**, 4267–4274.

Yoshimitsu, H., Hayashi, K., Shingu, K. *et al.* (1992). Two new cycloartane glycosides, thalictosides A and C from *Thalictrum thunbergii* D.C. *Chem. Pharm. Bull.*, **40**, 2465–2468.

Yoshimitsu, H., Hayashi, K., Kumabe, M. and Nohara, T. (1993). Two new cycloartane glycosides from Thalictri herba. *Chem. Pharm. Bull.*, **41**, 786–787.

Yosioka, I., Fujio, M., Osamura, M. and Kitagawa, I. (1966). A novel cleavage method of saponin with soil bacteria, intending to the genuine sapogenin: on *Senega* and *Panax* saponins. *Tetrahedron Lett.*, 6303–6308.

Yosioka, I., Matsuda, A., Imai, K., Nishimura, T. and Kitagawa, I. (1971). Saponin and sapogenol. IV. Seed sapogenols of *Aesculus turbinata* BLUME. On the configuration of hydroxyl functions in ring E of aescigenin, protoaescigenin, and isoaescigenin in relation to barringtogenol C and theasapogenol A. *Chem. Pharm. Bull.*, **19**, 1200–1213.

Yosioka, I., Hino, K., Matsuda, A. and Kitagawa, I. (1972). Saponin and sapogenol. VI. Sapogenol constituents of leaves of *Pittosporum tobira* AIT. *Chem. Pharm. Bull.*, **20**, 1499–1506.

Yosioka, I., Inada, A. and Kitagawa, I. (1974). Soil bacterial hydrolysis leading to genuine aglycone. VIII. Structures of a genuine sapogenol protobassic acid and a prosapogenol of seed kernels of *Madhuca longifolia* L. *Tetrahedron*, **30**, 707–714.

Younes, M.E.G. (1975). Chemical examination of local plants. XIV. Triterpenoids from the leaves of Egyptian *Calistemon lanceolatus*. *Aust. J. Chem.*, **28**, 221–224.

Young, H.S., Park, J.C. and Choi, J.S. (1987). Triterpenoid glycosides from *Rosa rugosa*. *Arch. Pharmacal Res.*, **10**, 219–222 (*Chem. Abstr.*, **108**, 218989).

Young, H.S., Park, J.C. and Choi, J.S. (1988). Analytical separation of isomeric saponins by LC/MS. *Kor. J. Pharmacog.*, **19**, 248–250.

Yu, B.-Y., Hirai, Y., Shoji, J. and Xu, G.-J. (1990). Comparative studies on the constituents of Ophiopogonis tuber and its congeners. VI. Studies on the constituents of the subterranean part of *Liriope spicata* var. *prolifera* and *L. muscari* (1). *Chem. Pharm. Bull.*, **38**, 1931–1935.

Yu, S.S. and Xiao, Z.Y. (1992). The structures of yiyeliangwanoside III and IV from the bark of *Nothopanax davidii* (FRANCE) HARMS. *Yaoxue Xuebao*, **27**, 42–47 (*Chem. Abstr.*, **116**, 211140).

Yu, S., Yu, D. and Liang, X. (1992). A new triterpene saponin from the root of *Aralia spinifolia*. *Chin. Chem. Lett.*, **3**, 289–290.

Yuchi, S. and Uchida, Y. (1988). Adzuki saponins from *Vigna angularis* as hypolipemics. Jap. Pat. 6,289,692 (*Chem. Abstr.*, **107**, 233370).

Yukananov, D.K., Sokolskii, I.N., Zinkevich, E.P. and Kuvaev, V.B. (1971). Plants of the family Caryophyllaceae studied to determine the presence of the triterpenoid saponin, gypsoside. *Rast. Resur.*, **7**, 386–390 (*Chem. Abstr.*, **75**, 121321).

Zacharius, R.M. and Osman, S.F. (1977). Glycoalkaloids in tissue culture of *Solanum* species. Dehydrocommersonine from cultured roots of *Solanum chacoense*. *Plant Sci. Lett.*, **10**, 283–287.

Zakharov, A.M., Zinkevich, E.P. and Bankovskii, A.I. (1968). Saponins of *Primula turkestanica*. *Khim. Prir. Soedin.*, 388–389.

Zakharov, A.M., Zakharova, O.I., Pakalns, D. and Sokolskii, I.N. (1971). Triterpenoid glycoside from *Primula macrocalyx*. *Khim. Prir. Soedin.*, 672.

Zeng, L., Zhang, R.-Y. and Wang, X. (1987). Studies on the constituents of *Zizyphus spinosus* HU. *Yaoxue Xuebao*, **22**, 114–120 (*Chem. Abstr.*, **107**, 12735).

Zhang, B., Pan, W., Feng, S. and Shi, G. (1985). Effects of saponins of *Panax quinquefolium* L. on experimental arrhythmia. *Shenyang Yaoxueyuan Xuebao*, **2**, 273–275 (*Chem. Abstr.*, **104**, 81750).

Zhang, F. and Chen, X. (1987). Effects of ginsenosides on sympathetic neurotransmitter release in pithed rats. *Zhongguo Yaoli Xuebao*, **8**, 217–220 (*Chem. Abstr.*, **107**, 32871).

Zhang, J. and Mao, Q. (1984). Studies on the chemical constituents of *Swertia cincta* BURKILL. *Yaoxue Xuebao*, **19**, 819–824 (*Chem. Abstr.*, **102**, 109805).

Zhang, R., Zhang, J. and Wang, M. (1986). Saponins from the root of *Glycyrrhiza uralensis* FISCH. *Yaoxue Huebao*, **21**, 510–515 (*Chem. Abstr.*, **105**, 187603).

Zhang, S. and Hu, Z. (1985). Effects of ginseng flower saponins and ginsenoside Re on experimentally-induced gastic ulcers. *Zhongyao Tongbao*, **10**, 331–332 (*Chem. Abstr.*, **104**, 512).

Zhang, Y., Shen, J., Song, J., Wang, Y. and Li, D. (1984). Analgesic and sedative effects of Astragalus saponin I. *Nanjing Yixueyuan Xuebao*, **4**, 225–227 (*Chem. Abstr.*, **104**, 122973).

Zhao, L., Chen, W. and Fang, Q. (1990). Triterpenoid saponins from *Anemone flaccida*. *Planta Medica*, **56**, 92–93.

Zhou, J. (1989). Some bioactive substances from plants of West China. *Pure Appl. Chem.*, **61**, 457–460.

Zhou, J., Wu, M.-Z., Taniyasu, S. *et al.* (1981). Dammarane-saponins of Sanchi-ginseng, roots of *Panax notoginseng* (BURK.) F.H. CHEN (Araliaceae): structures of new saponins, notoginsenosides-R_1 and -R_2, and identification of ginseno-sides-Rg_2 and -Rh_1. *Chem. Pharm. Bull.*, **29**, 2844–2850.

Zhou, X.-H., Kasai, R., Ohtani, K. *et al.* (1992). Oleanane and ursane glucosides from *Rubus* species. *Phytochemistry*, **31**, 3642–3644.

Zhu, Y.Z., Lu, S.H., Okada, Y., Takata, M. and Okuyama, T. (1992). Two new cycloartane-type glucosides, mongholicoside I and II, from the aerial part of *Astragalus mongholicus* BUNGE. *Chem. Pharm. Bull.*, **40**, 2230–2232.

Zinova, S.A., Isakov, V.V., Kalinovskii, A.I., Ulkova, Z.I., Ulanova, K.P. and Glebko, L.I. (1992). Four new triterpene glycosides from the roots of *Pulsatilla dahurica*. *Chem. Nat. Compd.*, **28**, 306–311.

Zollo, F., Finamore, E., Minale, L., Laurent, D. and Bargibant, G. (1986). Starfish saponins, 29. A novel steroidal glycoside from the starfish *Pentaceraster alveolatus*. *J. Nat. Prod.*, **49**, 919–921.

Zollo, F., Finamore, E. and Minale, L. (1987). Starfish saponins, 31. Novel polyhydroxysteroids and steroidal glycosides from the starfish *Sphaerodiscus placenta*. *J. Nat. Prod.*, **50**, 794–799.

Zollo, F., Finamore, E., Riccio, R. and Minale, L. (1989). Starfish saponins, part 37. Steroidal glycoside sulfates from starfishes of the genus *Pisaster*. *J. Nat. Prod.*, **52**, 693–700.

Zviadadze, L.D., Dekanosidze, G.E., Dzhikiya, O.D., Kemertelidze, E.P. and Shashkov, A.S. (1981). Triterpene glycosides of *Cephalaria gigantea*. II. Structure of giganteasides D and G. *Bioorg. Khim.*, **7**, 736–740 (*Chem. Abstr.*, **95**, 76861).

Index of Latin names

General index